Materials Horizons: From Nature to Nanomaterials

Series Editor

Vijay Kumar Thakur, School of Aerospace, Transport and Manufacturing, Cranfield University, Cranfield, UK

Materials are an indispensable part of human civilization since the inception of life on earth. With the passage of time, innumerable new materials have been explored as well as developed and the search for new innovative materials continues briskly. Keeping in mind the immense perspectives of various classes of materials, this series aims at providing a comprehensive collection of works across the breadth of materials research at cutting-edge interface of materials science with physics, chemistry, biology and engineering.

This series covers a galaxy of materials ranging from natural materials to nanomaterials. Some of the topics include but not limited to: biological materials, biomimetic materials, ceramics, composites, coatings, functional materials, glasses, inorganic materials, inorganic-organic hybrids, metals, membranes, magnetic materials, manufacturing of materials, nanomaterials, organic materials and pigments to name a few. The series provides most timely and comprehensive information on advanced synthesis, processing, characterization, manufacturing and applications in a broad range of interdisciplinary fields in science, engineering and technology.

This series accepts both authored and edited works, including textbooks, monographs, reference works, and professional books. The books in this series will provide a deep insight into the state-of-art of Materials Horizons and serve students, academic, government and industrial scientists involved in all aspects of materials research.

More information about this series at http://www.springer.com/series/16122

Ronaldo Ferreira do Nascimento ·
Vicente de Oliveira Sousa Neto ·
Pierre Basílio Almeida Fechine ·
Paulo de Tarso Cavalcante Freire
Editors

Nanomaterials and Nanotechnology

Biomedical, Environmental, and Industrial Applications

Springer

Editors
Ronaldo Ferreira do Nascimento
Department of Analytical Chemistry
and Physical Chemistry
Federal University of Ceará
Fortaleza, Ceará, Brazil

Pierre Basílio Almeida Fechine
Department of Analytical Chemistry
and Physical Chemistry
Universidade Federal do Ceará
Fortaleza, Ceará, Brazil

Vicente de Oliveira Sousa Neto
Department of Chemistry
State University of Ceará
Fortaleza, Ceará, Brazil

Paulo de Tarso Cavalcante Freire
Department of Physics
Federal University of Ceará
Fortaleza, Ceará, Brazil

ISSN 2524-5384 ISSN 2524-5392 (electronic)
Materials Horizons: From Nature to Nanomaterials
ISBN 978-981-33-6055-6 ISBN 978-981-33-6056-3 (eBook)
https://doi.org/10.1007/978-981-33-6056-3

This Springer imprint is published by the registered company Springer Nature Singapore Pte Ltd.
The registered company address is: 152 Beach Road, #21-01/04 Gateway East, Singapore 189721, Singapore

Contents

About the Editors

Ronaldo Ferreira do Nascimento is Professor at Federal University of Ceará (UFC), Brazil. He received his Ph.D. in Analytical Chemistry from Universidade de São Paulo (USP), Brazil, in 1997. He has experience in the field of analytical chemistry, with emphasis on the development of separation methods (gas chromatography and liquid chromatography), environmental chemistry (trace analysis) and adsorption (removal of pollutants from aqueous effluents using natural adsorbents).

Vicente de Oliveira Sousa Neto is Professor in the Faculty of Education, Sciences, and Letters (FECLESC) at the State University of Ceará (UECE), Brazil. He received his Ph.D. in Civil Engineering in the area of environmental sanitation by the Federal University of Ceará in 2012. He has experience in the area of environmental chemistry, with an emphasis on bioadsorption (treatment of wastewater using modified bioadsorbents and low cost), and chemistry of materials with emphasis on development of thin films for photovoltaics.

Pierre Basílio Almeida Fechine is Professor at Federal University of Ceará (UFC), Brazil, and has previously served as Head of the Department of Analytical Chemistry and Physical Chemistry. Prof. Fechine is a member of the Brazilian Academy of Sciences. He has experience in the field of chemistry, with emphasis on inorganic chemistry, working mainly in the following subjects: chemistry of materials, condensed state physics, dielectric and magnetic properties, biomaterials and nanoparticles.

Paulo de Tarso Cavalcante Freire is Professor at the Federal University of Ceará (UFC), and has previously served as Head of the Department of Physics and Dean of the Science Faculty at UFC. He received his Ph.D. in Physics from Universidade Estadual de Campinas in 1995. He has experience in the field of physics, with emphasis on Raman spectroscopy, nanomaterials, organic materials and high-pressure properties.

Chapter 1
Nanotechnology: Concepts and Potential Applications in Medicine

Luiziana Cavalcante Costa Fernandes, Karina Alexandre Barros Nogueira, Jéssica Roberta Pereira Martins, Elias Santos, Paulo George Cavalcante de Freitas, Beatriz Alexandre Barros Nogueira, Giovanni Loureiro Raspantini, Raquel Petrilli, and Josimar O. Eloy

1 Introduction

Nanotechnology is the science that studies the various applications of the nanoscale (1–1000 nm). Nanomedicine, a term that emerged in the 1990s, is the science that studies the potential use of the nanometric scale for healthcare, involving an interdisciplinarity field between research areas such as engineering, physics, chemistry, biology, pharmacy and medicine. Various nanomaterials/nanoparticles have been developed for use in medicine to assist in the diagnosis, prevention, control, monitoring, treatment of various diseases, providing alternative routes, increasing half-life, improving toxicity, reducing frequency of administration, with drug controlled release, features that provide unique patient benefits (Jain et al. 2014; Malinoski 2014; Nance 2019).

Paul Ehrlich has solidified the concept of the "magic bullet" which is nothing more than the ability of a substance to recognize a target and thereby provides

L. C. C. Fernandes · E. Santos · P. G. C. de Freitas · J. O. Eloy (✉)
Faculty of Pharmacy, Dentistry and Nursing, Department of Pharmacy, Federal University of Ceará, Fortaleza, Brazil
e-mail: josimar.eloy@ufc.br

K. A. B. Nogueira · J. R. P. Martins
Department of Chemical Engineering, Federal University of Ceará, Center of Technology, Fortaleza, Brazil

B. A. B. Nogueira
Department of Architecture, Urbanism and Design, Federal University of Ceará, Fortaleza, Brazil

G. L. Raspantini
School of Pharmaceutical Sciences of Ribeirão Preto, University of São Paulo, São Paulo, Brazil

R. Petrilli
Institute of Health Sciences,, University of International Integration of the Afro-Brazilian Lusophony- UNILAB, Redenção, Ceará, Brazil

R. F. do. Nascimento et al. (eds.), *Nanomaterials and Nanotechnology*,
Materials Horizons: From Nature to Nanomaterials,
https://doi.org/10.1007/978-981-33-6056-3_1

specific therapeutic action for that target. This concept currently encompasses three components (1) drug, (2) target-specific and (3) vehicle. The use of nanomedicine for this purpose was thought due to the physicochemical properties and modifications that can be introduced in nanoparticles for the treatment of a specific group of diseases, such as cancer (Fornaguera and García-Celma 2017).

Nanotechnology offers many advantages for drug delivery, such as (1) possibility of modification of size, shape and surface which impacts the transport, accumulation and recognition of nanoparticles in their targets, (2) allows the encapsulation of poorly water-soluble drugs, with low bioavailability, (3) protection of the drug against degradation, (4) allows combined therapy, since the nanocarrier can carry drugs with different therapeutic activities, which can act synergistically, (5) allows the nanoparticle to target the tumor/injured tissue, so that the nanoparticle surface undergoes a modification for binding of a ligand that is specific for a molecule that is overexpressed by tumor/tissue, (6) are capable of responding to specific endogenous stimuli such as pH, or exogenous, such as phototherapy, ultrasound, magnetic field, among others, reducing the non-specificity of treatment, (7) have the ability to accumulate in tumor tissues, this feature is possible due to the formation of new discontinuous blood vessels which do not have adequate blood flow and the lack of draining lymphatic vessels, named as enhanced permeation and retention (EPR) effect (8) can stay longer in the bloodstream and therefore allow sustained release of the nanoencapsulated drug due to the possibility of adding compounds on their surface such as polyethylene glycol (PEG) (Tran et al. 2017; Wolfram and Ferrari 2019).

There are many intrinsic characteristics of nanoparticles and nanotechnology that are applicable to medicine, such as the ability to circumvent tumor resistance, promote targeted treatment, reduce adverse effects, among other benefits, making nanoparticles successful in medicine. Doxil® was the first approved nanoencapsulated drug formulation, containing doxorubicin encapsulated in pegylated liposomes for cancer treatment. Doxorubicin has reduced therapeutic index and serious adverse effects, and the liposome improved pharmacokinetics and protected against side effects cardiotoxic. There are other nanostructured drugs marketed for other diseases, besides cancer. For instance, AmBisome® is an amphotericin B liposome and is the first line of treatment against fungal infections (Bavli et al. 2019; Cencig et al. 2012; Hussain et al. 2018). Throughout the chapter, we will cover the different types of nanoparticles, present some preparation and characterization strategies as well as discuss the applications of nanoparticles focused on scientific research and medical application, and finally, we will address treatments approved in the market.

2 Nanoparticles: Composition, Synthesis and Characterization

2.1 Lipid Nanoparticles

2.1.1 Liposomes

Liposomes were discovered by Bangham and are formed by amphiphilic molecules that organize in bilayer conformation in aqueous medium, the phospholipids that make up the bilayer, organize how their lipophilic tails interact with each other and the hydrophilic head interacts with water. Thus, by holding two compartments, the liposome is capable of carrying both hydrophilic drugs (encapsulated in the aqueous core) and lipophilic drugs (attached to the lipid bilayer). Some lipids that can be used for the development of these nanoparticles are phosphatidylethanolamine, phosphatidylcholine, phosphatidylserine, phosphatidylglycerol, phosphatidylinositol or glycolipids. Two main features of liposomes are their biocompatibility and biodegradability, since the lipids used to form the bilayer can be easily found in the human body (Carita et al. 2017; Bruch et al. 2019; Edwards and Baeumner 2006).

Essentially, the lipid bilayer first forms multilamellar vesicles (MLV); however, some methods have been developed to form unilamellar vesicles, which can be classified into giant unilamellar vesicles (GUV) and small and large unilamellar vesicles (SUV and LUV). Table 1 shows the conventional methods for preparing each of the vesicles (Patil and Jadhav 2014).

After formation of the giant vesicle, some techniques are employed for size reduction which are exemplified in Table 2 (Carita et al. 2017).

Over the years, structural modifications in liposomes have been allowed in order to improve the delivery of drugs that can be carried by these nanoparticles. Thus, different types of liposomes have been developed such as (Li et al. 2019):

1. Long-circulating liposomes: These are liposomes that have surface polymer molecules that are inert, such as oligosaccharides, glycoproteins, polysaccharides and synthetic polymers. Polyethylene glycol (PEG) is the most widely employed polymer used to coat the liposomes. This coating layer reduces the recognition of phagocytic cells and the capture of these nanoparticles by macrophages due to steric hindrance. Some disadvantages of this surface modification are slow blood clearance, low cell uptake and low cell selectivity.
2. Target-specific delivery: Tumor and normal cells express different types of receptors on their surface. These receptors are very specific and are responsible for tumor development. Based on this, many studies have been done to bind specific moieties on the liposome surface, making this nanoparticle specific to a cell/tumor target. Some of these ligands are antibodies, folic acid, polysaccharides, transferrin, among others. We will address these systems in more details in this chapter.

Table 1 Methods for synthesis of liposome vesicles

Vesicle types	Preparation methods
Multilamellar vesicles	
	Hydration of the lipid film under hydrodynamic flow: A film of phospholipids dried on a substrate is rehydrated under dynamic flows for a period of a few hours
	Solvent beads: an organic phase consisting of phospholipids and an aqueous phase are mixed in the presence of low vacuum, at the end we have an O/W emulsion containing solvent spheres with the lipids, the organic phase is removed by evaporation
	Proliposome hydration: they are formed by drying an organic phase composed of phospholipids with an encapsulated drug, wrapped in the phospholipid layer. It is possible to remove the organic phase by using vacuum rotary evaporator, fluidized bed or spray dryer
Giant unilamellar vesicles	
	Smooth hydration of lipid film: phospholipid dispersed in organic solvent is deposited on a substrate, subsequently hydrated for a period of time
	Electroformation: deposition of the lipid film on electrodes and subsequent hydration for a while under electric field
	Coalescence of small vesicles: induction of spontaneous coalescence of small vesicles by freeze/thaw cycles in electrolyte solution, use of oppositely charged phospholipids, addition of polyethylene glycol or fusogenic peptides and the addition of divalent cations in negatively charged phospholipids
Small and large unilamellar vesicles	
	Reverse phase evaporation: water hydration of phospholipids dissolved in an organic phase, forming an A/O emulsion, followed by evaporation of the organic phase
	Injection of organic solvent with phospholipids dissolved in an aqueous phase: injection of an organic solvent containing phospholipids into an aqueous buffer, leading to spontaneous formation of vesicles
	Detergent dialysis: phospholipid solubilization in aqueous phase detergent micelles with subsequent detergent removal by dialysis
	Reduction of size and lamellarity of MLVs: using techniques such as the French press, sonication, homogenization and membrane extrusion

3. Stimuli responsive: the in vivo microenvironment that liposomes are exposed to have different characteristics that can cause these nanoparticles to release encapsulated drug into specific tissues, so liposomes sensitive to these changes are also options for improving drug delivery. These stimuli-responsive liposomes can be sensitive to pH, presence of enzymes, temperature change, light sensitivity, among others.

Table 2 Methods to reduce liposome size

Method	
Sonication	It is based on high-energy cavitation through a sonicator
Membrane extrusion	The lipids are extruded, and under high pressure, they cross polycarbonate membranes with defined pore sizes, causing size and lamellarity reduction
Microfluidization	A lipid dispersion is added to a microfluidizer, which is pumped at high pressure through 1–5 μm channels. This dispersion is divided into two streams through two microchannels, after separation occurs the dispersion return and collision at high velocity. The process is repeated until a homogeneous dispersion is obtained
High-pressure homogenization	The lipid dispersion goes through a hole under high pressure and collides with a stainless steel wall. Continuous pumping of fluid occurs through the system to desired size

2.1.2 Microemulsion

Microemulsions are defined as thermodynamically stable and translucent, isotropically dispersed colloidal systems. They form spontaneously through thermodynamic self-assembly, with low surface tension and using low energy, through the interaction of at least three components which are water, oil and a surfactant. A fourth component can be employed, a co-surfactant. Microemulsions can be formed in the form of oil-in-water (O/W) or reverse microemulsions in the form of water in oil (W/O), because of their phases, capable of solubilizing hydrophilic and lipophilic substances, at room temperature. For the formation of this nanoparticle, it is necessary to find a specific concentration range between its components (Asgari et al. 2019; Vladisavljević 2019).

Microemulsions cannot be formed without the presence of a surfactant, which is the emulsifying agent of the formulation that binds the aqueous and oily phases together. To choose the ideal surfactant, one must understand the formulation as well as the drug that will be carried by it, so a parameter widely used by formulators is the hydrophilic–lipophilic balance (HLB). It analyzes the structure of the surfactant relating to its portion. Hydrophilic and hydrophobic can be calculated according to the following equation:

$$\text{HLB} = 7 + \sum(\text{number of hydrophilic groups}) + \sum(\text{number of lipophilic groups})$$

There are four classes of non-ionic, anionic, cationic and zwitterionic surfactants, the most widely used surfactants belong to the non-ionic class, they are Tween 80, Labrasol, Cremophor RH 40, Cremophor EL and Transcutol and Lecithin. Unlike microemulsions, nanoemulsions, which are known as miniemulsions, require energy and agitation for formation and lack thermodynamic stability (Callender et al. 2017).

During the process of synthesis of a microemulsion, the components present in the formulation collide, thus the particles are formed in two steps; the first step is nucleation inside the drop and the second step is aggregation to form the final particle. The surfactant controls the growth rate of the particles, so an adequate concentration of the particles allows the formation of the particles to occur evenly, so there is a suspension where the surfactant stabilizes the particles to prevent them from coiling (Eriksson et al. 2004).

2.1.3 Solid Lipid Nanoparticle/Nanostructured Lipid Carrier

They are heterogeneous lipid dispersions consisting of an inner lipid phase and an aqueous outer phase; such phases are stabilized by surfactants. The internal phase of such dispersions may be formed only of solid lipids at room temperature, such as solid lipid nanoparticles (SLN) or formed mainly of solid lipids mixed with liquid lipids or only liquid lipids, such as nanostructured lipid carriers (NLC). Both systems have advantages and are synthesized with biodegradable materials, and biocompatible excipients offer easy production on an industrial scale, can be used for various routes of administration, are capable of carrying high concentration of drugs, can carry both hydrophilic and lipophilic drugs, can bypass the reticuloendothelial system and are able to promote sustained drug delivery. A brief comparison between NLC and SLN is that the former has greater ability to carry drugs and also has better stability, since SLN is composed entirely of solid lipids that can crystallize, which leads to "expulsion" of the active substances and also undergo increase in particle size (Garcês et al. 2018; Gordillo-Galeano and Mora-Huertas 2018).

The stabilizing agents used for the preparation of the systems have the ability to decrease the interfacial energy between the lipid and aqueous phases, as well as improving the stability of the preparations during their storage and may also influence the crystalline form of the molecules and their kinetic behavior. Surfactants are the main substances used as stabilizing agents such as tweens, polaxamers, sodium lauryl Sulfate, cetrimonium bromide and soy lecithin. Furthermore, there are also co-surfactants in these formulations such as polyvinyl alcohol and propylene glycol (Gordillo-Galeano and Mora-Huertas 2018).

We can divide SLN and NLC preparation methods into three types, employing high energy, low energy and those that use organic solvents. Such methods are exemplified in Table 3 (Ganesan and Narayanasamy 2017; Geszke-Moritz and Moritz 2016).

2.1.4 Liquid Crystals

This delivery system has the characteristic to combine the symmetrical and mechanical properties present in a crystalline solid and in an isotropic liquid, representing an intermediate state between these two forms, in a crystal there is order as to positioning and orientation, while the liquid state is related to flowability. In a liquid crystal, the phases that form this structure are called mesophases and the compounds

Table 3 SLN/NLC preparation methods

Methods	Preparation technique
High energy	High-pressure homogenization: the liquid is pushed by high-pressure homogenizers through a micron-sized pore. The process can be performed in hot or cold conditions
	Hot homogenization: lipid and drug are melted and interact with an aqueous surfactant, with under homogenenization with a shear device, a preemulsion is formed. After further homogenization, it results in a colloidal emulsion, and then the droplets of the latter preparation are cooled and recrystallized until room temperature
	Cold homogenization: the lipid is melted, and the drug is incorporated into the matrix; this matrix is rapidly cooled forming a solid matrix that is milled by microparticle grinding. These microparticles are dispersed in a cold aqueous surfactant, and the latter preparation is homogenized under high pressure
	High shear and/or ultrasound homogenization: lipid is melted and dispersed in hot aqueous phase with speed-stirring surfactant to form an emulsion, then subjected to sonication to reduce particle size
Low energy	Microemulsion: lipids are melted and homogenized by gentle stirring with hot surfactant solution until microemulsion formation, and then the microemulsion is dispersed in cold water under moderate agitation, making the lipid droplets solid
	Membrane contactor: the lipid mixer fused together with the active ingredient of the formulation is pressed to cross the pores of a 0.05 nm diameter hydrophobic porous membrane into the aqueous phase of the formulation. Passage through the pores forms SLN or NLC droplets at the same time as the aqueous phase is being cooled
	Phase inversion temperature: this is the transformation of an O/W emulsion into an W/O type, a crucial factor for this technique is the change in surfactant properties at different temperatures. Hydrophilic–lipophilic balance (HLB) at 25 °C determines hydration of the hydrophilic part of the surfactants occurs, while dehydration of the ethoxy group occurs with increasing temperature. The phase inversion temperature occurs when the components of the aqueous and lipid phase surfactants are equal
	Coacervation technique: A fatty acid salt is homogeneously dispersed in the solution of the stabilizing agent with heating to the Krafft point of the constantly stirring fatty acid salt; thereafter, an ethanolic solution of the drug is slowly added under stirring until a single phase is obtained; shortly thereafter, a coacervation substance/acidifying solution is added to obtain a nanoparticle suspension
	Double emulsion: an aqueous solution is emulsified in a melted lipid mixture to form a W/O emulsion; then this first emulsion is dispersed in an aqueous emulsifier solution to form the final W/O/W emulsion

(continued)

Table 3 (continued)

Methods	Preparation technique
Organic solvents	Emulsification-solvent evaporation: occurs in three steps • Preparation of the organic phase • Preemulsification • Nanoemulsification After these steps, the formed nanodispersion is stirred so that the solvent is completely evaporated, and then the lipid material precipitates into the aqueous medium forming the nanoparticles
	Emulsification diffusion of the solvent: fully water immiscible organic solvents are saturated in water to ensure thermodynamic balance. The organic and aqueous phases are emulsified; the resulting emulsion is diluted in the aqueous medium, which leads to diffusion of the organic solvent and solidification of the nanoparticles that form the dispersed phase of the emulsion. Depending on the nature of the solvent, it may be removed by ultrafiltration or distillation
	Solvent injection: lipids are dispersed in water miscible solvents and quickly injected into an aqueous surfactant solution, resulting in the precipitation of the nanoparticle

are called mesogens. Upon temperature increase or addition of solvent, the formation of mesophases occurs, which form thermotropic and lyotropic liquid crystals, respectively (Singh 2000; Müller-Goymann 2004).

- Thermotropic liquid crystal: These crystals are formed upon temperature change; many of them migrate between phases and are called polymorphs. They can be classified into calamitic (consisting of rod-shaped molecules) or discotic (consisting of flat disk-shaped molecules). Depending on the degree of anisotropy, calamitic liquid crystals can be classified as phase (1) nematic, (2) smetic or (3) cholesteric, while discotic can be classified as (1) discotic or (2) columnar nematic (Villanueva-García et al. 2005; Lechuga-Ballesteros et al. 2003).
- Lyotropic liquid crystal: consists of the interaction of amphiphilic molecules with each other, which are arranged in different mesophases according to concentration and temperature. These mesophases are thermodynamically stable and are divided into lamellar, hexagonal and cubic (Chountoulesi et al. 2018).

  (a) Lamellar mesophase: They are made up of hydrophilic amphiphilic molecules that are shaped like cylinders, arranged in layers producing blades that intersect into polar and non-polar layers (Matjaž et al. 2019).
  (b) Cubic mesophase: It is the intermediate phase between lamellar mesophase and hexagonal mesophase. It has a unique structure consisting of a bicontinuous curved lipid bilayer, composed of three dimensions and two nanoaqueous channels, with a high interfacial area. This phase in its traditional form can cause contact irritation when applied to the skin. To circumvent this inconvenience, a dispersion into small particles has been formulated, forming the cubosomes that can be identified in three distinct phases which

are (1) double diamond structure (Pn3m), (2) body-centered cubic phase (Im3m) and (3) gyroid latitude (Ia3d) (Guo et al. 2010; Madheswaran et al. 2019).

(c) Hexagonal mesophase: These are closed structures formed by column-shaped micelles, formed by spontaneous ordering of amphiphilic lipids. According to the orientation of the hydrocarbons that make up the lipids, this phase can be classified into two types (1) normal or H1 hexosomes where the hydrocarbon chains are oriented toward the inner core or (2) inverse or H2 where the formation occurs of a water channel between the micelles (Guo et al. 2010; Madheswaran et al. 2019).

Better solubility of poorly soluble drugs, better drug loading, controlled release, high bioavailability, use in theranostics are some of the benefits that are intrinsic to liquid crystals. The lyotropic phase of the liquid crystal is of major interest for the delivery of drugs such as siRNA based on the lipid monoolein, especially forming cubic and hexagonal phases, or in the application of water-soluble zinc-phthalocyanine tetrasulfonate (ZnPcSO4) by photodynamic topical therapy in the treatment of cancer (Madheswaran et al. 2019; Gosenca, Bešter-Rogač, and Gašperlin 2013; Petrilli et al. 2016a, b; Praça et al. 2012).

2.2 *Polymeric Nanoparticles*

As previously mentioned, nanoparticles can be synthesized from various sources such as lipids, polymers, carbon or metals. When searching for the preparation of polymer-based drug carrier nanosystems, the range of possibilities is very wide. Thus, natural, synthetic or semi-synthetic polymers can be used; some examples of these type of material that are commonly used are: proteins, cellulose, chitosan, among the natural ones; aliphatic polyanhydrides, polystyrene as examples of the synthetic ones; agar–agar, alginates as examples of the semi-synthetic polymers (Severino et al. 2011).

The choice of which polymer will be employed is related among other factors to the physicochemical characteristics of the drug to be encapsulated, especially its polarity. For instance, hydrophilic drugs tend to be encapsulated by non-polar polymers in an organic phase that will allow the solubilization of the polymer and an easier encapsulation of the drug, since it is insoluble in the organic phase. Conversely, hydrophobic drugs tend to be encapsulated by polar polymers in an aqueous phase allowing, as in the previous one, a better solubilization of the polymer and easier encapsulation of the drug (Severino et al. 2011). In addition to the characteristics of the drugs, some negative characteristics present in the polymer should be considered prior to its selection for the delivery system, for example, natural polymers have varying purity degree between batches, which can be a problem when working with materials aimed for systemic delivery (Hans and Lowman 2002).

Polymeric nanoparticles have gained highlight in the scientific scenario due to some of their characteristics that allow a better therapeutic response to the detriment of other NPs, one of the main being their biocompatibility and, more important, its biodegradability, thanks to that characteristic of biosafety, once in the human body polymers will be degraded by metabolic process, such as hydrolysis, oxidation, conjugation and others, leaving the degradation products free to be eliminated by the commons pathways of excretion. Besides, it is possible to mention some other important advantageous characteristics of polymers, such as improved drug release control, good stability of loaded drug, by protecting of contact with oxidizing agents, proteases and other enzymes, This class of NPs has the possibility of wide variety of structural modifications that allow them to bind to different ligands, such as antibodies. These modifications lead to much more complex structures, allowing for targeted drug delivery (Crucho and Barros 2017; Soppimath et al. 2001; Gutjahr et al. 2016).

Polymeric NPs are a large class of nanocarriers that encompass both nanospheres and nanocapsules. Nanospheres have a structure formed by a polymeric matrix. The drug can be adsorbed throughout the polymeric matrix, both on its surface and on the inside; nanocapsules have a shallow nucleus in which occasionally the drug can be loaded, in some cases depending on the characteristics of the drug, such as its lipophilicity, it can only be found on the inside of the shell (Crucho 2015). Both forms are shown in Fig. 1.

It is important to emphasize that all factors of preparation of a nanoparticle must be properly known and understood, since they will dictate the final characteristics of the formed nanoparticles. It is worth mentioning that for NPs to release the drug at

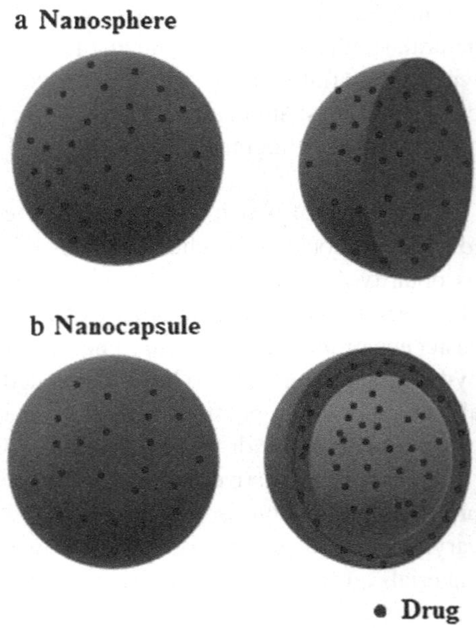

Fig. 1 3D representation of nanospheres (**a**) and nanocapsules (**b**) where the drug can be loaded in different portions within polymeric nanoparticles

the site of action, they must first be absorbed, and their absorption is related, among other parameters, to their size and surface loads and these parameters. Therefore, they are dependent on the preparation method, the physicochemical characteristics of the chosen polymer, its proportions and other factors that must be studied before preparing the formulation. Some examples of preparation methods are: the emulsification/solvent evaporation technique that was developed in 1979, and it is the pioneer method for preparation of polymeric nanoparticles. This method requires that the polymer must be diluted in a volatile solvent that will be emulsified in a second phase composed of an aqueous solution and a surfactant, the biphasic mixture must be subjected to a high-speed homogenization or some communing force that allows the production of an aqueous emulsion. Then, the polymeric nanoparticles will be produced by the evaporation of the solvent phase (Crucho and Barros 2017; Dash and Konkimalla 2012).

The nanoprecipitation method, one of the most common used technique developed to the prepare of polymeric nanoparticles, is an ease and cheap technique where the polymer is dissolved in an organic, which must be slowly dropped into a stirred a miscible aqueous solution containing a surfactant. The NPs will be formed thanks to the slowly evaporation of the solvent, allowing the formation of polymeric NPs with well-defined size and uniform distribution (Crucho and Barros 2017; Dash and Konkimalla 2012).

Table 4 shows a list of some of the main polymers used, followed by their method of preparation and some of the results obtained. From this, it is possible to observe the numerous variables are possible in the polymeric nanoparticle synthesis, in order

Table 4 Examples of studies reporting drug-loaded polymeric nanoparticles

Drug	Polymer	Preparation method	Size (nm)	Zeta potential (mV)	Encapsulation efficiency (%)	Reference
Retinoic acid	PLA	Nanoprecipitation	153–229	−10.4–(−29.4)	10.4–90.2	Almouazen et al. (2012)
Indomethacin	PCL	Emulsion diffusion	80–400	–	68–73	Limayem et al. (2004)
Ciprofloxacin	PLGA	Multiple emulsion-solvent evaporation	130–353	–	42.4–47.4	Jeong et al. (2008)
Curcumin	PLGA	Nanoprecipitation	150–235	−45.3–32.8	68.9–92.3	Klippstein et al. (2015)
Curcumin and camptothecin	PLGA	Oil-in-water emulsion-solvent evaporation	221–264	−10–(−20)	35.7–86	Xiao et al. (2015a)
Camptothecin	PLGA	Oil-in-water emulsion-solvent evaporation	258–286	−12.8–21.1	86.19–99.34	Xiao et al. (2015b)

*PLA: Poly(D,L)lactic acid; PCL: Poly(ε-caprolactone); PLGA: Poly(D,L-lactide-co-glycolide)

to achieve the ideal characteristics in NP, showing, as was stated at the beginning of the text, the great potential of polymeric nanoparticles for drug delivery.

2.3 *Inorganic Nanoparticles*

Previous sections reported the use of nanoparticles based on polymers, lipids, surfactants and other organic compounds (carbon-based compounds, usually containing C–H bonds). Such molecules are by far the most used platforms at the drug delivery systems, but as nanotechnology in medicine is not limited to that application, alternatives to those species also exist. Most inorganic nanoparticles possess magnetic, electric and optical characteristics that rely on their size, surface, shape and charge. Comparing to organic nanoparticles, they lack a complex and versatile chemical manipulability (possibilities to attach ligands, modulate drug release rate, limitations on drug association) and less progress has been made on unraveling the potentials of drug release on such systems; however, inorganic nanoparticles have been frequently studied for the diagnosis and even theranostics. Among the most known inorganic nanoparticles applied to biological systems, iron oxide nanoparticles, carbon nanotubes, gold nanoparticles and silver nanoparticles play a featured role. Many other inorganic nanoparticles are also well described in scientific literature, but their applications as biomedical aids are not well established yet—like nanoceramics, fullerenes and quantum dots. In this section, the use of such inorganic nanoparticles in medicine will be addressed.

2.3.1 Iron Oxide Nanoparticles

Iron is the most frequent transition metal in the Earth's crust and has been used for several applications since the Iron Age, more than 3000 years ago. It is an element avid for oxygen not only in biological systems, (e.g., as we see in hemoglobin) but often present many iron oxides are present in nature (e.g., as magnetite, maghemite and haematite) (Nedyalkova et al. 2017). Iron oxide nanoparticles are versatile and present in biomedical, health care, agriculture and food, environmental remediation, energy, defense, automotive, textiles and electronics industries. The most interesting and useful characteristics of such materials are the superparamagnetism, low toxicity and the easy separation methodology (Roca et al. 2019; Ali et al. 2016). These characteristics enable iron oxide NPs to be induced into magnetic resonance by self-heating when an external magnetic field is applied. These self-heating may harm certain cells or tissues, situation that needs to be handled with care; but also interesting when a cytotoxic activity is desired against certain tissues (like tumors). Still, one limitation of iron NPs is the extreme reactivity, which may cause stability and toxicity problems if the nanoparticle is not chemically stabilized by an external moiety (starches, surfactants and others) (Abd Elrahman and Mansour 2019; Dadfar et al. 2019; Valdiglesias et al. 2016).

The preparation of iron oxide magnetic nanoparticles may be done by physical (deposition of gas phase, electron beam lithography), chemical (e.g., chemical coprecipitation, electrochemical techniques) or microbiological techniques (microbial incubation). Most physical methods are unable to control particle size distribution—a critical parameter for the application in biological systems—and so not extensively used. Chemical methods are by far the most studied and comprehended, as they usually are simple, efficient and handle to manage the shape and other parameters of the NPs. Microbial methods are prominent due to their high reproducibility and low cost, but the elevated time consumption is generally a limiting factor (Ali et al. 2016; Dadfar et al. 2019).

2.3.2 Carbon Nanotubes

Carbon nanotubes (CNTs) are tridimensional tube-like hollow structures made of carbon atoms—essentially a graphene structure rolled into itself into narrow and long molecules (Xue 2016). Since its first description and discovery by Iijima (1991), this structure has drawn interest on materials science scientific community due to its particularly useful properties: unique structure, great electrical and conductive properties, extreme mechanical resistance and high specific surface area. However, one of the major drawbacks on carbon nanotubes in biomedical area is the high toxicity associated with those structures, which area strictly linked to the morphology, functional groups and dose administered. In general, CNTs toxicity is often due to the acceleration of oxidative stress generated by their contact with biological systems and the hydrophobicity associated with the non-polar structures; the hydrophobicity also impairs CNTs solubility in water and consequently in biological systems, generating an aggregation tendency. Surface modification approaches have been recently used to overcome such problems, turning the structure less cytotoxic and with potential to drug delivery (Simon et al. 2019; Xu et al. 2006).

2.3.3 Gold Nanoparticles

Among the several metallic nanoparticles applied in biomedicine, gold nanoparticles (GNPs) are among the most used systems due to their advantageous optical, chemical and catalytic characteristics (de Araújo et al. 2017; Pena-Pereira et al. 2017). Known to be inert and extremely stable against oxidation (as a common property of gold itself), GNPs may be prepared by several different methods with distinct outcomes in terms of particle structure and morphology. Turkevich first developed a method to prepare GNPs, which was further improved by Frens and Burst and is currently in use with some modifications up to nowadays (Adnan et al. 2016; Shah et al. 2014; Priyadarshini and Pradhan 2017).

As most of other nanoparticles, GNPs surface can be modified with different moieties as antibodies, small molecules, peptides and others, providing a selective targeting effect either by physical adsorption or by covalent coupling. GNPs can also be designed to react to external environmental alterations like pH, temperature, luminosity, specific frequencies, electricity, magnetic stimulation, electric current and even external heat application. Stimuli-responsive nanoparticles are designed to respond to specific intrinsic and extrinsic stimuli and deliver drugs to target site (Ajnai et al. 2014; Eloy et al. 2015). Such response may happen by physicochemical changes in the structure and surface of the GNPs, like swelling, charge switching or dissociation of moieties. Photothermal properties of GNPs are especially significant with nanorods, as those morphologies tend to facilitate the plasmonic heat generation, better producing heat with less energy (Yeh et al. 2012; Gormley et al. 2011). Scale-up methods, however, have not been extensively exploited on the preparation of gold nanoparticles.

2.3.4 Silver Nanoparticles

Silver nanoparticles (AgNPs) are also an extensively studied nanostructure system in medicine, especially in wound healing applications, antimicrobial approaches and drug delivery systems (Gunasekaran et al. 2011; Kumar et al. 2018; Rai et al. 2009; Malik and Mukherjee 2018). Silver is a soft and shiny metal, extensively used across centuries due to its disinfectant and wound healing properties. Silver presents both high thermal and electrical conductivity, interesting properties for inorganic materials that are candidates for nanodesign and enable development. However, as most of inorganic nanoparticles, several studies show potential cytotoxic activity against human cells when not properly coated with biocompatible polymers. General silver toxicity is well described in literature, and AgNPs tend to accumulate in the body (Courtois et al. 2019).

3 Targeted Nanoparticles

Aiming better therapeutic outcome with reduced systemic side effects, several nanoparticles are widely studied for use in tumor targeting, through two main mechanisms: (i) active targeting mechanisms based on receptor ligand affiliation (Fig. 2) or (ii) passive targeting based on the enhanced permeation and retention effect due to the increased permeability in the tumor region (EPR Effect), for example cancerous tissues and inflamed sites possess a leaky vasculature that increases liposome accumulation compared to use of the free drug (Anarjan 2019; Eloy et al. 2017b).

Active drug delivery nanoparticles can be divided into two main classes of nanoparticles: organic nanoparticles such as liposomes, micelles, niosomes, dendrimers, polymeric nanoparticles and inorganic nanoparticles such as metallic, magnetic, silica nanoparticles and quantum dots (Dissanayake et al. 2017; Anarjan 2019). Properties such as surface area and volume ratio are important to ensure high density of surface ligands and thus through specific interactions with receptors and

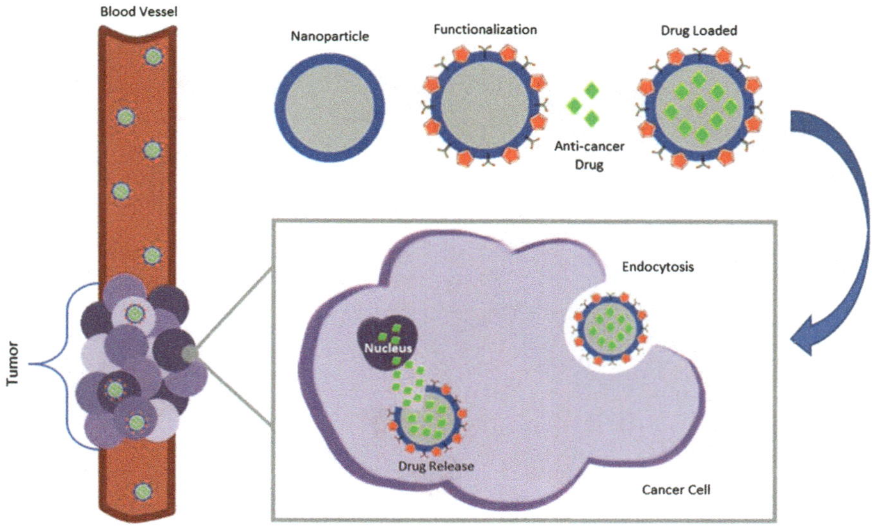

Fig. 2 Overview of molecular targeted therapy mechanism for cancer

nanocarriers on target cells to facilitate internalization of the nanoparticle—receptor complex through the endocytosis process (Anarjan 2019). Ideally, nanoparticles near 100 nm in size and with hydrophilic surface are also sought to reduce macrophage clearance and promote longer circulation times, thus increasing the ability to reach the target (Brannon-Peppas and Blanchette 2012).

Different targeting strategies can be used with versatile carrier nanoparticles to optimize selective target delivery. Any failure to maintain the concentration of anti-tumor agent at adequate levels in the tumor tissue may result in the development of drug resistance and tumor cell growth, so the choice of nanocarrier and the ligand used are relevant (Anarjan 2019; Eloy et al. 2017b). Several targeting ligands may be used as shown in Fig. 3; these have different functions and characteristics, varying according to target and acting on cell surface antigens, signal transduction receptors responsible for regulating cell cycle progression, growth factors, angiogenesis, metastasis and cell death. The targeting strategy involves many ligands such as peptide, antibodies, saccharide, transferrin, folate, attached covalently or non-covalently to the surface of nanoparticles using different strategies (Petrilli et al. 2014; Lee et al. 2018; Eloy et al. 2017b).

3.1 Antibody Targeted Nanocarriers

Briefly, antibodies are proteins usually in the form of "Y" that perform two major functions in the immune system, acting in the antigen elimination process and able to recognize and bind antigens. In the structure, the antibody has three main parts; the

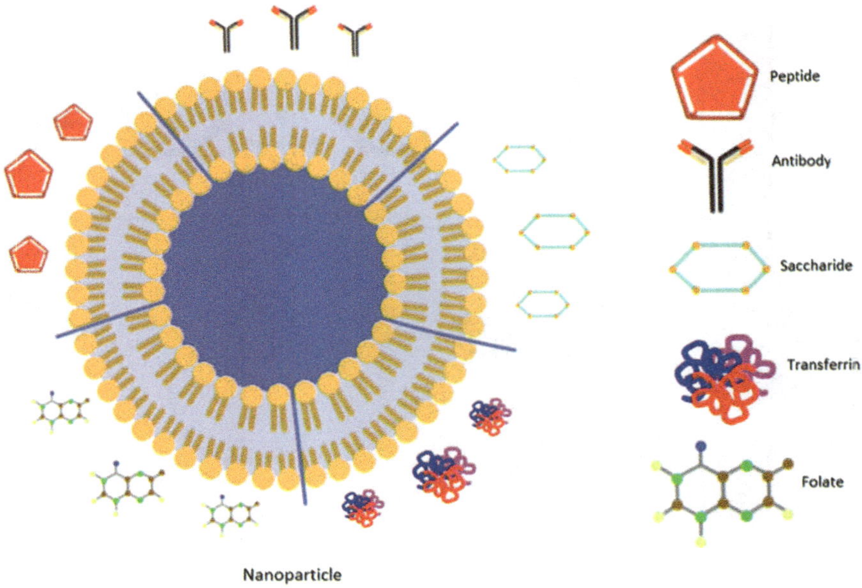

Fig. 3 Schematic representation of different ligands that can be attached to the surface of nanoparticles for targeted delivery

portion of crystalline fragment (Fc) enables complementation by binding to target cell receptors, triggering antigen elimination functions. It is the double portion of fragments antigen binding (Fab) responsible for antigen recognition and specificity. Thus, antibody or also called immunoglobulin or antibody fragments such as Fab or single-chain variables fragment (scFv) become an important factor in the active targeting using nanoparticles (Eloy et al. 2017b; Anarjan 2019).

Anti-vascular endothelial growth factor (VEGF) and anti-epidermal growth factor receptor (EGFR) in cancer cells are the most widely used antibodies in active targeting systems. For the conjugation of the antibody to the nanoparticle surface usually binding molecules such as polyethylene glycol (PEG) are used, acting as spacers for the linkage of maleimide and antibody, resulting in the exposition of antibodies in the surface of immunoliposomes (Petrilli et al. 2018; Wang et al. 2016; Petrilli et al. 2014; Anarjan 2019).

Different strategies can be used to attach antibodies or its fragments to the surface of nanoparticles. Among them, the most commonly applied is the reaction of thiolated antibody to maleimide, this occurs through covalent interactions when antibodies containing sulfhydryl groups after thiolation form a thioether linkage through maleimide-containing lipids. For targeting purposes, subtherapeutic doses of the antibody are conjugated to nanoparticles. For instance, Petrilli et al. 2016a, b developed a cetuximab immunoliposomes, where Traut's reagent was used for thiolation of cetuximab with 1 mg/mL (Petrilli et al. 2016a, b; Eloy et al. 2017b).

3.2 *Peptide Targeted Nanocarriers*

Briefly, peptides are identified as charged carriers such as cell penetrating peptides (CPPs) or also known as protein transduction domain (PTD), characterized by having a short chain, with about 10–30 amino acids. Through single or multiple endocytic mechanisms, they have the ability to cross cell membranes together with a conjugated biological charge such as drugs, nucleic acids, proteins and nanoparticles (Petrilli et al. 2014; Dissanayake et al. 2017). Peptides have advantages because of their low toxicity, small size, ease of synthesis in large quantities, favorable pharmacokinetics, high affinity for specific receptors, and peptide sequence can be altered for load adjustments, hydrophobicity, stability, affinity and solubility (Dissanayake et al. 2017; Petrilli et al. 2014).

A variety of CPPs are currently known, but they usually consist in a non-amphipathic arginine-rich CPP such as transcriptional activator peptide (TAT) that demonstrated to be capable of translocate into cells of the viral genome and activate the transcription. Another frequently studied CPP is penetratin (PNT), obtained from an Antennapedia homeodomain. The mechanism of internalization of CPPs has not yet been fully elucidated, but studies of internalization in cell lines are performed under conditions to prevent active transport of CPP and its translocation by endocytotic pathways. Zorko and Langel (2005) suggest that CPP internalization occurs in several steps starting with the interaction of CPP with the cell surface, through electrostatic interaction to charged phospholipids, followed by endocytosis, peptidolytic degradation of CPP in the cell and finally the release of degradation products of this process happens (Zorko and Langel 2005; Dissanayake et al. 2017).

Another common group of peptides is arginine–glycine–aspartic acid (RGD) which is constantly used for cell adhesion in extracellular matrix proteins as an integrin ligand expressed, for example, in activated tumor vasculature endothelial cells. Integrins such as $\alpha v\beta 3/\alpha v\beta 5/\alpha v\beta 6/\alpha v\beta 8/\alpha vb8/\alpha IIb\beta 3/\alpha 8\beta 1/\alpha 5\beta 1$ are receptors of heterodimeric transmembrane glycoproteins that have non-covalent α and β complexes and high affinity with RGD. Integrins are then targeted against cancer cell angiogenesis and metastasis. The $\alpha v\beta 3$ integrin, during angiogenesis, is overexpressed on the endothelial cell surface in order to favor the growth and survival of new vessels, so RGD nanoparticles block αv integrin function and may produce an antiangiogenic effect inhibiting tumor growth (Anarjan 2019; Petrilli et al. 2014).

3.3 *Folate Targeted Nanocarriers*

The folate receptor is a glycoprotein, overexpressed in solid tumor types including ovarian, brain, breast, kidney, lung, making this ligand a potential target for nanoparticles containing anticancer drugs (Farran et al. 2019; Anarjan 2019). Due to its high affinity, small size, ease of modification, low immunogenicity, storage stability, the conjugation of folic acid has improved the targeting to cancer cells that overexpress folate receptors (Lu and Low 2002; Petrilli et al. 2014; Anarjan 2019).

The mechanism suggests the folate receptor is organized by glycosylphosphatidylinositol into membrane-rich receptor complexes and that a folate receptor (FR) is overexpressed in cancer cells. Basically, folic acid is covalently attached to the surface of nanoparticles, which allows specific recognition and drug delivery depends on the receptor-based endocytosis (Ceborska 2017).

3.4 Transferrin Targeted Nanocarriers

Human transferrin is a glycoprotein that is mostly produced by the liver, that acts on the delivery of iron through circulation to cells, which is important for cell growth (Kawabata 2019; Anarjan 2019). Due to the high demand for iron, there is a high expression of the transferrin receptor (TfR) on the surface of cancer cells. Thus, transferrin-functionalized nanoparticles can work directing the release of drugs in malignant cells and using molecules with ability to contrast receptor function. Wei et al 2016 propose nanoparticles capable to pass through the impermeable blood–brain barrier. They obtained more accumulation in brain tissues with nanoparticles using transferrin receptor than the non-targeted nanoparticles. Besides, in vivo experiments demonstrated for tumor therapy, prolonged survival time in mice with intracranial U87 glioma treated using nanoparticles with transferrin (Wei et al. 2016).

One promising approach is the conjugation of nanoparticles with transferrin for delivery of specific site genes, drugs and dyes for diagnostics, because the use of transferrin as a moiety favors the entry of nanoparticle into tumor cells by receptor-mediated endocytosis. In one study by Singh and collaborators, using polymeric micelles containing docetaxel functionalized with transferrin showed the effectiveness of nanoparticle with transferrin as a targeting moiety compared to free drugs. The suggested mechanism involves TfR-mediated endocytosis, followed by the recycling of TfR receptor from the cytosol back to the membrane (Singh et al. 2016; Anarjan 2019; Petrilli et al. 2014).

3.5 Saccharide Targeted Nanocarriers

Carbohydrates can be used as biodegradable nanoparticle material for controlled release and directing effect due to specific carbohydrate protein interactions. By taking advantage of the interaction between carbohydrates on the surface of a nanocarrier and overexpressed proteins in cancer cells, it is possible to direct drugs to the target (Seidi et al. 2018).

Due to the expression of asialoglycoprotein receptors in hepatocytes in the liver, galactose, mannose, fucose ligands may be used in specific targeting materials due to the ability to selectively bind to galactose-terminated glycoproteins for example.

In addition, there is evidence that saccharide binding not only leads to the targeting effect, but also improves the physical stability of the complexes that permits small formations with neutral zeta potential. Also low immunogenicity compared to antibody and peptide targeting moieties is expected, making their in vivo use promising (Seidi et al. 2018; Petrilli et al. 2014; Merdan et al. 2002).

4 In Vitro Studies for Nanoparticles Evaluation

4.1 Cytotoxicity and Cellular Uptake

The previous sections of this chapter introduced some of the main nanoparticles types. The preparation and characterization methods previously discussed can potentially influence nanoparticles performance. Basically, in vitro studies can be used for screening and precede the in vivo studies. Thus, a variety of in vitro assays such as cell viability, uptake and apoptosis detection can be applied to evaluate the toxicity and selectivity of nanoparticles for target cells (Yazdimamaghani et al. 2019). Usually, the reduction of mitochondrial function, loss of selective cellular permeability, as well changes in cell morphology and replication are indicative of cytotoxicity (Borenfreund and Puerner 1985).

In the next sections, cytotoxicity studies will be presented and different available methodologies will be discussed. Furthermore, confocal and flow cytometry analysis will be described as methods for evaluating cellular internalization of nanoparticles.

4.1.1 Cytotoxicity

Cytotoxicity testing is a type of biological and screening evaluation that uses in vitro tissue cells to observe cell growth and morphological effects. In this context, most cytotoxicity studies assess cell viability after different treatments. Thus, the concentration of the cytotoxic agent, influence of encapsulation and time of exposure can be established. To this end, generally one or more concentrations of cytotoxic agent may be applied to the cells, and the viability percentage is analyzed with one or more exposure conditions. They may therefore have the quantitative response of individual toxic compounds in different systems or of various compounds in individual systems (Eisenbrand et al. 2002). In the study by Petrilli et al. (2017), 5-FU immunoliposomes were evaluated against skin cancer cells using 120 h treatment, showing a potent cytotoxic effect. The study obtained an immunoliposome designed to improve the cytotoxicity of administered drugs while maintaining a particle size of less than 180 nm (Petrilli et al. 2017).

Most cell types readily absorb nanoparticles. For instance, Epple (2018) in his review paper, focusing on inorganic calcium phosphate, presented and discussed a series of studies with nanoparticles. Only very small nanoparticles can penetrate the cell membrane themselves, which is not the case with larger particles, being absorbed by pinocytosis, endocytosis or phagocytosis. Thus, the critical size depends on the properties of the nanoparticles, the cell type and the uptake mechanism. Finally, the particle shape, spherical or rod-like, plays a minimal role in the cytotoxic biologic response (Epple 2018).

Neuhaus et al. (2016) and his collaborators have developed a DNA-coated organic calcium phosphate nanoparticle. In the study, the authors synthesized and stabilized the nanoparticle by DNA coating. The nanoparticle obtained was characterized as a spherical structure with size between 30 and 140 nm, with some agglomeration and zeta potential between +20 and +25 mV. The authors concluded that particle size and cell type are important for cytotoxicity (Neuhaus et al. 2016).

Different methods can be used to evaluate cellular cytotoxicity. Tetrazolium salts are generally some of the most commonly applied compounds for investigating cytotoxicity, proliferation and cell viability (Funk et al. 2007). Some of these methods will be presented and discussed below, such as MTT, MTS and XTT. MTT is widely applied in homogenous cells measuring mitochondrial activity where the net positive charge in tetrazolium salts easily penetrates eukaryotic cells, chemical electron transfer reactions are also involved in this process, such as NADH that transfer electrons to MTT. MTS and XTT are also composed of tetrazolium; however, the net negative charge of salt presents difficulty in penetrating the cell, requiring intermediate electron acceptors that easily transfer electrons from the plasma membrane or cytoplasm which in turn reduce tetrazolium to formazan (Sittampalam et al. 2016).

Methyl Thiazolyl Tertrazolium (MTT) Assay

The methyl thiazolyl tertrazolium (MTT) assay is currently one of the most applied tests for in vitro evaluation of cytotoxicity. It analyzes mitochondrial viability by measuring formazan crystal formation generated by tetrazolium reduction (Fig. 4). The result is measured though the absorbance compared to a control, usually measured at 570 nm, after formazan solubilization in adequate solvent (Sjögren and Sletten 2000).

Mitochondrial dehydrogenase from living cells can cleave the tetrazole ring by transferring electrons and converting soluble yellow MTT to a purple crystalline formazan. Formazan although insoluble in water is soluble in dimethyl sulfoxide and other organic solvents. The number of cells and their activity has a positive correlation with the amount of crystals formed, and the measurement of the colorimetric absorbance value; thus the optical density will reflect on the metabolic activity and the number of surviving cells (Li et al. 2015).

This method for cell viability evaluation is simple and precise; the quantification is easy and can be evaluated and quantified by photometric plate readers. The salts formed are directly related to membrane permeability. On the other hand, the assay has disadvantages such as underestimation of living cells because it is dependent on any physiological changes in cells, such as alterations in lysosomal and mitochondrial

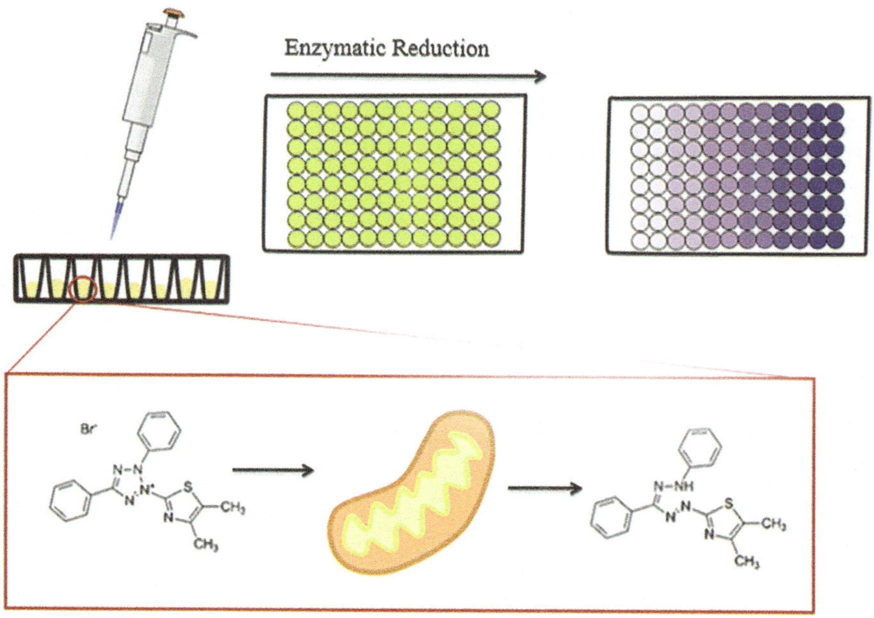

Fig. 4 Methyl thiazolyl tertrazolium (MTT) assay

activities. Another disadvantage is the dependence on cell conditions such as cell density (Stockert et al. 2018).

The assay has several time-consuming steps, when mitochondrial cells undergo drug-mediated stress and inhibitor off-target effects interfere with the MTT reduction rate, resulting in inconsistent results. In this context, to avoid misinterpretation of the result, it is important to perform also other non-metabolic trials (Stepanenko and Dmitrenko 2015). Other compounds that underestimate the test result is single-wall carbons, nanotubes, antioxidants, flavones, flavanones, polyphenols, copper and silver nanoparticles (Holder et al. 2012; Stockert et al. 2018). Some applications of MTT assay in recent papers can be found in Table 5.

XTT

XTT [2,3-bis (2-methoxy-4-nitro-5-sulfophenyl) -2H-tetrazolium-5-carboxanilide], as well as MTT, is a widely explored approach to measure the cell viability. Although both assays are widely used they have different solubility. XTT requires an electron coupling reagent for optimal yield, and formazan salts formed are soluble and consequently can be used in real-time assays (Berridge et al. 2005; Funk et al. 2007).

This assay also has interferents, as constituents of the cell medium and other substances such as cellular enzymes, as well as bovine or human albumin preparations reduce tetrazolium salts. For instance, MTT and XTT were evaluated for their

Table 5 Examples on application of MTT assays in some recent papers

Study aim	Nanoparticle	Drug	Cell culture and incubation time	Reference
Inhibition of H1N1 influenza virus infection by zinc oxide nanoparticles	Inorganic nanoparticle	Oseltamivir	MDCK-SIAT1 cells incubated for 24 h	Ghaffari et al. (2019)
Use of hybrid nanoparticles as drug carriers for in vivo cancer therapy	Inorganic nanoparticle	Cisplatin	MDA-MB-231 cells cell line incubated for 24 h	Abedi et al. (2019)
Effects of mesoporous silica nanoparticles (MSNs) on HeLa cells and astrocytes	Inorganic nanoparticles	Aminofluorescein dye and diethylenetriaminopentaacetic acid	HeLa cells and astrocytes incubated during 24 h	Fisichella et al. (2009)
5-FU liposome and immunoliposomes effect in skin cancer cells	Lipid nanoparticle	5-fluorouracil	Cutaneous carcinoma cells A431 for 72 and 120 h	Eloy et al. (2017a)
Evaluation of paclitaxel and rapamycin liposomes in breast cancer	Lipid nanoparticle	Paclitaxel and rapamycin	4T1 breast cancer incubated for 24 h	Eloy et al. (2016)

accuracy in the presence of serum albumin in the study by Funk et al. (2007). Serum albumin was diluted at different concentrations and plated into 96 wells, to which 50 μL of XTT was added, passing 4 h while stirring at 37 ° C. At the end of the incubation time, the samples were read using a 480 nm absorbance plate reader. The same conditions were adopted for MTT testing. At the end of the incubation period, 100 μL isopropanol was added to each well of the plate; absorbance was read at a wavelength of 540 nm. The authors concluded that when there is albumin or other free proteins in XTT and MTT assays, attention should be paid to overestimating cell numbers and underestimating the potential cytotoxic effects tested (Funk et al. 2007).

MTS

MTS 5-[3-(carboxymethoxy) phenyl]-3-(4,5-dimethyl-2-thiazolyl)-2-(4-sulfophenyl)-2H-tetrazolium internal salt salts are an intermediate electron acceptor for reducing and forming formazans that can be used in real time assays. In contrast,

MTS emits a weak signal; the ability of MTS to be rapidly reduced suggests that these tetrazolium dyes can be applied in simple microplate assays for measuring NADH and NADPH. MTS other than MTT has its positive charge counterbalanced by a negatively charged sulfonate in a phenyl ring; in addition, it can be used for kinetic study in viable cell measuring assays performed sequentially several times without extracellular formazan crystal formation (Berridge et al. 2005). The amount of sulfonated formazans delivered in cell culture and colorimetric detection is directly proportional to the number of living cells (Dunigan et al. 1995; Stockert et al. 2018).

Mofokeng (2019) and colleagues developed a water-soluble alanine-capped CuS nanoparticle using the MTS test to measure cell viability in human cervical carcinoma cells. At the end of the incubation process, the medium was removed from the wells and the MTS dye added for 4 h; finally, the MTS reagent was replaced with 100 μL of an isopropanol HCl solution and incubated for 1 h. The results were obtained by a microplate reader at 570 nm. Gold nanoparticles were used as negative control in the study due to previous reports of low cytotoxicity. The study results show that synthesized CuS nanoparticles induce a reduction in cell viability (Mofokeng et al. 2019)

Other Methods

The trypan blue exclusion assay is also used in the cytotoxicity study. Stepanenko and Dmitrenko (2015) conducted a cytotoxicity study observing the MTT test in comparison with the trypan blue cell exclusion test. The assay was performed with U251/T98G and C6 cells. Viable cells exclude trypan blue due to membrane integrity, non-viable cells are easily stained and counted with the aid of a microscope with a hemocytometer. In the results obtained in the study, it was clear that trypan blue showed significant over/underestimation of cell viability compared to MTT assay is possible due to cell line and experimental parameters (Stepanenko and Dmitrenko 2015).

The trypan blue test makes explicit some disadvantages that may influence the underestimation of the result, such as this dye stains only necrotic or cells with apoptosis not marking cells that started the apoptotic process and that still has intact membrane. Additionally, cells subjected to senescence or failed mitosis where they are unlikely to be stained (Stepanenko and Dmitrenko 2015)

Another method for measuring viable cells is the resazurin assay. Resazurin is metabolized by mitochondrial enzymes that reduce non-fluorescent blue resazurin to highly fluorescent pink resorufin. Breznan (2015) studying the nonspecific interaction of carbon nanotubes performed the cytotoxicity test with the resazurin assay to assess the in vitro impact of this nanoparticle. Human lung (A549) and murine macrophage (J774A.1) epithelial cells were incubated for 24 h with the nanoparticles. At the end of the incubation time, the supernatant was discarded and added to resazurin for fluorescence microplate monitoring (Breznan et al. 2015)

The test is simple, fast, versatile, economical and shows a high degree of correlation with cytotoxicity evaluated by other methods, since resazurin cell reduction occurs by mitochondrial enzyme and the cell-metabolized reagent is not toxic to the cell. It can be reused in cytotoxicity research saving time and money in cell culture, unlike the MTT that needs to kill the cell to get the result. Although the test also has some disadvantages such as the reduction of resazurin by antioxidant compounds present in the cell culture medium, another point observed was the non-fluorescent and colorless final product metabolism with potential for underestimation of cell viability (O'Brien et al. 2000; Breznan et al. 2015).

4.1.2 Cellular Uptake

Understanding how a nanoparticle interaction process with the biological system occurs is important for the development of new nanotechnology-based drugs. This interaction is dynamic, complex and has several parameters. Nanocarriers first are in the extracellular medium either in vitro or in vivo, then find the target cell plasma membrane consisting of a selectively permeable lipid bilayer and molecules, after this passage through the membrane, enter the intracellular compartment by a process called endocytosis, encompassing nanocarriers in the invaginations of the membrane that formed vesicles called endosomes (Donahue et al. 2019; Seo et al. 2017).

These routes that promote nanocarrier crossing through the membrane can be divided into two general routes. The first pathway is based on endocytosis which is divided into four different mechanisms: macropinocytosis, clathrin-dependent endocytosis, caveolin-dependent endocytosis and clathrin and caveolin independent endocytosis. The other pathway is direct access of nanocarriers to the cytoplasm using biochemical or physical means that is divided into four distinct mechanisms: direct entry by translocation, lipid fusion, electroporation and microinjection. The different input mechanisms are directly correlated with the physical chemical properties and type of nanocarrier such as surface geometry, size and chemistry. Usually, the mechanisms of caveolae are more commonly found upon the administration of nanoparticles with size limits near 100 nm (Donahue, Acar, and Wilhelm 2019; Seo et al. 2017; Lu et al. 2002).

Understanding how nanocarriers are internalizing and interacting with cell membranes, intracellular compartments and nuclear membrane molecules has helped the design and evaluation of more effective nanoparticles (Donahue, Acar and Wilhelm 2019; Seo et al. 2017).

For cellular uptake studies, both flow cytometry and confocal studies can be performed. Flow cytometry allows morphological characterization of the cell as size, expression of molecular targets, determination of viability and fluorescence measurement of nanoparticle internalization (Romero et al. 2010). It is a quantitative analysis technique that enables the simultaneous measurement of the physical, chemical and biological characteristics of suspended cells (Bajgelman 2019). In this methodology, a beam of light with constant wavelength intercepts the sample generating signals that are captured, transformed, stored and further processed by the software. The optical system of this equipment is formed by monochrome lasers and filters that direct the light signals to a detector. The sample to be analyzed is injected into the center of the flow chamber which must be filled with a saline solution called sheath fluid (Biosciences 2002).

Combinations of lasers, filters and compensation technologies enable the simultaneous identification of different surface cell markers, making flow cytometry a powerful analytical tool (Bajgelman 2019). In addition to analyzing cell events, flow cytometry can assess cell viability upon nanoparticles application.

Confocal microscopy is a technique that provides the magnification and contrast of the microscopic image and creates three-dimensional images. This technique can also be used for real-time testing. In general, the equipment emits a light that reflects and passes through a small opening illuminating a small point in the region or substance observed by the apparatus. The reflected light is captured by a detector and creates an image at the detected focal point. A disadvantage of the technique is that microanatomic structures can only be observed to a depth of 350 μm (Rito and Pineiro-Maceira 2009).

For instance, Petrilli (2018) evaluated the internalization of liposomes and immunoliposomes in confocal microscopy using Dio (3,3′-dioctadecyloxacarbocyanine perchlorate) as a fluorescent dye as a lipid bilayer marker. For flow cytometry analysis, propidium iodide was used in addition to Dio and stains the dead cell nucleus. The two markers complement each other in the study, as the results can demonstrate the amount of living cells that captured the liposome or immunoliposome (Petrilli et al. 2018a).

Nanoparticles produced by emulsion techniques with polyethyleneimine or bovine serum can be studied accompanied by their internalization by the cell. Romero et al. 2010 studied HepG2 cells after being exposed to polymeric nanoparticles, with the aid of confocal microscopy and flow cytometry. These techniques were used to observe the stability and internalization of the nanoparticle and the region where it was located, finally obtaining comparable or better cell viability than the control. Finally, the test showed low toxicity in the environment for the internalized nanoparticle by the cell. In another recent study, Eloy et al. 2017a, b observed liposome and immunoliposome uptake in the triple negative breast cancer cell line 4T1 and HER2 positive breast cancer SKBR3 cells. Basically, the cell nuclei were stained with DAPI dye and the lipid nanoparticles were stained with Dio. With the aid of confocal microscopy, it was possible to observe that cells incubated with immunoliposomes had higher fluorescence intensity than cells treated with liposomes, which is probably related to the specific binding to HER2 receptors (Romero et al. 2010; Eloy et al. 2017a, b).

5 In Vivo Studies of Nanoparticles

5.1 *Pharmacokinetics of Nanoparticles*

Nanocarriers are designed to increase drug absorption, bioavailability, enable controlled release, protect the drug from physiological degradation and promote the drug selectively to the target. In oncology, these features offer many benefits, considering that in traditional chemotherapy, low-molecular-weight drugs can diffuse through the endothelial wall into healthy tissues, causing toxic effects. Thus, nanocarriers cause antineoplastic drugs to accumulate in tumor tissue and reach a high level of cytotoxicity, preventing the rest of the body from being exposed to this effect (Ernsting et al. 2013).

Nanoparticles change the pharmacokinetics profile of drugs, significantly affecting their efficacy and toxicity. To study the pharmacokinetic profile of nanocarriers, it is necessary to measure drug concentrations in different tissues until it is eliminated from the body. Among the most studied parameters to determine the biodistribution of these compounds are: Cmax (maximum concentration), Cl (clearance), $t_{1/2}$ (plasma half-life), area under the curve (AUC) and MRT (mean time permanence of the molecule in the body). When a drug-loaded nanoparticle is developed, the aim is to increase Cmax, $t_{1/2}$, AUC, MRT and reduce Cl. Drug accumulation in the target tissue is expected to generate an increased therapeutic effect and reduce the risk of unwanted toxicity in non-diseased tissues (Li and Huang 2008).

When a hydrophilic drug is injected intravenously, it is rapidly cleared by renal filtration. In the case of hydrophobic drugs, the half-life is longer because of the increased level of plasma protein binding and the liver biotransformation process to generate a hydrophilic metabolite that can be renally eliminated. When encapsulated into nanoparticles, drugs are protected from this enzymatic biotransformation mechanism prior to release and because of the increased carrier size renal clearance is also avoided; this is because compounds can only cross the renal glomerulus if they have a smaller than approximately 5.5 nm (Li and Huang 2008).

In tumors, abnormal tumor vascularization generates a suitable environment for nanoparticle penetration. The accelerated angiogenesis process in the tumor environment generates an exaggerated proliferation of endothelial cells. Thus, cell spacings are generated that can be from 100 nm to 2 μm depending on the type of tumor, forming large pores in the tumor vasculature, increasing vascular permeability and allowing macromolecules to pass through. Lymphatic drainage at the tumor is defective because of high interstitial pressure in the tumor nucleus, causing particle retention. With irregularities in the tumor vessels and poor flow in the lymphatic circulation, the enhanced permeability and retention (EPR) effect is generated, which is widely exploited for the development of antitumor drug-containing nanoparticles (Khawar et al. 2015; Morshed et al. 2018).

For nanoparticles to enter the tumor environment to take advantage of the EPR effect, they must remain in the circulation for a sufficient period of time for tumor accumulation to occur. However, these particles may be targeted by the reticuloendothelial system (RES), which is part of the immune system and may perform phagocytosis. To overcome this obstacle, a strategy was prepared in the preparation of nanoparticles, conjugating a hydrophilic polymer that mimics the surface of the erythrocytes. For this purpose, initially the monosialoganglioside (GM1) was used. Currently, a widely used polymer is polyethylene glycol (PEG), which is incorporated into the nanoparticle to decrease recognition by serum proteins as complement system components, avoiding opsonization and also any interaction with cells of the immune system. The polymer forms an edge that generates steric hindrance on the nanocarrier surface. Sadzuka et al. (1998) demonstrated that liposome pegylation allowed a threefold decrease in RES component removal and a sixfold increase in the area under the curve, thereby increasing the drug's internalization rate by threefold (Immordino et al. 2006).

The EPR effect observed in solid tumors ensures that nanocarriers accumulate in the tumor area and the use of PEG to increase half-life allows the passive targeting of nanoparticles (Immordino et al. 2006).

In addition to modifying the surface with PEG, there are other parameters that need to be adjusted to increase the half-life of nanoparticles. Particle size is one of the determining factors. Several authors show that nanocarriers with a diameter around 100 nm present an increase in plasma half-life and a reduction in phagocyte uptake. To examine the pattern of nanoparticle biodistribution in vivo, nanoparticles can be labeled with a radioisotope and be injected in mice, and dose the drug into different tissues after a few hours of administration. Liu et al. (1998) found that after 4 h of liposome administration in mice, 60% of nanoparticles with a diameter between 100 and 200 nm were in circulation, while only 20% of those with sizes smaller than 50 nm and greater than 250 nm were detected in the blood (Li and Huang 2008).

Particles that are smaller than 50 nm tend to be trapped in the liver, as fenestrations in the hepatic sinusoids are 50–100 nm of space. Conversely, large particles with sizes around 400 nm are captured by the spleen. Data presented by Moreira et al. (2001) are in agreement with this observation, reporting that particles with a size around 120 nm have 10–20 times more tumor penetration. Even controlling the particle size, in some cases, although the nanoparticles may have good tumor penetration, they can be retained for a short time, so many studies have been conducted to actively target nanoparticles to increase selectively, by adhering specific ligands on nanocarriers such as antibodies, integrins and receptors (Li and Huang 2008; de Oliveira et al. 2012; Bahrami et al. 2017; Ernsting et al. 2013).

Another factor that has an impact on nanoparticle pharmacokinetics is surface charge, which is measured as zeta potential. Particles with a charge lower than −10 mV are rapidly eliminated by the reticuloendothelial system (RES), while positively charged particles greater than 10 mV aggregate with plasma proteins, whereas those with a value close to neutrality have a longer plasma half-life and lower clearance. Positively charged nanocarriers can lead to pulmonary embolism and vessel obstruction due to the state of aggregation, which is considered a toxic effect. However, it was shown by Levchenko et al. (2002) and Zhang et al. (2005) that the insertion of PEG on the surface decreases the charge effect, i.e., the polymer does reduce the clearance by the RES as it protects the surface charge and can be used to improve the cationic particle biodistribution (Ernsting et al. 2013).

5.2 Barriers to Nanoparticles Penetration in the Tumor Microenvironment

Herein, it was previously shown how it is possible to modify the physicochemical characteristics of nanoparticles so that they have a prolonged distribution and thus are able accumulate in the tumor, but there are barriers that prevent the penetration of tumor tissues and may decrease the therapeutic efficacy of the drug carried. The tumor microenvironment contains a dense extracellular matrix (ECM), high pressure exerted by interstitial fluid, macrophages and activated fibroblasts in constant interaction with tumor cells (Sun et al. 2017; Niu et al. 2018; Ernsting et al. 2013).

The tumor has an inflammatory environment with a process of fibrosis, cancer cell proliferation, fibroblast overactivation and other cells that cause uncontrolled production of extracellular matrix elements, generating exacerbated pressure in the interstitial fluid, impeding flow and hindering the penetration of nanoparticles in cells located at deep areas of the tumor (Ernsting et al. 2013; Sun et al. 2017; Niu et al. 2018).

In light of the above, several researchers have proposed strategies to increase tumor penetration of nanoparticles, such as administration of VEGF-specific antiangiogenic drugs (bevacizumab), use of a prostaglandin inhibitor (imatinib) that reduces interstitial pressure as well as insertion of ECM degrading enzymes (hyaluronidases and collagenases) in nanoparticles to disrupt the fibers that make up the extracellular matrix (Ernsting et al. 2013; Sun et al. 2017; Niu et al. 2018).

Another example is the application of losartan directly to the tumor, as this drug may decrease collagen formation by negative feedback for TGF-β, i.e., it is possible to inhibit the synthesis of ECM components (Diop-Frimpong et al. 2011). Cun et al (2016) demonstrated that it was possible to use a gelatin sensitive to degradation by metalloproteinases (enzyme overexpressed in tumor environment) in nanoparticles, which on entering the tumor reduced their size and increased their penetration without affecting biodistribution (Cun et al. 2016).

5.3 Evaluation of the in Vivo Effectiveness of Nanoparticulate Anticancer Drugs

As previously addressed, in addition to improved pharmacokinetics and for nanocarrier-based treatment to be effective there are other important factors such as intratumor penetration and retention. Tumors have been shown to exhibit properties such as EPR effect, dense extracellular matrix and high interstitial pressure that affect these processes. However, other variables should be evaluated, such as tumor cell multiplication rate; tumor volume, location and type; blood flow at the site, composition and stiffness of the ECM. To assess nanoparticle penetration and tumor efficacy, there are models that attempt to create a histology similar to that found in the tumor environment (Hare et al. 2017; Niu et al. 2018).

Preclinical data allow for important decisions such as dose–effect and effectiveness. Current in vivo models can simulate the physiopathology and diversity of tumors associated with human subjects. An example is the model of xenograft implants of human tumors in immunosuppressed mice. For example, the insertion of high tumorigenic human mammary carcinoma cells in athymic mice, which do not have mature T lymphocytes, can develop mammary tumors in these animals. These models mimic the complexity of human tumor cells, but differ in size and physiology. However, a disadvantage is that it is not possibility of studying all the interaction between the neoplasia and the individual that occurs when the tumor naturally develops in that organism (Pantaleão and Luchs 2010; Hare et al. 2017).

To verify whether the process of tumor cell insertion in animals will develop cancer with growth similar to the pathology that occurs in humans, techniques have been developed, such as orthotopic and subcutaneous xenograft models. Du et al (2014) conducted a comparison experiment between the two animal models and human gallbladder carcinoma. In this experiment, it was established for the orthotopic model that an incision of the animal's abdominal cavity would be made and a cell suspension would be injected directly into the gallbladder. For the other model, this injection would occur subcutaneously. The study demonstrated that the orthotopic xenograft model showed a development more similar to human gallbladder carcinoma, since the cells were injected directly into the target tissue and provided the formation of a microenvironment similar to the human tumor, presenting more invasive capacity than in the human tumor. In a subcutaneous xenograft model, in this second model there was no formation of metastasis in the lymph nodes and the tumor was covered by a wrap. The first model mentioned presented liver metastasis in the lymph nodes and greater invasive power, better suited to reproduce the tumor growth that occurs in humans with gallbladder carcinoma (Du et al. 2014).

In order to conduct the in vivo efficacy evaluation experiment by the xenograft model of human tumors, the animal's tumor is measured to evaluate tumor growth and then the tumor can be excised for further analysis such as histopathological examination of the affected tissue is possible. With this, tumor growth inhibition or shrinkage results can be obtained, revealing the enhanced cytotoxicity of the drug loaded in the nanoparticle. For instance, Eloy et al (2016) evaluated the efficacy of paclitaxel and rapamycin-containing liposomes for breast cancer treatment, revealing that there was a synergistic effect of two drugs, both in vitro experiments with human breast cancer cell line and in vivo experiments by the orthotopic xenographic model, demonstrating that there was a reduction in tumor growth. Furthermore, Baksi et al (2018) evaluated the anticancer efficacy of chitosan polymeric nanoparticles carrying quercetin (flavonoid present in several plant species) for breast and lung cancer. In this study, it was used the orthotopic xenographic model for breast tumor development and the subcutaneous model to insert the lung cancer cell line. After the treatment period, there was a reduction in tumor volume for both types of cancer(Eloy et al. 2016; Baksi et al. 2018).

6 Clinical Applications

Cancer is a heterogeneous disease and affects millions of people in the world. Among its classic treatment alternatives, we can mention chemotherapy, radiotherapy, surgery and immunotherapy. However, frequently the treatment is unsuccessful, because of tumor resistance during treatment, development of short- and long-term side effects. Thus, nanotechnology arises as a strategy to circumvent these and other drawbacks, enabling targeted delivery, enhancing accumulation of drugs in the tumor, thus reducing adverse effects and improving treatment efficacy and efficiency (Arranja et al. 2017; Hussain et al. 2018). Table 6 illustrates nanomedicines available on the market for the treatment of mainly cancer (Wolfram and Ferrari 2019; Fornaguera and García-Celma 2017; Tran et al. 2017).

Table 6 Nanostructured drug delivery systems available in the clinic

Drug name	Type of nanoformulation	Indication
Doxil	Doxorubicin hydrochloride pegylated liposomal	Solid tumor/breast cancer, ovarian, multiple myeloma
Myocet	Doxorubicin-citrate liposome	Solid tumor
DaunoXome	Daunorubicin citrate liposome	Solid tumor/first-line cytotoxic therapy for treatment of advanced Kaposi's sarcoma (KS)
Abraxane	Albumin nanoparticle-bound paclitaxel	Solid tumor/in combination with gemcitabine is the first line of treatment for metastatic pancreatic adenocarcinoma
Onivyde	Liposomal irinotecan	Metastatic pancreatic adenocarcinoma
Vyxeos	Daunorubicin + liposome-encapsulated cytarabine	Acute myeloid leukemia (AML)
AmBisome	Liposomal amphotericin B	Opportunistic and/or endemic and systemic fungal infections
Brentuximab	Antibody-drug conjugate	Untreated classic stage III or IV Hodgkin's lymphoma (cHL) in combination with chemotherapy
Marqibo	Liposomal vincristine sulfate	Lymphoblastic leukemia and melanoma
Genexol-PM	Polymeric nanoparticles containing paclitaxel	Breast, lung and ovarian cancer
Vyxeos	Liposome-encapsulated combination of daunorubicin–cytarabine	Acute myeloid leukemia (AML) with newly diagnosed therapy-related or myelodysplasia-related AML

Doxil® is pegylated liposome-encapsulated doxorubicin and was the first nanoencapsulated drug approved by the Food and Drug Administration (FDA) in 1995 and the European Agency for the Evaluation of Medicines (EMA). Doxorubicin has low therapeutic index and serious adverse effects, with cardiotoxicity being the most pronounced. Doxorubicin liposomes improved the pharmacokinetics of the drug and protected against cardiotoxic and other side effects, enhancing quality of life of patients after chemotherapy. However, some side effects of Doxil® are present, such as palmar-plantar erythrodysesthesia (PPE) popularly known as hand and foot syndrome which is characterized by the sensitivity, redness and peeling of the skin, and infusion or pseudo-allergy reactions that are related to activation of the complement system generating symptoms such as flushing, headaches, redness that are more related to the first infusion of the drug. Since its approval, Doxil® has been indicated for the treatment of ovarian cancer and Kaposi multiple myeloma sarcoma (Alibolandi et al. 2017; Bavli et al. 2019).

Another interesting product based on nanotechnology for cancer treatment is Abraxane®, which consists of a 130 nm nanoparticle of albumin-bound paclitaxel. This formulation was developed to decrease the toxicity associated with the commercial solution formulation, Taxol®, which contains Cremophor EL/ethanol, used as a solubilizer for the drug, but this excipient has been shown to trigger side effects, mainly hypersensitivity. Abraxane® has been shown to avoid side effects, as well as presenting a more pronounced antitumor effect, since as it is bound to human albumin, better tumoral vascular permeability is observed (Gardner et al. 2008; Cysteine et al. 2012; Shigematsu et al. 2015).

Not only cancer is benefited from drug encapsulation in nanocarriers. AmBisome® is a liposomal formulation of amphotericin B and is the first line of treatment for invasive fungal infections such as leishmaniasis viscera because it has a high affinity for ergosterol, a predominant sterol in the parasite membrane. Amphotericin B liposomes have been shown to have an excellent therapeutic index, which allows higher doses to be administered providing greater treatment efficacy (Ven et al. 2012; Cencig et al. 2012).

7 Conclusion

Herein, some main concepts on nanotechnology were presented and discussed. We described briefly some of the main types of nanoparticles that can be applied both to the treatment and diagnosis of different diseases. Basically, lipid, polymeric and inorganic nanoparticles can be used to different aims, but some important aspects such as size, composition and biocompatibility should be considered for the therapeutic outcome and administration route. Also, different methods of preparation were described, with highlight to some available methods for nanoparticles size controlling. Surface-modified nanoparticles with different moieties aiming for targeted therapy were presented and discussed. Also, in vitro and in vivo methods commonly used for the efficacy and toxicity evaluation were considered in this chapter. In conclusion, some available marketed products were presented. In this context, it is expected for the next years the development and approval of many types of other nanoparticles for medical application.

Acknowledgements This work was supported by the National Council for Scientific and Technological Development (CNPq) (grants # 409352/2018-7; #409362/2018-2).

Reference

Abd Elrahman AA, Mansour FR (2019) Targeted magnetic iron oxide nanoparticles: preparation, functionalization and biomedical application. J Drug Deliv Sci Technol 52(January):702–712. https://doi.org/10.1016/j.jddst.2019.05.030

Abedi M, Sadat S, Abedanzadeh M, Borandeh S (2019) Citric acid functionalized silane coupling versus post-grafting strategy for dual PH and saline responsive delivery of cisplatin by Fe 3 O 4/Carboxyl functionalized mesoporous SiO_2 hybrid nanoparticles: a-synthesis. Mater Sci Eng C 104(June):109922. https://doi.org/10.1016/j.msec.2019.109922

Adnan NNM, Cheng YY, Ong NMN, Kamaruddin TT, Rozlan E, Schmidt TW, Duong HTT, Boyer C (2016) Effect of gold nanoparticle shapes for phototherapy and drug delivery. Polym Chem 7(16):2888–2903. https://doi.org/10.1039/c6py00465b

Ajnai G, Chiu A, Kan T, Cheng CC, Tsai TH, Chang J (2014) Trends of gold nanoparticle-based drug delivery system in cancer therapy. J Exp Clin Med 6(6):172–178. https://doi.org/10.1016/j.jecm.2014.10.015

Alibolandi M, Abnous K, Mohammadi M, Hadizadeh F, Sadeghi F, Taghavi S, Reza M, Ramezani M (2017) Extensive preclinical investigation of polymersomal formulation of doxorubicin versus Doxil-mimic formulation. J Controlled Release 264(August):228–236. https://doi.org/10.1016/j.jconrel.2017.08.030

Almouazen E, Bourgeois S, Boussaïd A, Valot P, Malleval C, Fessi H, Nataf S, Brianc S (2012) Development of a nanoparticle-based system for the delivery of retinoic acid into macrophages. 430:207–15. https://doi.org/10.1016/j.ijpharm.2012.03.025

Anarjan FS (2019) Active targeting drug delivery nanocarriers: Ligands. Nano-Structures Nano-Objects 19:100370

Arranja AG, Pathak V, Lammers T, Shi Y (2017) Tumor-targeted nanomedicines for cancer theranostics. Pharmacol Res 115:87–95. https://doi.org/10.1016/j.phrs.2016.11.014

Asgari S, Saberi AH, McClements DJ, Lin M (2019) Microemulsions as nanoreactors for synthesis of biopolymer nanoparticles. Trends Food Sci Technol 86:118–130. https://doi.org/10.1016/j.tifs.2019.02.008

Attarad A, Hira Zafar MZ, ul Haq I, Phull AR, Ali JS, Hussain A (2016) Synthesis, characterization, applications, and challenges of iron oxide nanoparticles. Nanotechnol Sci Appl 9:49–67. https://doi.org/10.2147/NSA.S99986

Bahrami B, Hojjat-Farsangi M, Mohammadi H, Anvari E, Ghalamfarsa G, Yousefi M, Jadidi-Niaragh F (2017) Nanoparticles and targeted drug delivery in cancer therapy. Immunol Lett 190(April):64–83. https://doi.org/10.1016/j.imlet.2017.07.015

Bajgelman MC (2019) Chapter 8. principles and applications of flow cytometry. In: Data processing handbook for complex biological data sources. Elsevier Inc. https://doi.org/10.1016/B978-0-12-816548-5.00008-3

Baksi R, Singh DP, Borse SP, Rana R, Sharma V, Nivsarkar M (2018) In vitro and in vivo anti-cancer efficacy potential of quercetin loaded polymeric nanoparticles. Biomed Pharmacother 106(July):1513–1526. https://doi.org/10.1016/j.biopha.2018.07.106

Bavli Yaelle, Winkler Ilan, Mae Bing, Roffler Steve, Cohen Rivka, Szebeni Janos (2019) Doxebo (doxorubicin-free Doxil-like liposomes) is safe to use as a pre-treatment to prevent infusion reactions to PEGylated nanodrugs. J Controlled Release 306(June):138–148. https://doi.org/10.1016/j.jconrel.2019.06.007

Berridge MV, Herst PM, Tan AS (2005) Tetrazolium dyes as tools in cell biology : new insights into their cellular reduction. 11(05):127–52. https://doi.org/10.1016/S1387-2656(05)11004-7

Biosciences BD (2002) Introduction to flow cytometry : a learning guide. https://www.bdbiosciences.com/en-us

Borenfreund E, Puerner JA (1985) Toxicity determined in vitro by morphological alterations and neutral red absorption. Toxicol Lett 24(2–3):119–124. https://doi.org/10.1016/0378-4274(85)90046-3

Brannon-Peppas Lisa, Blanchette James O (2012) Nanoparticle and targeted systems for cancer therapy. Adv Drug Deliv Rev 64:206–212

Breznan D, Das D, MacKinnon-Roy C, Simard B, Kumarathasan P, Vincent R (2015) Non-specific interaction of carbon nanotubes with the resazurin assay reagent: impact on in vitro assessment of nanoparticle cytotoxicity. Toxicol In Vitro 29(1):142–147. https://doi.org/10.1016/j.tiv.2014.09.009

Bruch GE, Fernandes LF, Bassi BLT, Alves MTR, Pereira IO, Frézard F, Massensini AR (2019) Liposomes for drug delivery in stroke. Brain Res Bull 152(June):246–256. https://doi.org/10.1016/j.brainresbull.2019.07.015

Callender SP, Mathews JA, Kobernyk K, Wettig SD (2017) Microemulsion utility in pharmaceuticals: implications for multi-drug delivery. Int J Pharm 526(1–2):425–442. https://doi.org/10.1016/j.ijpharm.2017.05.005

Carita AC, Eloy JO, Chorilli M, Lee RJ, Leonardi GR (2017) Recent advances and perspectives in liposomes for cutaneous drug delivery. Curr Med Chem 25(5):606–635. https://doi.org/10.2174/0929867324666171009120154

Ceborska M (2017) Folate appended cyclodextrins for drug, DNA, and SiRNA delivery. Eur J Pharm Biopharm 120(September):133–145. https://doi.org/10.1016/j.ejpb.2017.09.005

Cencig S, Coltel N, Truyens C, Carlier Y (2012) Evaluation of benznidazole treatment combined with nifurtimox, posaconazole or AmBisome® in mice infected with Trypanosoma cruzi strains. Int J Antimicrob Agents 40(6):527–532. https://doi.org/10.1016/j.ijantimicag.2012.08.002

Chountoulesi M, Pippa N, Pispas S, Chrysina ED, Forys A, Trzebicka B, Demetzos C (2018) Cubic lyotropic liquid crystals as drug delivery carriers: physicochemical and morphological studies. Int J Pharm 550(1–2):57–70. https://doi.org/10.1016/j.ijpharm.2018.08.003

Courtois P, Rorat A, Lemiere S, Guyoneaud R, Attard E, Levard C, Vandenbulcke F (2019) Ecotoxicology of silver nanoparticles and their derivatives introduced in soil with or without sewage sludge: a review of effects on microorganisms, plants and animals. Environ Pollut 253:578–598. https://doi.org/10.1016/j.envpol.2019.07.053

Crucho CIC (2015) Stimuli-responsive polymeric nanoparticles for nanomedicine. ChemMedChem 10(1):24–38. https://doi.org/10.1002/cmdc.201402290

Crucho CIC, Barros MT (2017) Polymeric nanoparticles: a study on the preparation variables and characterization methods. Mater Sci Eng C 80:771–784. https://doi.org/10.1016/j.msec.2017.06.004

Cun X, Ruan S, Chen J, Zhang L, Li J, He Q, Gao H (2016) A dual strategy to improve the penetration and treatment of breast cancer by combining shrinking nanoparticles with collagen depletion by losartan. Acta Biomater 31:186–196. https://doi.org/10.1016/j.actbio.2015.12.002

Cysteine R, Yang Y, Niu X, Zhang Q, Hao L, Ding Y (2012) The efficacy of abraxane on osteosarcoma xenografts in nude mice and expression of secreted protein, acidic. Am J Med Sci 344(3):199–205. https://doi.org/10.1097/MAJ.0b013e31823e62e5

Dadfar SM, Roemhild K, Drude NI, von Stillfried S, Knüchel R, Kiessling F, Lammers T (2019) Iron oxide nanoparticles: diagnostic, therapeutic and theranostic applications. Adv Drug Deliv Rev 138:302–325. https://doi.org/10.1016/j.addr.2019.01.005

Dash TK, Konkimalla VB (2012) Polymeric modification and its implication in drug delivery : poly- ε-caprolactone (PCL) as a model polymer. https://doi.org/10.1021/mp3001952
de Araújo RF, de Araújo AA, Pessoa JB, Freire Neto FP, da Silva GR, Leitão Oliveira ALCS, de Carvalho TG et al (2017) Anti-inflammatory, analgesic and anti-tumor properties of gold nanoparticles. Pharmacol Rep 69(1):119–129. https://doi.org/10.1016/j.pharep.2016.09.017
de Oliveira LC, Taveira JF, Souza LG, Marreto RN, Lima ME, Taveira FS (2012) Aplicações das nanopartículas lipídicas no tratamento de tumores sólidos: revisão de literatura. Rev Bras Cancerologia 58(45):695–70112
Diop-Frimpong B, Chauhan VP, Krane S, Boucher Y, Jain RK (2011) Losartan inhibits collagen I synthesis and improves the distribution and efficacy of nanotherapeutics in tumors. Proc Natl Acad Sci USA 108(7):2909–2914. https://doi.org/10.1073/pnas.1018892108
Dissanayake S, Denny WA, Gamage S, Sarojini V (2017) Recent developments in anticancer drug delivery using cell penetrating and tumor targeting peptides. J Controlled Release 250:62–76
Donahue N, Acar H, Wilhelm S (2019) Concepts of nanoparticle cellular uptake, intracellular trafficking, and kinetics in nanomedicine. Adv Drug Deliv Rev
Du Qiang, Jiang Lei, Wang Xiao Qian, Pan Wei, She Fei Fei, Chen Yan Ling (2014) Establishment of and comparison between orthotopic xenograft and subcutaneous xenograft models of gallbladder carcinoma. Asian Pac J Cancer Prev 15(8):3747–3752. https://doi.org/10.7314/APJCP.2014.15.8.3747
Dunigan DD, Waters SB, Owen TC (1995) Aqueous soluble tetrazolium/formazan MTS as an indicator of NADH- and NADPH-dependent dehydrogenase activity. Biotechniques 19(4):640–649
Edwards KA, Baeumner AJ (2006) Analysis of liposomes. Talanta 68(5):1432–1441. https://doi.org/10.1016/j.talanta.2005.08.031
Eisenbrand G, Pool-Zobel B, Baker V, Balls M, Blaauboer BJ, Boobis A, Carere A et al (2002) Methods of in vitro toxicology. Food Chem Toxicol 40(2–3):193–236. https://doi.org/10.1016/S0278-6915(01)00118-1
Eloy J, Petrilli R, Lopez R, Lee R (2015) Stimuli-responsive nanoparticles for SiRNA delivery. Curr Pharm Des 21(29):4131–4144. https://doi.org/10.2174/1381612821666150901095349
Eloy JO, Petrilli R, Topan JF, Antonio HMR, Barcellos JPA, Chesca DL, Serafini LN, Tiezzi DG, Lee RJ, Marchetti JM (2016) Co-loaded paclitaxel/rapamycin liposomes: development, characterization and in vitro and in vivo evaluation for breast cancer therapy. Colloids Surf B 141:74–82. https://doi.org/10.1016/j.colsurfb.2016.01.032
Eloy JO, Petrilli R, Chesca DL, Saggioro FP, Lee RJ, Maldonado J (2017a) Anti-HER2 immunoliposomes for co-delivery of paclitaxel and rapamycin for breast cancer therapy. Eur J Pharm Biopharm 115:159–167
Eloy JO, Petrilli R, Trevizan L, Chorilli M (2017b) Immunoliposomes: a review on functionalization strategies and targets for drug delivery. Colloids Surf B 159:454–467
Epple M (2018) Review of potential health risks associated with nanoscopic calcium phosphate. Acta Biomater 77:1–14. https://doi.org/10.1016/j.actbio.2018.07.036
Eriksson S, Nylén U, Rojas S, Boutonnet M (2004) Preparation of catalysts from microemulsions and their applications in heterogeneous catalysis. Appl Catal A 265(2):207–219. https://doi.org/10.1016/j.apcata.2004.01.014
Ernsting MJ, Mami M, Roy A, Li S (2013) Factors controlling the pharmacokinetics, biodistribution and intratumoral penetration of nanoparticles. J Control Realese 23(1):1–7. https://doi.org/10.1038/jid.2014.371
Farran B, Pavitra E, Kasa P, Peela S, Seeta G, Raju R (2019) Cytokine and growth factor reviews folate-targeted immunotherapies: passive and active strategies for cancer. Cytokine Growth Factor Rev 45(January):45–52. https://doi.org/10.1016/j.cytogfr.2019.02.001
Fisichella M, Dabboue H, Bhattacharyya S, Saboungi M, Salvetat J, Hevor T, Guerin M (2009) Mesoporous silica nanoparticles enhance MTT formazan exocytosis in HeLa cells and astrocytes. Toxicol In Vitro 23(4):697–703. https://doi.org/10.1016/j.tiv.2009.02.007

Fornaguera C, García-Celma MJ (2017) Personalized nanomedicine: a revolution at the nanoscale. J Personalized Med 7(4):14–21. https://doi.org/10.3390/jpm7040012

Funk D, Schrenk HH, Frei E (2007) Serum albumin leads to false-positive results in the XTT and the MTT assay. Biotechniques 43(2):178–186. https://doi.org/10.2144/000112528

Ganesan P, Narayanasamy D (2017) Lipid nanoparticles: different preparation techniques, characterization, hurdles, and strategies for the production of solid lipid nanoparticles and nanostructured lipid carriers for oral drug delivery. Sustain. Chem Pharm 6(May):37–56. https://doi.org/10.1016/j.scp.2017.07.002

Garcês A, Amaral MH, Lobo JMS, Silva AC (2018) Formulations based on Solid Lipid Nanoparticles (SLN) and Nanostructured Lipid Carriers (NLC) for cutaneous use: a review. Eur J Pharm Sci 112(November 2017):159–67. https://doi.org/10.1016/j.ejps.2017.11.023

Gardner ER, William D, Figg WD (2008) Quantitative determination of total and unbound paclitaxel in human plasma following Abraxane treatment. 862:213–18. https://doi.org/10.1016/j.jchromb.2007.12.013

Geszke-Moritz M, Moritz M (2016) Solid lipid nanoparticles as attractive drug vehicles: composition, properties and therapeutic strategies. Mater Sci Eng C 68:982–994. https://doi.org/10.1016/j.msec.2016.05.119

Ghaffari H, Tavakoli A, Moradi A, Tabarraei A, Bokharaei-Salim F, Zahmatkeshan M, Farahmand M, Javanmard D, Jalal Kiani S, Esghaei M (2019) Inhibition of H1N1 influenza virus infection by zinc oxide nanoparticles : another emerging application of nanomedicine. 4:1–10

Gordillo-Galeano A, Mora-Huertas CE (2018) Solid lipid nanoparticles and nanostructured lipid carriers: a review emphasizing on particle structure and drug release. Eur J Pharm Biopharm 133(October):285–308. https://doi.org/10.1016/j.ejpb.2018.10.017

Gormley AJ, Greish K, Ray A, Robinson R, Gustafson JA, Ghandehari H (2011) Gold nanorod mediated plasmonic photothermal therapy: a tool to enhance macromolecular delivery. Int J Pharm 415(1–2):315–318. https://doi.org/10.1016/j.ijpharm.2011.05.068

Gosenca M, Bešter-Rogač M, Gašperlin M (2013) Lecithin based lamellar liquid crystals as a physiologically acceptable dermal delivery system for ascorbyl palmitate. Eur J Pharm Sci 50(1):114–122. https://doi.org/10.1016/j.ejps.2013.04.029

Gunasekaran T, Nigusse T, Dhanaraju MD (2011) Silver nanoparticles as real topical bullets for wound healing. J Am Coll Clin Wound Spec 3(4):82–96. https://doi.org/10.1016/j.jcws.2012.05.001

Guo C, Wang J, Cao F, Lee RJ, Zhai G (2010) Lyotropic liquid crystal systems in drug delivery. Drug Discovery Today 15(23–24):1032–1040. https://doi.org/10.1016/j.drudis.2010.09.006

Gutjahr A, Phelip C, Coolen AL, Monge C, Boisgard AS, Paul S, Verrier B (2016) Biodegradable polymeric nanoparticles-based vaccine adjuvants for lymph nodes targeting. Vaccines 4(4):1–16. https://doi.org/10.3390/vaccines4040034

Hans ML, Lowman AM (2002) Biodegradable nanoparticles for drug delivery and targeting. 6(September):319–27

Hare JI, Lammers T, Ashford MB, Puri S, Storm G, Barry ST (2017) Challenges and strategies in anti-cancer nanomedicine development: an industry perspective. Adv Drug Deliv Rev 108:25–38. https://doi.org/10.1016/j.addr.2016.04.025

Holder AL, Goth-goldstein R, Lucas D, Koshland CP (2012) Particle-induced artifacts in the MTT and LDH viability assays

Hussain Z, Arooj M, Malik A, Hussain F, Safdar H, Khan S (2018) Nanomedicines as emerging platform for simultaneous delivery of cancer therapeutics : new developments in overcoming drug resistance and optimizing anticancer efficacy. 46:1015–1025

Iijima S (1991) Helical microtubules of graphitic carbon. Lett Nat 353:412–414

Immordino ML, Dosio F, Cattel L (2006) Stealth liposomes: review of the basic science, rationale and clinical applications, existing and potential. 116–120

Jain K, Mehra NK, Jain NK (2014) Potentials and emerging trends in nanopharmacology. Curr Opin Pharmacol 15(1):97–106. https://doi.org/10.1016/j.coph.2014.01.006

Jeong Y-I, Na H-S, Seo D-H, Kim D-G, Lee H-C, Jang M-K, Na S-K, Roh S-H, Kim S-I, Nah J-W(2008) Nanoparticles and its antibacterial activity. 352:317–323. https://doi.org/10.1016/j.ijpharm.2007.11.001

Kawabata H(2019) Transferrin and transferrin receptors update. Free Radical Biol Med 133(June 2018):46–54. https://doi.org/10.1016/j.freeradbiomed.2018.06.037

Khawar IA, Kim JH, Kuh HJ (2015) Improving drug delivery to solid tumors: priming the tumor microenvironment. J Controlled Release 201:78–89. https://doi.org/10.1016/j.jconrel.2014.12.018

Klippstein R, Wang JTW, El-Gogary RI, Bai J, Mustafa F, Rubio N, Bansal S, Al-Jamal WT, Al-Jamal KT (2015) Passively targeted curcumin-loaded pegylated PLGA nanocapsules for colon cancer therapy in vivo. Small 11(36):4704–4722. https://doi.org/10.1002/smll.201403799

Kumar SSD, Rajendran NK, Houreld NN, Abrahamse H (2018) Recent advances on silver nanoparticle and biopolymer-based biomaterials for wound healing applications. Int J Biol Macromol 115(2017):165–75. https://doi.org/10.1016/j.ijbiomac.2018.04.003

Lechuga-Ballesteros D, Abdul-Fattah A, Stevenson CL, Bennett DB (2003) Properties and stability of a liquid crystal form of cyclosporine—the first reported naturally occurring peptide that exists as a thermotropic liquid crystal. J Pharm Sci 92(9):1821–1831. https://doi.org/10.1002/jps.10444

Lee YT, Tan YJ, Oon CE (2018) Molecular targeted therapy: treating cancer with specificity. Eur J Pharmacol 834(June):188–196

Li S-D, Huang L (2008) Reviews pharmacokinetics and biodistribution of nanoparticles

Li W, Zhou J, Xu Y (2015) Study of the in vitro cytotoxicity testing of medical devices (review). 617–620. https://doi.org/10.3892/br.2015.481

Li M, Chunyang D, Guo N, Teng Y, Meng X, Sun H, Li S, Peng Y, Galons H (2019) Composition design and medical application of liposomes. Eur J Med Chem 164:640–653. https://doi.org/10.1016/j.ejmech.2019.01.007

Limayem I, Charcosset C, Fessi H (2004) Purification of nanoparticle suspensions by a concentration/diafiltration process. 38:1–9. https://doi.org/10.1016/j.seppur.2003.10.002

Lu Y, Low PS (2002) Folate-mediated delivery of macromolecular anticancer therapeutic agents. Adv Drug Deliv Rev 54:675–693

Madheswaran T, Kandasamy M, Bose RJC, Karuppagounder V (2019) Current potential and challenges in the advances of liquid crystalline nanoparticles as drug delivery systems. Drug Discovery Today 24(7):1405–1412. https://doi.org/10.1016/j.drudis.2019.05.004

Malik P, Mukherjee TK (2018) Recent advances in gold and silver nanoparticle based therapies for lung and breast cancers. Int J Pharm 553(1–2):483–509. https://doi.org/10.1016/j.ijpharm.2018.10.048

Malinoski FJ (2014) The nanomedicines alliance: an industry perspective on nanomedicines. Nanomed Nanotechnol Biol Med 10(8):1819–20. https://doi.org/10.1016/j.nano.2014.07.003

Matjaž MG, Škarabot M, Gašperlin M, Janković B (2019) Lamellar liquid crystals maintain keratinocytes' membrane fluidity: an AFM qualitative and quantitative study. Int J Pharm 572:118712. https://doi.org/10.1016/j.ijpharm.2019.118712

Merdan T, Kopeček J, Kissel T (2002) Prospects for cationic polymers in gene and oligonucleotide therapy against cancer. Adv Drug Deliv Rev 54:715–758. https://doi.org/10.1016/S0169-409X(02)00046-7

Mofokeng TP, Moloto MJ, Shumbula PM, Tetyana P (2019) Synthesis, characterization and cytotoxicity of alanine-capped CuS nanoparticles using human cervical carcinoma HeLa cells. Anal Biochem 580(June):36–41. https://doi.org/10.1016/j.ab.2019.06.008

Morshed N, Jahan N, Penheiro DE(2018) Polymeric nanoparticles for targeted delivery in cancer treatment: an overview. Int J Pharm Sci Rev Res 52(1):101–11. http://globalresearchonline.net/journalcontents/v52-1/19.pdf%0A http://ovidsp.ovid.com/ovidweb.cgi?T=JS&PAGE=reference&D=emexa&NEWS=N&AN=624202638

Müller-Goymann CC (2004) Physicochemical characterization of colloidal drug delivery systems such as reverse micelles, vesicles, liquid crystals and nanoparticles for topical administration. Eur J Pharm Biopharm 58(2):343–356. https://doi.org/10.1016/j.ejpb.2004.03.028

Nance E (2019) Careers in nanomedicine and drug delivery. Adv Drug Deliv Rev. https://doi.org/10.1016/j.addr.2019.06.009

Nedyalkova M, Donkova B, Romanova J, Tzvetkov G, Madurga S, Simeonov V (2017) Iron oxide nanoparticles—in vivo/in vitro biomedical applications and in silico studies. Adv Coll Interface Sci 249(February):192–212. https://doi.org/10.1016/j.cis.2017.05.003

Neuhaus B, Tosun B, Rotan O, Frede A, Westendorf AM, Epple M (2016) Nanoparticles as transfection agents: a comprehensive study with ten different cell lines. RSC Advances 6(22):18102–18112. https://doi.org/10.1039/c5ra25333k

Niu Y, Zhu J, Li Y, Shi H, Gong Y, Li R, Huo Q, Ma T, Liu Y (2018) Size shrinkable drug delivery nanosystems and priming the tumor microenvironment for deep intratumoral penetration of nanoparticles. J Controlled Release 277:35–47. https://doi.org/10.1016/j.jconrel.2018.03.012

O'Brien J, Wilson I, Orton T, Pognan F (2000) Investigation of the Alamar Blue (resazurin) fluorescent dye for the assessment of mammalian cell cytotoxicity. Eur J Biochem 267(17):5421–5426. https://doi.org/10.1046/j.1432-1327.2000.01606.x

Pantaleão C, Luchs A (2010) Câncer e modelos experimentais de tumores murinos cancer and experimental models of mouse tumor. Rev Inst Adolfo Lutz 69(4):439–445

Patil YP, Jadhav S (2014) Novel methods for liposome preparation. Chem Phys Lipid 177:8–18. https://doi.org/10.1016/j.chemphyslip.2013.10.011

Pena-Pereira F, Lavilla I, Bendicho C (2017) Unmodified gold nanoparticles for in-drop plasmonic-based sensing of iodide. Sens Actuators, B 242:940–948. https://doi.org/10.1016/j.snb.2016.09.161

Petrilli R, Eloy J, Marchetti J, Lopez R, Lee R (2014) Targeted lipid nanoparticles for antisense oligonucleotide delivery. Curr Pharm Biotechnol 15(9):847–855. https://doi.org/10.2174/1389201015666141020155834

Petrilli R, Eloy JO, Praça FSG, Del Ciampo JO, Fantini MAC, Fonseca MJV, Bentley MVLB (2016a) Liquid crystalline nanodispersions functionalized with cell-penetrating peptides for topical delivery of short-interfering RNAs: a proposal for silencing a pro-inflammatory cytokine in cutaneous diseases. J Biomed Nanotechnol 12(5):1063–1075. https://doi.org/10.1166/jbn.2016.2211

Petrilli R, Eloy J, Lopez R, Lee R (2016b) Cetuximab immunoliposomes enhance delivery of 5-FU to skin squamous carcinoma cells. Anticancer Agents Med Chem 17(2):301–308. https://doi.org/10.2174/1871520616666160526110913

Petrilli R, Eloy J, Lopez R, Lee R (2017) Cetuximab immunoliposomes enhance delivery of 5-FU to skin squamous carcinoma cells. Anticancer Agents Med Chem 17(2):301–308. https://doi.org/10.2174/1871520616666160526110913

Petrilli R, Eloy JO, Saggioro FP, Chesca DL, Claro M, de Souza MVS, Dias LLP, daSilva R, Lee J, Lopez RFV (2018) Skin cancer treatment effectiveness is improved by iontophoresis of EGFR-targeted liposomes containing 5-FU compared with subcutaneous injection. J Controlled Release 283(June):151–162. https://doi.org/10.1016/j.jconrel.2018.05.038

Praça FS, Garcia WS, Medina G, Petrilli R, Bentley MVLB (2012) Liquid crystal nanodispersions enable the cutaneous delivery of photosensitizer for topical PDT: fluorescence microscopy study of skin penetration. Curr Nanosci 8(4):535–540. https://doi.org/10.2174/157341312801784203

Priyadarshini E, Pradhan N (2017) Gold nanoparticles as efficient sensors in colorimetric detection of toxic metal ions: a review. Sens Actuators, B 238:888–902. https://doi.org/10.1016/j.snb.2016.06.081

Rai M, Yadav A, Gade A (2009) Silver nanoparticles as a new generation of antimicrobials. Biotechnol Adv 27(1):76–83. https://doi.org/10.1016/j.biotechadv.2008.09.002

Rito C, Pineiro-Maceira J (2009) Microscopia confocal reflectante aplicada ao diagnóstico do. 636–642

Roca AG, Gutiérrez L, Gavilán H, Brollo MEF, Veintemillas-Verdaguer S, del Puerto Morales M (2019) Design strategies for shape-controlled magnetic iron oxide nanoparticles. Adv Drug Deliv Rev 138:68–104. https://doi.org/10.1016/j.addr.2018.12.008

Romero G, Estrela-Lopis I, Zhou J, Rojas E, Franco A, Espinel CS, Gonza A, Gao C, Donath E, Moya SE (2010) Surface engineered poly (lactide-co-glycolide) nanoparticles for intracellular delivery : uptake and cytotoxicity s a confocal raman microscopic study. 2993–2999

Seidi F, Jenjob R, Phakkeeree T, Crespy D (2018) Saccharides, oligosaccharides, and polysaccharides nanoparticles for biomedical applications. J Controlled Release 284(June):188–212. https://doi.org/10.1016/j.jconrel.2018.06.026

Seo S-J, Chen M, Wang H, Sil M, Leong KW, Kim H-W (2017) Nano today extra- and intra-cellular fate of nanocarriers under dynamic interactions with biology. Nano Today 14:84–99

Severino P, Santana MHA, Souto EB, de Ciências F (2011) Polímeros usados como sistemas de transporte de princípios ativos

Shah M, Badwaik V, Kherde Y, Waghwani HK, Modi T, Aguilar ZP (2014) Table of contents 1. (8):1320–1344

Shigematsu H, Kadoya T, Masumoto N, Sasada T, Emi A, Ohara M, Kajitani K, Okada M (2015) The efficacy and safety of preoperative chemotherapy with triweekly abraxane and cyclophosphamide followed by 5-fluorouracil, epirubicin, and cyclophosphamide therapy for resectable breast cancer: a multicenter clinical trial. Clin Breast Cancer 15(2):110–116. https://doi.org/10.1016/j.clbc.2014.09.010

Simon J, Flahaut E, Golzio M (2019) Overview of carbon nanotubes for biomedical applications. Materials 12(4):1–21. https://doi.org/10.3390/ma12040624

Singh S (2000) Phase transitions in liquid crystals. Phys Rep 324(2–4):107–269. https://doi.org/10.1016/S0370-1573(99)00049-6

Singh R, Sharma G, Agrawal P, Sonali BP, Koch B, Muthu M (2016) Transferrin receptor targeted PLA-TPGS micelles improved efficacy and safety in docetaxel delivery. Int J Biol Macromol 83:335–344

Sittampalam G, Coussens N, Arkin M, Auld D, Austin C, Bejcek B, Glicksman M et al (2016) Assay guidance manual. Assay Guidance Manual Md:305–36. https://doi.org/PMID:22553881

Sjögren G, Sletten G (2000) Cytotoxicity of dental alloys, metals, and ceramics assessed by millipore filter, agar overlay, and MTT tests

Soppimath KS, Aminabhavi TM, Kulkarni AR, Rudzinski WE (2001) Biodegradable polymeric nanoparticles as drug delivery devices. J Controlled Release. https://doi.org/10.1067/mpr.2000.107227

Stepanenko AA, Dmitrenko VV (2015) Pitfalls of the MTT assay: direct and off-target effects of inhibitors can result in over/underestimation of cell viability. Gene 574(2):193–203. https://doi.org/10.1016/j.gene.2015.08.009

Stockert JC, Horobin RW, Colombo LL, Blázquez-Castro A (2018) Tetrazolium salts and formazan products in cell biology : viability assessment, Fl uorescence imaging, and labeling perspectives. Acta Histochem no February

Sun Q, Ojha T, Kiessling F, Lammers T, Shi Y (2017) Enhancing tumor penetration of nanomedicines. Biomacromol 18(5):1449–1459. https://doi.org/10.1021/acs.biomac.7b00068

Tran S, DeGiovanni P-J, Piel B, Rai P (2017) Cancer nanomedicine: a review of recent success in drug delivery. Clin Transl Med 6(1). https://doi.org/10.1186/s40169-017-0175-0

Valdiglesias V, Fernández-Bertólez N, Kiliç G, Costa C, Costa S, Fraga S, Bessa MJ, Pásaro E, Teixeira JP, Laffon B (2016) Are iron oxide nanoparticles safe? Current knowledge and future perspectives. J Trace Elem Med Biol 38:53–63. https://doi.org/10.1016/j.jtemb.2016.03.017

Van de Ven H, Paulussen C, Feijens PB, Matheeussen A, Rombaut P, Kayaert P, Van Den Mooter G (2012) PLGA Nanoparticles and nanosuspensions with amphotericin B: potent in vitro and in vivo alternatives to fungizone and am bisome. J Controlled Release 161(3):795–803. https://doi.org/10.1016/j.jconrel.2012.05.037

Villanueva-García M, Gutiérrez-Parra RN, Martínez-Richa A, Robles J (2005) Quantitative structure-property relationships to estimate nematic transition temperatures in thermotropic liquid crystals. J Mol Struct (Theochem) 727(1–3 SPEC. ISS.):63–69. https://doi.org/10.1016/j.theochem.2005.02.033

Vladisavljević GT (2019) Preparation of microemulsions and nanoemulsions by membrane emulsification. Colloids Surf, A 579:123709. https://doi.org/10.1016/j.colsurfa.2019.123709

Wang Y, Huang H-Y, Yang L, Zhang Z, Ji H (2016) Cetuximab-modified mesoporous silica nanomedicine specifically targets EGFR-mutant lung cancer and overcomes drug resistance. Nature Publishing Group no May:1–10. https://doi.org/10.1038/srep25468

Wei L, Guo XY, Yang T, Min Zhi Y, Chen DW, Wang JC (2016) Brain tumor-targeted therapy by systemic delivery of SiRNA with transferrin receptor-mediated core-shell nanoparticles. Int J Pharm 510(1):394–405. https://doi.org/10.1016/j.ijpharm.2016.06.127

Wolfram J, Ferrari M (2019) Clinical cancer nanomedicine. Nano Today 25:85–98. https://doi.org/10.1016/j.nantod.2019.02.005

Xiao B, Han MK, Viennois E, Wang L, Zhang M, Si X, Merlin D (2015a) Hyaluronic acid-functionalized polymeric nanoparticles for colon cancer-targeted combination chemotherapy. Nanoscale 7(42):17745–17755. https://doi.org/10.1039/c5nr04831a

Xiao B, Zhang M, Viennois E, Zhang Y, Wei N, Baker MT, Jung Y, Merlin D (2015b) Inhibition of MDR1 gene expression and enhancing cellular uptake for effective colon cancer treatment using dual-surface-functionalized nanoparticles. Biomaterials 48:147–160. https://doi.org/10.1016/j.biomaterials.2015.01.014

Xu ZP, Zeng QH, Lu GQ, Yu AB (2006) Inorganic nanoparticles as carriers for efficient cellular delivery. Chem Eng Sci 61(3):1027–1040. https://doi.org/10.1016/j.ces.2005.06.019

Xue Y (2016) Carbon nanotubes for biomedical applications. Industrial applications of carbon nanotubes. Elsevier Inc., Amsterdam. https://doi.org/10.1016/B978-0-323-41481-4.00011-3

Yazdimamaghani M, Moos PJ, Dobrovolskaia MA, Ghandehari H (2019) Genotoxicity of amorphous silica nanoparticles: status and prospects. Nanomed: Nanotechnol Biol Med 16(xxxx):106–25. https://doi.org/10.1016/j.nano.2018.11.013

Yeh YC, Creran B, Rotello VM (2012) Gold nanoparticles: preparation, properties, and applications in bionanotechnology. Nanoscale 4(6):1871–1880. https://doi.org/10.1039/c1nr11188d

Zorko M, Langel U (2005) Cell-penetrating peptides: mechanism and kinetics of cargo delivery. Adv Drug Deliv Rev 57:529–545

Chapter 2
Biopolymers for Eco-Safe Remediation

Vicente de Oliveira Sousa Neto, Antonio Joel Ramiro de Castro, Cícero Pessoa de Moura, Guilherme Augusto Magalhães Júnior, Rafael Ribeiro Portela, Gilberto Dantas Saraiva, and Ronaldo Ferreira do Nascimento

1 Introduction

The removal of pollutants, as well as the prevention of environmental pollution, has been an important strategy of the green economy. Currently, with the implementation of stricter environmental laws regarding the release of pollutants, much research has focused on the use of biodegradable materials in the prevention and remediation of environmental damage. Environmental compensation is key to sustainable development. Sites polluted by industrial activity, the use of pesticides and fertilizers, or the release of other pollutants must be cleaned to rebuild or return them to their natural state.

Social and economic demands require us to be environmentally conscious. Prior to modern environmental regulation, many companies simply threw hazardous materials into the environment. They poured chemicals and other pollutants onto unused land or into lakes, rivers, and streams. The sites would also be polluted by accidents

V. de Oliveira Sousa Neto (✉) · G. D. Saraiva
Laboratory of Study and Research in Pollutants Removal by Adsorption—LERPAD, Department of Chemistry, Laboratory of Synthesis and Characterization of Materials—LASCAM, Department of Physics, State University of Ceará (UECE-FECLESC), Rua José de Queiroz Pessoa, N° 2554, Quixadá, Ceará 63900-000, Brazil
e-mail: vicente.neto@uece.br

C. P. de Moura · G. A. M. Júnior · R. R. Portela · R. F. do Nascimento
Federal Institute of Education, Science and Technology of Ceará (IFCE), Quixadá Campus, Quixadá, Ceará 63902-580, Brazil

A. J. R. de Castro · G. D. Saraiva
Federal University of Ceará - Quixadá Campus(UFC), Av. José de Freitas Queiroz, 5003, Quixadá, Ceará 63902-580, Brazil

A. J. R. de Castro · R. F. do Nascimento
Department of Analytical and Physical Chemistry, Federal University of Ceará (UFC), Campus do Pici, Bloco 940, Fortaleza, Ceará 60455-970, Brazil

R. F. do. Nascimento et al. (eds.), *Nanomaterials and Nanotechnology*, Materials Horizons: From Nature to Nanomaterials, https://doi.org/10.1007/978-981-33-6056-3_2

or improperly functioning equipment. Polluted sites (cleaned and rebuilt) are known as brownfield sites.

1.1 Remediation

Conventional technologies have been used to treat all types of pollutants such as organic and toxic waste. There are many options in which choice is linked (Fig. 1) to the cost and efficiency of the process. New technologies are in development for the remediation of contaminants in the air, water, and soil (Ye et al. 2019). Figure 1 shows some environmental remediation approaches.

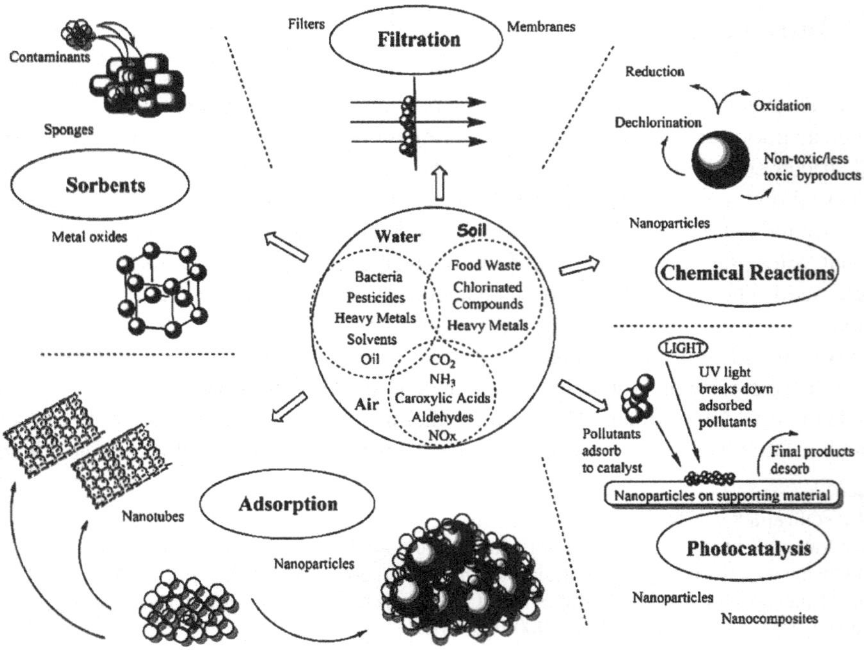

Fig. 1 Environmental remediation approaches. This is figure by Guerra et al. (2018), is from an open-access article distributed under the Creative Commons Attribution License which permits unrestricted use, distribution, and reproduction in any medium, provided the original work is properly cited

1.2 International Environmental Legislation

The global pollution is attributed primarily to increased population density and anthropogenic activity through the misuse of environmental resources and improper waste disposal (Kowalska et al. 2018). Table 1 shows the European annual waste generation from four aggregated industrial sectors and gross value added.

This chapter will be approaching the application of biopolymers such as cellulose, chitosan, alginate, and others in the remediation of water and soil contaminated by toxic metals or pesticides. Initially, a general exposure of these contaminants and their effects on ecosystems will be given. After the biopolymers will be addressed with emphasis on their properties, characteristics, and applications in environmental remediation.

2 Metal Toxic Levels: Soil, Water

Although European industry continues to be a significant source of pollutant release into the environment, statistical data indicate that, in general, pollution releases have declined over the past decade. Sustainability has been a fundamental component of the competitiveness of the national and international markets in the globalized world.

The reduction of pollutant emissions can be attributed to a greater environmental awareness of the great political leaders, a more informed society and concerned with future generations. This naturally results in a more stringent and environmentally restrictive legislation. Environmental regulation combined with the enhancement of

Table 1 European annual waste generation from four aggregated industrial sectors and gross value added

Year	Extractive industry	Manufacturing industry	Energy supply	Waste	Total industry gross value added (except construction)	Manufacturing industry gross value added
	(in million tonnes)				(in billion euros)	
2010	67,121	25,492	9018	16,429	221,087	177,621
2012	73,269	24,978	1011	19,424	223,777	181,533
2014	70,319	25,136	9833	23,127	227,645	187,506
2016	63,298	25,028	9719	25,361	241,885	201,174

Data sources

Eurostat statistics on waste (env_wasgen) provided by Statistical Office of the European Union (Eurostat)

Gross value added for industry provided by Statistical Office of the European Union (Eurostat)

Link:https://commons.wikimedia.org/wiki/File:European_annual_waste_generation_from_four_aggregated_industrial_sectors_and_gross_value_added.jpg new technologies focused on reducing pollutants has led to reduced release of water, soil, and pollutants in Europe.

Toxic metals in the environment bring serious damage to nature, public health, and consequently the economy. The negative impacts of the disposal of these pollutants are a constant concern (Edelstein and Ben-Hur 2018). Nature-released metal pollutants are predominantly sourced from agriculture, industry, mining, transportation, fuel consumption, residual organic matter, sewage water and include iron, copper, zinc, nickel, cobalt, cadmium, titanium, strontium, tin, vanadium, nickel, molybdenum, mercury, lead, etc. The discharge of industrial effluents containing high toxic metals is usually associated with intense industrial activity (Li et al. 2019).

Surface water eutrophication is characterized by a variety of indicators, including nitrogen and phosphorus concentrations, chemical oxygen demand, and water transparency (Yao et al. 2019). Industrial pollutant releases to water include compounds that contain nutrients that can cause eutrophication, such as nitrogen (referred to as total nitrogen) and phosphorous (total phosphorus). In Fig. 2 is shown water pollutant emissions and gross value added for the European industry from 2007 to 2017. According to Fig. 2, whose data were obtained from the European Environment Agency, the release of toxic metals has been decreasing over the years although

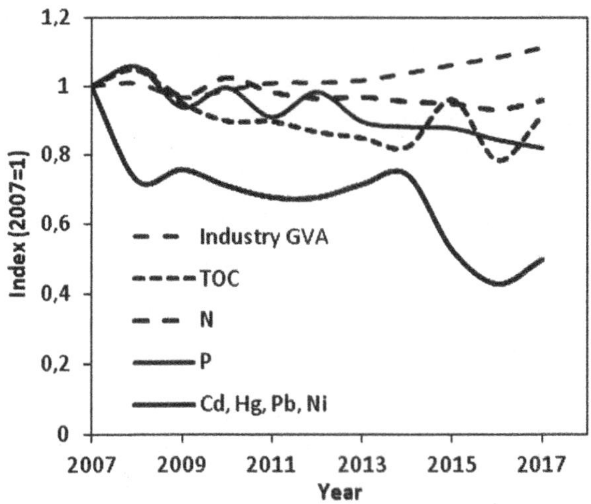

Fig. 2 Releases of water pollutants and gross value added for industry (European Environment Agency, EEA-33). This figure by Vicente Neto is licensed under the Creative Commons Attribution 4.0 International. Link https://commons.wikimedia.org/wiki/File:Releases_of_water_pollutants_and_gross_value_added_for_industry.jpg. *Notes* Data obtained from European Environment Agency: https://www.eea.europa.eu/data-and-maps/indicators/industrial-pollution-in-europe-3/assessment accessed on 09.15.2019. EEA-33 refers to the 33 member countries of the European Environment Agency including the EU-28 Member States, the four European Free Trade Agreement countries (Iceland, Liechtenstein, Norway, and Switzerland) as well as Turkey. The E-PRTR does not contain data for Turkey, which is not covered by the E-PRTR Regulation. Gross value added (GVA) is used as a proxy for the economic activity of industry in Europe. The metric used accounts for inflation based on 2010 values

with occasional peaks in 2019, 2014, and 2017. All these peaks are lower than the reference year 2007.

Soil contamination in Europe is, among other things, linked to industrial activity. Waste transfers from industrial facilities in the EU have remained relatively stable in the last decade.

3 Pesticides

The development of agricultural chemicals with direct application in pest and weed control has enabled an increase in food production in the world. In this context, this economic segment has brought positive impacts to both developed and developing countries.

However, the unregulated use of these pesticides results in food residues, leading to concern about the possible adverse effects of these chemicals on human health. This concern is related to data that assumes that exposure to pesticide residues through the diet is up to five times the magnitude of exposure through other routes, such as air and drinking water (Knezevic et al. 2012; Claeys et al. 2011). Figure 3 shows the use of pesticides (%) in the world by continents, and Table 2 shows the pesticides use (average 2014–2017) by countries according to Food and Agriculture Organization (FAO).

Pesticides, as well as any chemical product, have physicochemical properties such as solubility, partition coefficient (in soils/liquids), vapor pressure, and melting point. These properties affect how they behave in the environment under natural conditions.

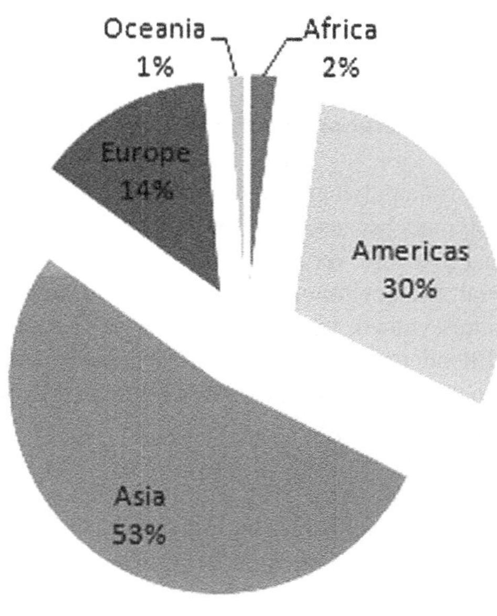

Fig. 3 Pesticides in the world by continent. Average 1990–2017. This figure by Vicente Neto is licensed under the Creative Commons Attribution 4.0 International. Link: https://commons.wikimedia.org/wiki/File:Pesticides_in_the_world_by_continent._Average_1990_%E2%80%93_2017.jpg. *Notes* Data obtained from Food and Agriculture Organization (FAO) http://www.fao.org/faostat/en/#data/RP/visualize

Table 2 Top 10 Countries: Pesticides use. Average 2014–2017

Country	Unit	2014	2015	2016	2017
China	Tonnes	181,573,279	177,242,036	177,258,889	177,363,440
United States of America		40,777,920	40,777,920	40,777,920	40,777,920
Brazil		352,336	395,646	377,176	377,176
Argentine		207,706	20,380,698	19,990,795	19,600,893
Canada		76,314	7,531,830	9,083,870	9,083,870
Ukraine		78,201	78,201	78,201	78,201
France		7,490,960	6,653,102	7,195,122	7,058,870
Malaysia		4,919,943	6,330,763	6,728,838	6,728,838
Span		61,067	59,018	61,895	6,089,640
Italy		59,422	63,322	60,239	56,641

Notes Data obtained from Food and Agriculture Organization (FAO)
Link: http://www.fao.org/faostat/en/#data/RP/visualize

Pesticides can be classified by their use and chemical structure. Generally, the most common classifications for use are insecticides, fungicides, herbicides, rodenticides, fumigants, and insect repellents (WHO 2006). The classification of pesticides according to their functional groups (chemical structure) follows the list: organophosphates, pyrethroids, carbamates, and organochlorines (Table 3).

4 Biopolymers

4.1 Chitosan

4.1.1 Chitosan: Source and Properties

Chitosan polymers are recognized as an important alternative to the use of synthetic polymers because they are biologically biodegradable and have a modifiable chemical structure, very functional that allows a variety of applications of medical, industrial, and environmental interest. Chitosan is a derivative of chitin, which makes it a semi-synthetic aminopolysaccharide of natural origin, and it is obtained industrially (Ellendersen et al. 2018) from the controlled deacetylation of the chitin. This one is a natural polymer abundantly found in crustacean carapaces and as fungal cell wall component (Fig. 4).

Chitosan is biodegradable which has an ability to remove toxic metals from aqueous systems. This affinity for heavy metals is related to the free amino and hydroxyl groups present in their structure. Chitosan is also considered a renewable material (Sivakami et al. 2013). These features make chitosan a good material to be applied to environmental remediation.

Table 3 Pesticides and a brief summary of the mode of action

Classification	Pesticide	Mechanism	Reference
Organophosphates	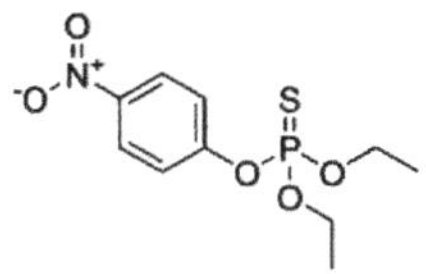 Parathion 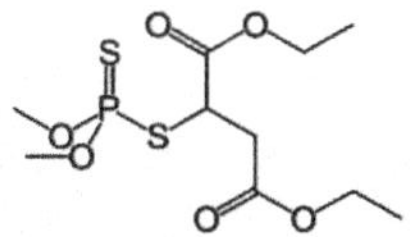 Malathion 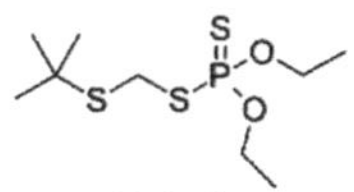 Terbufos 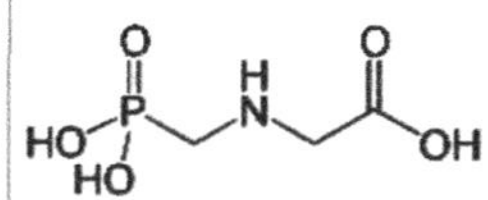 Glyphosate	Organophosphates act on the enzyme acetylcholinesterase(Ach). They inhibit Ach by irreversible process Glyphosate interferes with the shikimate pathway, which produces the aromatic amino acids phenylalanine, tyrosine, and tryptophan in plants and microorganisms	Metcalf (2000) Steinrucken and Amrhein (1980)
Pyrethroids	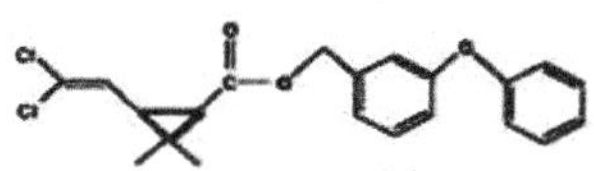 Permethrin	Pyrethrones act to keep the sodium channels open in the neural membranes of insects	Coats (1990)

(continued)

Table 3 (continued)

Classification	Pesticide	Mechanism	Reference
	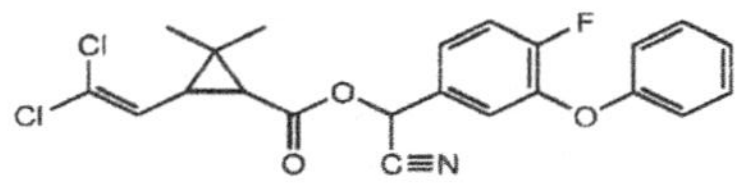 Cyfluthrin	Its mode of action is characterized by interference with nerve signaling by inhibition of the membrane sodium channel systems in the target organism	Elyazar et al. (2011)
	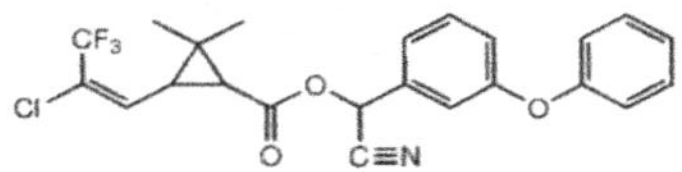 l-cyihalothrin	The binding of the insecticide to the protein keeps the sodium channels in their open state, so the nerves cannot repolarize, thereby paralyzing the organism	Soderlund et al. (2002)
	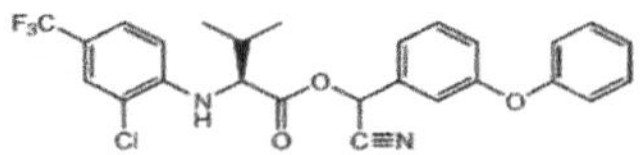 τ–Fluvalinate	Acts to inhibit sodium channel modulators	USEPA (1996)
Carbamates	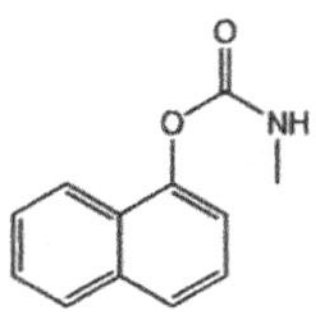 Carbaryl 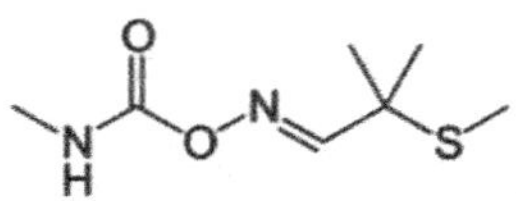 Adilcarb	These insecticides kill insects by reversibly inactivating the enzyme acetylcholinesterase	Fukuto (1990)

(continued)

Table 3 (continued)

Classification	Pesticide	Mechanism	Reference
Organochlorines	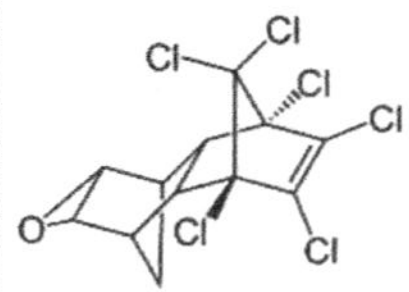 Dieldrin	The mechanism of action is the insecticide binding at the GABAA site in the gamma-aminobutyric acid (GABA) chloride ionophore complex, which inhibits chloride flow into the nerve	Coats (1990)
	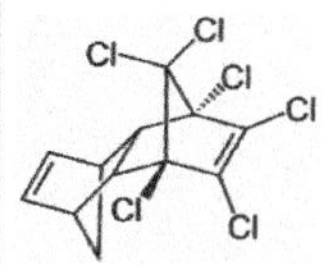 Aldrin	(1) Ability of Aldrin to inhibit brain calcium ATPases (2) Ability of Aldrin to block gamma-aminobutyric acid (GABA) activity	Mehrotra et al. (1988), Glotfelty (1978)
	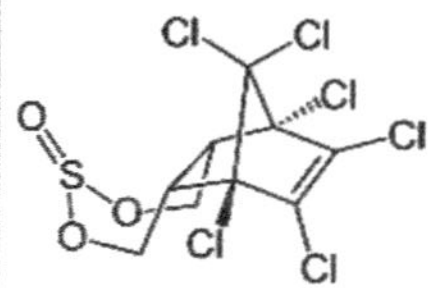 Endossulfan	Act by altering the electrophysiological and associated enzymatic properties of nerve cell membranes, causing a change in the kinetics of Na+ and K+ ion flow through the membrane	Hayes and Laws (1991)

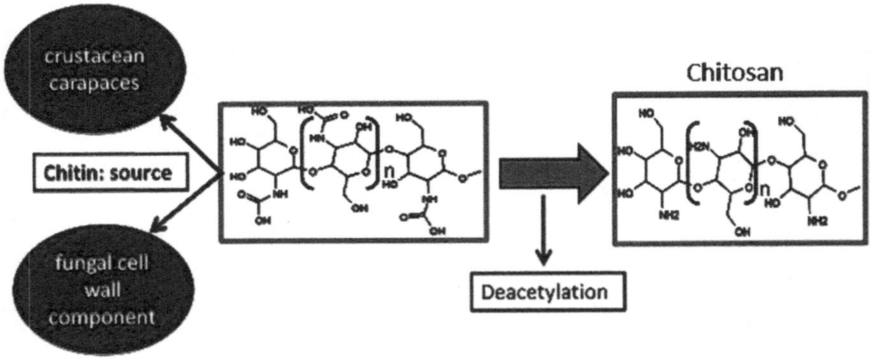

Fig. 4 Chitin and chitosan. This figure by Vicente Neto is licensed under the Creative Commons Attribution 4.0 International. Link https://commons.wikimedia.org/wiki/File:Chitin_and_chitosan.jpg

4.1.2 Modification and Remediation Applications

Chitosan has low solubility in an aqueous medium. This feature limits its application for several purposes. Therefore, several kinds of research have been developed to improve the solubility of chitosan in the aqueous medium (Barros et al. 2015).

Chitosan has two highly reactive functional groups in its structure. Such groups are hydroxyl (–OH) and amino ($-NH_2$). Structurally, as shown in Fig. 3, chitosan has a secondary hydroxyl at C3 and a primary hydroxyl at C-6. The amino functional group is found at C-2. Both groups are responsible for the adsorption capacity of chemical species due to the good reactive capacity of these functional groups under natural conditions (Barros et al. 2015, Carvalho et al. 2018).

The following topic was listed some recent techniques that have been successfully applied to remove pesticides by adsorption using chitosan modified as adsorbent.

Schiff's Base Derivative of Chitosan

Schiff bases have been introduced into the structure in chitosan via reaction with the amino group in carbon (C-2) in order to improve its chelating capacity. The Schiff base has a carbonyl group that is very active in the formation of metal complexes (Gavalyan 2016).

Shankar et al. (2020), modified chitosan and studied its ability to remove pentachlorophenol pesticides. In that work, chitosan was subjected to two stages of functionalization. Initially, it was cross-linked with 2-hydroxy-1-naphthaldehyde to obtain CHTAC. Then, it was subjected to an additional graft with $CuCl_2$ to synthesize CHTAC.

According to the authors, the modification of chitosan both with a Schiff base containing aromatic groups (1st step) and the introduction of a copper complex

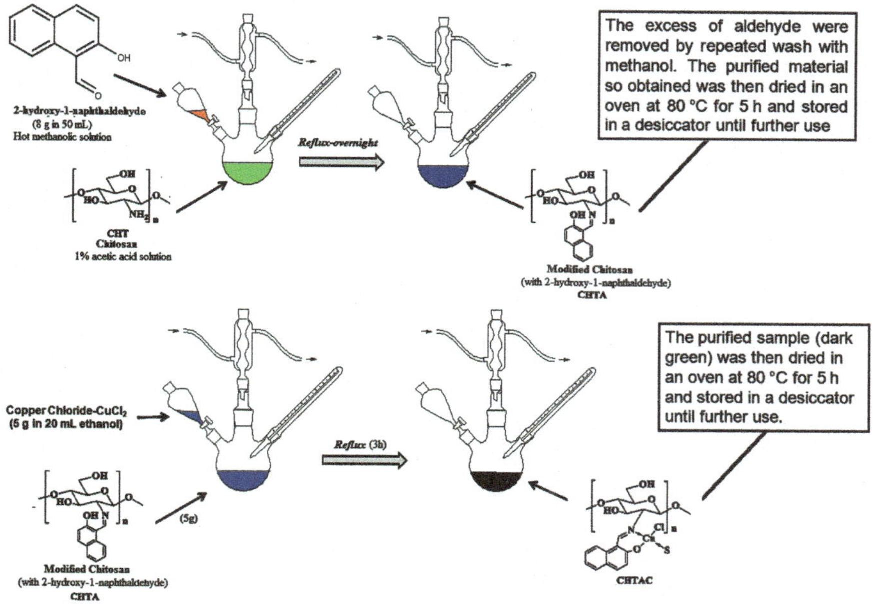

Fig. 5 Functionalization in two steps proposed by Shankar et al. (2020). Modification of chitosan using 2-hydroxy-1-naphthaldehyde and $CuCl_2$. This figure by Vicente Neto is licensed under the Creative Commons Attribution 4.0 International. Link https://commons.wikimedia.org/wiki/File:Chitosan-modified.jpg

(2nd step) was a strategy whose objective was to improve the affinity of organic compounds with the adsorbent. This possible due to metal sites electron-deficient to adsorbent surface that can improve its interaction with electron-rich molecules such as pesticides containing aromatic groups (Fig. 5).

Modified Magnetic Chitosan Nanoparticles

Undoubtedly, organophosphorus pesticides (OPPs) are one of the organic pollutants more damaging to environmental aquatic. However, this class of pesticides is commonly used in agriculture due to its efficiency against insects (Bandforuzi and Hadjmohammadi 2019).

OPPs and their residues are acetylcholinesterase inhibiting substances and therefore are highly toxic to both aquatic life and human health when absorbed by the organisms. Exposure to high levels of pesticide tests and residues can cause acute food poisoning.

Bandforuzi and Hadjmohammadi (2019), investigated the application of magnetic chitosan nanoparticles (MCNPs) modified with mixed sodium dodecyl sulfate (SDS) hemimicelles for both removing and trace quantification of three organophosphate pesticides in agricultural wastewater and environmental water samples. According

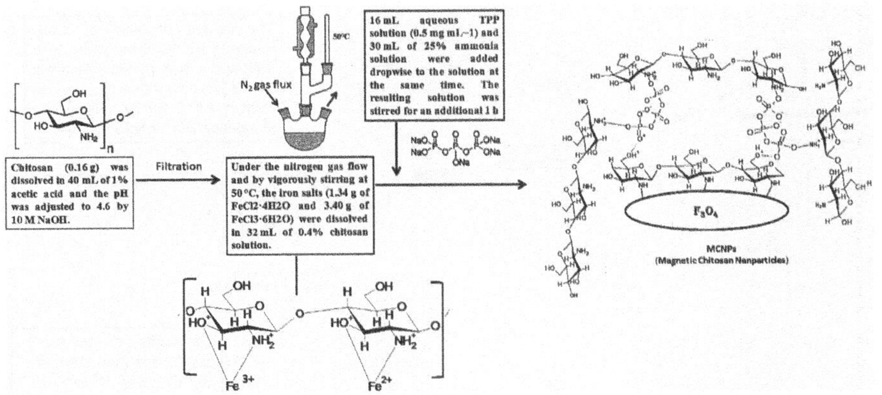

Fig. 6 Schema of the preparation of MCNPs. This figure by Vicente Neto is licensed under the Creative Commons Attribution 4.0 International. Link https://commons.wikimedia.org/wiki/File:Chitosan-hibrid_magnetic.jpg

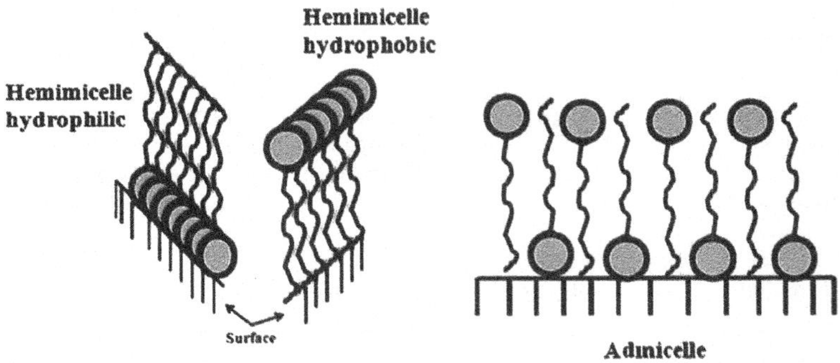

Fig. 7 Hemimicelle and admicelles. This figure by Vicente Neto is licensed under the Creative Commons Attribution 4.0 International. Link https://commons.wikimedia.org/wiki/File:Hemimicelle_e_Admicelle.jpg

to the authors, the introduction of magnetic cores was via co-precipitation of Fe(II)/Fe(III) (Fe_3O_4, nanoform) based on mixed hemimicelle of sodium dodecyl sulfate. In this method, the formation of mixed hemimicelles made up of SDS on the surface of MCNPs leads to solubilization and the retention of target analytes by strong hydrophobic and electrostatic interactions. In Fig. 6 is illustrated schema of the preparation of MCNPs.

New hybrid-magnetic material presented magnetic sensitivity that allows a separation (by an external magnetic field) and recovery of the adsorbent for reuse can be easily separated from the reaction mixture by an external magnet. In adsorption study, Bandforuzi and Hadjmohammadi (2019) used SDS surfactant to make up hemimicelles or admicelles (Fig. 7). The principle of adsorption, in this case, is associated

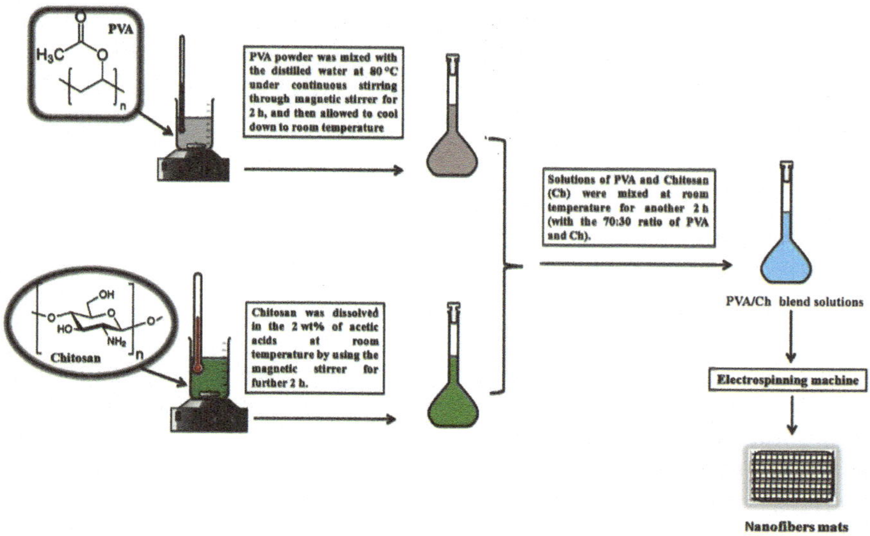

Fig. 8 Preparation of nanofibers mats from chitosan/PPV blend solution. This figure by Vicente Neto is licensed under the Creative Commons Attribution 4.0 International. Link https://commons.wikimedia.org/wiki/File:Preparation_of_nanofibers_mats.jpg

formation of the aggregates of surfactant monomers on the solid ionic surface while being adsorbed on it (Esumi and Ueno 2003).

Bandforuzi and Hadjmohammadi (2019) based on experimental evidence indicated that magnetic chitosan-coated with SDS was able to extract and stabilize pesticides from the environmental water, groundwater, and wastewater. This ability was attributed to strong interactions, both hydrophobic and electrostatic in nature, between mixed hemimicelles and analytes. Such interactions were responsible for the high efficiency in the magnetic solid-phase extraction (MSPE). The authors concluded that the adsorbent has potential to removing pesticides.

Composite Nanofibers Membranes of Chitosan

Chitosan through its amino and hydroxyl groups is a versatile reactant. This way, under specific conditions, chitosan is able to participate in diverse chemical reactions to produce a functional derivative with an adequate and specific purpose to industrial application (Teo et al. 2019).

Karim et al. (2019), developed a kind of membrane nanofiber from polyvinyl alcohol (PVA) and chitosan (Chi) for removing toxic of Pb(II) and Cd(II) ions from wastewater samples (Karim et al. 2019). Figure 8 illustrated the steps to obtain chitosan/PVA blend solution according to Karim et al. (2019). Blend solution was injected into the electrospinning machine. After electrospinning, the nanofibers mats were collected and prepared to be used.

According to Karim et al. (2019), chitosan/PPV was an effective adsorbent for the removal of Pb(II) and Cd(II) by adsorption process in aqueous medium and in experimental conditions of competition in the presence of several metal ions. Table 4 summarizes some environmental applications of chitosan.

4.2 Cellulose

4.2.1 Source and Properties

Cellulose is a natural biopolymer largely found in nature. It has several properties such as biodegradability, hydrophilicity, potential as a sorbent, non-toxicity, susceptible to chemical modification via hydroxyl groups, good mechanical properties; it is a renewable, and non-meltable polymer, which has low solubility in most solvents due to hydrogen bonding and crystallinity (Suhas et al. 2016; Sanchez-Salvador et al. 2019). From the chemical point of view, cellulose is a β(1-4)-linked chain of glucose molecules. Cellulose is a polysaccharide in which the carbon content can reach up to 50% of the lignocellulosic biomass. The hydroxyl groups of cellulose are strategic ways for functionalization because they are determinants in the reactivity of this natural polymer. Structurally, hydroxyls can contribute to resistance and rigidity of the cellulose due to their ability to form hydrogen bonds. Cellulose is a natural polymer that has a molecular formula ($C_6H_{10}O_5$) whose linear chain is composed of repetitive units of Cellobiose. This compound is a disaccharide formed by two units of glucose. Figure 9 shows an illustration of the cellulose composition.

4.2.2 Cellulose Derivatization

The most common cellulose derivatization reactions are esterification, etherification, oxidation, co-polymerization graphitization, and cross-linking.

Esterification and etherification: synthetic routes of wide industrial application with the purpose of giving cellulose greater solubility in solvents such as water, acetone, and alcohol (Gupta et al. 2019).

Oxidation: In the vast majority of cases, oxidation of cellulose has as main objective the conversion of hydroxyl groups into carbonyl groups or carboxyl groups.

Gupta et al. (2019) obtained modified cellulose from cellulose simultaneous oxidation and esterification using Octenyl Succinic Anhydride (OSA) and sodium hypochlorite (NaOCl) mixtures. In Fig. 10 shows a schematic illustration summarizing the synthetic route to obtain the derivative cellulose.

According to Gupta et al. (2019), the method has the advantage due to the fact that the simultaneous oxidation and succinylation of cellulose let hight degree of substitution of carbonyl groups in the interior of cellulose. The modified cellulose showed an adsorption capacity for the Cu(II) equal to 5.4 mg/g in 15 min at neutral pH. The derivative obtained also has the potential to be used as a low-cost adsorbent

Table 4 Environmental applications of natural and synthetic polymer based on chitosan materials

	Functional group introduced	Pollutant	Mechanism	Reference
Chitosan	–	NH_4^+	Adsorption process	Bernardi et al. (2018)
	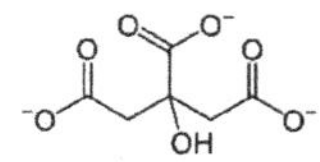 Citrate: used as an auxiliary complexing agent	Cu(II)	Copper sorption by chitosan in the presence of citrate ions	Guzman et al. (2003)
	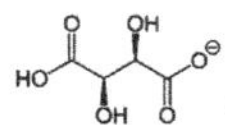 Tartrate 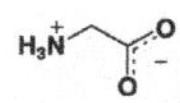 Glycine 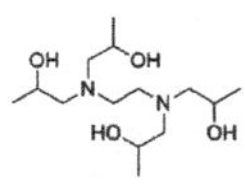 Quadrol All ligantes used as an auxiliary complexing agent	Cu(II)	Copper sorption by chitosan in the presence of citrate ions	Gyliene et al. (2009)
Chitosan-thymine conjugate	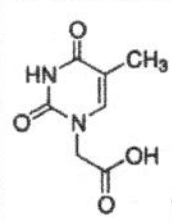 Thymine-1-acetic acid	pyrethroid pesticides	Flocculation performance	Ghimici and Dinu (2019)
Mercaptoacetyl chitosan	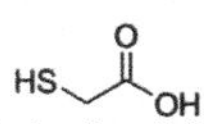 Thioglycolic acid (TGA	Cu (II) turbidity	Flocculation performance	Chang et al. (2009)
Salicylaldehyde functionalized chitosan nanoparticles	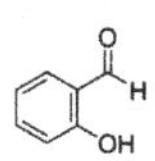 Salicylaldehyd	Pb(II), Cu(II), and Cd(II)	Adsorbents have electron donor groups such as amine, hydroxyl, and imine, which can interact with the metal ions to form complex a	Hussain et al. (2020)

(continued)

Table 4 (continued)

	Functional group introduced	Pollutant	Mechanism	Reference
Carboxylate functionalized chitosan co-polymer	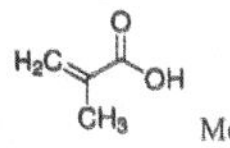 Methacrylic acid	Zn(II), Cd(II), Pb(II) and Cu(II)	Active functional groups are involved in the adsorptive removal process	Dev et al. (2020)
Chitosan/glutaraldehyde	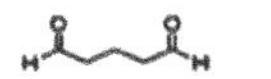 Glutaraldehyd: used as cross-linker	Cr(VI)	Interaction with cross-linked chitosan. It is chelation	Rojas et al. (2005)

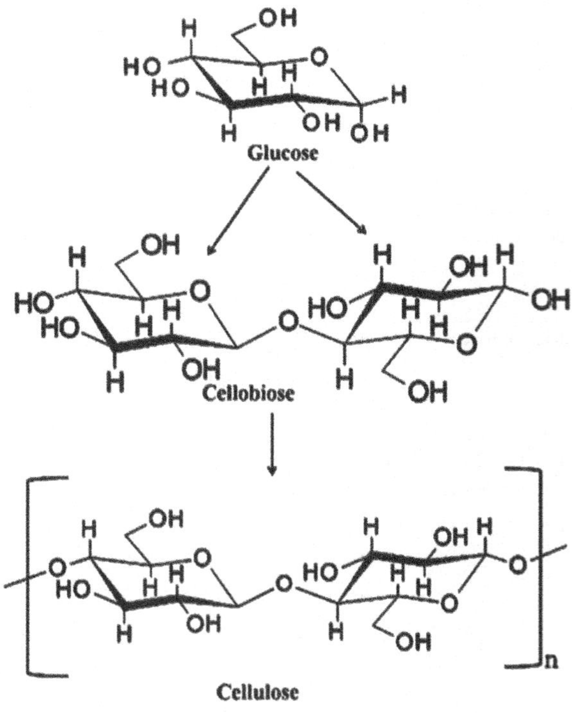

Fig. 9 Illustration of the cellulose composition. This figure by Vicente Neto is licensed under the Creative Commons Attribution 4.0 International. Link https://commons.wikimedia.org/wiki/File:Illustration_of_the_cellulose_composition.jpg

for the removal of other toxic metals in aqueous media. From thermodynamic studies, it was observed that ΔG° to the adsorption of Cu(II) ions over modified cellulose is a spontaneous process and the reuse study suggests that the regeneration of the adsorbent can be carried out on a commercial scale reducing the economic impact and energy cost. The SEM morphology and thermal study showed that the modified

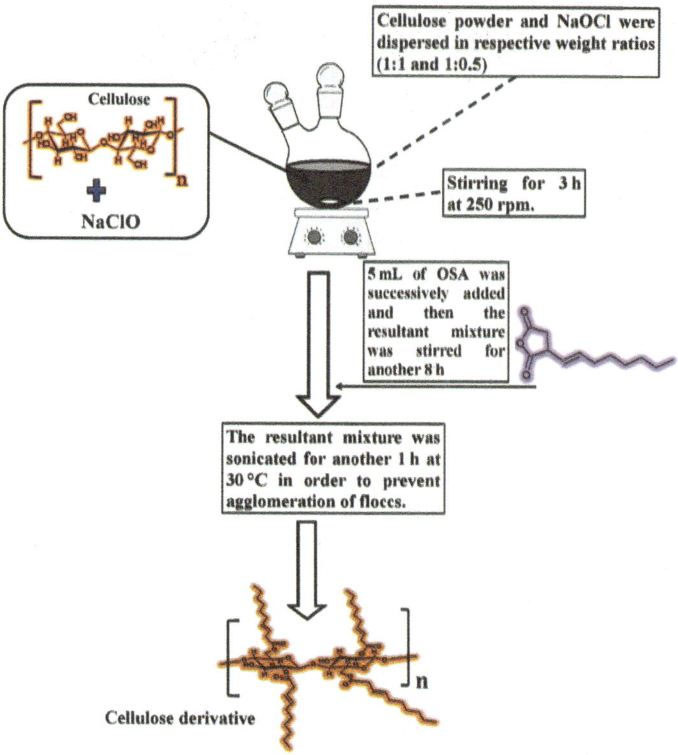

Fig. 10 Schematic illustration summarising the synthetic route to obtain the derivative cellulose. This figure by Vicente Neto is licensed under the Creative Commons Attribution 4.0 International. Link https://commons.wikimedia.org/wiki/File:Schematic_illustration_summarising_the_synthetic_route_to_obtain_the_derivative_cellulose.jpg

cellulose had a porous structure that enhanced the metal uptake capacity and that the succinylated cellulose exhibited lower thermal stability than native cellulose (Gupta et al. 2019).

Yuan et al. (2019) developed a method to produce a new material able to removing neonicotinoid insecticides. The new material was obtained from cellulose and in the synthetic route was used methacrylic acid as a modification reagent. The modified cellulose-based complex particle (MCCP) was produced through hydrothermal carbonization with methacrylic acid in the stirring and sand bath circumstance. Then, activated modified carbon-based porous particle (AMCCP) was prepared by treating MCCP with potassium hydroxide at high temperature (Fig. 11).

Experimental data discussed by Yuan et al. (2019) suggest that meta acrylic acid is important in hydrothermal synthesis because cellulose with methacrylic acid can form a microspherical surface very homogeneous with respect to synthesis in the absence of methacrylic acid. Another aspect in hydrothermal synthesis is associated to catalyze cellulose in different oligomers and glucose by hydrolysis. These

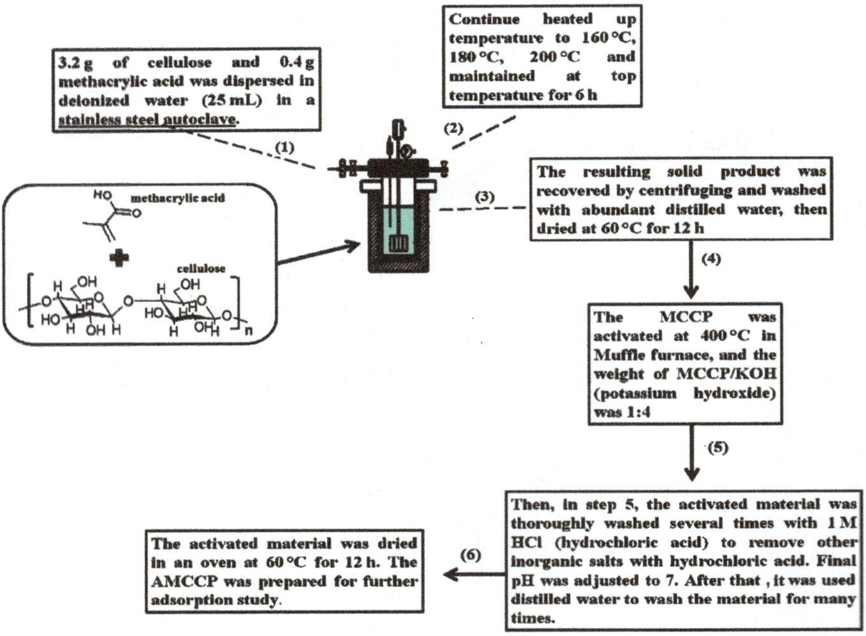

Fig. 11 Schematic illustration: preparation of the modified cellulose-based complex particle. This figure by Vicente Neto is licensed under the Creative Commons Attribution 4.0 International. Link: https://commons.wikimedia.org/wiki/File:Preparation_of_the_modified_cellulose-based_complex_particle.jpg

catalyzes are the hydronium ions generated due to the auto-ionization of water. Clearly, the authors found out that using methacrylic acid during hydrothermal treatment at 200 °C with stirring results in a good performance in removing neonicotinoids.

According to the authors, the adsorption capacity of acetamiprid for the adsorbent was 99.206 mg g at a temperature of 303 K. The adsorbent also showed high efficiency in the reuse study, being able to remove more than 95% of the adsorbate after fifth cyclical recycling. Table 5 summarizes some environmental applications of cellulose.

4.3 *Carrageenan*

4.3.1 Chemistry of Carrageenans

Carrageenan is a natural polymer that is extracted from red algae. It is a sulfated polysaccharide (Fig. 12) of high molecular weight whose repetitive units of the polymer consist of alternating units of D-galactose and 3,6-anhydro-galactose (3,6-AG) joined by α-1,3 and β bonds-1,4-glycoside. The carrageenan extraction method

Table 5 Environmental applications of natural and synthetic polymer-based cellulose materials

	Functional group	Pollutant	Mechanism	Reference
Silanized cellulose	Hexadecyltrimethoxysilane (HDTMS) 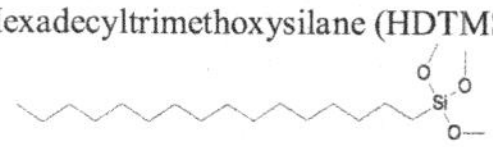	Oil	Superhydrophilicity and superoleophobicity can permeate oil while completely repelling water	Wang et al. (2016)
Cellulose/PVDF–HFP	Polyvinylidene fluoride-co-hexafluoropropylene membranes 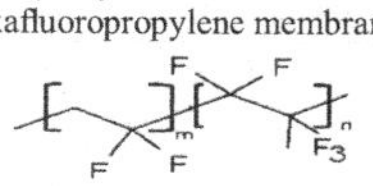	Oil	The resulting membrane exhibits superhydrophilicity and underwater supoeroleophobicity and is successfully applied for selective separation of water from oil	Ahmed et al. (2014)
Microcrystalline cellulose, MCC/DTPA-PEIA)	Diethylenetriaminepentaacetic acid (DTPA) - polyethylenimine (PEI) 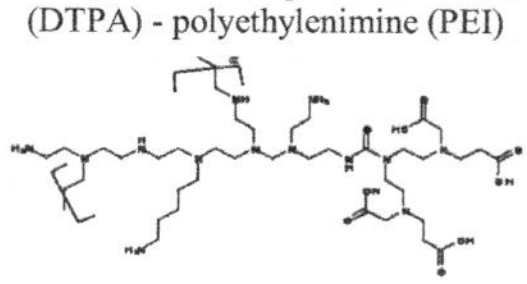	Anionic and cationic heavy metal ion	High density of amino and carboxyl groups for the simultaneous complete removal of anionic and cationic heavy metal ions	Zhou et al. (2020)
Microcrystalline cellulose MCC/TEPAA/	Tetraethylenepentamine,TEPA 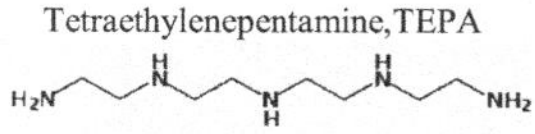	Cr(VI)	High density of amino groups via crosslink by epichlorohydrin (ECH)	Xue et al. (2019)

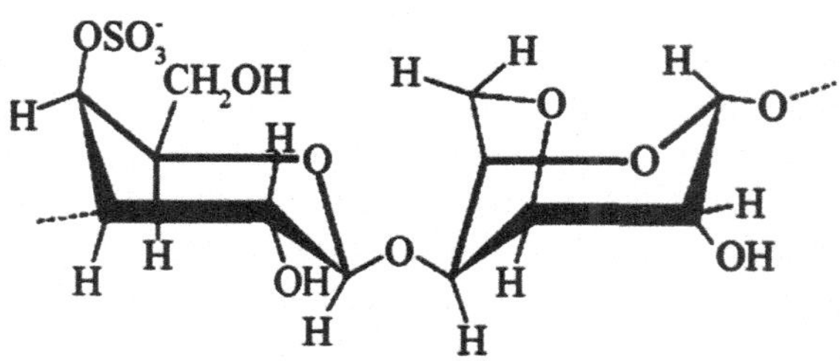

Fig. 12 Schematic representation of a κ-carrageenan macromolecule (Torres et al. 2019). This article is an open access article distributed under the terms and conditions of the Creative Commons Attribution (CC BY) license

influences the molecular weight of this natural polymer (Torres et al. 2019; Weiner 1991).

In the carrageenan, the HSO_3^- functional groups, as well as the hydroxyl groups, are potential sites for heavy metal binding, from the ability that found to other natural polymers such as chitin, cellulose, starch, and alginate. Ali et al. (2019) investigated the adsorption capacity to carrageen and derivative carrageen with cellulosic nanomaterials to metal removing. In this study were used three different types of cellulosic nanomaterials: cellulose nanocrystals (CNC), cellulose nanofibers (CNF), microfibrillated cellulose (MFC). Both CNC and CNF were prepared from dissolved bagasse pulp. CNF was obtained by a mechanical disintegration process after 2,2,6,6-tetramethyl-1-piperidinyloxy (TEMPO)-mediated oxidation of dissolved cellulose pulp. The bio-polymeric matrixes were characterized with transmission electron microscopy (SEM), Fourier transform infrared spectroscopy (FT-IR), and X-ray diffraction. Adsorption study suggested that the removal efficiency of metal ions increases with increasing concentration of carrageenan beads and it also was observed that the adsorbent with better adsorption capacity has the highest content of carboxylated groups in its structure. In addition, to Isogai and co-workers, this method produces a CNF with high content of the negatively charged carboxylate groups onto their surface (Isogai et al. 2011). Figure 13 illustrates the experimental procedures used to obtain both CNC and carrageenan modified with nanocellulose-based material Car/F-CNC.

Figure 14 illustrates the procedures to obtain both T-CNF and carrageenan modified with T-CNF-based material Car/T-CNF. The same figure also illustrates obtaining the Car/CNC from Car/T-CNF.

Figure 15 shows the preparation of T-CNF and Car/T-CNF according to Ali et al. (2019). Carrageenan has been used extensively in environmental applications, which are summarized in Table 6.

5 Conclusion

In this chapter was summarized the development of material based on biopolymers (chitosan, cellulose, and carrageenans) involving the treatment technologies for the removal of environmental pollutants. These biopolymers could be potential tools for environmental remediation. These compounds are high molecular weight molecules with repeated sequences, which may become reactive sites suitable for modifications, creating an opportunity for chemical functionalization. Depending on the contained functional groups, biopolymers such as chitosan, cellulose, and carrageenan can bind metals or soil particles and can form cross-linking networks with other polymers. Biopolymer networks can bind to metals on one side, and soil particles on the other side, trapping the contaminants in very stable complexes. This chapter intends to make a general approach on the potential applications of biopolymers previously mentioned in the remediation of water, soil, and air in an ecologically safe way and illustrates the procedures to obtain both modified.

Thus, the exploration of the chemistry of carrageenans, a natural polymer extracted from red algae, revealed that adsorption studies suggested that the removal efficiency of metal ions increases with increasing concentration of carrageenan beads and also was observed that the adsorbent with better adsorption capacity has the highest content of carboxylated groups in the structure carrageenan modified. Cellulose, a natural biopolymer largely found in nature, has several properties such as

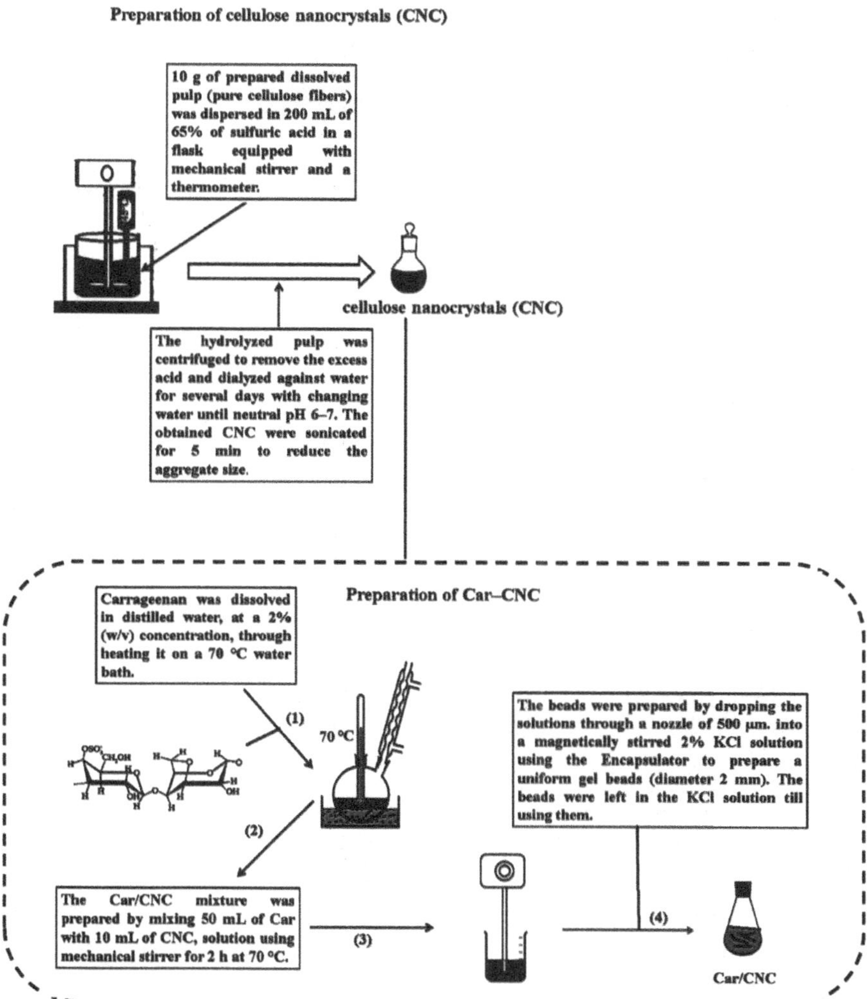

Fig. 13 Preparation of CNC and car/CNC according to Ali et al. (2019). This figure by Vicente Neto is licensed under the Creative Commons Attribution 4.0 International. Link https://commons.wikimedia.org/wiki/File:Car-CNC.jpg

biodegradability, hydrophilicity, and potential as a sorbent, non-toxicity, susceptible to chemical modification via hydroxyl groups. Its derivatives have good mechanical properties, it is a renewable, and non-meltable polymer, which has low solubility in most solvents due to hydrogen bonding and crystallinity. Thus, was summarized the environmental applications of natural and synthetic polymer-based cellulose materials. Chitosan polymers very well recognized as an important alternative to the use of synthetic polymers, because they are biologically biodegradable and have

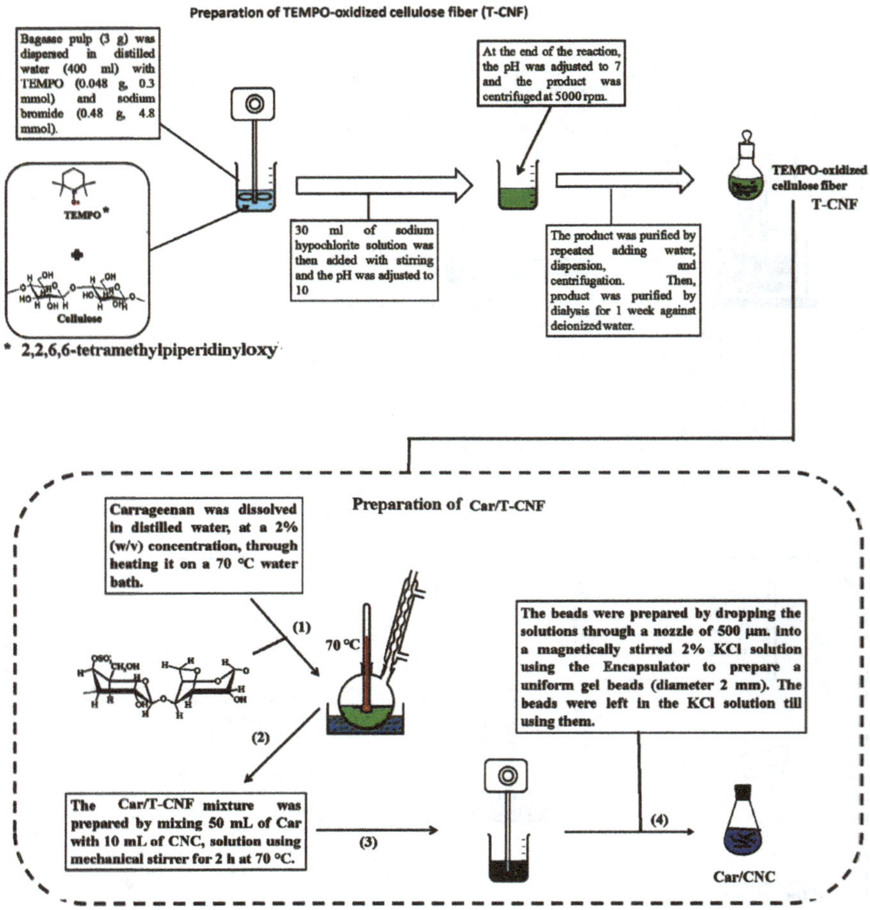

Fig. 14 Preparation of T-CNF and car/T-CNF according to Ali et al. (2019). This figure by Vicente Neto is licensed under the Creative Commons Attribution 4.0 International. Link https://commons.wikimedia.org/wiki/File:Preparation_of_T-CNF_and_Car-T-CNF.jpg

a modifiable chemical structure. Thus, is very functional for a variety of applications of medical, industrial, and environmental interest. In addition, and was showed that a functional derivative based on the m*odified magnetic chitosan nanoparticles and composite nanofibers membranes of chitosan is an* important alternative for removal of organophosphorus pesticides which are one of the organic pollutants more damaging to environmental aquatic.

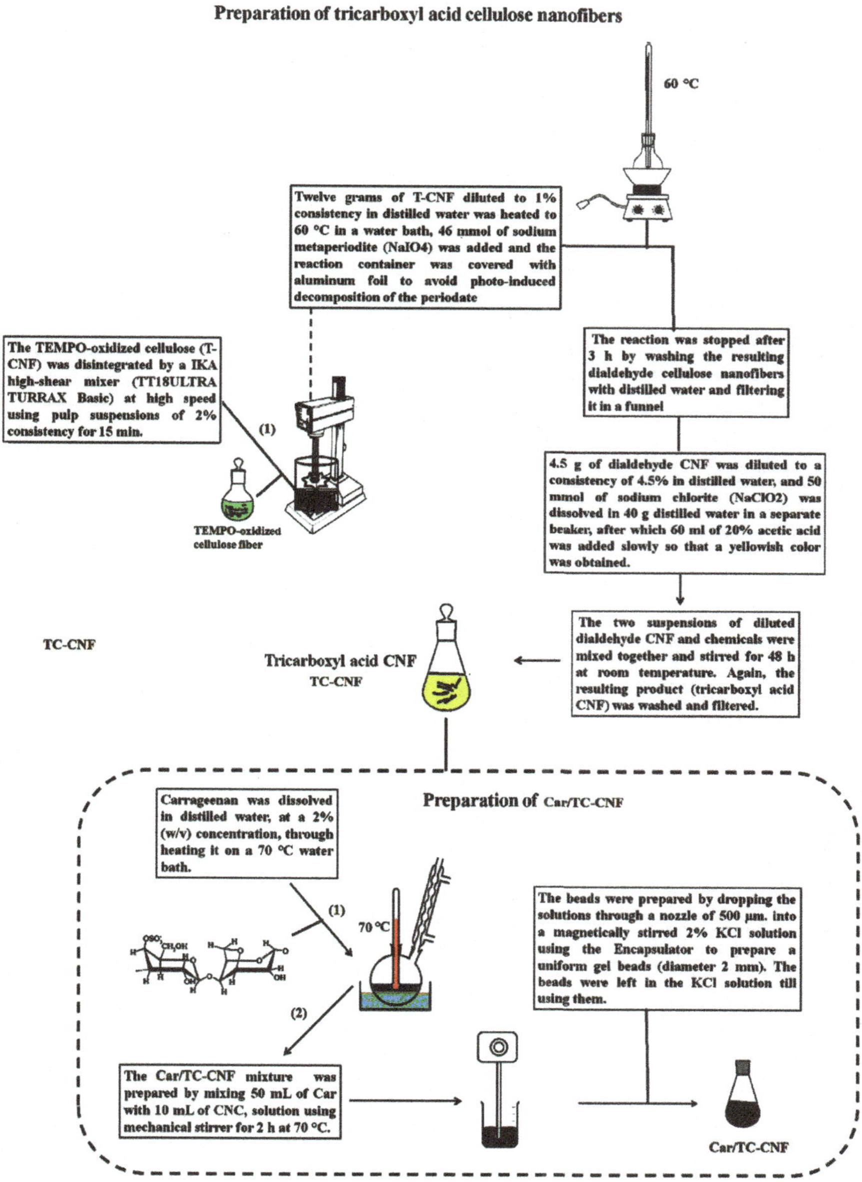

Fig. 15 Preparation of TC-CNF and car/TC-CNF according to Ali et al. (2019). This figure by Vicente Neto is licensed under the Creative Commons Attribution 4.0 International. Link https://commons.wikimedia.org/wiki/File:Preparation_of_TC-CNF_and_Car-TC-CNF.jpg

Table 6 Applications of natural and synthetic polymer-based carrageenan (CG) materials

	Functional group introduced	Pollutant	Mechanism	Reference
Carrageenan	Polyethyleneimine, PEI 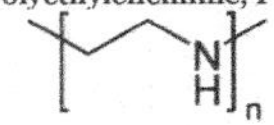	Fe(II)/Fe(III), Mg(II)	Carrageenan beads were treated with PEI to prepare polymeric beads containing NH_2 and OH groups	Ali et al. (2017)
	poly(glycidyl methacrylate),PG 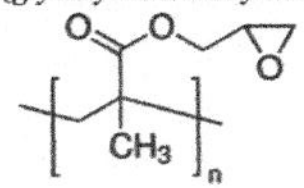	Methylene blue	The repeating units of carrageenan and PG in the blend hydrogel consist of reactive sulfate and epoxide functional groups that are favorable for binding with cationic dye	Lapwanit et al. (2018)
laponite/Carrageenan modified membranes	Hydrolyzed polyacrylonitride (h-PAN) membranes via layer-by-layer (LbL) self-assembly 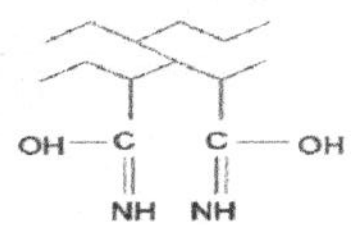	oil	With the presence of laponite nanoclay and CGN multilayers, the obtained LbL membranes exhibit superhydrophilic properties	Prasannan et al. (2020)
Carrageenan/glutaraldehyde	Glutaraldehyde 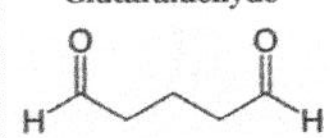	Metoprolol (MTPL), Pharmaceutical	Hydrogen bonds between polymer matrix and MTPL	Nanaki et al. (2015)

(continued)

Table 6 (continued)

	Functional group introduced	Pollutant	Mechanism	Reference
Graft copolymer (k-CG-g-vinylsulfonic acid	Poly(vinyl sulfonic acid) 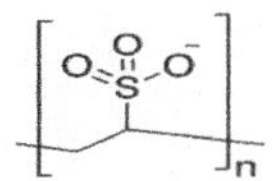	(i) Ni(II), Zn(II), Pb(II)	Incorporation of more functional groups of poly(vinylsulfonic acid), number of sorption sites increases	Yadav et al. (2012)
N-(hydroxymethyl) acrylamide-g-/k-CG	N-(hydroxymethyl) acrylamide 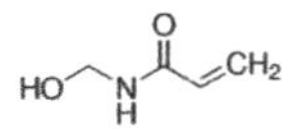	Ni(II),Zn(II), Pb(II	The incorporation of more functional groups of poly *N*-(hydroxymethyl) acrylamide, number of sorption sites increase	Verma et al. (2014)
κ-carrageenan-g-poly(methacrylic acid)	Poly(methacrylic acid) 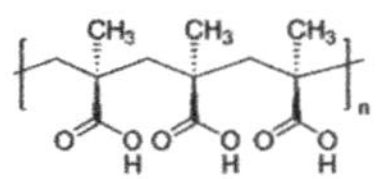	Cationic dye	The electrostatic interactions between the –COO^- and positively charged cationic dye	Gholami et al. (2016)

References

Ahmed FE, Lalia BS, Hilal N, Hashaikeh R (2014) Underwater superoleophobic cellulose/electrospun PVDF–HFP membranes for efficient oil/water separation. Desalination 344:48–54

Ali KA, Hassan ME, Elnashar MMM (2017) Development of functionalized carrageenan, chitosan and alginate as polymeric chelating ligands for water softening. Int J Environ Sci Technol 14:2009–2014

Ali KA, Wahba MI, Abou-Zeid RE, Kamel S (2019) Development of carrageenan modified with nanocellulose-based materials in removing of Cu^{2+}, Pb^{2+}, Ca^{2+}, Mg^{2+}, and Fe^{2+}. Int J Environ Sci Technol 16:5569–5576

Bandforuzi SR, Hadjmohammadi MR (2019) Modified magnetic chitosan nanoparticles based on mixed hemimicelle of sodium dodecyl sulfate for enhanced removal and trace determination of three organophosphorus pesticides from natural waters. Anal Chim Acta 1078:90–100

Barros FC, Sousa Neto VO, Carvalho TV, Vieira RS, Silva GM, Nascimento RF (2015) Recent development of chitosan nanocomposites with multiple potential uses. Advanced structured materials, 1st edn., vol 74. Springer, India, pp 497–531

Bernardi F, Zadinelo IV, Alves HJ, Meurer F, dos Santos LD (2018) Chitins and chitosans for the removal of total ammonia of aquaculture effluents. Aquaculture 483:203–212

Carvalho TV, Barreto LV, Angelim AL, Costa SP, Bezerra WM, Melo FEA, Craveiro AA, Sampaio GMMS, Saraiva GD, Sousa Neto VO, Melo VM, Nascimento RF (2018) Chitosan and its derivatives: synthesis strategy and applications. Advances in organic synthesis, 1st edn., vol 1. Bentham Science, Dubai, pp 121–174

Chang Q, Zhang M, Wang J (2009) Removal of Cu^{2+} and turbidity from wastewater by mercaptoacetyl chitosan. J Hazard Mater 169:621–625

Claeys WL, Schmit J-F, Bragard C, Maghuin-Rogister G, Pussemier L, Schiffers B (2011) Exposure of several Belgian consumer groups to pesticide residues through fresh fruit and vegetable consumption. Food Control 2010(22):508–516

Coats JR (1990) Mechanisms of toxic action and structure-activity relationships for organochlorine and synthetic pyrethroid insecticides. Environ Health Perspect 87:255–262

Dev VV, Baburaj G, Antony S, Arun V, Krishnan KA (2020) Zwitterion-chitosan bed for the simultaneous immobilization of Zn(II), Cd(II), Pb(II) and Cu(II) from multi-metal aqueous systems. J Clean Prod 255:120309

Edelstein M, Ben-Hur M (2018) Heavy metals and metalloids: sources, risks and strategies to reduce their accumulation in horticultural crops. Sci Hortic 234:431–444

Ellendersen LSN, Milinsk MC, Feroldi M, Zadinelo IV, Santos LD, Muniz GIB, Gasparrini LJ, Alves HJ (2018) Biopolymer foam for remediation of aquatic environments contaminated with particulates and heavy metals. J Environ Chem Eng 6:6131–6138

Elyazar IRF, Hay SI, Baird JK (2011) Chapter 2—Malaria distribution, prevalence, drug resistance and control in Indonesia. In: Rollinson D, Hay SI (eds) Advances in parasitology. Academic Press, Cambridge, vol 74, pp 41–175

Esumi K, Ueno M (2003) Structure-performance relationships in surfactants. CRC Press, Boca Raton

Fukuto TR (1990) Mechanism of action of organophosphorus and carbamate insecticides. Environ Health Perspect 87:245–254

Gavalyan VB (2016) Synthesis and characterization of new chitosan-based Schiff base compounds. Carbohydr Polym 145:37–47

Ghimici L, Dinu IA (2019) Removal of some commercial pesticides from aqueous dispersions using as flocculant a thymine-containing chitosan derivative. Sep Purif Technol 209:698–706

Gholami M, Vardini MT, Mahdavinia GR (2016) Investigation of the effect of magnetic particles on the Crystal Violet adsorption onto a novel nanocomposite based on κ-carrageenan-g-poly(methacrylic acid). Carbohydr Polym 136:772–781

Glotfelty DE (1978) The atmosphere as a sink for applied pesticides. J Air Pollut Control Assoc 28(9):917–921

Guerra FD, Attia MF, Whitehead DC, Alexis F (2018) Nanotechnology for environmental remediation: materials and applications. Molecules 23:1760

Gupta AD, Pandey S, Jaiswal VK, Bhadauria V, Singh H (2019) Simultaneous oxidation and esterification of cellulose for use in treatment of water containing Cu(II) ions. Carbohydr Polym 222:114964

Guzman J, Saucedo I, Revilla J, Navarro R, Guibal E (2003) Copper sorption by chitosan in the presence of citrate ions: influence of metal speciation on sorption mechanism and uptake capacities. Int J Biol Macromol 33:57–65

Gyliene O, Binkiene R, Butkiene R (2009) Sorption of Cu(II) complexes with ligands tartrate, glycine and quadrol by chitosan. J Hazard Mater 171:133–139

Hayes WJ Jr, Laws ER Jr (1991) Handbook of pesticide toxicology. Academic Press Inc, Cambridge, pp 816–822

Hussain MS, Musharraf SG, Bhanger MI, Malik MI (2020) Salicylaldehyde derivative of nano-chitosan as an efficient adsorbent for lead(II), copper(II), and cadmium(II) ions. Int J Biol Macromol 147:643–652

Isogai A, Saito T, Fukuzumi H (2011) TEMPO-oxidized cellulose nanofibers. nanoscale 3(1):71–85

Karim MR, Aijaz MO, Alharth NH, Alharbi HF, Al-Mubaddel FS, Awual MdR (2019) Composite nanofibers membranes of polyhaHavard(vinyl alcohol)/chitosan for selective lead(II) and cadmium(II) ions removal from wastewater. Ecotoxicol Environ Saf 169:479–486

Knezevic Z, Serdar M, Ahel M (2012) Risk assessment of the intake of pesticides in Croatian diet. Food Control 23:59–65

Kowalska JB, Mazurek R, Gąsiorek M, Zaleski T (2018) Pollution indices as useful tools for the comprehensive evaluation of the degree of soil contamination—a review. Environ Geochem Health 40:2395–2420

Lapwanit S, Sooksimuang T, Trakulsujaritchok T (2018) Adsorptive removal of cationic methylene blue dye by kappa-carrageenan/poly(glycidyl methacrylate) hydrogel beads: preparation and characterization. J Environ Chem Eng 6(5):6221–6230

Li Y, Bai P, Yan Y, Yan W, Shi W, Xu R (2019) Removal of Zn^{2+}, Pb^{2+}, Cd^{2+}, and Cu^{2+} from aqueous solution by synthetic clinoptilolite. Microporous Mesoporous Mater 273:203–211

Mehrotra BD, Ravichandra Reddy S, Desaiah D (1988) Effect of subchronic dieldrin treatment on calmodulin-regulated Ca^{2+} pump activity in rat brain. J Toxicol Environ Health 25(4):461–469

Metcalf RL (2000) "Insect Control". Ullmann's encyclopedia of industrial chemistry. Wiley-VCH, Weinheim

Nanaki SG, Kyzas GZ, Tzereme A, Papageorgiou M, Kostoglou M, Bikiaris DN, Lambropoulou DA (2015) Synthesis and characterization of modified carrageenan microparticles for the removal of pharmaceuticals from aqueous solutions. Colloids Surf B 127:256–265

Prasannan A, Udomsin J, Tsaia H-C, Wanga C-F, Laia J-Y (2020) Robust underwater superoleophobic membranes with bio-inspired carrageenan/laponite multilayers for the effective removal of emulsions, metal ions, and organic dyes from wastewater. Chem Eng J 391:123585

Rojas G, Silva J, Flores JA, Rodriguez A, Ly M, Maldonado H (2005) Adsorption of chromium onto cross-linked chitosan. Sep Purif Technol 44:31–36

Sanchez-Salvador JL, Balea A, Monte MC, Blanco A, Negro C (2019) Pickering emulsions containing cellulose microfibers produced by mechanical treatments as stabilizer in the food industry. Appl Sci 9:359

Shankar A, Kongot M, Saini VK, Amit Kumar A (2020) Removal of pentachlorophenol pesticide from aqueous solutions using modified chitosan. Arab J Chem 13:1821–1830

Sivakami MS, Gomathi T, Venkatesan J, Jeong HS, Kim SK, Sudha PN (2013) Preparation and characterization of nano chitosan for treatment wastewaters. Int J Biol Macromol 57(2013):204–212

Soderlund DM, Clark JM, Sheets LP, Mullin LS, Piccirillo VJ, Sargent D, Stevens JT, Weiner ML (2002) Mechanisms of pyrethroid neurotoxicity: implications for cumulative risk assessment. Toxicology 171(1):3–5

Steinrucken HC, Amrhein N (1980) The herbicide glyphosate is a potent inhibitor of 5-enolpyruvyl-shikimic acid-3-phosphate synthase. Biochem Biophys Res Commun 94:1207–1212

Suhas, Gupta VK, Carrott PJM, Singh R, Chaudhary M, Kushwaha S (2016) Cellulose: a review as natural, modified and activated carbon adsorbent. Bioresour Technol 216:1066–1076

Teo SH, Islam A, Chan ES, Thomas Choong SY, Alharthi NH, Taufiq-Yap YH, Awual MR (2019) Efficient biodiesel production from Jatropha curcas using $CaSO_4/Fe_2O_3$-SiO_2 core-shell magnetic nanoparticles. J Clean Prod 208:816–826

Torres FG, Troncoso OP, Pisani A, Gatto F, Bardi G (2019) Natural polysaccharide nanomaterials: an overview of their immunological properties. Int J Mol Sci 20:5092

USEPA (1996) Reregistration eligibility decision for profenofos

Verma SK, Pandey VS, Yadav M, Behari K (2014) Grafting of N-(hydroxymethyl) acrylamide on to κ-carrageenan: synthesis, characterization and applications. Carbohydr Polym 102:590–597

Wang X, Xu S, Tan Y, Du J, Wang J (2016) Synthesis and characterization of a porous and hydrophobic cellulose-based composite for efficient and fast oil–water separation. Carbohydr Polym 140:188–194

Weiner ML (1991) Toxicological properties of carrageenan. Agents Actions 32:46–51

World Health Organization (2006) Pesticides and their application: or the control of vectors and pests of public health importance, (6th edn.), World Health Organization

Xue F, He H, Zhu H, Huang H, Wu Q, Wang S (2019) Structural design of a cellulose-based solid amine adsorbent for the complete removal and colorimetric detection of Cr(VI). Langmuir 35(39):12636–12646

Yadav M, Sand A, Mishra MM, Tripathy J, Pandey VS, Behari K (2012) Synthesis, characterization and applications of graft copolymer (κ-carrageenan-g-vinylsulfonic acidInt. J Biol Macromol 50(3):826–832

Yao J, Wang G, Xue B, Wang P, Hao F, Xie G, Peng Y (2019) Assessment of lake eutrophication using a novel multidimensional similarity cloud model. J Environ Manage 248:109259

Ye J, Chen X, Chen C, Bate B (2019) Emerging sustainable technologies for remediation of soils and groundwater in a municipal solid waste landfill site—a review. Chemosphere 227:681–702

Yuan M, Liu X, Li C, Yu J, Zhang B, Ma Y (2019) A higher efficiency removal of neonicotinoid insecticides by modified cellulose-based complex particle. Int J Biol Macromol 126:857–866

Zhou H, Zhu H, Xue F, He H, Wang S (2020) Cellulose-based amphoteric adsorbent for the complete removal of low-level heavy metal ions via a specialization and cooperation mechanism. Chem Eng J 385:123879

Chapter 3
Magnetite-Zeolite Nanocomposite Applied to Remediation of Polluted Aquatic Environments

Carla B. Vidal, Breno A. dos Santos, Antônia Mayza M. França, Raquel A. Bessa, Adonay R. Loiola, and Ronaldo Ferreira do Nascimento

1 Introduction

The use of nanomaterials, described as structured materials of sizes between 1 and 100 nm, as adsorbents for specific application in the treatment of water and wastewater has been studied for the past few decades. This is mainly due to the adsorptive properties of these materials, such as, large surface areas due to their small size that promotes an increase in adsorption capacity and provides good catalytic activity. Another important property of nanomaterials is their mobility in aqueous media, making them potential adsorbents when applied to the removal of different classes of pollutants, such as: toxic metals ions, organic and inorganic compounds, and also bacteria (de Oliveira Sousa Neto et al. 2019; Lu and Astruc 2020).

Among the studied nanomaterials, special attention has been given to nanostructured magnetic adsorbents, especially those containing iron oxide nanosized particles to solve problems related to the environment, is one of the new technologies that has received considerable attention in recent years, due to its particular ability to be easily separated from water by an external magnetic field after the adsorption process. Besides, magnetic nanoparticles have unique physicochemical, magnetic, and optical properties that are of great importance in diverse applications. Cobalt,

C. B. Vidal
Department of Chemistry and Biology, Federal University of Technology – Paraná (UTFPR), Curitiba, PR 81280-340, Brazil

B. A. dos Santos · A. M. M. França · R. F. do Nascimento (✉)
Department of Analytical and Physical Chemistry, Federal University of Ceará (UFC), Campus do Pici, Bloco 940, S/N, Fortaleza, CE 60455-970, Brazil
e-mail: ronaldo@ufc.br

R. A. Bessa · A. R. Loiola
Department of Organic and Inorganic Chemistry, Federal University of Ceará, Campus do Pici, Fortaleza, CE 60440-900, Brazil

R. F. do. Nascimento et al. (eds.), *Nanomaterials and Nanotechnology*,
Materials Horizons: From Nature to Nanomaterials,
https://doi.org/10.1007/978-981-33-6056-3_3

iron, and iron oxides nanoparticles, such as maghemite (γ-Fe_2O_3) and magnetite (Fe_3O_4), have been widely considered as the most suitable materials for production of magnetic nanocomposites due to their significant magnetic properties (Bercoff et al. 2009; de Andrade Bessa et al. 2017; Lu and Astruc 2020).

Composites based on nanosized particles of magnetite and natural or synthetic microsized zeolite have been widely applied in photodegradation and dye removal studies (Shirani et al. 2014; Ghanbari et al. 2016), application as a contrast agent in magnetic resonance (Atashi et al. 2017), metal ions preconcentration and determination (Naghizadeh et al. 2017), application as anti-cancer drug carriers (Sağir et al. 2016) and metal ions adsorption (Pizarro et al. 2015; Zhao et al. 2015; Mthombeni et al. 2016; Sharifi and Baghdadi 2016), as they show great efficiency in water purification and separation of toxic pollutants, since supported nanomagnetite prevents risks of the nanoparticle's contamination (Ghanbari et al. 2016).

This chapter aims to address nanomaterials based on magnetite-zeolite nanocomposites. Special attention was given to the synthesis and characterization of these materials, as well as its application to remediation of polluted aquatic environments.

2 Magnetite Nanocomposites

The composition of magnetite is represented by $(Fe^{3+})_{tet}[Fe^{2+}Fe^{3+}]_{oct}O_4^{-2}$, with ratio $Fe^{2+}/Fe^{3+} = 0.5$. The crystalline structure of magnetite is an inverse spinel with a unit cell consisting of 32 oxygen ions in a face-centered cubic structure and a cell parameter of 0.839 nm. In this crystalline structure, Fe^{2+} and half of the Fe^{3+} occupy octahedral sites and the other half of the Fe^{3+} occupies tetrahedral sites (Fig. 1).

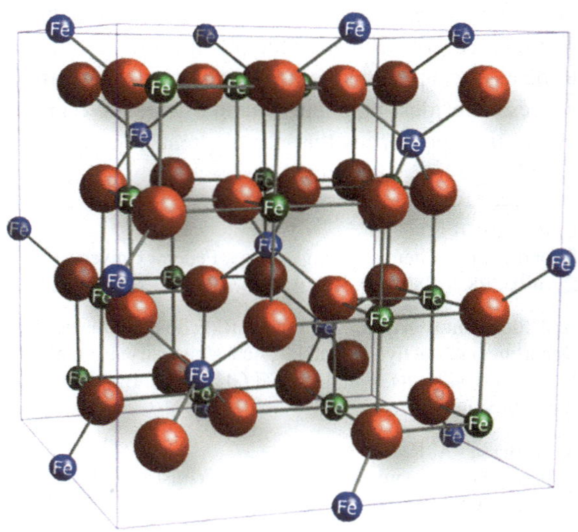

Fig. 1 Ball and stick model of the unit cell of magnetite. Octahedral Fe ions are in green, tetrahedral Fe ions are in blue, oxygen ions are in red. Reproduced with permission from ACS Publications (Usman et al. 2018)

The natural magnetic behavior of magnetite is ferrimagnetism, since the electron spins of Fe ions are antiparallel with different magnitudes, generating the magnetic moment. The oxidation numbers of iron (Fe (II) and Fe (III)) contribute to the magnetic properties of magnetite making it very viable for biomedical applications (supply of medicines, separation, sensors) and electromagnetic materials (Chen et al. 2015; Scapim et al. 2017), as well as magnetic separations.

Generally, conventional separation processes, such as filtration or centrifugation, require high time and costs, since the separation of adsorbent materials becomes difficult with the use of large volumes of solution. Magnetic separation technology emerges as an alternative to this problem, since magnetic adsorbents can be separated easily and quickly when exposed to an external magnetic field (Javanbakht et al. 2016).

Figure 2 shows the magnetic attraction of magnetic nanoparticles (blackpowder), which are strongly attracted to approaching magnet compared to pure zeolites (white powders) and the zeolite-magnetite nanocomposites (brown powders). It can be seen that nanocomposites have similar characteristics to the magnetic nanoparticles regarding the attraction to the magnet, unlike pure zeolite that showed no attraction to the magnet.

Figure 3 shows the magnetic attraction behavior of a magnetite-based adsorbent supported on synthetic zeolite-4A also in aqueous media, highlighting the displacement of the magnetized particles to the magnet after approximately 5 min of exposure to the external magnetic field, thus proving its efficiency in the magnetic separation process, commonly observed in magnetic zeolite composites (Barquist and Larsen 2010; Fungaro 2011; Yamaura and Fungaro 2013).

Magnetic adsorbents have two separation techniques that are reduced to adsorption and magnetic attraction. Thus, the removal of pollutants is promoted by the adsorption

Fig. 2 Photograph presenting the response to the magnet of the **a** zeolite A, **b** zeolite A composite, **c** magnetic nanoparticles, **d** zeolite P, and **e** zeolite P composite. Reproduced with permission from Elsevier (de Andrade Bessa et al. 2017)

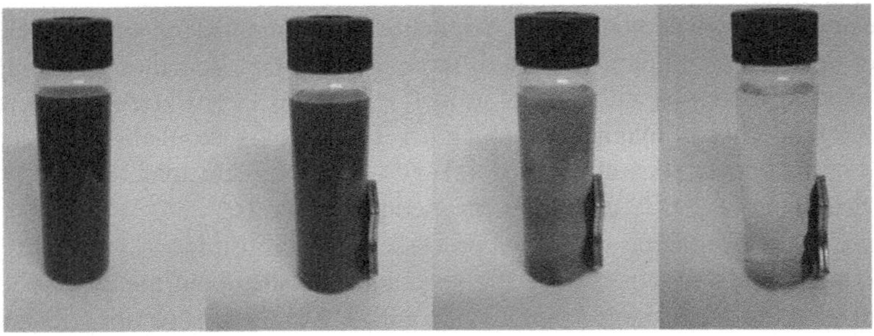

Fig. 3 Magnetic attraction using magnetite-zeolite nanocomposite in aqueous media

process, and then, the particles of the magnetic adsorbent are agglomerated and attracted instantly by an external magnetic field, being separated from the liquid medium. Such process dispenses the use of other technologies commonly used in solid/liquid separation (Fungaro et al. 2010).

3 Magnetite-Zeolite Nanocomposite

It is expected that magnetite-zeolite composites present a combination of the many properties of the zeolites with the magnetic properties of the magnetite. In this sense, it is important to employ proper synthesis methods so that such properties are not compromised. A compilation of the most used synthesis strategies, accompanied by the description of some techniques can help to evaluate important aspects related to structure, morphology, composition, and magnetic response of these materials are presented in the following sections.

3.1 Nanocomposite Synthesis

The synthesis of magnetite-zeolite composites can be achieved by several different approaches. The most common method consists of incorporating magnetite nanoparticles over zeolite surface, but it is also possible to synthesize magnetite nanoparticles within zeolite crystals, which is usually carried out simultaneously to the zeolite synthesis. In any case, crystallization of magnetite is a tricky procedure. As it is known, magnetite contains equal parts of Fe^{2+} and Fe^{3+}, so the preparation of Fe_3O_4 can be based either on an equimolar mixture of Fe^{2+} and Fe^{3+} or on a Fe^{2+} solution source, in which Fe^{2+} is partially converted to Fe^{3+}. Small variations in the synthesis conditions can lead iron to its higher oxidation state and therefore the formation of Fe_2O_3, which in turn can exist in many different varieties, such as γ-Fe_2O_3 (maghemite) and α-Fe_2O_3 (hematite), the latter being the most stable form, but of little use in magnetic zeolite composites (Dar and Shivashankar 2014). The control of the magnetic particle size is also important to be taken into consideration during the preparation of the zeolite composite, once the stabilization of the final product and the maintenance of some properties of the zeolite such as surface area depend at large on it (Luo et al. 2015).

3.1.1 Magnetic Nanoparticles Over Zeolite Surface

The vast majority of magnetic zeolite composites are produced via the impregnation of magnetic nanoparticles, usually iron oxides, over the external surfaces of zeolitic crystals. This method presents several advantages such as simplicity and costs relatively low. On the other hand, the exposure of these particles on the zeolite surfaces can lead them to oxidation and consequent the loss of their magnetic properties.

One example of magnetite/zeolite composite was presented by Jahangirian et al. (2013), that produced zeolite 3A/magnetic iron oxide nanoparticles composite via a simple precipitation method, involving very small iron oxide particles (3.55 nm) obtained as the result of the vigorous stirring applied during the precipitation in basic medium. Mollahosseini et al. (2015) reported the preparation of Fe_3O_4 nanoparticles by employing an electrochemical method in which iron electrodes were used as iron source. Those particles were incorporated to a commercial clinoptilolite zeolite using tetramethylammonium chloride solution medium, following a heating treatment. Zendehdel et al. (2019) used zeolite Y nanocrystals to prepare a magnetic composite. The zeolite was dispersed in a Fe^{3+}/Fe^{2+} solution and, under N_2 atmosphere, ammonia solution was added dropwise so that magnetite nanoparticles could be precipitated. A similar approach was adopted by Kouli et al. (2020) the preparation of a magnetic zeolite X composite, but with the additional use of microwave heating. The stabilization of the magnetic particles over the zeolite surface was achieved in the preparation of a magnetite/zeolite Y composite by the modification of zeolite Y with chlorosulfonicacid, as described by Kalhor and Zarnegar (2019). A green method to prepare zeolite/magnetic iron oxide nanocomposite involved the use of deoxygenated aqueous solution coupled to ultrasonic irradiation (Jahangirian et al. 2013).

Several works use raw materials such as fly ash, kaolin, and diatom for zeolite synthesis as an alternative to overcome the high costs involved in the production of magnetic zeolite composites.

Fly ash are thin particles in powder form composed mainly of silica (SiO_2), alumina (Al_2O_3), iron oxide (Fe_2O_3), and calcium oxide (CaO) (Kumar et al. 2012). It is produced as a residue from the coal burning, particularly in thermal power plants, and can present a great compositional variety, which is an important limiting factor toward zeolite synthesis. Nevertheless, numerous works report its use as an attractive material to produce magnetic zeolite composites. Magnetic sodalite and hydroxysodalite from fly ash, using the hydrothermal method, was prepared by Fungaro et al. (2010). Fly ash was also used to obtain magnetic Na-P1 zeolite composite following a relatively simple precipitation method, as described by Aono et al. (2013). A mixture of fly ash and a waste material recovered from the caustic leaching of bauxite in the production of alumina was used by Belviso et al. (2015) to produce magnetic zeolite in a one-step synthesis. This method presented the disadvantage of multiple zeolite co-crystallization. Kaolin was used by Liu et al. (2013) as the main Si and Al source in the synthesis of magnetic zeolite A. Diatom, another natural compound that can be used to generate magnetic zeolite, was used in the synthesis of the mordenite zeolite impregnated with magnetite nanoparticles.

Natural zeolites are widely employed in the preparation of magnetic composites, although the control of purity is almost always a challenging task, as they usually are found in deposits where other zeolitic phases are present. Clinoptilolite is found as one of the most used natural zeolites applied for this purpose (Badeenezhad et al. 2019). Mthombeni et al. (2015) for example describes it in a magnetic composite form applied to removal of Cr(VI) from aqueous solution. For its preparation, a solution containing Fe^{2+}/Fe^{3+} was added dropwise to a mixture containing the zeolite and NaOH, in a proper concentration. Additionally, this magnetic composite was modified by the impregnation of pyrrole. Mordenite coated with a layer of magnetite nanoparticles showed practical in the adsorption of arsenate from aqueous systems (Pizarro et al. 2015), and also in refinery oily wastewater purification (Hesas et al. 2019).

An ingenious method involving the use of chitosan, allowed that magnetically cylindrically shaped zeolite L crystals, adopted a precise orientation in the polymeric matrix, as illustrated in Fig. 4 (Fibikar et al. 2015).

Several works show that important properties of zeolites in form of magnetic composites, in particular adsorptive properties, can be improved by means of surface modification, or functionalization. Shirani et al. (2015) prepared a magnetic zeolite NaY by the co-precipitation method via modification with 2-(3,4-dihydroxyphenyl)-1,3-dithiane, which, despite the costs related to functionalization, considerably increased the removal capacity of the toxic heavy metals ions Cd^{2+} and Cu^{2+} from soil and water samples. Barquist and Larsen (2010) promoted the amino-functionalized of nanocrystalline magnetic zeolite NaY by using the NH_4^+ ion exchanged form of the zeolite mixed with $FeCl_3$ and $FeSO_4$ solutions, in N_2 atmosphere, being thermally treated up to 723 K, and then treated with aminopropyltriethoxisilane dissolved in toluene. Mollahosseini et al. (2015) employed an electrochemical method in which iron electrodes, in tetramethylammonium chloride solution medium, under heating, were used to obtain Fe_3O_4 nanoparticles. Those particles were incorporated to a commercial clinoptilolite zeolite to producing a magnetic composite. A cysteine-modified clinoptilolite-magnetite nanocomposite obtained by co-precipitation method via hydrothermal route was reported by Sharifi and Baghdadi (2016). A magnetic composite based on clinoptilolite zeolite and modified by cetyltrimethyl ammonium bromide and dithizone was described by Amiri-Yazani et al. (2019). Baile et al. (2018) used hexadecyltrimethylammonium bromide (HDTMABr) to modify a magnetic composite of ZSM-5 decorated with iron oxide magnetic nanoparticles.

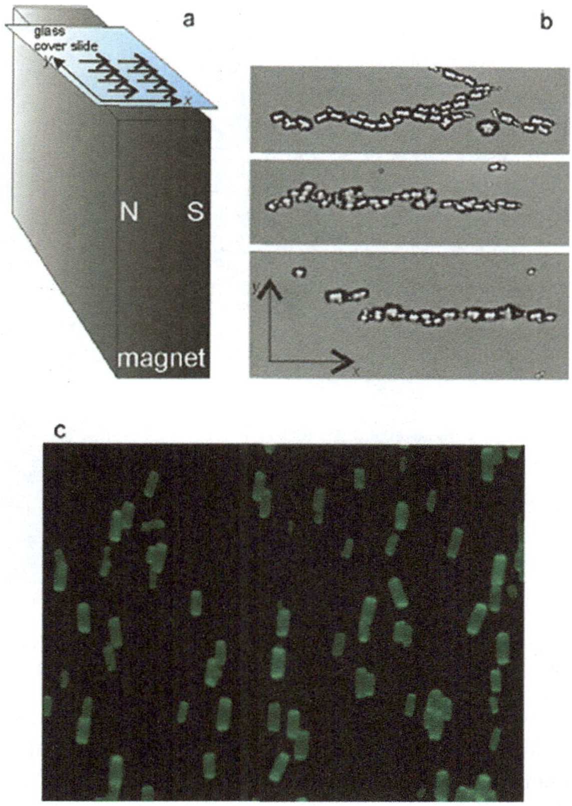

Fig. 4 **a** Schematic illustration of the experiment layout. N = north, S = south. **b** Optical wide-field microcopy images. **c** Fluorescence microscopy image of zeolite L with intercalated acridine molecule aligned unidirectionally. Reproduced with permission from Chemistry Europe (Fibikar et al. 2015)

3.1.2 Magnetic Nanoparticles Within Zeolite Crystals

Magnetic zeolite composites formed by magnetite nanoparticles inside zeolitic crystals represent a group of materials with potential for many applications. The preparation of these materials tends to be more complex. The magnetite particles have to be small enough (<100 nm) to be accommodated within the zeolite crystals and the zeolite must be synthesized in the presence of magnetite particles, making the syntheses hard to be accomplished or susceptible to result in a mixture of crystalline phases.

In spite of the many prerequisites that must be attained to efficiently prepare zeolite crystals containing magnetic particles inside the zeolite crystals, several synthesis methods have been proposed. Highly dispersed and magnetically separable silicalite-1 nanozeolites (MZCNPs) were obtained by Shan et al. (2006) by adding superparamagnetic magnetite nanoparticles into a tetraethyl orthosilicate nanozeolite solution precursor. This approach allowed magnetite nanoparticles and zeolite nanoparticles to be tightly combined before the hydrothermal treatment demanded for the growth of zeolite crystals. This way, the magnetite nanoparticles remained occluded

Fig. 5 **a** SEM and **b–e** TEM images of MZCNPs samples, showing the presence of magnetic nanoparticles within the zeolite crystals. Reproduced with permission from ACS (Shan et al. 2006)

in the zeolite particles. This effect can be easily observed in transmission electron micrographs presented in Fig. 5.

By acid treatment of natural zeolite clinoptilolite together with Fe^{2+} and Fe^{3+} solutions, followed by addition of NH_4OH dropwise under nitrogen atmosphere, at a temperature range of 60–90 °C, Bosînceanu and Sulițanu (2008) prepared magnetic clinoptilolite containing very small $FeO(OH)/Fe_3O_4$ particles (<2 nm). The synthesis of magnetic zeolite P containing Fe_3O_4 particles was reported by Cao et al. (2008). In that work, Fe_3O_4 particles were mixed to an aged solution of sodium silicate, sodium aluminate, and sodium hydroxide, and then hydrothermally treated to generate the magnetic zeolite. Magnetic ZSM-5 zeolite was synthesized by hydrothermal process, adding magnetic Fe_3O_4 particles during the zeolite crystallization process (Cao et al. 2013). Nabiyouni et al. (2015) reported the incorporation of magnetite nanoparticles into the pores of as-synthesized zeolite Y. The preparation of magnetic ZSM-5 was reported by Atashi et al. (2017). In this work, Fe_3O_4 nanoparticles were first prepared by precipitation of Fe^{2+} and Fe^{3+} with ammonia solution and then dispersed in a sodium aluminate and silicic acid solution, followed by hydrothermal treatment. In an attempt to produce magnetic zeolite Al-ZSM-5, El-Din et al. (2011) synthesized Fe_3O_4 nanoparticles from iron chloride solution containing nanozeolites, but the resulting materials consisted of a mixture of phases.

Some methods involving the functionalization of the zeolites can help to improve some properties such as selectivity and reactivity. Chen et al. (2015) proposed

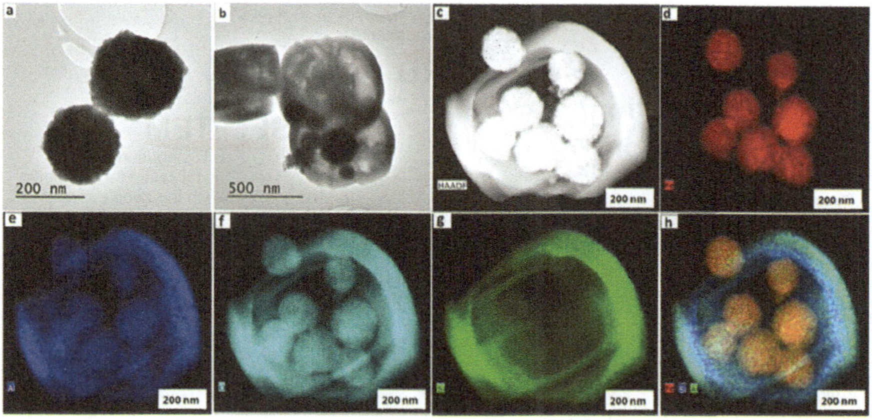

Fig. 6 **a** HR-TEM image of magnetite spheres at 200 nm; **b** HR-TEM image of γ-Fe_2O_3-HZM-5; **c** HAADF image of γ-Fe_2O_3 microspheres encapsulated into the HZSM-5 grain; **d**–**h** elemental mapping of Fe, Al, O, Si, and Fe/Si/Al, respectively. Reproduced with permission from Royal Society of Chemistry (Lima et al. 2016)

a method to generate a magnetic organic-functionalized ZSM-5 zeolite by first coating Fe_3O_4 with mesoporous silica gel and then synthesizing a subsequent layer of zeolite ZSM-5, which was also functionalized with carboxylatocalix[4]arenes (CC[4]A). In the preparation of a magnetically recoverable ZSM-5 zeolite, Lima et al. (2016) adopted the strategy of first synthesizing Fe_3O_4 nanoparticles via solvothermal method with iron chloride mixed with trisodium citrate, sodium acetate, and ethylene glycol. Next, these particles were functionalized then with poly(diallyldimethylammonium chloride), and then added to a zeolite precursor solution which gave rise to the maghemite/zeolite ZSM-5 composite, as represented in the Fig. 6.

The effect of the Fe_3O_4 content in Fe_3O_4/zeolite NaA nanocomposites for magnetic resonance image contrast was studied by Gharehaghaji et al. (2018). Interestingly, the samples with the lowest iron oxide content showed the most promising results. The work of Divband et al. (2018) describes the preparation of magnetic zeolite A from a gel containing a mixture of zeolite nanocrystals and Fe_3O_4 nanoparticles. This potential of this composite as a drug delivery system was evaluated.

3.2 *Nanocomposite Characterizations*

To evaluate the products obtained from the zeolite syntheses, characterization methods should provide information about their structure, morphology, chemical composition, and ability to sorb and retain molecules. When a magnetic zeolite is formed, some information about the magnetite nanoparticles on the nanocomposite

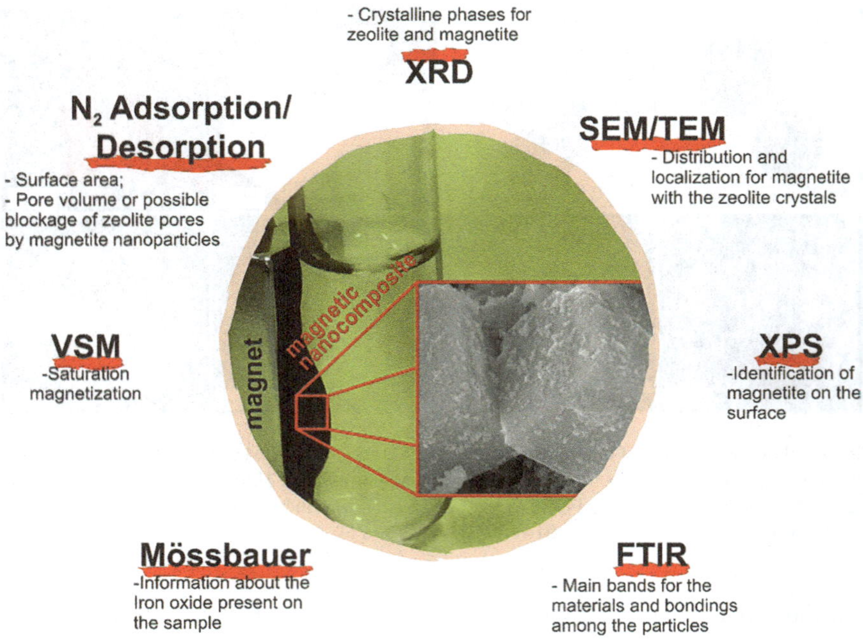

Fig. 7 Characterization techniques used for magnetic zeolites and main information obtained from their study

is also required and elucidation about their structure, magnetization, and oxidation of iron species is valuable. This section aims to present the main techniques used for characterizing those nanocomposites. Figure 7 summarizes the main techniques and main properties obtained from them.

XRD

X-ray powder diffraction is routinely applied to identify the phases present in a sample. Powder diffraction pattern databases are available and can be easily matched using peaks positions and intensities. Due to the crystalline nature of both zeolites and magnetite, the XRD pattern for a given magnetite/zeolite composite provides information about the degree of crystallinity and even the amount of the phases present. If the structure is related to one compound that is already known, structure details can be determined using profile refinement methods such as Rietveld method (Wright 2008).

Weight percentage (wt%) for main and secondary phases is trustful after a set of adjustments in parameters, as presented by de Andrade Bessa et al. (2017) for magnetic particles containing more than 90% of magnetite and the composites formed with zeolites A and P. Another study performed by Lima et al. (2016) used ZSM-5 zeolite in a nanocomposite, presents the average crystallite size ca. 16 nm for the Fe_3O_4 particles from Scherrer's equation; another interesting property that can be calculated from the XRD results acquired.

The eventual presence of undesired secondary phases can lead the composites to present loss of performance in the intended application. Analyzing the peaks positions and intensities, it is possible to identify other different phases that can be present to that sample, as observed by Fungaro et al. (2010) with the zeolite used as a mixture of sodalite, hydroxysodalite and hydrated sodium and aluminum silicate, while the magnetic particle can be assigned to magnetite and maghemite mixture. This comparison is possible by means of the identification of the main peaks for magnetite at 2θ: 30.17°, 35.46°, 43.38°, 53.69°, 57.23°, and 62.77° using Cu-Kα radiation, which correspond to (200), (311), (400), (422), (511), and (440) Bragg reflections, respectively (Sharifi and Baghdadi 2016; Amiri-Yazani et al. 2019).

FTIR

Infrared spectroscopy is a widely available technique, applied extensively in the study of microporous solids for giving some structural insights about the zeolite framework and the magnetic particle used on the nanocomposite. This information is provided by stimulating the vibration or rotation of two or more atoms in the probed matter and a typical spectrum is collected between 400 and 4000 cm^{-1}, which enables most of the fundamental bands from framework and hydroxyls to be measured. Also, it may be possible to deduce some structural information on a new zeolite for which X-rays diffraction data does provide enough information for analyzing it, particularly via the interaction of probe molecules with acid/base adsorption sites of the zeolite (Breck 1984; Lercher and Jentys 2007; Wright 2008).

The zeolite framework vibrations bands depend on both the structure type and the composition. However, for qualitative interpretation of the IR spectrum, it is traditionally separated into regions characteristic for classes of vibrations: 3200–3800 cm^{-1} for fundamental hydroxyl stretches, 400–1200 cm^{-1} for framework vibrations and adsorbed water in the region of 3600 and around 1600 cm^{-1}. Table 1 presents the general assignments of lattice vibrations for zeolite and the main assignment observed for magnetite structure.

Fungaro et al. (2010) in a sample with Fe_3O_4 and γ-Fe_2O_3 observed two characteristic bands to each of them. This was also reported by Cabrera et al. (2008) that the presence of a band at 630 cm^{-1} could be an indicative of maghemite formation to the magnetic nanoparticles' synthesis. It is also common to compare the wave numbers observed for the zeolite structure before and after the nanocomposite formation where

Table 1 General assignments of lattice vibrations for zeolite and the main one for magnetite structure

Assignments		Region (cm^{-1})
Zeolite internal tetrahedra	Asymmetrical stretch	950–1250
	Symmetrical stretch	650–720
	T–O–T bending	420–500
Zeolite external linkages	Double ring	500–650
	Symmetrical stretch	750–820
	Asymmetrical stretch	1050–1150
Magnetite	Fe–O vibration	570–590

the maintenance of such bands in the same wave numbers suggests that the structure is not modified in the nanocomposite, as observed by Jahangirian et al. (2013) using a composite formed by clinoptilolite and magnetite nanoparticles.

TEM/SEM

The electron microscopy techniques are probably the most common in characterization studies for materials chemistry. Regarding specifically magnetic zeolites, the observation of the crystals forming the nanocomposite can elucidate how they are spatially arranged and explain some behavior observed in application steps. For example, Cao et al. (2013) show SEM images for nanocomposites and a trend on decreasing the ZSM-5 crystals sizes when the synthesis is performed with Fe_3O_4 nanoparticles and when analyzing the N_2 adsorption/desorption results, it can be seen that the particles are also blocking the pores, affecting surface area.

Commonly, the nanoparticles are distributed over the zeolite crystals and enhance not only the magnetic separation process, but also the adsorptive characteristics. Both natural and synthetic zeolites usually show the particles dispersed over the surface, as reported by Liu et al. (2013), Pizarro et al. (2015), de Andrade Bessa et al. (2017). Sometimes the magnetic particles can be at the grain boundaries between the polycrystalline zeolite (Shan et al. 2006; Aono et al. 2013), inside the pores (El-Din et al. 2011; Mthombeni et al. 2015), or in a core-shell structure (Lima et al. 2016) depending on how the synthesis was performed.

Mössbauer

Mössbauer spectroscopy has been extensively used to study magnetic materials for being element-specific and ^{57}Fe is a convenient active isotope to this analysis. It supplies information about the iron valence and the type of coordination (Schwertmann and Cornell 2000). The bulk Fe_3O_4 at room temperature is composed of two ferri/ferromagnetic sextets corresponding to Fe^{3+} in tetrahedral and Fe^{3+} and Fe^{2+} in octahedral sites. When the broad sextet becomes a paramagnetic doublet, there is a characteristic of superparamagnetic particles (Neto et al. 2017).

Singh et al. (2016) reported the accommodation of magnetic nanoparticles inside and outside the pores which is seen with the Mössbauer results as doublet and sextet, suggesting that the particles inside the pores present a small particle size, which is associated with the presence of a superparamagnetic doublet, while the sextet comes from the larger particles over the surface. This mixed spectrum was also observed by Lima et al. (2016) due to the presence of iron(III) oxide nanoparticles, (sizes below 15 nm). The composition to the iron oxide present on the sample is obtained using fitting parameters for the mixed valence $Fe^{3+/2+}$ in octahedral and Fe^{3+} in tetrahedral sites that can be ascribed to the corresponding iron oxide pattern. Pizarro et al. (2015) were able to use these results to state the purity for magnetite described in their work, while Oliveira et al. (2004) found the profile for maghemite and goethite.

VSM

Vibrating sample magnetometer analysis is commonly used to measure the magnetic properties of materials as a function of magnetic field, temperature, and time. The usual response obtained from this technique is the saturation magnetization that can be compared to the values observed for bulk magnetite. The addition of the zeolites to the magnetic matrix promotes the decrease of those values reaching extremely low values, such as of 0.85 and 0.87 emu/g, as observed by El-Din et al. (2011) and Cao et al. (2013), respectively. Shan et al. (2006) observed 8.9 emu/g to a composite with 20% wt. NP that is attracted by the magnet within 90 s. Values up to 52 emu/g can be found for this property in a nanocomposite with ca. 100 nm (Atashi et al. 2017).

Another information acquired for the magnetic materials is the temperature dependence of magnetization with zero-field-cooled (ZFC) and field-cooled (FC) curves, which allow one to identify the superparamagnetic behavior. Magnetic nanoparticles with this characteristic present graphically a flat profile of the FC curves at low temperatures and the ZFC curves do not show any maximum in the measured temperature range. Thus, there is indication that the superparamagnetic blocking temperatures for the samples are above 300 K (Kolen'ko et al. 2014; Neto et al. 2017). de Andrade Bessa et al. (2017) performed this study to magnetic composites that presented typical features of an ordered magnetic system, whereas a relatively weak magnetic field is enough to split the curves even without for the pure NPs.

XPS

X-ray photoelectron spectroscopy (XPS) enables the surface analysis by irradiating a sample with monoenergetic soft X-rays and limited penetrating power in a solid in a way that they interact with the surface region causing the electrons to be emitted by the photoelectric effect. Each element has a unique spectrum and some information regarding their identification at the surface, chemical states, quantification, and spatial distribution can be obtained from an XPS spectrum (Watts and Wolstenholme 2003).

Applied to studies using magnetic zeolites, this technique can infer the magnetic particle achieved with the synthesis or surface modification. Teng et al. (2003) and Kokate et al. (2013) suggest the use of XPS to distinguish between maghemite and magnetite since they present specific binding energies. They identified the doublet Fe $2p_{3/2}$ and $2p_{1/2}$ as related to Fe_3O_4, with binding energies of 711.9 and 725.8 eV, and 710.6 and 724.3 eV for γ-Fe_2O_3. XPS can also be used qualitatively to identify the magnetic nanoparticles over the zeolite surface as performed by Chen et al. (2015). This way, this technique can be complementary to XRD as well as can indicate the presence of particles on the zeolite surface.

N_2 Adsorption/desorption

Gas adsorption is a well-established tool for characterization of texture in porous solids and fine powders. An accurate characterization of microporous materials is particularly important for optimizing their applications since surface area and porosity are interesting properties that design catalytic and adsorptive processes. The manometric method is the most suitable technique used for physisorption isotherms and the official recommendation according to IUPAC for assessment of those properties. The adsorptive used is crucial for this and liquid nitrogen (77 K) is widely used, but for several reasons, liquid argon (87 K) is also considered and recommended particularly for micropore size analysis (Thommes 2007; Thommes et al. 2015; Cychosz et al. 2017).

Important information about where the particles are located on magnetic zeolites/composite can be inferred by N_2 adsorption/desorption isotherms. Afshin et al. (2020) found higher surface area and total pore volume for the nanocomposite than the precursor zeolite. Amiri-Yazani et al. (2019) and Mollahosseini et al. (2015) performing different studies with clinoptilolite also reported the same behavior when forming magnetic composite with magnetite. The authors suggest that the enhancement on specific surface area and pore volume is due to deposition of iron oxide nanoparticles on the zeolite surface leading to the formation of secondary porosity. Magnetic zeolites in which the magnetite nanoparticles are located both over and within the zeolite crystals present a 50% decrease in surface area as observed from Cao et al. (2013), as well as pore blockage to pores from 3 to 4.5 nm. Thus, the use of this technique for observation in such properties can bring positive information about the proposed materials.

3.3 Adsorption Process Using Nanocomposites

The rapid industrial development, advancement of technology, urbanization and activities focused on agriculture have led to an increase in environmental pollution, due to the large amount of waste and effluents generated and discarded in the environment (Moura et al. 2011; Vidal et al. 2012, 2015, 2020; Melo et al. 2014; Raulino et al. 2018; de Oliveira Sousa Neto et al. 2019; Santana et al. 2020). Consequently, the contamination of water resources and soils is mainly due to the carelessness of industries in treating their effluents before disposing it. These effluents, after being discharged with their characteristic pollutants, cause changes in the quality of the receiving bodies and, consequently, their pollution (Buarque et al. 2019; de Oliveira et al. 2020).

A variety of impacts on the environment and health of living things are caused due to the stability, high solubility, and extensive migration activity of pollutants present in wastewater. Thus, it is necessary to remove and recover them from industrial effluents (Moura et al. 2011; Martins et al. 2014; Raulino et al. 2014; de Quadros Melo et al. 2016; Vidal et al. 2016; Silva et al. 2018; Fang et al. 2018; Burakov et al. 2018; Santana et al. 2020).

Among the various technologies developed for the removal of pollutants from water and wastewater, the most efficient methods for treating residual water are based on adsorption processes (Fang et al. 2018). Adsorption is considered an efficient and flexible process, presenting easy operating conditions, producing quality effluents, high capacity for interaction with metals and performance in a wide pH range (Fu and Wang 2011; Ihsanullah et al. 2016). In addition, low-cost adsorbent are widely used in the adsorption process such as natural clays (Kushwaha et al. 2020), industrial waste (Bensalah et al. 2020), agricultural waste (Özsin et al. 2019), zeolites

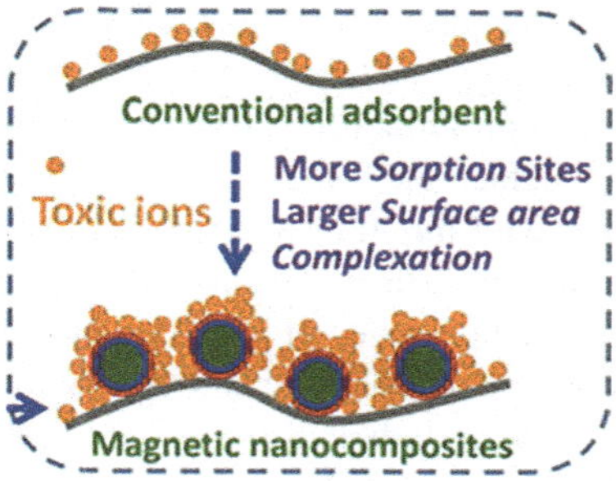

Fig. 8 Schematic adsorption mechanisms. Reproduced with permission from Elsevier (Zhu et al. 2013)

(Gaffer et al. 2017) and metal oxide (Wang et al. 2020). Although these adsorbents have been shown to be very effective in removing pollutants from wastewater, they suffer an inherent disadvantage related to the difficulty of separating them from the solution. Thereby, the use of magnetic nanocomposites in the adsorption process has been gaining a lot of attention, due to its ability to easily and quickly separate adsorbents used in suspension by means of an external magnetic field (Mehta et al. 2015; Reddy and Yun 2016). Several studies highlight the efficiency of magnetic zeolite nanocomposites in removing pollutants from aqueous media (Periyasamy et al. 2018; Mirjavadi et al. 2019).

Supporting nanomagnetite particles on the natural zeolite has economical, technological, and environmental advantages. The starting materials, the synthetic magnetite, and the natural or synthetic silicate (zeolite) are relatively affordable, making the composite attractive to be used in industrial scale. From the technological point of view, the use of composites formed with the magnetic iron oxide allows the adsorbent to be magnetically removed from the water medium. Environmentally, the proposed adsorbing magnetic system represents a prospective procedure to clean larger natural bodies of water being affected with chemical contamination (Pizarro et al. 2015). Figure 8 shows a schematic adsorption mechanisms of metal ions on nanocomposites compared to conventional adsorbents.

Pizarro et al. (2015) studied composites based on nanosized particles of magnetite and natural microsized zeolite in order to adsorb arsenate from aqueous systems. The authors synthesized a well-crystallized nanomagnetite with an averaged particle size of 50 nm when supported on zeolite. According to them, the nanomagnetite enhances the arsenate adsorption of zeolite. The composite prepared with a nanomagnetite zeolite showed a similar adsorption capacity as that observed for pure magnetite. The smaller the mean particle size of zeolite the lower the mass proportion of nanomagnetite in the composite required to reach the maximum adsorptive efficiency. The authors also highlighted that supported nanomagnetite prevents risks of the nanoparticle's contamination.

Salem Attia et al. (2013) synthesized magnetic nanoparticles coated zeolite for the adsorption of pharmaceuticals and personal care products (PPCP) from aqueous solution using batch and column studies. They found that magnetic nanoparticles adsorbed more than 95% of PPCP such as Ibuprofen, Naproxen, Diclofenac-Na, and Gemfibrozil from acidic aqueous solution in 10 min at low pH values, but less efficiency under basic conditions. The contact time, pH dependence, and initial PPCP concentration affect adsorption, depending on the nature of the organic pollutant.

In fact, the fast-kinetic adsorption has been observed for some researches and this behavior might be attributed to the abundance of adsorption sites or free sites on the external surface of the adsorbent (Safinejad et al. 2017). Besides, nanosized particles and the presence of hydroxyl groups on their surface lead to the rapid adsorption of compounds (organic and inorganic) (Barreto et al. 2011; Safinejad et al. 2017).

Some studies also attribute this phenomenon to the adsorption's driving force, which is generally associated with the difference in concentration between the solution and the solid/liquid interface, thus resulting in a high removal rate in the first moments of adsorption (Ibrahim et al. 2018; Shalaby et al. 2018). Table 2 shows the comparison of adsorption equilibrium time for metal ions using magnetic nanoparticles or only pure zeolites as adsorbents.

Ren et al. (2017) used an amino-functionalized superparamagnetic $CoFe_2O_4@SiO_2$ ($CoFe_2O_4@SiO_2–NH_2$) core–shell nanospheres (NPs) composed of amorphous silica (SiO_2) to remove metal ions from aqueous solutions. According to them, monodisperse $CoFe_2O_4@SiO_2$ nanospheres grafted with more amino groups had a greater adsorption capacity and higher removal efficiency for metal ions (Cd(II): 199.9 mg g^{-1}, 99.96%; Cu(II): 177.8 mg g^{-1}, 88.05%; Pb(II): 181.6 mg g^{-1}, 90.79%).

According to the authors, the effects of the pH, initial concentrations, reaction temperature, and time on the adsorption of heavy metal ions by $CoFe_2O_4@SiO_2–NH_2$ were analyzed systematically. Owing to the superparamagnetic properties with

Table 2 Comparison of adsorption equilibrium time for metal ions in different materials

Adsorbent	Equilibrium time (min)	Metal ions	Conditions	Reference
Zeolite X	120	$Pb^{2+}/Cd^{2+}/Cu^{2+}$	pH 5, 21 ± 2 °C, 0.03 g adsorbent	Apiratikul and Pavasant (2008)
Magnetite zeolite	120	Ga^{3+}/In^{3+}	pH 5, 25 ± 1 °C, 0.02 g adsorbent	Zhao et al. (2014)
Nanomagnetite zeolite	120	Cs^{+}/Sr^{2+}	pH 7, 28 ± 2 °C, 0.2 g adsorbent	Ibrahim et al. (2018)
Zeolite NaP1	90	$Pb^{2+}/Cd^{2+}/Cu^{2+}/Zn^{2+}/Ni^{2+}$	pH 5, 22 ± 1 °C, 0.2 g adsorbent	Visa (2016)
Magnetite zeolite	20	Zn^{2+}/Cu^{2+}	pH 6, 27 °C, 0.1 g adsorbent	Shalaby et al. (2018)
Magnetite graphene	5	Cr^{4+}	pH 7, 1 g L^{-1}	Zhu et al. (2013)

a high saturation magnetization value (32.92 emu g^{-1}) of $CoFe_2O_4@SiO_2–NH_2$, the metal-loaded nanospheres can be quickly removed from an aqueous solution (30 s) by magnetic separation. Moreover, they found that the nanospheres exhibited good reusability for up to five cycles, confirming that these magnetic nanospheres could be a potential adsorbent for the effective and regenerable removal of metals ions from aqueous solutions (Fig. 9).

de Andrade Bessa et al. (2017) reported an experimental investigation on the synthesis of nanocomposites based on magnetite and zeolites A and P, using kaolin as the main SiO_2 and Al_2O_3 sources, by hydrothermal route. The authors found average size of the magnetite nanoparticles of ca. 50 nm as determined by TEM analyses. The magnetic characterization confirmed the ferrimagnetic behavior of the magnetite nanoparticles and of the composites, as well as verified that the magnetic properties of the nanoparticles are not affected by the zeolites in the composite formation. According to the authors, the results evidenced that high-quality zeolite composites with magnetite nanoparticles can be reached by considering the employed low-cost method, placing this route as an attractive alternative for water softening reaching removal levels of about 97% in the first application times.

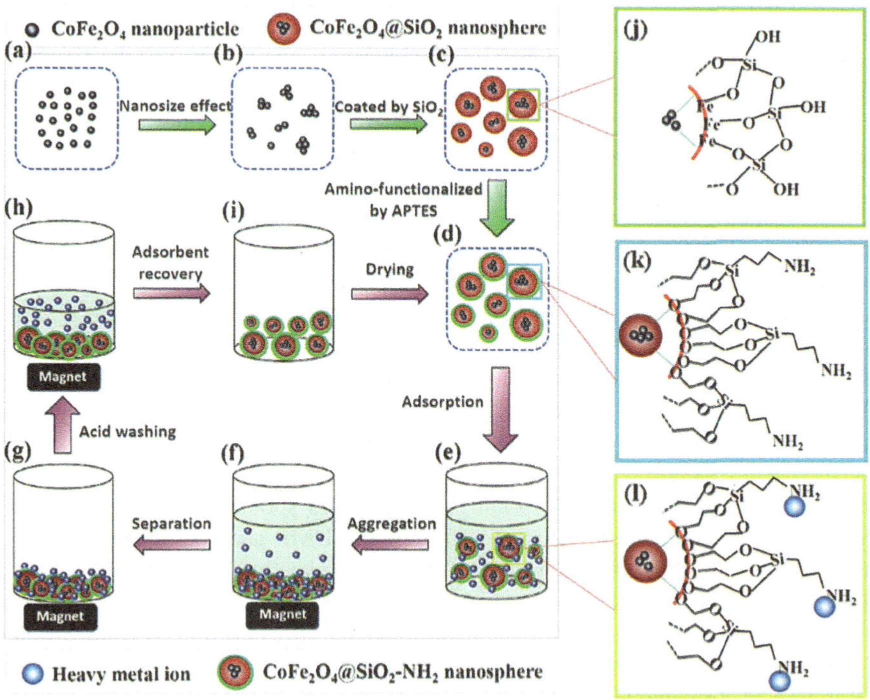

Fig. 9 Schematic illustration of the synthetic procedure of $CoFe_2O_4@SiO_2–NH_2$ core–shell nanospheres (**a**–**d**, **j** and **k**) and the adsorption and regeneration process of heavy metals (**d**–**i** and **l**). Reproduced with permission from Royal Society of Chemistry (Ren et al. 2017)

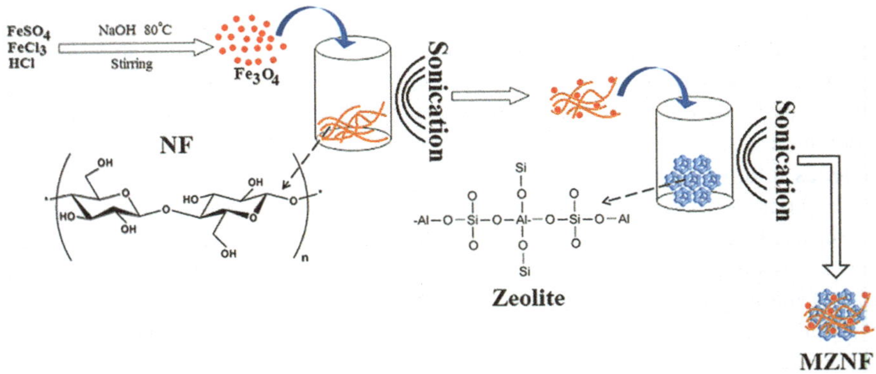

Fig. 10 Preparation procedure of MZNF. Reproduced with permission from Springer Nature (Mirjavadi et al. 2019)

Mirjavadi et al. (2019) studied the zinc adsorption behavior in zeolite/cellulose magnetic nanofibers (MZNF). The authors showed that zeolite played an important role in the prepared nanocomposite due to its large surface area. In addition, the time of 30 min, the temperature of 30 °C and at pH 7 were highlighted as ideal conditions for the adsorption of Zn^{2+}, with 96% removal and a maximum adsorption capacity of 9.45 mg L^{-1}. The authors also reported that the presence of competing ions did not influence the efficiency of the adsorption process, exhibiting an adsorption higher than 93% and zeolite/cellulose magnetic nanofibers showed efficiency greater than 50% after 6 adsorption-desorption cycles. Thus, they concluded that the MZNF obtained may be a promising candidate for the removal of Zn^{2+} in wastewater. Figure 10 presents a summary of the synthetic process for obtaining magnetic zeolite/cellulose nanofibers (Mirjavadi et al. 2019).

Ahmadi et al. (2016) synthesized the magnetic zeolite nanocomposite with an average diameter from 90 to 100 nm by the chemical co-precipitation method and they used it for adsorption of dimethyl phthalate (DMP) from aqueous solution. According to the authors, the higher the adsorbent dosage and the contact time, the greater adsorption efficiency. However, they also found that efficiency decreases by increasing the pH and the initial DMP concentration. Thermodynamic analyses have shown that the adsorption process occurs spontaneously and is inherently endothermic. According to the authors, the DMP adsorption efficiency did not change after 10 batch adsorption-desorption cycles, indicating the prospect of potential application of the synthesized adsorbent in the water treatment.

Mesdaghinia et al. (2017) investigated the in situ growth of Fe_3O_4 nanoparticles on the surface of the zeolite through a co-precipitation process, resulting in the formation of a magnetic zeolite/Fe_3O_4 nanocomposite (MZNC). The synthesized material was used in the adsorption and degradation of phthalate esters (PAEs) in the presence of H_2O_2. According to transmission electron microscopy (TEM) analysis, the structure formed by Fe_3O_4 is spiraled with an average size of approximately

14 nm. However, MZNCs have a cubic structure, almost uniform and interlaced with an average diameter of 80–100 nm.

According to the authors, the maximum removal by adsorption was 95.4% and 41.14% with the equilibrium time of 30 min, pH 3, 0.2 g L^{-1} of adsorbent and initial concentration of 10 and 50 mg L^{-1}, respectively. The authors pointed out that after 10 consecutive adsorption-desorption cycles, the reuse of MZNC reduced slightly. Therefore, it can be speculated that MZNC is capable of being applied repeatedly for DEP adsorption without much loss in the initial adsorption capacities and its removal efficiency. In addition, more than 70% of adsorbed DEP can be desorbed/recovered in the presence of methanol in the tenth cycle, which can be reused in different cases, as industrial applications.

Piri et al. (2019) carried out the synthesis of magnetic zeolite/hydroxyapatite nanocomposites (MZeo-HAP) using the microwave assisted method to remove anionic dyes reactive orange 5 (RO5), reactive orange 16 (RO16), and Congo red (CR) in aqueous solutions by adsorption in batch (Fig. 11). The nanocomposites of MZeo-HAP had a high surface area of 103.56 m^2 g^{-1}, increasing the number of active sites available for the dye molecules to adsorb. Maximum dye removal was achieved at pH 2 with an equilibrium contact time of 30 min. At the dye concentration of 80 mg L^{-1}, the highest RO5, RO16, and CR adsorption capacity was 92.45, 88.31, and 104.05 mg g^{-1}, respectively.

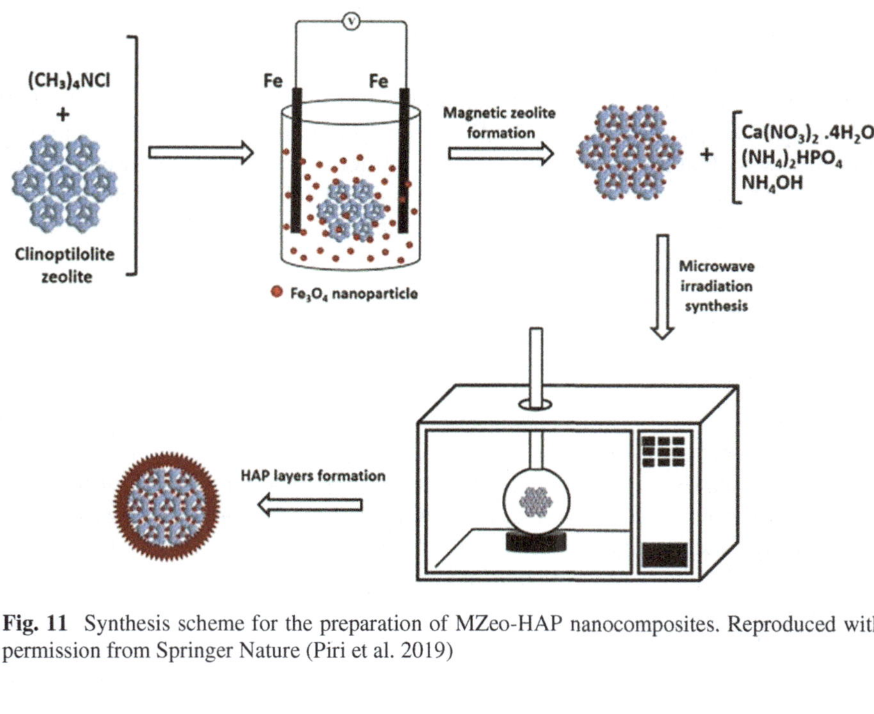

Fig. 11 Synthesis scheme for the preparation of MZeo-HAP nanocomposites. Reproduced with permission from Springer Nature (Piri et al. 2019)

According to the authors, the presence of zeolite in the composite is one of the positive points of this adsorbent, because besides being economical it has influential functional groups. Zeolite as a silica-based material also reinforces the HAP crystal. In addition, the use of magnetic nanoparticles in MZeo-HAP provides a better situation for a simple separation of the spent adsorbent, applying an external magnetic field compared to other adsorbents. Regarding the thermodynamic parameters ($\Delta G°$, $\Delta S°$ and $\Delta H°$) estimated in the adsorption study, the authors suggested an endothermic process, in which the maximum adsorption capacity of the dyes occurred at the highest temperature studied (40 °C).

Jorfi et al. (2020) investigated the efficiency of Fe_3O_4 magnetic nanoparticles coated with natural zeolite (Fe_3O_4@Z), for removing cadmium from aqueous solution. After synthesis, the authors observed a uniform distribution of nanocomposite particles with a size range of 16.63–29.03 nm in zeolite before Cd^{2+} adsorption. In addition, they reported that the maximum cadmium adsorption using a dosage of 0.6 g L^{-1}, pH = 6, a contact time of 60 min and a concentration of 20 mg L^{-1} was 42.9%. The authors also revealed that in the recovery study, HNO_3 1 mol L^{-1} was considered the best eluent, after four cycles, recovering 38% of Cd^{2+} ions. Finally, they concluded that the Fe_3O_4@Z nanocomposite can be applied as an efficient adsorbent for removing cadmium from aquatic environments.

Afshin et al. (2020) used a zeolite/Fe_3O_4 nanocomposite in the adsorption of the cationic dye blue 41 from aqueous solutions in a wide range of concentrations. The authors report that the efficiency of the adsorption process increased by increasing the reaction time, pH and the amount of adsorbent. However, when there was an increase in the initial concentration of the dye, a significant decrease in the adsorption efficiency was observed. Under optimal conditions of pH = 9, initial dye concentration of 100 mg L^{-1}, adsorbent dose of 3 g L^{-1}, and reaction time of 60 min, the removal efficiency was obtained 71.4%. According to the authors, the results showed that the zeolite/Fe_3O_4 nanocomposite could be used as an adsorbent of high efficiency and availability, ecologically correct and economical, in order to remove the dye from the wastewater of different industries.

4 Conclusions

This chapter provided a revision on synthesis and characterization of magnetite-zeolite nanocomposites and its application as adsorbents to remediation of polluted aquatic environments. For instance, magnetite-zeolite nanocomposites are used as electrochemical supercapacitors, sensors, magnetic controlling systems, and adsorbents and catalysts to remove organic and inorganic compounds from water through adsorption process. High-quality zeolite composites with magnetite nanoparticles can be reached by low-cost hydrothermal method. The presence of zeolite in the composite plays an important role in the prepared nanocomposite due to its large surface area and it is also positive for adsorption proposes because it is economical, and it has influential functional groups. Zeolite supported nanomagnetite prevents risks of the nanoparticle's contamination. The use of magnetic nanoparticles provides simple separation of the spent adsorbent, just applying an external magnetic field compared to other adsorbents.

References

Afshin S, Rashtbari Y, Vosoughi M et al (2020) Removal of basic blue-41 dye from water by stabilized magnetic iron nanoparticles on clinoptilolite zeolite. Rev Chim 71:218–229. https://doi.org/10.37358/RC.20.2.7919

Ahmadi E, Kakavandi B, Azari A et al (2016) The performance of mesoporous magnetite zeolite nanocomposite in removing dimethyl phthalate from aquatic environments. Desalin Water Treat 57:27768–27782. https://doi.org/10.1080/19443994.2016.1178174

Amiri-Yazani T, Zare-Dorabei R, Rabbani M, Mollahosseini A (2019) Highly efficient ultrasonic-assisted pre-concentration and simultaneous determination of trace amounts of Pb (II) and Cd (II) ions using modified magnetic natural clinoptilolite zeolite: Response surface methodology. Microchem J 146:498–508. https://doi.org/10.1016/j.microc.2019.01.050

Aono H, Tamura K, Johan E et al (2013) Preparation of composite material of Na-P1-type zeolite and magnetite for Cs decontamination. Chem Lett 42:589–591. https://doi.org/10.1246/cl.130123

Apiratikul R, Pavasant P (2008) Sorption of Cu^{2+}, Cd^{2+}, and Pb^{2+} using modified zeolite from coal fly ash. Chem Eng J 144:245–258. https://doi.org/10.1016/j.cej.2008.01.038

Atashi Z, Divband B, Keshtkar A et al (2017) Synthesis of cytocompatible Fe_3O_4@ZSM-5 nanocomposite as magnetic resonance imaging contrast agent. J Magn Magn Mater 438:46–51. https://doi.org/10.1016/j.jmmm.2017.04.062

Badeenezhad A, Azhdarpoor A, Bahrami S, Yousefinejad S (2019) Removal of methylene blue dye from aqueous solutions by natural clinoptilolite and clinoptilolite modified by iron oxide nanoparticles. Mol Simul 45:564–571. https://doi.org/10.1080/08927022.2018.1564077

Baile P, Vidal L, Aguirre MÁ, Canals A (2018) A modified ZSM-5 zeolite/Fe_2O_3 composite as a sorbent for magnetic dispersive solid-phase microextraction of cadmium, mercury and lead from urine samples prior to inductively coupled plasma optical emission spectrometry. J Anal At Spectrom 33:856–866. https://doi.org/10.1039/C7JA00366H

Barquist K, Larsen SC (2010) Chromate adsorption on bifunctional, magnetic zeolite composites. Microporous Mesoporous Mater 130:197–202. https://doi.org/10.1016/j.micromeso.2009.11.005

Barreto ACH, Santiago VR, Mazzetto SE et al (2011) Magnetic nanoparticles for a new drug delivery system to control quercetin releasing for cancer chemotherapy. J Nanoparticle Res 13:6545–6553. https://doi.org/10.1007/s11051-011-0559-9

Belviso C, Agostinelli E, Belviso S et al (2015) Synthesis of magnetic zeolite at low temperature using a waste material mixture: fly ash and red mud. Microporous Mesoporous Mater 202:208–216. https://doi.org/10.1016/j.micromeso.2014.09.059

Bensalah H, Younssi SA, Ouammou M et al (2020) Azo dye adsorption on an industrial waste-transformed hydroxyapatite adsorbent: kinetics, isotherms, mechanism and regeneration studies. J Environ Chem Eng 8:103807. https://doi.org/10.1016/j.jece.2020.103807

Bercoff PG, Bertorello HR, Saux C et al (2009) Magnetic properties of Co-impregnated zeolites. J Magn Magn Mater 321:3813–3820. https://doi.org/10.1016/j.jmmm.2009.07.046

Bosînceanu R, Sulitanu N (2008) Synthesis and characterization of FeO(OH)/Fe_3O_4 nanoparticles encapsulated in zeolite matrix. J Optoelectron Adv Mater 10:3482–3486

Breck DW (1984) Zeolite molecular sieves. Wiley, New York

Buarque PMC, de Lima RBP, Vidal CB et al (2019) Enhanced removal of emerging micropollutants by applying microaeration to an anaerobic reactor. Eng Sanit Ambient 24:667–673. https://doi.org/10.1590/s1413-4152201920190030

Burakov AE, Galunin EV, Burakova IV et al (2018) Adsorption of heavy metals on conventional and nanostructured materials for wastewater treatment purposes: a review. Ecotoxicol Environ Saf 148:702–712. https://doi.org/10.1016/j.ecoenv.2017.11.034
Cabrera L, Gutierrez S, Menendez N et al (2008) Magnetite nanoparticles: electrochemical synthesis and characterization. Electrochim Acta 53:3436–3441. https://doi.org/10.1016/j.electacta.2007.12.006
Cao J, Liu X-W, Fu R, Tan Z (2008) Magnetic P zeolites: synthesis, characterization and the behavior in potassium extraction from seawater. Sep Purif Technol 63:92–100. https://doi.org/10.1016/j.seppur.2008.04.015
Cao J, Chang G, Guo H, Chen J (2013) Synthesis and characterization of magnetic ZSM-5 zeolite. Trans Tianjin Univ 19:326–331. https://doi.org/10.1007/s12209-013-1912-0
Chen L, Zhou Q, Xiong Q et al (2015) Shape-evolution and growth mechanism of fepolyhedrons. Adv Mater Sci Eng 2015. https://doi.org/10.1155/2015/763124
Cychosz KA, Guillet-Nicolas R, García-Martínez J, Thommes M (2017) Recent advances in the textural characterization of hierarchically structured nanoporous materials. Chem Soc Rev 46:389–414. https://doi.org/10.1039/C6CS00391E
Dar MI, Shivashankar SA (2014) Single crystalline magnetite, maghemite, and hematite nanoparticles with rich coercivity. RSC Adv 4:4105–4113. https://doi.org/10.1039/C3RA45457F
de Andrade Bessa R, de Sousa Costa L, Oliveira CP et al (2017) Kaolin-based magnetic zeolites A and P as water softeners. Microporous Mesoporous Mater 245:64–72. https://doi.org/10.1016/j.micromeso.2017.03.004
de Oliveira Sousa Neto V, Freire TM, Saraiva GD et al (2019) Water treatment devices based on zero-valent metal and metal oxide nanomaterials
de Oliveira FF, Moura KO, Costa LS et al (2020) Reactive adsorption of parabens on synthesized micro- and mesoporous silica from coal fly ash: pH effect on the modification process. ACS Omega 5:3346–3357. https://doi.org/10.1021/acsomega.9b03537
de Quadros Melo D, de Oliveira Sousa Neto V, de Freitas Barros FC et al (2016) Chemical modifications of lignocellulosic materials and their application for removal of cations and anions from aqueous solutions. J Appl Polym Sci 133. https://doi.org/10.1002/app.43286
Divband B, Rashidi MR, Khatamian M et al (2018) Linde type A and nano magnetite/NaA zeolites: cytotoxicity and doxorubicin loading efficiency. Open Chem 16:21–28. https://doi.org/10.1515/chem-2018-0001
El-Din TAS, Elzatahry AA, Aldhayan DM et al (2011) Synthesis and characterization of magnetite zeolite nano composite. Int J Electrochem Sci 6:6177–6183
Fang L, Li L, Qu Z et al (2018) A novel method for the sequential removal and separation of multiple heavy metals from wastewater. J Hazard Mater 342:617–624. https://doi.org/10.1016/j.jhazmat.2017.08.072
Fibikar S, Luppi G, Martínez-Junza V et al (2015) Manipulation and orientation of zeolite L by using a magnetic field. ChemPlusChem 80:62–67. https://doi.org/10.1002/cplu.201402252
Fu F, Wang Q (2011) Removal of heavy metal ions from wastewaters: a review. J Environ Manage 92:407–418. https://doi.org/10.1016/j.jenvman.2010.11.011
Fungaro DA (2011) Adsorption of anionic dyes from aqueous solution on zeolite from fly ash-iron oxide magnetic nanocomposite. J At Mol Sci 2:305–316. https://doi.org/10.4208/jams.032211.041211a
Fungaro DA, Yamaura M, Graciano JEA (2010) Remoção de íons Zn^{2+}, Cd^{2+}E Pb^{2+} de soluções aquosas usando compósito magnético de zeólita de cinzas de carvão. Quim Nova 33:1275–1278. https://doi.org/10.1590/S0100-40422010000600011
Gaffer A, Al Kahlawy AA, Aman D (2017) Magnetic zeolite-natural polymer composite for adsorption of chromium (VI). Egypt J Pet 26:995–999. https://doi.org/10.1016/j.ejpe.2016.12.001
Ghanbari D, Sharifi S, Naraghi A, Nabiyouni G (2016) Photo-degradation of azo-dyes by applicable magnetic zeolite Y-Silver–CoFe2O4 nanocomposites. J Mater Sci: Mater Electron 27:5315–5323. https://doi.org/10.1007/s10854-016-4430-8

Gharehaghaji N, Divband B, Zareei L (2018) Nanoparticulate NaA zeolite composites for MRI: effect of iron oxide content on image contrast. J Magn Magn Mater 456:136–141. https://doi.org/10.1016/j.jmmm.2018.02.013

Hesas RH, Baei MS, Rostami H et al (2019) An investigation on the capability of magnetically separable Fe_3O_4/mordenite zeolite for refinery oily wastewater purification. J Environ Manage 241:525–534. https://doi.org/10.1016/j.jenvman.2018.09.005

Ibrahim HA, Abdel Moamen OA, Monem NA, Ismail IM (2018) Assessment of kinetic and isotherm models for competitive sorption of Cs+ and Sr^{2+} from binary metal solution onto nanosized zeolite. Chem Eng Commun 205:1274–1287. https://doi.org/10.1080/00986445.2018.1446004

Ihsanullah, Abbas A, Al-Amer AM et al (2016) Heavy metal removal from aqueous solution by advanced carbon nanotubes: critical review of adsorption applications. Sep Purif Technol 157:141–161. https://doi.org/10.1016/j.seppur.2015.11.039

Jahangirian H, Shah Ismail MH, Jelas Haron M et al (2013) Synthesis and characterization of zeolite/Fe_3O_4 nanocomposite by green quick precipitation method. Dig J Nanomater Biostructures 8:1405–1413

Javanbakht V, Ghoreishi SM, Habibi N, Javanbakht M (2016) A novel magnetic chitosan/clinoptilolite/magnetite nanocomposite for highly efficient removal of Pb(II) ions from aqueous solution. Elsevier B.V

Jorfi S, Shooshtarian MR, Pourfadakari S (2020) Decontamination of cadmium from aqueous solutions using zeolite decorated by Fe_3O_4 nanoparticles: adsorption modeling and thermodynamic studies. Int J Environ Sci Technol 17:273–286. https://doi.org/10.1007/s13762-019-02350-2

Kalhor M, Zarnegar Z (2019) Fe_3O_4/SO_3H@zeolite-Y as a novel multi-functional and magnetic nanocatalyst for clean and soft synthesis of imidazole and perimidine derivatives. RSC Adv 9:19333–19346. https://doi.org/10.1039/c9ra02910a

Kokate M, Garadkar K, Gole A (2013) One pot synthesis of magnetite–silica nanocomposites: applications as tags, entrapment matrix and in water purification. J Mater Chem A 1:2022–2029. https://doi.org/10.1039/C2TA00951J

Kolen'ko YV, Bañobre-López M, Rodríguez-Abreu C et al (2014) Large-scale synthesis of colloidal Fe_3O_4 nanoparticles exhibiting high heating efficiency in magnetic hyperthermia. J Phys Chem C 118:8691–8701. https://doi.org/10.1021/jp500816u

Kouli ME, Banis G, Savvidou MG et al (2020) A study on magnetic removal of hexavalent chromium from aqueous solutions using magnetite/zeolite-x composite particles as adsorbing material. Int J Mol Sci 21. https://doi.org/10.3390/ijms21082707

Kumar B, Garg R, Singh U (2012) Utilization of flyash as filler in Hdpe/flyash polymer composites: a review. Int J Appl Eng Res 7:1679–1682

Kushwaha A, Rani R, Patra JK (2020) Adsorption kinetics and molecular interactions of lead [Pb(II)] with natural clay and humic acid. Int J Environ Sci Technol 17:1325–1336. https://doi.org/10.1007/s13762-019-02411-6

Lercher JA, Jentys A (2007) Infrared and Raman spectroscopy for characterizing zeolites. In: Čejka J et al (ed) Studies in surface science and catalysis. Elsevier, pp 495–XIII

Lima TM, Lima CGS, Rathi AK et al (2016) Magnetic ZSM-5 zeolite: a selective catalyst for the valorization of furfuryl alcohol to γ-valerolactone, alkyl levulinates or levulinic acid. Green Chem 18:5586–5593. https://doi.org/10.1039/C6GC01296E

Liu H, Peng S, Shu L et al (2013) Magnetic zeolite NaA: synthesis, characterization based on metakaolin and its application for the removal of Cu^{2+}, Pb^{2+}. Chemosphere 91:1539–1546. https://doi.org/10.1016/j.chemosphere.2012.12.038

Lu F, Astruc D (2020) Nanocatalysts and other nanomaterials for water remediation from organic pollutants. Coord Chem Rev 408:213180. https://doi.org/10.1016/j.ccr.2020.213180

Luo L, Dai C, Zhang A et al (2015) Facile synthesis of zeolite-encapsulated iron oxide nanoparticles as superior catalysts for phenol oxidation. RSC Adv 5:29509–29512. https://doi.org/10.1039/C5RA02194D

Martins AVPR, Ramos JET, Coelho JA et al (2014) Metal-impregnated carbon applied as adsorbent for removal of sulphur compounds using fixed-bed column technology. Environ Technol (United Kingdom) 35. https://doi.org/10.1080/09593330.2013.868530

Mehta D, Mazumdar S, Singh SK (2015) Magnetic adsorbents for the treatment of water/wastewater—a review. J Water Process Eng 7:244–265. https://doi.org/10.1016/j.jwpe.2015.07.001

Melo DQ, Vidal CB, da Silva AL et al (2014) Removal of Cd^{2+}, Cu^{2+}, Ni^{2+}, and Pb^{2+} ions from aqueous solutions using tururi fibers as an adsorbent. J Appl Polym Sci 131. https://doi.org/10.1002/app.40883

Mesdaghinia A, Azari A, Nodehi RN et al (2017) Removal of phthalate esters (PAEs) by zeolite/Fe_3O_4: investigation on the magnetic adsorption separation, catalytic degradation and toxicity bioassay. Elsevier B.V

Mirjavadi ES, Tehrani RMA, Khadir A (2019) Effective adsorption of zinc on magnetic nanocomposite of Fe_3O_4/zeolite/cellulose nanofibers: kinetic, equilibrium, and thermodynamic study. Environ Sci Pollut Res 26:33478–33493. https://doi.org/10.1007/s11356-019-06165-z

Mollahosseini A, Toghroli M, Kamankesh M (2015) Zeolite/Fe_3O_4 as a new sorbent in magnetic solid-phase extraction followed by gas chromatography for determining phthalates in aqueous samples. J Sep Sci 38:3750–3757. https://doi.org/10.1002/jssc.201500510

Moura CP, Vidal CB, Barros AL et al (2011) Adsorption of BTX (benzene, toluene, o-xylene, and p-xylene) from aqueous solutions by modified periodic mesoporous organosilica. J Colloid Interface Sci 363. https://doi.org/10.1016/j.jcis.2011.07.054

Mthombeni NH, Onyango MS, Aoyi O (2015) Adsorption of hexavalent chromium onto magnetic natural zeolite-polymer composite. J Taiwan Inst Chem Eng 50:242–251. https://doi.org/10.1016/j.jtice.2014.12.037

Mthombeni NH, Mbakop S, Ochieng A, Onyango MS (2016) Vanadium (V) adsorption isotherms and kinetics using polypyrrole coated magnetized natural zeolite. J Taiwan Inst Chem Eng 66:172–180. https://doi.org/10.1016/j.jtice.2016.06.016

Nabiyouni G, Shabani A, Karimzadeh S et al (2015) Synthesis, characterization and magnetic investigations of Fe_3O_4 nanoparticles and zeolite-Y nanocomposites prepared by precipitation method. J Mater Sci: Mater Electron 26:5677–5685. https://doi.org/10.1007/s10854-015-3118-9

Naghizadeh M, Taher MA, Behzadi M, Moghaddam FH (2017) Preparation a novel magnetic natural nano zeolite for preconcentration of cadmium and its determination by ETAAS. Environ Nanotechnol Monit Manag 8:261–267. https://doi.org/10.1016/j.enmm.2017.10.001

Neto DMA, Freire RM, Gallo J et al (2017) Rapid sonochemical approach produces functionalized Fe_3O_4 nanoparticles with excellent magnetic, colloidal, and relaxivity properties for MRI application. J Phys Chem C 121:24206–24222. https://doi.org/10.1021/acs.jpcc.7b04941

Oliveira LCA, Petkowicz DI, Smaniotto A, Pergher SBC (2004) Magnetic zeolites: a new adsorbent for removal of metallic contaminants from water. Water Res 38:3699–3704. https://doi.org/10.1016/j.watres.2004.06.008

Özsin G, Kılıç M, Apaydın-Varol E, Pütün AE (2019) Chemically activated carbon production from agricultural waste of chickpea and its application for heavy metal adsorption: equilibrium, kinetic, and thermodynamic studies. Appl Water Sci 9:56. https://doi.org/10.1007/s13201-019-0942-8

Periyasamy S, Gopalakannan V, Viswanathan N (2018) Hydrothermal assisted magnetic nano-hydroxyapatite encapsulated alginate beads for efficient Cr(VI) uptake from water. J Environ Chem Eng 6:1443–1454. https://doi.org/10.1016/j.jece.2018.01.007

Piri F, Mollahosseini A, Khadir A, Milani Hosseini M (2019) Enhanced adsorption of dyes on microwave-assisted synthesized magnetic zeolite-hydroxyapatite nanocomposite. J Environ Chem Eng 7:103338. https://doi.org/10.1016/j.jece.2019.103338

Pizarro C, Rubio MA, Escudey M et al (2015) Nanomagnetite-zeolite composites in the removal of arsenate from aqueous systems. J Braz Chem Soc 26:1887–1896. https://doi.org/10.5935/0103-5053.20150166

Raulino GSC, Vidal CB, Lima ACA et al (2014) Treatment influence on green coconut shells for removal of metal ions: pilot-scale fixed-bed column. Environ Technol (United Kingdom) 35. https://doi.org/10.1080/09593330.2014.880747

Raulino GSC, Silva LSD, Vidal CB et al (2018) Role of surface chemistry and morphology in the reactive adsorption of metal ions on acid modified dry bean pods (*Phaseolus vulgaris L.*) organic polymers. J Appl Polym Sci 135. https://doi.org/10.1002/app.45879

Reddy DHK, Yun Y-S (2016) Spinel ferrite magnetic adsorbents: alternative future materials for water purification? Coord Chem Rev 315:90–111. https://doi.org/10.1016/j.ccr.2016.01.012

Ren C, Ding X, Fu H et al (2017) Core-shell superparamagnetic monodisperse nanospheres based on amino-functionalized $CoFe_2O_4$@SiO_2 for removal of heavy metals from aqueous solutions. RSC Adv 7:6911–6921. https://doi.org/10.1039/c6ra27728d

Safinejad A, Goudarzi N, Chamjangali MA, Bagherian G (2017) Effective simultaneous removal of Pb(II) and Cd(II) ions by a new magnetic zeolite prepared from stem sweep. Mater Res Express 4:16. https://doi.org/10.1088/2053-1591/aa9738

Sağir T, Huysal M, Durmus Z et al (2016) Preparation and in vitro evaluation of 5-flourouracil loaded magnetite-zeolite nanocomposite (5-FU-MZNC) for cancer drug delivery applications. Biomed Pharmacother 77:182–190. https://doi.org/10.1016/j.biopha.2015.12.025

Salem Attia TM, Hu XL, Yin DQ (2013) Synthesized magnetic nanoparticles coated zeolite for the adsorption of pharmaceutical compounds from aqueous solution using batch and column studies. Chemosphere 93:2076–2085. https://doi.org/10.1016/j.chemosphere.2013.07.046

Santana GB, Nóbrega DC, Oliveira JT et al (2020) Aplicação da semente de moringa (*Moringa oleífera*) como coagulante natural no tratamento de efluente de indústria de tintas no Ceará. Rev Tecnol 41:1–17. https://doi.org/10.5020/23180730.2020.9889

Scapim LCM, Borges SB, de Paula JN et al (2017) Síntese E Caracterização De Nanomagnetita Pelo Processo De Coprecipitação. J Eng Exact Sci 3:1182–1191. https://doi.org/10.18540/jcecvl3iss8pp1182-1191

Schwertmann U, Cornell RM (2000) Iron oxides in the laboratory: preparation and characterization. Wiley-VCH, Germany

Shalaby T, Eissa M, El Kady M, Abd El-Gaber S (2018) Geochemistry of El-Salam Canal and the adjacent groundwater in north Sinai, Egypt: an application to a water treatment process using magnetic zeolite nanoparticles. Appl Water Sci 8. https://doi.org/10.1007/s13201-018-0741-7

Shan W, Yu T, Wang B et al (2006) Magnetically separable nanozeolites: promising candidates for bio-applications. Chem Mater 18:3169–3172. https://doi.org/10.1021/cm060530f

Sharifi M, Baghdadi M (2016) Enhanced selectivity and capacity of clinoptilolite for Cd^{2+} removal from aqueous solutions by incorporation of magnetite nanoparticles and surface modification with cysteine. Water Sci Technol 73:2284–2293. https://doi.org/10.2166/wst.2016.016

Shirani M, Semnani A, Haddadi H, Habibollahi S (2014) Optimization of simultaneous removal of methylene blue, crystal violet, and fuchsine from aqueous solutions by magnetic NaY zeolite composite. Water Air Soil Pollut 225. https://doi.org/10.1007/s11270-014-2054-2

Shirani M, Akbari A, Hassani M (2015) Adsorption of cadmium(<scp>ii</scp>) and copper(<scp>ii</scp>) from soil and water samples onto a magnetic organozeolite modified with 2-(3,4-dihydroxyphenyl)-1,3-dithiane using an artificial neural network and analysed by flame atomic absorption spec. Anal Methods 7:6012–6020. https://doi.org/10.1039/C5AY01269D

Silva LS, Raulino GSC, Vidal CB et al (2018) Peculiar properties of LTA/FAU synthetic composite zeolite and its effect on Cu^{2+}adsorption: factorial experimental design. Desalin Water Treat 107. https://doi.org/10.5004/dwt.2018.22165

Singh LH, Pati SS, Coaquira JAH et al (2016) Magnetic interactions in cubic iron oxide magnetic nanoparticle bound to zeolite. J Magn Magn Mater 416:98–102. https://doi.org/10.1016/j.jmmm.2016.05.003

Teng X, Black D, Watkins NJ et al (2003) Platinum-maghemite core−shell nanoparticles using a sequential synthesis. Nano Lett 3:261–264. https://doi.org/10.1021/nl025918y

Thommes M (2007) Textural characterization of zeolites and ordered mesoporous materials by physical adsorption. In: Čejka J et al (ed) Studies in surface science and catalysis. Elsevier, pp 495–XIII

Thommes M, Kaneko K, Neimark AV et al (2015) Physisorption of gases, with special reference to the evaluation of surface area and pore size distribution (IUPAC Technical Report). Pure Appl Chem 87:1051–1069. https://doi.org/10.1515/pac-2014-1117

Usman M, Byrne JM, Chaudhary A et al (2018) Magnetite and green rust: synthesis, properties, and environmental applications of mixed-valent iron minerals. Chem Rev 118:3251–3304. https://doi.org/10.1021/acs.chemrev.7b00224

Vidal CB, Raulino GSC, Barros AL et al (2012) BTEX removal from aqueous solutions by HDTMA-modified Y zeolite. J Environ Manage 112. https://doi.org/10.1016/j.jenvman.2012.07.026

Vidal CB, Feitosa AV, Pessoa GP et al (2015) Polymeric and silica sorbents on endocrine disruptors determination. Desalin Water Treat 54. https://doi.org/10.1080/19443994.2014.880377

Vidal CB, Melo DQ, Raulino GSC et al (2016) Multielement adsorption of metal ions using Tururi fibers (*Manicaria Saccifera*): experiments, mathematical modeling and numerical simulation. Desalin Water Treat 57. https://doi.org/10.1080/19443994.2015.1025441

Vidal C, Barbosa P, Pessoa G et al (2020) Multiresidue determination of endocrine disrupting compounds in sewage treatment plants (SPE-HPLC-DAD). J Braz Chem Soc. https://doi.org/10.21577/0103-5053.20200127

Visa M (2016) Synthesis and characterization of new zeolite materials obtained from fly ash for heavy metals removal in advanced wastewater treatment. Powder Technol 294:338–347. https://doi.org/10.1016/j.powtec.2016.02.019

Wang L, Shi C, Wang L et al (2020) Rational design, synthesis, adsorption principles and applications of metal oxide adsorbents: a review. Nanoscale 12:4790–4815. https://doi.org/10.1039/C9NR09274A

Watts JF, Wolstenholme J (2003) An introduction to surface analysis by XPS and AES. John Wiley & Sons Ltd, England

Wright PA (2008) Structure determination experimental techniques. In: Microporous framework solids. The Royal Society of Chemistry, pp 79–147

Yamaura M, Fungaro DA (2013) Synthesis and characterization of magnetic adsorbent prepared by magnetite nanoparticles and zeolite from coal fly ash. J Mater Sci 48:5093–5101. https://doi.org/10.1007/s10853-013-7297-6

Zendehdel R, Ansari S, Sedghi R, Jafari MJ (2019) Magnetic nano-zeolite Y as a novel fluidized bed for air decontamination. Int J Environ Sci Technol 16:1261–1268. https://doi.org/10.1007/s13762-017-1588-4

Zhao FY, Li YL, Li LH (2014) Preparation and characterization of magnetite nanoparticles. Appl Mech Mater 618:24–27. https://doi.org/10.4028/www.scientific.net/AMM.618.24

Zhao F, Zou Y, Lv X et al (2015) Synthesis of $CoFe_2O_4$-zeolite materials and application to the adsorption of gallium and indium. J Chem Eng Data 60:1338–1344. https://doi.org/10.1021/je501039u

Zhu J, Wei S, Chen M et al (2013) Magnetic nanocomposites for environmental remediation. Adv Powder Technol 24:459–467. https://doi.org/10.1016/j.apt.2012.10.012

Chapter 4
Chitosan Nanoparticle: Alternative for Sustainable Agriculture

André Luiz Barros de Oliveira, Francisco Thálysson Tavares Cavalcante, Katerine da Silva Moreira, Paula Jéssyca Morais Lima, Rodolpho Ramilton de Castro Monteiro, Bruna Bandeira Pinheiro, Kimberle Paiva dos Santos, and José Cleiton Sousa dos Santos

1 Introduction

Chitosan is a natural polysaccharide, obtained from the alkaline deacetylation of chitin. After cellulose, it is the most important organic compound in nature and can easily be found as a by-product of the fishing industry, such as the crustacean exoskeleton and other animals (Doan et al. 2019). In the wake, chitosan is one of the most frequently cited polymers in scientific research (Dongre 2019) dealing with a wide range of applications, such as in the food industry, cosmetics, agriculture, and environmental protection, quality monitoring and wastewater treatment, as adsorbents in oil clarification, among others (Casadidio et al. 2019). Furthermore, chitosan has been strongly indicated as a suitable functional material given its unique set of properties as an extremely abundant, biocompatible, non-toxic, and renewable natural biodegradable polymer, which makes it still ecologically interesting and economically viable material (Song et al. 2018). It is noteworthy that due to the presence of amino groups in its chemical structure, chitosan has distinct biological functions and versatility of modifications and formulations (Dongre 2018), offering the production of a wide variety of beneficial derivatives, which have aroused the interest of several researchers in the last decade (Kumar et al. 2019) (Fig. 1).

A. L. B. de Oliveira · F. T. T. Cavalcante · K. da Silva Moreira · P. J. M. Lima ·
R. R. de Castro Monteiro · B. B. Pinheiro · K. P. dos Santos
Departamento de Engenharia Química, Universidade Federal do Ceará, Campus do Pici, Fortaleza, CE CEP 60455-760, Brazil

J. C. S. dos Santos (✉)
Instituto de Engenharias e Desenvolvimento Sustentável, Universidade da Integração Internacional da Lusofonia Afro-Brasileira, Rua José Franco de Oliveira, s/n, Redenção, CE CEP 62790-970, Brazil
e-mail: jcs@unilab.edu.br

R. F. do. Nascimento et al. (eds.), *Nanomaterials and Nanotechnology*,
Materials Horizons: From Nature to Nanomaterials,
https://doi.org/10.1007/978-981-33-6056-3_4

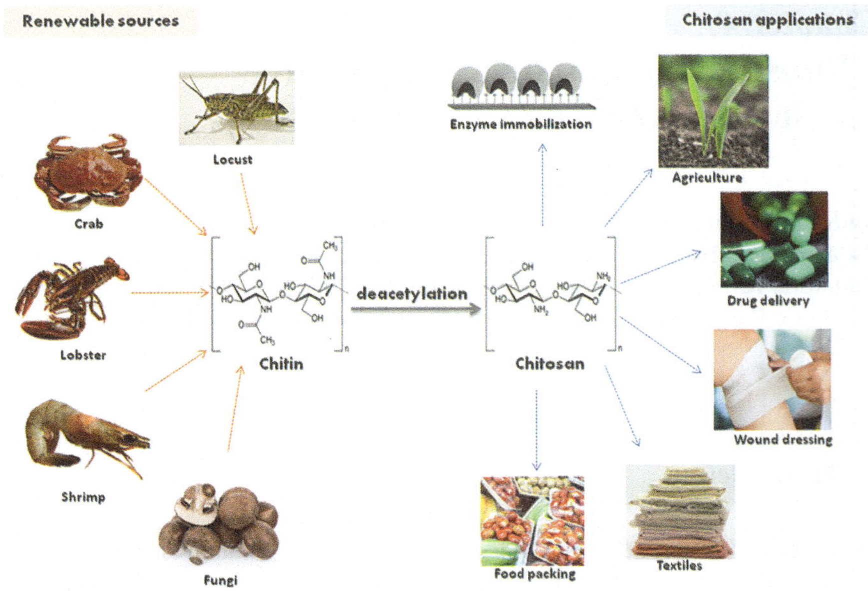

Fig. 1 Chitosan obtained from renewable sources and some applications

Studies on polymeric nanoparticles in the pharmaceutical and biomedical areas have been increasingly deepened due to their controlled release properties and to reach specific drug action sites (Lombardo et al. 2019). In this context, chitosan nanoparticles present a slow and continuous release of compounds. They have been also used with the function of controlled delivery of agrochemicals and genetic materials (Itodo 2019). For the more, the use of nanomaterials in agriculture is in line with sustainable development, using, for example, nanofertilizers and nanopesticides in the soil to control nutrients, protect from pests and microbes, decontaminate, pollute waters, and consequently increase productivity (Joshi et al. 2019).

In this chapter, it is given a comprehensive review of the advantages and recent developments in the formulation of chitosan nanoparticles as an alternative for sustainable agriculture. This chapter presents a review of chitosan and chitosan nanoparticles, and the importance and versatility of applications. Some bad variability of these nanoparticles under specific conditions and possible solutions is also presented. Finally, it highlights the importance of nanotechnology development as an alternative for sustainable agriculture, presenting the advantages of using chitosan nanoparticles, development, and perspectives.

2 Chitosan Properties

Chitosan is a useful biomaterial derived from partial deacetylation of chitin in alkaline solutions with characteristics that make it a linear polysaccharide (Yadav et al. 2019) of great interest for a significant number of applications such as nontoxic, biocompatibility, biodegradability, biosorption, and bioactivity (Abo Elsoud and El Kady 2019); these specific properties make it a unique molecule. For this reason, it has been proposed as a potentially attractive material for many uses and also, of great economic and environmental importance (Batista et al. 2019).

Chitosan is a linear polyamine copolymer (Sánchez-Machado et al. 2019) of β-(1 → 4)-linked 2-amino-2-deoxy-D-glucose (D-glucosamine) and 2-acetamido-2-deoxy-D-glucose (*N*-acetyl-D-glucosamine) units (Sweidan et al. 2011). Its molecular structure has an amino group (C2) and 2 hydroxyl groups (C3 and C6), which form intermolecular hydrogen bonds that determine the stability of the polymer (Varelis et al. 2019). The chemical structure of chitosan is shown in Fig. 2. Chemically, chitosan can be modified through amino and hydroxyl groups. Physical and chemical properties depend principally on its molecular weight and degree of deacetylation (Abdelgawad and Hudson 2019).

One of the proprieties of chitosan is its high hydrophobicity and it is insoluble in water. However, it dissolves in aqueous solutions of organic acids such as acetic, formic, and citric, as well as inorganic acids such as dilute hydrochloric acid resulting in viscous solutions (Crini et al. 2019).

Its application is extensive and can be used in agriculture (defensive mechanisms and fertilizer for plants) (Choudhary et al. 2019), biotechnology (Ma et al. 2019), medicine (de Masi and Tonazzini 2019), pharmacy (immunological, antitumor, hemostatic, and anticoagulant) (dos Santos Rodrigues et al. 2019; Li et al. 2019), food industry (dietary fibers, cholesterol-lowering) (Harkin et al. 2019), sauce preservative, fungicide and bactericide (Devlieghere et al. 2004), and cosmetics (skin exfoliator, acne treatment, hair moisturizer, toothpaste) (Dowling 2019). The development of innovative chitosan is driven by the fact that the polymer can be obtained from renewable sources (Ahmad et al. 2019) as shown in Fig. 3.

The use possibilities of this polymer are increased due to the fact that chitosan can be prepared in different forms, such as viscosity-controlled solutions, gels, film forming, microspheres, and nanoparticles (Islam et al. 2019); for this reason, chitosan

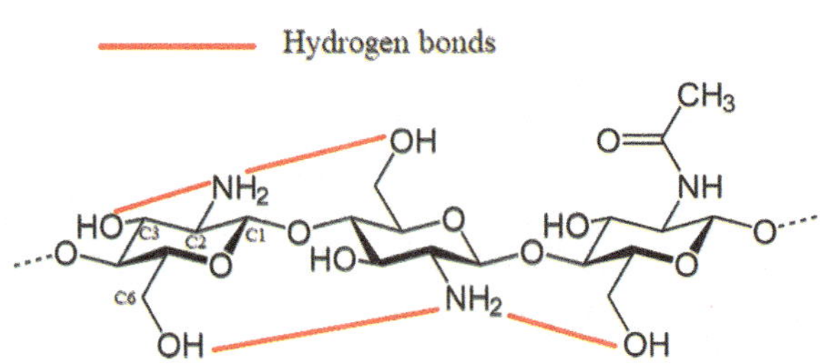

Fig. 2 Chemical structure of chitosan

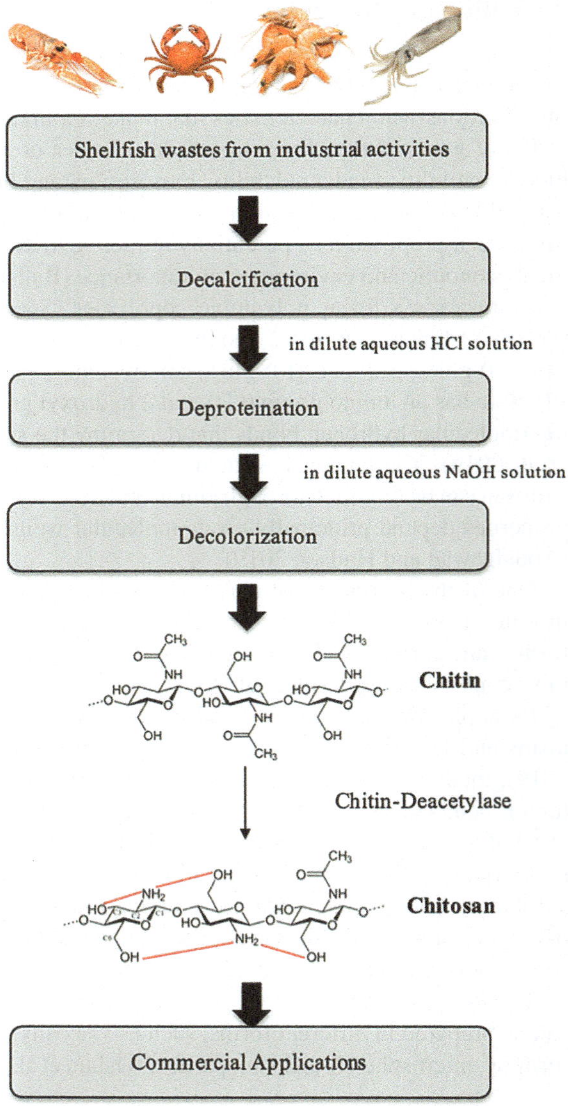

Fig. 3 Chitosan obtained from renewable sources and some applications

has been used as packaging material, especially as edible films and coatings for fruits and vegetables to extend the shelf life of foodstuffs (Parreidt et al. 2018).

To diversify the chitosan application, several strategies have been adopted to improve their properties like blending, chemical modifications, complexation, crosslinking, and graft copolymerization (El-Hefian et al. 2014). Furthermore, chitosan is passive to several chemical modifications (immobilization of chelating agents, quaternization, carboxylation, acylation, sulfonation, amination, and polyelectrolytic complex formation) once it has a high proportion of reactive amino groups distributed in the polymeric matrix (Zhou et al. 2019).

In addition, chitosan is also considered as a polyelectrolyte (Brar and Kaur 2018). Polyelectrolytes are macromolecules that have ionizable groups along their entire chain (Atta-ur-Rahman 2018). These ionizable groups are classified according to their functionality and can be anionic or cationic. In this case, the amino clusters present in the chitosan polymer chain have reactivity and, at pH less than 6.5 (mean acid), are capable of binding to hydrogen ions, presenting a positive global charge (Ng et al. 2016). Thus, chitosan is considered a cationic polyelectrolyte, with additional properties such as adsorption capacity and chelating ability (Crini et al. 2019). In view of these characteristics, chitosan reveals tremendous versatility of application in completely different areas, including from wastewater treatment (Goh et al. 2019) used as a flocculating agent in the treatment of aqueous effluents, removal of metal ions, eco-friendly polymer, and odor reduction.

3 Nanomaterials

In general, a particle is considered a nanometer to the materials found at the nanoscale ($1\ nm = 10^{-9}\ m$) (Kolahalam et al. 2019). The properties of materials change as their size approaches the nanoscale, and the percentage of atoms on the surface of a material becomes significant. In other words, nanomaterials have a higher area to volume ratio, so specific surface effects become more important giving different characteristics for certain applications compared to the same materials without nanometer dimensions (Gopakumar et al. 2019).

The synthesis of nanoparticles can be done through bottom-up or top-down methods (Demirci et al. 2019). The bottom-up approach includes the reduction of material components (up to atomic level) with further self-assembly processes leading to the formation of nanostructures (Kumar et al. 2017). Inversely, the top-down approach uses larger (macroscopic) initial structures, which can be externally controlled in the processing of nanostructures (Dincer 2018). On the other hand, nanomaterials occur naturally or synthesized with various structures. The nanomaterials are divided into diverse categories depending on their morphology, size, and physicochemical properties. The categories can be carbon-based NPs, ceramics NPs, lipid-based NPs, metal NPs, and also polymeric NPs (Khan et al. 2017). Besides, depending on their construction, they are classified as composites (Saleh 2016).

Another possibility for obtaining nanocomposite materials is by combining carbon nanoparticles and a polymer matrix. The main nanocomposite characteristics lie in their multifunctionality, a possibility to make combinations of unique characteristics, unreachable with traditional materials (Fu et al. 2019a, b). The use of biopolymers, like chitosan, in the formation of biomaterials is particularly interesting, as organic or inorganic nanoparticles serve as modifiers of the structure of the nanocomposites (Sukhova et al. 2019).

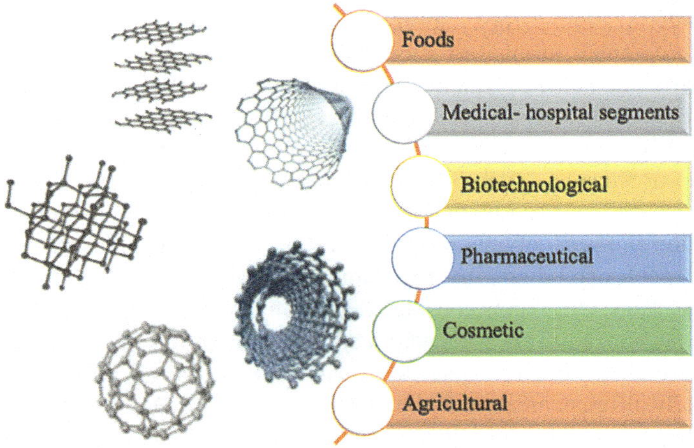

Fig. 4 Application of nanomaterials in various segments

Nanomaterials have various technological applications used in several segments, and they are shown in Fig. 4. (Jiang et al. 2019; Weber and Bechelany 2019; Yu et al. 2019). Due to the nanometric dimensions, the catalytic reaction is considered very efficient since the catalysis process is strongly conditioned to the atoms present on the material surface (Gopakumar et al. 2019). In other words, the great potential of nanomaterials is related to the high catalytic activity exhibited by these materials due to the high surface/volume ratio of nanoparticles (Mauricio et al. 2018).

One of the most important nanomaterials studied in recent years is chitosan-based nanoparticles (Bai et al. 2019). Basically, the main mechanisms for the preparation of chitosan nanoparticles depend on a chemical or physical cross-linking process (Kosheleva et al. 2019), formation of polyelectrolyte complexes, and self-assembly of hydrophobically modified chitosan (Quiñones et al. 2018). Also, other possible processes like reverse micelle, desolvation, precipitation/coacervation, and emulsion-droplet coalescence are available in the literature (Hasnain and Nayak 2019). In this context, chitosan has been used as a carrier in polymeric nanoparticles for drug delivery through various routes of administration, once chitosan nanoparticles present a high ability to control drug and vaccine release rate (Mohammed et al. 2017). In addition, among the many possibilities of nanoparticles' applications in the agricultural area, we can mention the use of chitosan nanoparticles as antivirus in some plants, seed coating biocompatibility, replacement or amount fertilizers reduction, herbicides and insecticides (Duhan et al. 2017) used for production, reducing environmental impacts without losing efficiency, due to proven, biodegradability, non-toxicity and adsorption (Itodo 2019).

4 Synthesis of Chitosan Nanoparticles

To transform residual chitosan into a product with a broad range of commercial applications, it is necessary to synthesize nanoparticles from this material (Anand et al. 2018). In this way, there are several techniques to produce them, and the first was described in 1994, by emulsification and cross-linking process (Asiri et al. 2018). Since then, several synthesis techniques have been described and each of them can form particles of different sizes and properties and may suit the most diverse application needs (Grenha 2012; Huang et al. 2010; Mitra et al. 2001; Qi et al. 2004; Wang et al. 2011; Yu et al. 2010) as shown in Fig. 5. The insertion of materials to the chitosan matrix, reactants concentrations, and the time of the production reaction are factors that influence the properties of the particles formed, and this could be manipulated for agricultural applications. The techniques most employed in the literature will be briefly described (Divya and Jisha 2018).

4.1 *Covalent Crosslinking*

The synthesis of chitosan microparticles technique consists of cross-linking the functional amine group of chitosan with the aldehyde group of the cross-linking agent,

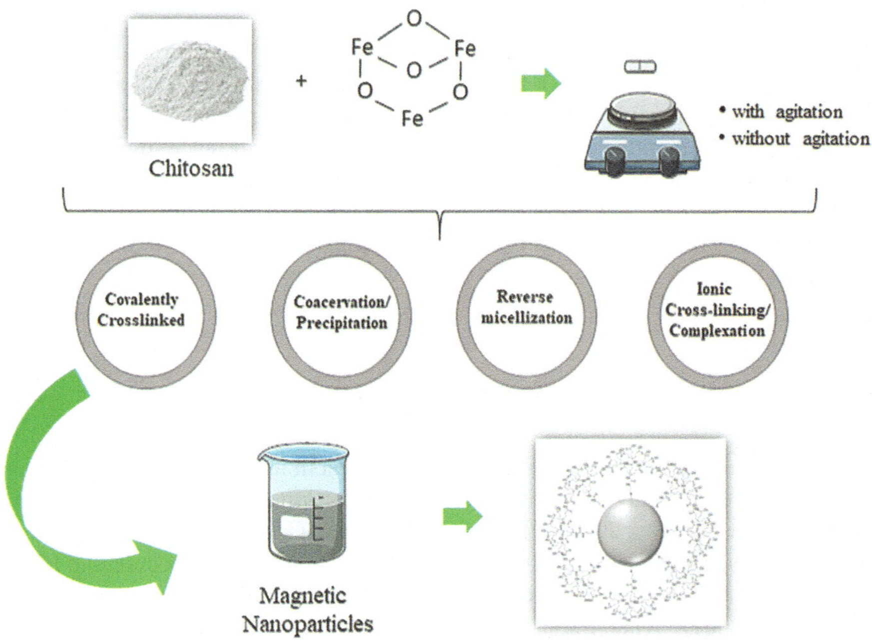

Fig. 5 Since then, several synthesis techniques

by covalent bonds (Divya and Jisha 2018). In this technique, a water-in-oil emulsion containing chitosan is prepared in an aqueous mixture. Droplets of this solution are stabilized using a surfactant compound. Then, the cross-linking agent is added, which will harden the droplets, which will be filtered, washed, and dried (Alizadeh et al. 2019). The size of the particles will depend on the agitation speed of the medium and the cross-linking agent used (Yadav et al. 2017). However, this method has the disadvantage of depending on a powerful cross-linking agent (such as glutaraldehyde), which may react with the active site of the substance of interest (Noori et al. 2019).

4.2 Coacervation/Precipitation

This method uses the principle of the insolubility of the chitosan in alkaline pH but precipitates in solutions of this type (Garg et al. 2019; Tokumitsu et al. 1999). Two chitosan emulsions are made, one made in oil and the other in an aqueous alkaline solution, such as NaOH. Both emulsions are mixed at high speed, allowing coalescence and precipitating solidified particles of chitosan.

4.3 Reverse Micellization

Reverse micelles are mixtures of thermodynamically stable water, oil, and surfactant, and this concept can be used to produce chitosan nanoparticles (Lone et al. 2018). First, a water-in-oil microemulsion is prepared using a surfactant dissolved in an organic solvent. Then, the chitosan aqueous solution, the substance of interest (as a drug or an enzyme), and a cross-linking agent are then added under constant stirring to avoid any turbidity. The substance of interest added to the nanoparticles can be obtained after evaporation, washing, and centrifugation (Mitra et al. 2001; Farias et al. 2014).

4.4 Ionic Cross-Linking/Complexation

The cationic nature of chitosan can be used for reactions with negatively charged polymers. Another property of this material is the formation of a gel with certain anions (Divya and Jisha 2018). These combined processes can be used for nanoparticle synthesis, as well as the individual gelation process, which occurs due to the formation of inter- and intramolecular cross-linking of chitosan with these anions (Janes et al. 2001). When the process occurs with anions of small molecules, it is called gelation (Grenha 2012). On the other hand, when it occurs with anionic macromolecules, it is called complexation. Both processes have the advantage of occurring in mild conditions and do not require a chemical cross-linking agent; avoiding undesired interactions with the compounds of interest (Agnihotri et al. 2004; Grenha 2012) is therefore the most found technique in the literature (Ali et al. 2011; Anitha et al. 2009; Corsi et al. 2003; Fernández-Urrusuno et al. 1999; Gan and Wang 2007; Mohammadpour Dounighi et al. 2012; Pant and Singh 2018; Qi et al. 2004; Saharan et al. 2013; Xu and Du 2003).

4.5 *Chemically Modified Chitosan*

The chemical modification of chitosan is performed for specific cases of aggregation of compounds, which may improve the interaction between them and the particles, and increase the carrying capacity or continuous release of these compounds. Different chemical groups, like bioactive molecules, can be incorporated into chitosan nanoparticles, as shown by many studies (Prabaharan and Mano 2004). For example, Calvo et al. (2002) coated chitosan nanoparticles with poly(lactic-co-glycolic acid), increasing the protein delivery into intestinal and nasal mucosae. Like it was described, many processes of synthesis of chitosan nanoparticles can be performed, and it is necessary to choose the best technique and modification (if required) for each, and this is consequently applied for agricultural purposes.

5 Chitosan Nanoparticles' Applications

5.1 *Pharmaceuticals*

As will be discussed below, chitosan nanoparticles are applied in a broad range of pharmaceuticals industry (Baghdan et al. 2018; Luangtana-anan et al. 2019; Pan et al. 2019; Panão Costa et al. 2019). There are two types of loading methods of the substance desired: during the preparation of the particles, called incorporation, or after the formation of the particles, called incubation (Chandra Hembram et al. 2016). The efficiency of the loading will depend on the method of preparation of the particles and the physicochemical properties of the substance of interest (Agnihotri et al. 2004; Esfandiarpour-Boroujeni et al. 2017).

5.1.1 Drug/Protein/Gene Delivery Devices

As already cited, chitosan nanoparticles have some interesting properties to be applied in drug delivery systems such as biodegradability and mucoadhesive character, promoting a gradual releasement of substances (Mitra et al. 2001; Wang et al. 2011). Taking advantage of these features, many devices were successfully applied

(Divya and Jisha 2018). In this context, chitosan can load a broad range of drugs including antivirus, antiallergic, and hormone drugs (Wang et al. 2011). Ibrahim et al. (2015) entrapped brimonidine (intra-ocular pressure reducer) in chitosan nanoparticles for ocular delivery (Ibrahim et al. 2015). The amount of drug released was controlled, without any burst effect, following a slow release for at least 24 h, indicating a potential for prolonged delivery of the substance (Ibrahim et al. 2015).

Protein drugs can be easily degraded by enzymes in vivo and have poor permeability and stability. Chitosan nanoparticles can be applied to protect the protein from enzyme degradation in vivo and promote contact between drug and biomembrane, improving bioavailability (Wang et al. 2011). Luangtana-anan et al. (2019) made a systematic study about chitosan nanoparticles obtained by ionic cross-linking, observing how the loading and releasement of proteins could be predicted and manipulated, using bovine serum albumin as a model (Luangtana-anan et al. 2019). They cite an initial rapid release of proteins as the major drawback of using these particles as carriers, but that can be controlled by changing the loading concentrations or varying the preparation protocols, changing the chemical interactions of the medium (Luangtana-anan et al. 2019).

In another paper, Al-Qadi et al. (2012) applied a dry powder system consisting of microencapsulated protein-loaded chitosan nanoparticles for insulin delivery to the deep lung in rats. They observed a great potential of this delivery for macromolecules and local therapy of lung diseases. In another application, Pan et al. (2019) studied the antibacterial activity of pure chitosan when compared with chitosan nanoparticles (Pan et al. 2019). The interaction between the positively charged particles and negatively charged biological membranes inhibited the growth of various microorganisms, and this interaction was higher in chitosan nanoparticles, bringing better results than pure chitosan with the same molecular weight (Pan et al. 2019).

Chen et al. (2013) applied chitosan–DNA nanoparticles for an intranasal vaccine against *Streptococuus mutans*, a cariogenic bacteria (Chen et al. 2013). The vaccine was successfully tested, inducing an effective immune response in the tested animals (Chen et al. 2013). In another application, Xu et al. (2011) made their particles for another intranasal vaccine, but against *Streptococcus pneumoniae* (respiratory pathogen). They also obtained positive results, immunizing the tested animals against pneumococcal infections (Xu et al. 2011).

As a non-virus gene carrier, chitosan has excellent biodegradability and biocompatibility, which has led to its increasing application in gene and drug delivery (Wang et al. 2011). Panão Costa et al. (2019) used chitosan–casein nanoparticles for gene delivery. They obtained great transfection results in COS-7 cells (Panão Costa et al. 2019). In the same context, Baghdan et al. (2018) observed excellent transfection and gene expression results with chorioallantoic membrane models, with liposomes coating their chitosan–DNA nanoparticles, bringing new insights into the non-viral gene therapy (Baghdan et al. 2018).

Chitosan nanoparticles are being recurrently studied to overcome their major problems for drug delivery: rapid releasement of substances and biocompatibility of their modifications and its derivatives. Specific studies of the systems of interest are necessary to solve such problems (Chuan et al. 2019; Urimi et al. 2019; Wang et al. 2011).

5.2 Biomedical

5.2.1 Scaffold Fabrication for Tissue Regeneration

One of the biggest problems in the world is replacing a diseased or injured organ with biocompatible materials (Sivashankari and Prabaharan 2019). Currently, in medicine, this replacement depends on organ transplants, heart valves, metal implants, etc. However, these methods have faced great difficulties due to the lack of tissue donors, rejection, and corrosion of metallic implants (Chen and Thouas 2015; Fisher et al. 2013; Oryan and Sahvieh 2017). In recent decades, tissue engineering has attracted the attention of researchers seeking to restore and regenerate damaged or lost human organs and tissues using a variety of biologically compatible techniques (Hasnain and Nayak 2019; Sivashankari and Prabaharan 2019). Tissue engineering takes into consideration three aspects called the "Tissue Engineering Triad," which are growth factors, cells, and scaffolds/dies (O'Brien 2011). Tissue engineering focuses primarily on the production and optimization of scaffolds, as it provides the necessary and appropriate environment for tissue cell development (Dhandayuthapani et al. 2011).

Scaffolds are used to support tissues or organs during damage recovery or regeneration. Compound-based tissue engineering and scaffolds are well-studied subjects in the age of tissue engineering (Ortega et al. 2015). Scaffolds can operate as a conductor for bioactive molecules, such as cells to cure defects and growth factors. After tissue formation in the scaffold's matrix, the matrix gradually deteriorates for the healing process to be completed (Dhandayuthapani et al. 2011). Currently, biomaterials obtained from natural and synthetic polymers are widely applied for scaffold fabrication (Sivashankari and Prabaharan 2019). Natural polymers, such as collagen, gelatin, chitosan, alginate, and cellulose, are biocompatible and enable interaction between cells and polymers, leading to increased cell proliferation (Dhandayuthapani et al. 2011). Other materials such as ceramics, metal nanoparticles, and carbon-based nanomaterials are being used as fillers in tissue engineering scaffolds because of their bioactivity and support for stem cell development, proliferation, and differentiation in various lineages (O'Brien 2011; Sivashankari and Prabaharan 2019).

Scaffolds are used to support tissues or organs during damage recovery or regeneration. Compound-based tissue engineering and scaffolds are well-studied subjects in the age of tissue engineering (Kumar et al. 2019). Scaffolds can operate as a conductor for bioactive molecules, such as cells to cure defects and growth factors. After tissue formation in the scaffold's matrix, the matrix gradually deteriorates for the healing process to be completed. Currently, biomaterials obtained from natural and synthetic polymers are widely applied for scaffold fabrication (Sivashankari and Prabaharan 2019). Natural polymers, such as collagen, gelatin, chitosan, alginate, and cellulose, are biocompatible and enable interaction between cells and polymers, leading to increased cell proliferation (Dhandayuthapani et al. 2011). Other materials such as ceramics, metal nanoparticles, and carbon-based nanomaterials are being used as fillers in tissue engineering scaffolds because of their bioactivity and support for stem cell development, proliferation, and differentiation in various lineages (Sivashankari and Prabaharan 2019).

Regeneration or restructuring of the injured nerve is a challenging clinical aspect (Kumar et al. 2019). Porous conductive scaffolding can be used to recover peripheral nerves. In studies developed by Baniasadi et al. (2015), the authors developed conductive porous structures including conductive polyaniline/graphene nanoparticles in a chitosan/gelatin matrix. The researchers calculated various nerve regeneration parameters such as porosity, controlled biodegradability, electrical conductivity, and mechanical property (Baniasadi et al. 2015). The study showed that mechanical properties and electrical conductivity are directly proportional, while biodegradability and porosity are inversely proportional to the inclusion of graphene and polyaniline content in the scaffolding matrix (Baniasadi et al. 2015).

5.2.2 Barrier of Chitin-Based Dressing to Microbial Infections

Healing recovers the sweetness of the injured tissue and prevents the disruption of homeostasis in organisms (Kermanizadeh et al. 2018). Dressings are primarily aimed at inhibiting bleeding and protecting against environmental irritation, water, and electrolyte disturbances (Huang et al. 2010). The skin should be covered with a bandage after being injured. Dressings can be divided into three categories: biological, synthetic, and biological-synthetic (Jayakumar et al. 2011).

Curative organics mark disadvantages high as antigenicity, low competitiveness, and the risk of cross-contamination (Jayakumar et al. 2011). Synthetic dressings with a shelf life once influence emotion and have a high risk of pathogenic transmissions. Biological-synthetic dressings are bilayer and consist of polymers and high biological materials (Matsuda et al. 1990; Suzuki 2000).

In recent years, researchers have been dedicated to producing a synthesized and modified dressing from biocompatible materials (Schoukens 2019). Production is focused on the use of biologically derived materials, such as hydrocolloids, alginates, collagen, chitosan, chitin, chitosan, or chitin derivatives, which can accelerate healing processes at the molecular, cellular, and systemic levels (Jayakumar et al. 2011).

The authors (Abdelgawad et al. 2014) presented the superiority of the chitosan and polyvinyl alcohol (PVA) nanofiber mats loaded with silver nanoparticles that were crosslinked by glutaraldehyde over environmental agents (Abdelgawad et al. 2014). The study on drug release in the 7 days immersion test showed increased cross-linking time resulting from the decline in Ag+ ion release. Antibacterial testing revealed that the number of bacterial colonies was reduced as chitosan increased in the system (Abdelgawad et al. 2014).

In their research, the authors (Annur et al. 2015) incorporated silver nanoparticles into chitosan nanofiber structures since silver nanoparticles were known for their antibacterial properties (Annur et al. 2015). The study showed that 1% by weight of chitosan nanofibers combined with $AgNO_3$ showed a 30% reduction in fiber diameter after post-plasma treatment. Antibacterial research shows that 2% by weight of

$AgNO_3$ combined with chitosan nanofibers (0.38 mm) has a larger zone of inhibition than chitosan nanofibers (0.01 mm).

Thus, chitosan nanoparticles applied to the medical field have a high potential. As noted earlier, they can be applied in the manufacture of scaffolding for tissue and organ regeneration and restoration, an area that has faced problems with relying on transplants and implants. They also play an important role in the manufacture of synthetic dressings as they reduce the action of bacterial agents.

5.3 Cosmetic

In recent years, the cosmetics industry has grown steadily, moving billions of dollars worldwide (Chiari-Andréo et al. 2019). To ensure this advance, innovative formulations and products are needed, as well as improved compositions, providing benefits to consumers. Nanoparticles have been considered a promising tool for solving various pharmaceutical problems as well as cosmetic problems (Chiari-Andréo et al. 2019). Nanoparticles may be applied for controlled drug release and targeted delivery of substances, which makes them of great interest to the cosmetic industry. The controlled drug release keeps the substance chemically stable, safe, and effective, improving its solubility, film formation, besides the combination of various active substances, and other uses (Brigger et al. 2012; Müller et al. 2017).

Chitosan-derived nanoparticles make them promising candidates in the cosmetics field because of their high water solubility characteristics (K. Mouryaa et al. 2010; Kumar et al. 2004, 2014; Saewan and Jimtaisong 2015). Chitosan has similarities to extracellular matrix polysaccharides; that is, they have water solubility, anionic functionality, high hydrodynamic volumes, and cation-binding characteristics (Argüelles-Monal et al. 2018; Muzzarelli 1988).

5.3.1 Active Cosmetic Product Transporters

Chitosan and its derivatives have been widely investigated for use as a carrier for pharmaceuticals and cosmetics (Chiari-Andréo et al. 2019; Jimtaisong and Saewan 2014). The use of chitosan for encapsulation of active products has gained interest because of its mucosal adhesiveness, biodegradability, biocompatibility, and non-toxicity. The advantages of encapsulating active components in a polymer matrix involve protecting the surrounding environment from controlled release and processing (Jimtaisong and Saewan 2014).

In their work, Harris et al. (2011) prepared chitosan nanoparticles and microspheres by the chitosan hydrochloride and sodium tripolyphosphate ion gelation method for encapsulating an antioxidant from yerba mate extracts (Harris et al. 2011). The spray drying methodology was used to produce microspheres, and the resulting substances allowed to control the release of natural antioxidants, thus being a promising system in food and cosmetic applications (Harris et al. 2011). In another work, researchers (Morganti et al. 2008) investigated the use of chitin as an active carrier for cosmetic products. In the study, chitin nanofibrils were aggregated with antioxidant substances (ectoin, lutein, and melatonin). Chitin nanofibrils improved the penetration of active substances through the skin layer. Because of its free radical scavenging activity, the combination of antioxidant agents and chitin can be employed to prevent the harmful effects of solar radiation that is responsible for wrinkles and photoaging (Morganti et al. 2008).

5.3.2 Protective Emulsions Protective Ultraviolet Radiation

The effects of exposure of skin to solar radiation are mainly caused by ultraviolet radiation from the sun's rays (Ntohogian et al. 2018a). Although UV-C (100–290 nm) radiation is completely filtered by the ozone layer, UV-A (320–400 nm) and UV-B (290–320 nm) radiations cause numerous skin conditions, such as skin cancer, photo-discovery, sunburn, skin degeneration, and photosensitivity (Agbai et al. 2014; Cadet et al. 2015). In order to prevent these adverse reactions, sunscreens that contain filter substances with a strong protective effect against solar radiation are widely used by the population (Ntohogian et al. 2018b). In this context, cosmetic products such as lotions, oils, creams, and emulsions are used for sunscreen, while chemical or physical materials are often selected as sunscreen agents (Mutalik et al. 2015). These agents can act by absorbing, dispensing, or intensifying solar radiation, respectively (Serpone et al. 2007). Commercially available sunscreen products present various problems such as erythema, edema, and irritation (Mutalik et al. 2015). In order to solve these problems, researchers are looking for solutions in nanotechnology, developing formulas in which appropriate carriers act as carriers of manometric sunscreens, such as zinc and titanium dioxide, which are the most effective and acceptable products for nanotechnology and production of sunscreens (Mutalik et al. 2015; Newman et al. 2009). One of the most efficient biocompatible materials to act as a nanocharger in the manufacture of sunscreen products is chitosan (Ntohogian et al. 2018b)

Researchers (Ntohogian et al. 2018a) evaluated the preparation of protective emulsions from chitosan nanoparticles with annatto, filtered annatto, turmeric, and ultra-filtered turmeric (Ntohogian et al. 2018b). Ion gelation methodology was used to prepare chitosan nanoparticles. The results obtained by the authors showed that all encapsulated materials obtained good thermal stability and decor stability. However, all prepared emulsions had low cytotoxicity and good storage stability for up to 90 days, and minimum sun protection was observed with SPF sun protection factor values ranging from 2.15 to 4.85 (Ntohogian et al. 2018b).

5.4 Foods

The mechanical and antibacterial characteristics of chitosan films are enhanced by the addition of metal nanoparticles and metal oxides such as silver, zinc and TiO_2, zinc oxide, and nano clay (Othman 2014). Studies show the effectiveness of using chitosan nanoparticles as a protective barrier against food spoilage, improving antibacterial properties, and they are good for reducing environmental problems and extending food shelf life as it improves the storage environment of them (Sarojini et al. 2019). Also, a thin layer of biodegradable material can be formed into a film and, without changing the original ingredients or processing method, can be used as a food wrap (Cazón et al. 2017).

In this regard, a major current concern is the reduction of environmental problems associated with food storage and demand for biodegradable materials (Cazón et al. 2017). Thus, films derived from biopolymers have been widely used, also considering their edible characteristics (Ortega et al. 2019; Youssef and El-Sayed 2018). In the food industry, edible active biological films and coatings offer many benefits due to their edibility, biocompatibility with human tissues, barrier properties against pathogenic microorganisms, non-toxicity, aesthetic appearance, non-polluting, and low cost (Vásconez et al. 2009). Edible and biodegradable films present themselves as an alternative to synthetic packaging materials due to the effectiveness of the water barrier, its ability to prevent moisture loss, loss of flavor, or to reduce oxidation of some foods (Aider 2010). However, these polymers have low tensile strength and high absorption, a fact that can be overcome by mixing them with other polymeric materials that enhance the desired properties of the film, such as polybutylene succinate (Threepopnatkul et al. 2014).

Due to bioadhesiveness, biodegradability, and biocompatibility, chitosan is commonly used as food stabilizers and thickeners (El Knidri et al. 2018). When used alone as a coating material for nanoparticle assembly, chitosan has some limitations due to its pH sensitivity (Zhao et al. 2011). The formation of a chitosan polyelectrolyte complex layer on the surface of the nano-delivery system that controls release and confers stability can overcome these obstacles (Liu et al. 2016).

Another important application of the use of chitosan nanoparticles in the food industry is prebiotics (Călinoiu et al. 2019). They bring host health benefits, provide beneficial bacterial growth, and also inhibit the growth of pathogenic bacteria (Roberfroid 2000). Prebiotics can be obtained by chemical or enzymatic synthesis, where enzyme-catalyzed synthesis becomes a more attractive alternative because it allows the control of the regioselectivity and stereochemistry of reaction end products (Hashem et al. 2016). In this context, Nguyen et al (2019) (Nguyen and Yang 2017) obtained excellent results using the enzyme immobilized on chitosan-coated magnetic nanoparticles, revealing the potential application in the production of prebiotic lactulose-based galactooligosaccharides.

Thus, chitosan nanoparticles have great potential in the food industry, highlighting some of the mentioned advantages of biocompatibility, biodegradability, antimicrobial activity, and film-forming capacity, which consequently reduces environmental problems (Muxika et al. 2017).

5.5 *Textiles*

Chitosan nanoparticles also have a wide application in the textile industry (Shahid-ul-Islam et al. 2013; Muxika et al. 2017; Meramo-Hurtado et al., 2019). Recently, a new area of research has emerged in the development of new textile surfaces, exploring the potential of green synthesized metal nanoparticles (Durán et al. 2007). In this context, Shahid-ul-Islam et al. (2019) reported the production of silver nanoparticles on the flax surface in the presence of chitosan, with the aim of simultaneously conferring color, antibacterial, and antioxidant activity, since silver is known as a reliable choice to confer such functional properties. Considering the serious environmental risk presented by dyes, from colored wastewater that can block both sunlight penetration and the dissolution of oxygen essential for aquatic life, chitosan nanoparticles act in the treatment of these waters and are used as sorbents for heavy metal ions or sorption of various textile dyes (Khataee and Kasiri 2010).

In this regard, chitosan nanoparticles can also be used as matrices for photocatalyst immobilization in photocatalytic degradation processes (Khataee and Kasiri 2010). According to Henderson (2011) (Henderson 2011), the most commonly used semiconductor as a photocatalyst, because of its cheapness, non-toxic, photochemically stable, very abundant, and water-insoluble under most environmental conditions, is titanium dioxide (TiO_2). The biggest advantage of immobilizing these nanoparticles in support matrices such as chitosan is that liquid–solid separation is much easier and cost-effective (Shoaebargh and Karimi 2014).

In recent years, a very important issue for the textile industries is the use of anti-odor textiles in various types of clothing such as sportswear, socks, shoes, and underwear, which has increased due to the growing of consumer health awareness, as textiles are close to the microorganisms of the environment and the skin (Bashari et al. 2018; Chandrasekar et al. 2014). The heat and humidity present in the skin contribute to the growth of bacteria that can cause the appearance of blemishes, unpleasant odors, allergies, and skin infections (Callewaert et al. 2014). Therefore, it is essential to prevent the bacteria from multiplying to control the unpleasant odor, so the chitosan nanoparticles present as an antimicrobial agent preventing the proliferation of bacteria that cause these odors, which can be controlled by applying antimicrobial finish to tissues (Bashari et al. 2018).

Another important feature for the use of chitosan nanoparticles in textile industries, which has attracted great interest in recent years, relates to their potential use as ultraviolet (UV) protection agents, since chitosan has a great ability to form metal complexes with various metals such as zinc and copper, due to their amino and hydroxyl groups, forming the chitosan–ZnO or chitosan/copper complexes (AbdElhady 2012). AbdElhady (2012) presented the results of chitosan/zinc oxide nanoparticles preparation and characterization with the objective of conferring UV and antimicrobial protection to cotton fabric and found that with increasing concentration of ZnO/chitosan nanoparticles there was a significant improvement in UV protection of finished cotton fabric (AbdElhady 2012).

In another approach, Hebeish et al. (2013) (Hebeish et al. 2013) also used chitosan nanoparticles as a green finish in the multifunction cotton fabrics, where the results also showed that with increasing the concentration of chitosan nanoparticles together with the copper complex, that is, after the formation of chitosan/metal complex, there was an increase in UV protection of cotton fabric. Thus, better UV protection of the treated cotton fabrics was obtained with the high surface area to volume ratio in chitosan nanoparticles providing a greater amount of more accessible free amino groups that will interact with copper metal ions (Hebeish et al. 2013). In another study, Lu et al. (2019) prepared hydroxypropyl chitosan (HCS) nanoparticles by ion-gel technology with the aim of applying to *Antheraea pernyi* silk tissue by a conventional process of dry cure, strengthening the wrinkle-resistant properties. In the anti-growth treatment of cotton fabric, Huang et al. (2008) studied the effect of using low molecular weight chitosan (LWCS).

5.6 Agriculture

Besides the classical application of chitosan, such as pharmaceutical and biomedical uses, it has been used in agriculture due to its antimicrobial activities against fungi, bacteria, and viruses and its action as an elicitor of plant defense mechanisms (Cota-Arriola et al. 2013).

In agriculture, the control of pathogens generally relies on fertilizers and pesticides of chemical nature (Xing et al. 2015). Nevertheless, excessive use of chemical fertilizers and pesticides harms biological diversity, natural and agricultural systems, and public health; besides, it leads to the development of resistant pests (Sun et al. 2012). Thus, natural-based pesticides and fertilizers are an environmentally friendly alternative that mitigate the negative impacts on human health and on the environment of synthetic pesticides (Xing et al. 2015). Beyond that, nano-based smart delivery systems have been employed to combat crop pathogens and to enhance the efficiency of pesticides at lower dosage rates (Parisi et al. 2015).

In this sense, chitosan nanoparticles have been used for active ingredient delivery due to its characteristics, such as biocompatibility, biodegradability, high permeability, cost-effectiveness, non-toxicity, and excellent film-forming ability (Kashyap et al. 2015; Shukla et al. 2013). Furthermore, chitosan nanoparticles can easily absorb to plant surface, prolonging the contact of the active ingredient and the target absorptive surface (Nagpal et al. 2010). In agricultural systems, the release of an active ingredient from chitosan nanoparticles is by diffusion and/or degradation, and it depends upon the particle's features (morphology, size, and density, for example) and active ingredient environment (pH, solvent polarity, and presence of enzymes, for example) (Martínez-Ruvalcaba et al. 2009). Thus, the literature reports a wide application of chitosan nanoparticles for agricultural purposes.

Chitosan nanoparticles have been used as a matrix for the release of pesticides for crop production (Itodo 2019). Amphiphilic carboxymethyl chitosan with ricinoleic acid (R-CM-chitosan) was synthesized, and it was used as a carrier for the pesticide azadirachtin (Aza) by Feng and Peng (2012). For their study, R-CM-chitosan efficiently restrained Aza degradation in a natural environment, as the loading efficiency reached up to 56%; besides, Aza/R-CM-chitosan provided a better controlled release of the drug (11 days) compared to the control groups. In addition, chitosan nanoparticles have been used for nutrients delivery in order to reduce the consumption of fertilizers and to minimize the environmental impact (Kashyap et al. 2015). Furthermore, the design of tailored fertilizers allows releasing formulation for specific food production systems (Muxika et al. 2017). In another approach, Oliveira et al. (2016) encapsulated nitric oxide donors in chitosan nanoparticles to mitigate the deleterious effects of salinity on the growth of maize plants (Oliveira et al. 2016). Results showed that encapsulated nitric oxide donors were more effective in preventing salt-induced effects than those non-encapsulated (Cipolatti et al. 2016).

In addition to the applications mentioned above, chitosan nanoparticles have been used in agriculture for herbicide delivery (dos Silva et al. 2011), determination of heavy metals in soils (Hamed et al. 2016), and gene delivery for plant modification (Saharan and Pal 2016). Therefore, chitosan nanoparticles have great potential in agriculture, once it may not only improve plant nutrition and, consequently, growth, but it might mitigate the environmental impact of pesticides, fertilizers, and herbicides.

5.7 *Paper*

Chitosan has been used in the paper industry due to its antimicrobial and mechanical properties. In fact, because paper contaminated by bacteria allows the spread of diseases, it has been given attention to the development of an antibacterial paper (Ma et al. 2010). Furthermore, once the molecular structure of chitosan is very similar to that of cellulose, it is possible to promote the formation of strong bonding, enhancing the strength of paper sheets (Muxika et al. 2017). In addition, this polymer could be used to confer strength to paper against moisture and packaging for food wrapping (Hamed et al. 2016). To enhance the properties of paper, chitosan is added in its furnish or chitosan, and its derivative coatings are used (Diab et al. 2015).

Ma et al. (2010) used chitosan nanoparticles for preparing antibacterial paper. Their results showed that the antibacterial activity of chitosan nanoparticles was improved with the increase in its concentration (Ma et al. 2010). Raghavendra et al. (2017) investigated the effect of chitosan silver nanoparticles on a traditional paper used for food packaging. As a result, the combination of chitosan and silver nanoparticles improved mechanical, oil resistance, air resistance, and antibacterial properties of the paper (Oliveira et al. 2016). Similarly, Habibie et al. (2016) studied the effect of chitosan on the physical and mechanical properties of paper. They observed that water absorption was decreased while the strength and smoothness of the paper were increased.

5.8 Enzyme Immobilization/Applications

Due to the excellent properties such as high selectivity and specificity presented by enzymes, it is considered remarkable biocatalyst (dos Santos et al. 2015a, b; Reis et al. 2019). These properties allow them to perform a variety of specific chemical reaction conditions (dos Santos et al. 2014; Suescun et al. 2015), which makes them even more attractive in various industrial applications, for example, in the food, textile, paper, and agricultural areas, among others (de Oliveira et al. 2018; Virgen-Ortíz et al. 2019). As it presents a protein nature, a large portion of the enzymes has a great instability under certain reaction conditions, as well as high production and purification costs (Fernandez-Lopez et al. 2016; Manoel et al. 2015b; Villalba et al. 2016). However, an excellent way to circumvent these problems is to employ immobilization techniques (Galvão et al. 2018; Rueda et al. 2016). Also, the advantages of immobilized enzyme are the possibility of reuse of the biocatalyst and (Bezerra et al. 2017; Manoel et al. 2016) ease of separation of the catalyst and reaction product (Pinheiro et al. 2018), thus reducing costs in the process (Rios et al. 2018).

In general terms, immobilization consists of confining the protein to an aqueous-insoluble solid support and organic solvents, and can be used alone or in combination with other protein stabilization techniques, considered one of the most efficient tools for altering specificity selectivity, activity, and stability of enzymes (de Oliveira et al. 2018; Lima et al. 2017; Rueda et al. 2015). Enzymes can be immobilized by different methods (Radt 2018; Sheldon and van Pelt 2013), such as encapsulation in polymeric membranes; confinement in polymeric matrices; adsorption on hydrophobic insoluble materials or ion exchange resins (Rodrigues et al. 2019); encapsulation; covalent bonding to an insoluble matrix (dos Santos et al. 2017; Pinheiro et al. 2019; Reis et al. 2019), or by crosslinking (Ortiz et al. 2019). The choice of immobilization protocols and the support for immobilization in order to improve enzyme activity will depend on the characteristics of the enzyme, and the type and conditions of the catalytic process (de Souza et al. 2016; Manoel et al. 2015a; Santos et al. 2015). Moreover, an inadequate enzyme immobilization may affect negatively enzyme performance (Melo et al. 2017).

Lipases, obtained from several sources, can be immobilized onto various kinds of supports (organic solvents and also in nonconventional media) (de Oliveira et al. 2018; dos Santos et al. 2015a, b). One of the most natural biopolymers used for enzyme immobilizations is the chitosan (Bonazza et al. 2018), suitable to some of its specific proprieties like low cost of acquiring, diverse forms, high protein affinity, nontoxicity, physiological inertness, and hydrophilicity, among others (de Oliveira et al. 2018; Neto et al. 2005). The possibilities of application of chitosan as support to enzymes' immobilization are extended mainly due to its interesting properties derived from the high area/volume ratio (Meryam Sardar 2015).

5.8.1 Nanoparticles of Chitosan for Enzyme Immobilization

Nanoparticles of chitosan are widely used as support for immobilization enzymes, such as lipases (Kuo et al. 2012; Liu et al. 2011; Monteiro et al. 2019), ®-galactosidases (Klein et al. 2012), laccase (Fang et al. 2009; Kalkan et al. 2012; Sadighi and Faramarzi 2013), proteases (Wang et al. 2014), lactase, and glucose oxidase (Fig. 6). This potential as support is due to chitosan being renewable, biodegradable, and water-insoluble (Aranaz et al. 2009; Krajewska 2004; Shen et al. 2016). In the chitosan, the structure has the presence of hydroxyl and amino groups which allow the activation of it with reagents such as glutaraldehyde, divinyl sulfone, genipin, and hyaluronic acid (de Oliveira et al. 2018; Lu et al. 2019; Ragab et al. 2019; Tamura et al. 2019). This activation enables the enzyme to be covalently immobilized (Ju et al. 2012). In this context, various modifications in the structure of chitosan nanoparticles were made at the beginning of the years, resulting in several "supports" enabling a wide variety of applications as in agriculture (Kashyap et al. 2015), water and waste treatment (Goh et al. 2019), food and beverages (Harkin et al. 2019), cosmetics and toiletries (Chiari-Andréo et al. 2019), and biopharmaceutics (Alizadeh et al. 2019). That modification in its chemical structure produces a material which retains less water and is more chemically resistant in acidic media (Berger et al. 2004).

Applications of enzymes on chitosan nanoparticles have been used in a large variety of food processes and have grown over the years. As an illustration of this variety, it can be used as support for ⟨-galactosidase immobilization (Liu et al. 2011) in food applications and ®-D-galactosidase for the synthesis of (GOS) (Nguyen et al.

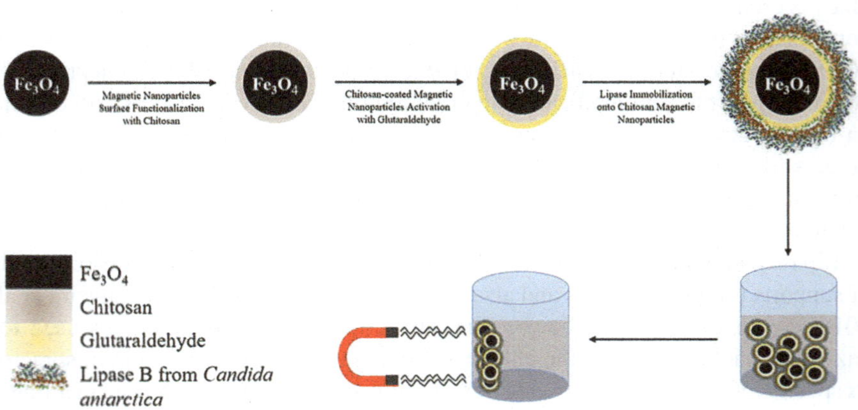

Fig. 6 Immobilization proposal of lipase B from *Candida antarctica* onto chitosan-coated magnetic nanoparticles activated with glutaraldehyde

2019; Pan et al. 2009). Another application widely used in the food industry is the enzyme ®-galactosidases immobilized on chitosan-coated magnetic nanoparticles for lactose hydrolysis (Chen and Duan 2015). The biocatalyst hydrolyzes the lactose present in whey and produces glucose and galactose. The hydrolyzed lactose solution is sweet syrup that has higher sweetening power than lactose and is used in confectionery, baking, soft, and dairy industries (Kosseva et al. 2009).

Some enzymes, such as pepsin an endopeptidase used to degrade proteins, can be used in the production of cheese (Altun and Cetinus 2007). Cheese has a wide variety, and some of them require longer ripening to develop the flavor, texture, and aroma characteristics of mature cheese (Agulló et al. 2003). The immobilization of enzyme surges as a way to solve problems during cheese making as loss of enzyme in whey, distribution, reduced yield, and reduced quality cheese (Anjani et al. 2007). Chitosan nanoparticles are one of the supports used for the production of this biocatalyst due to being simplest (Monteiro et al. 2019).

5.8.2 Biomedical Applications and Biosensors

Some diseases can be treated by using chitosan nanoparticles and specific enzymes, as the case of widespread skin allergies which can be treated using enzymes such as therapeutic agents (Leonida et al. 2019). Aiming to solve the impossibility of protein absorption into the skin caused by the stratum corneum, that allows the absorption of only small lipophilic molecules, and also to increase the bioavailability, catalase and histaminase were encapsulated into nanochitosan (Leonida et al. 2019). Benefic results of stability and activity maintenance and release kinetics were obtained, suggesting the posterior use of these produced nanobiocomposites in creams and gels formulation. In this context, nanochitosan as film form was reported for potential therapeutic use in cutaneous leishmaniasis, which induces the scar tissue generation and, in combination with glucantime, decreases the lesion extent and parasite loads (Bahrami et al. 2015). The wound healing dressing can also be obtained by the formation of novel materials from the mixture of chitosan and another substance with different properties, aiming the generation of a material with enhanced characteristics. For example, the mixture of chitosan, polyvinyl pyrrolidone (PVP), and silver oxide nanoparticles produced a film with good antibacterial activity and more wound healing benefits than cotton gauge and even 100% chitosan dressings (Archana et al. 2015). Besides that, the addition of specific enzymes into the wound dressing can improve the treatment results, as some proteolytic enzymes are, which can be used to digest the necrotic tissue of patients with skin wounds (Dabiri et al. 2016).

Another application for chitosan in biomedical applications is in the production of biosensors. Enzyme immobilization on a solid matrix is necessary to retain the enzyme's specific biological function (Wang et al. 2002). In this way, immobilization of urease in chitosan Fe_3O_4 nanoparticles (Ali et al. 2013; Kaushik et al. 2009) can be used for removal of urea from blood in artificial kidneys, blood detoxification, or dialysate regeneration system of artificial kidneys (Chellapandian and Krishnan 1998). Urease catalyzes the conversion of urea to carbon dioxide and ammonia and the determination of urea in blood serum by the development of urea biosensors. Consequently, the development of urea biosensors receives much interest since the urea is monitored in blood as an indicator of renal function (Magalhães 1998).

Chitosan is also used as a glucose biosensor because it can promote electron transfer between Glucose oxidase and electrode. Several authors have been used chitosan nanoparticles as a matrix for glucose oxidase immobilization (Anusha et al. 2015; Chaichi and Ehsani 2016). Glucose oxidase has a great catalytic ability and because of this feature has been extensively used to monitor blood glucose levels in diabetics (Liu et al. 2005).

As already said, chitosan and its derived nanoparticles have many applications in the biomedical area such as drug delivery, antibacterial agent, and wound healing. Similarly, chitosan nanoparticles' use can be applied as an alternative to sustainable agriculture (Choudhary et al. 2019) as a carrier for controlled delivery of genetic materials for plant transformation and also of pesticides, micronutrients, and fertilizers (Fu et al. 2019a, b).

5.8.3 Environmental Applications

Biosensors are not only used for biomedical applications; they can be used to remove phenols from water. Biosensors are not only used for biomedical applications; they can be used to remove phenols from water. They might be used as a tool for monitoring phenolic compounds in wastewater from industrial processes such as petroleum refining, coal conversion, resins, chemical, textiles, and plastics (Wang et al. 2002). Phenols are taken up through the skin, and part of it is detoxified and secreted in the urine. Laccase is an oxidoreductase that can oxidize phenols present in water, and essays have been done to immobilize laccase in support for an easier recovery after the process. One of the most prominent supports for this situation is magnetic chitosan nanoparticles (Monteiro et al. 2019). There are in the scientific literature a high number of studies reporting the use of chitosan nanoparticles in water treatment, as removal of metal ions, eliminate synthetic polymers, reduce odors, and flocculant to clarify water (Aydemir and Güler 2015; Koyani and Vazquez-Duhalt 2016).

Chitosan nanoparticles also have been used for other oxidoreductase and hydrolase immobilizations aiming to apply in bioremediation processes. The oxidoreductases immobilized in chitosan nanoparticles can eliminate endocrine disruptors and pharmaceuticals from water and decolorate diverse industrial dyes (Alarcón-Payán et al. 2017).

Among the vast varieties of application of chitosan nanoparticles, one is the immobilization of xylanase to convert hemicelluloses present as residual carbohydrates in the cellulosic pulp industry (Liu et al. 2014). This biocatalyst can hydrolyze the main compound of hemicelluloses, the xylan, by broking β-1,4 glycosidic linkages to produce short-chain xylooligosaccharides (Manrich et al. 2010).

5.8.4 Drug Delivery Applications

Another promising application of chitosan nanoparticle is its potential in drug delivery systems (DDSs). DDSs are engineered technologies created to replace the conventional application of drugs that are characterized by limited effectiveness, poor biodistribution, and a lack of selectivity (Unsoy et al. 2014). In the case of DDSs, the therapeutic agents are delivered in a targeted and control manner protecting the vital tissues, minimizing the undesirable side effects, and protecting the drug from rapid degradation (Kermanizadeh et al. 2018). For this reason, the use of suitable materials is essential once they can reduce the unwanted toxic side effects and improve the therapeutic effect (Shagholani et al. 2015).

Chitosan was used to coat magnetic nanoparticles and used for lysosomal enzymatic degradation (Rastegari et al. 2017). Another work already used magnetic chitosan nanoparticles for delivery of doxorubicin that is a strong cytotoxic compound to the normal tissues used for cancer therapy (Unsoy et al. 2014). Some studies had developed a new method for drug delivery using electrospinning chitosan nanofibers to deliver drugs for cancer therapy. Electrospinning is a method to fabricate ultrafine fibers with diameters between nanometers and micrometers that have a high surface area, high porosity, and very small pore size (Sedghi et al. 2017).

5.8.5 Other Applications

Chitosan nanoparticles can also be a visual indicator of food quality, indicating the degree of ripening once this depending on pH and CO_2 present on the packaging (Suh et al. 2016). The variation in pH and CO_2 makes changes in the turbidity of chitosan suspensions. When the suspension of chitosan is opaque, that indicates the low quantity of CO_2 concentrations (early ripening); however, if the suspension becomes transparent, that means high CO_2 levels, once chitosan is soluble in acid (Lee et al. 2015; Suh et al. 2016).

Due to the antimicrobial activity, polycationic and chelation properties of the chitosan, useful for heavy metal trapping, studies involving nanochitosan as potential food additive have been performed. The results presented no signs of toxicity or inflammation by chitosan in the analyzed organisms, and that material can also be used for delivery not only to proteins but also to other nutrients (Darwesh et al. 2018; Norton et al. 2014; Souza et al. 2014). Still, in food applications, nanochitosan can be used in the production of edible coatings used in the storage of fresh fruits or meats, which allied to the addition of specific enzymes—the result may be even more profitable. For example, nanochitosan with and without copper addition showed required properties for shelf-life extension, such as weight loss control, firmness maintenance, and polyphenol oxidase/peroxidase activity suppression (Eshghi and Hashemi 2014). That result may be improved with the addition of enzymes with antimicrobial activity such as lactoperoxidase, which was studied in the packaging of fish and ameliorated the trout fillets preservation, by decreasing the number of bacteria and other undesirable micro-organisms (Jasour et al. 2014).

6 Future Trends

In the last decade, nanotechnologies have gained attention and revolutionized mainly the application of medicine by presenting drug-carrying properties, forming a slow and continuous release system (Zottel et al. 2019) (Fig. 7). Due to its great importance and versatility of application, the use of chitin-based nanoparticles has been widely studied in several areas in search of improvement and development of new applications (Khattak et al. 2019).

Several recent researches available in the literature (Islam et al. 2018; Liang et al. 2017; Ye et al. 2018) in the pharmaceutical field reported higher absorption and bioavailability of tea polyphenols by chitosan nanoparticles. Tea polyphenols find a variety of applications in the food, pharmaceutical, and medical industries, with health benefits such as antioxidant and antibacterial activities, used for cancer prevention, anti-radiation, and immune booster (Khan and Mukhtar 2019). However, the low absorption rate of tea polyphenols precludes their bioactivity in vivo (Ye et al. 2018). Future trends are expected in the development of tea polyphenol-encapsulated chitosan nanoparticles suitable for oral administration to increase the absorption of bioactive compounds (Zeng et al. 2019)

In the biomedical field, significant efforts are made to develop innovative dressings, sequential drug release, indicated skin regeneration treatments, and wound repair (Jadhav et al. 2016), due to their high antibacterial power. For having multifunctional properties of drug-laden nanoparticles, they are important and promising

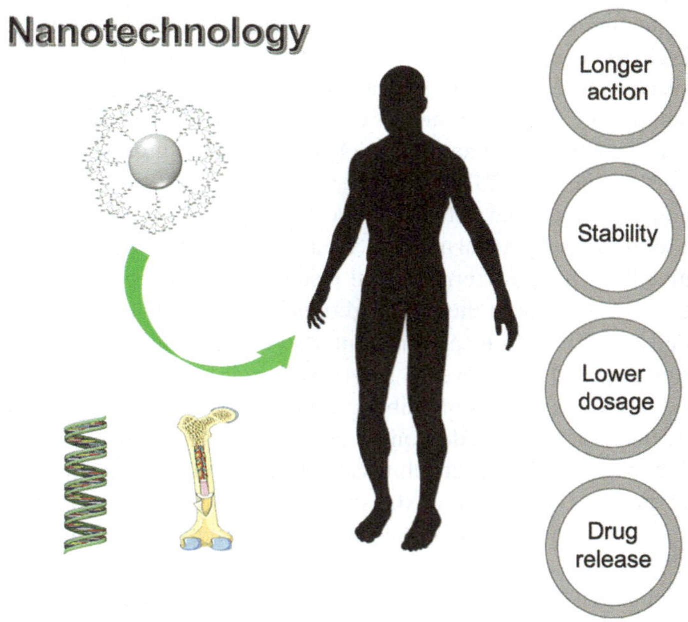

Fig. 7 Nanotechnologies application of medicine area

alternatives for wound healing. On the other hand, in agriculture, there is a great need for in-depth knowledge regarding the potential application of chitosan for the encapsulation of active ingredients in agriculture (Choudhary et al. 2019).

Chitosan-based nanomaterials present promising fertilizer carriers in various crops to improve nutrients, reduce environmental pollution, minimize nutrient losses in fertilization, and increase yield through pest and nutrient management. It is noteworthy that the effectiveness of these nanomaterials is better than current pesticides and synthetic fertilizers and that in terms of low cost, concentration, and eco-safety (Choudhary et al. 2019; Shang et al. 2019). Due to its potentiality, chitosan nanomaterials may become a product widely used in the agricultural area in the near future, in order to become an alternative for sustainable agriculture.

7 Conclusions

Chitosan, the second most abundant biopolymer after cellulose, is obtained by chitin deacetylation and has been widely used in several areas, since they have distinct properties such as biodegradability, biocompatibility, non-toxicity, and renewable characteristics that make them environmentally sustainable. The development of nanotechnology and nanomaterials has allowed new formulations of chitosan, like nanoparticles, making their applicability significantly higher, as it is associated with specific properties of chitosan with the characteristics presented by nanomaterials, for example, small dimensions, variety shapes, improved specific surface, and high adsorption capacity. Thus, they began to exhibit other important properties such as antioxidant, antibacterial, analgesic, fungistatic, and hemostatic. For those reasons, nanomaterials based on chitosan become of great commercial interest in biomedicine, pharmaceuticals, cosmetics, personal care, agriculture, and some other areas.

In agriculture, chitosan has shown the potential to be applied in agriculture for its antimicrobial activity on a wide variety of phytopathogens, favoring increased crop production. However, there is still a knowledge gap, which requires further research to strengthen the ability to use chitosan in agriculture as an alternative to replace or complement current methods.

It can be concluded that the use of marine polysaccharides, such as chitosan, is a potential instrument for innovative technologies that improve quality and environmental sustainability.

Acknowledgements We gratefully acknowledge the financial support of Brazilian Agencies for Scientific and Technological Development, Fundação Cearense de Apoio ao Desenvolvimento Científico e Tecnológico (FUNCAP), project number BP3-0139-00005.01.00/18, Conselho Nacional de Desenvolvimento Científico e Tecnológico (CNPq), project number 422942/2016-2, Coordenação de Aperfeiçoamento de Ensino Superior (CAPES).

References

Abdelgawad A, Hudson S (2019) Chitosan nanoparticles: polyphosphates cross-linking and protein delivery properties. Int J Biol Macromol 136:133–142

Abdelgawad AM, Hudson SM, Rojas OJ (2014) Antimicrobial wound dressing nanofiber mats from multicomponent (chitosan/silver-NPs/polyvinyl alcohol) systems. Carbohydr Polym 100:166–178

AbdElhady MM (2012) Preparation and characterization of chitosan/zinc oxide nanoparticles for imparting antimicrobial and UV protection to cotton fabric. Int J Carbohydr Chem 2012:1–6

Abo Elsoud MM, El Kady EM (2019) Current trends in fungal biosynthesis of chitin and chitosan. Bull Natl Res, Cent, p 43

Agbai ON, Buster K, Sanchez M, Hernandez C, Kundu RV, Chiu M, Roberts WE, Draelos ZD, Bhushan R, Taylor SC, Lim HW (2014) Skin cancer and photoprotection in people of color: a review and recommendations for physicians and the public. J Am Acad Dermatol 70:748–762

Agnihotri SA, Mallikarjuna NN, Aminabhavi TM (2004) Recent advances on chitosan-based micro- and nanoparticles in drug delivery. J Control Release

Agulló E, Rodríguez MS, Ramos V, Albertengo L (2003) Present and future role of chitin and chitosan in food. Macromol Biosci 3:521–530

Ahmad M, Manzoor K, Ikram S (2019) Chitosan nanocomposites for bone and cartilage regeneration. In: Applications of nanocomposite materials in dentistry. Elsevier, pp 307–317

Aider M (2010) Chitosan application for active bio-based films production and potential in the food industry: review. LWT—Food Sci Technol 43:837–842

Alarcón-Payán DA, Koyani RD, Vazquez-Duhalt R (2017) Chitosan-based biocatalytic nanoparticles for pollutant removal from wastewater. Enzyme Microb Technol 100:71–78

Ali SW, Rajendran S, Joshi M (2011) Synthesis and characterization of chitosan and silver loaded chitosan nanoparticles for bioactive polyester. Carbohydr Polym 83:438–446

Ali A, Alsalhi MS, Atif M, Ansari AA, Israr MQ, Sadaf JR, Ahmed E, Nur O, Willander M (2013) Potentiometric urea biosensor utilizing nanobiocomposite of chitosan-iron oxide magnetic nanoparticles. In: Journal of physics: conference series. p 012024

Alizadeh L, Zarebkohan A, Salehi R, Ajjoolabady A, Rahmati-Yamchi M (2019) Chitosan-based nanotherapeutics for ovarian cancer treatment. J Drug Target 27:839–852

Al-qadi S, Grenha A, Carrión-recio D, Seijo B, Remuñán-lópez C (2012) Microencapsulated chitosan nanoparticles for pulmonary protein delivery: in vivo evaluation of insulin-loaded formulations. J Control Release 157:383–390

Altun GD, Cetinus SA (2007) Immobilization of pepsin on chitosan beads. Food Chem 100:964–971

Anand M, Sathyapriya P, Maruthupandy M, Hameedha Beevi A (2018) Synthesis of chitosan nanoparticles by TPP and their potential mosquito larvicidal application. Front Lab Med 2:72–78

Anitha A, Rani VVD, Krishna R, Sreeja V, Selvamurugan N, Nair SV, Tamura H, Jayakumar R (2009) Synthesis, characterization, cytotoxicity and antibacterial studies of chitosan, O-carboxymethyl and N, O-carboxymethyl chitosan nanoparticles. Carbohydr Polym 78:672–677

Anjani K, Kailasapathy K, Phillips M (2007) Microencapsulation of enzymes for potential application in acceleration of cheese ripening. Int Dairy J 17:79–86

Annur D, Wang ZK, Der Liao J, Kuo C (2015) Plasma-synthesized silver nanoparticles on electrospun chitosan nanofiber surfaces for antibacterial applications. Biomacromol 16:3248–3255

Anusha JR, Raj CJ, Cho B-B, Fleming AT, Yu K-H, Kim BC (2015) Amperometric glucose biosensor based on glucose oxidase immobilized over chitosan nanoparticles from gladius of Uroteuthis duvauceli. Sens Actuators B Chem 215:536–543

Aranaz I, Mengíbar M, Harris R, Paños I, Miralles B, Acosta N, Galed G, Heras Á (2009) Functional characterization of chitin and chitosan. Curr Chem Biol 3:203–230

Archana D, Singh BK, Dutta J, Dutta PK (2015) Chitosan-PVP-nano silver oxide wound dressing: in vitro and in vivo evaluation. Int J Biol Macromol 73:49–57

Argüelles-Monal W, Lizardi-Mendoza J, Fernández-Quiroz D, Recillas-Mota M, Montiel-Herrera M (2018) Chitosan derivatives: introducing new functionalities with a controlled molecular architecture for innovative materials. Polymers (Basel) 10:342

Asiri SM, Khan FA, Bozkurt A (2018) Synthesis of chitosan nanoparticles, chitosan-bulk, chitosan nanoparticles conjugated with glutaraldehyde with strong anti-cancer proliferative capabilities. Artif Cells Nanomedicine Biotechnol 46:S1152–S1161

Atta-ur-Rahman FRS(2018) Advances in organic synthesis. Bentham Science, Sharjah

Aydemir T, Güler S (2015) Characterization and immobilization of Trametes versicolor laccase on magnetic chitosan–clay composite beads for phenol removal. Artif Cells Nanomedicine Biotechnol 43:425–432

Baghdan E, Pinnapireddy SR, Strehlow B, Engelhardt KH, Schäfer J, Bakowsky U (2018) Lipid coated chitosan-DNA nanoparticles for enhanced gene delivery. Int J Pharm 535:473–479

Bahrami S, Esmaeilzadeh S, Zarei M, Ahmadi F (2015) Potential application of nanochitosan film as a therapeutic agent against cutaneous leishmaniasis caused by L. major. Parasitol Res 114:4617–4624

Bai Y, Shen B, Zhang S, Zhu Z, Sun S, Jun G, Li B, Wang Y, Zhang R, We F (2019) Storage of mechanical energy based on carbon nanotubes with high energy density and power density. Adv Mater 31:1800680

Baniasadi H, Ramazani SA, Mashayekhan S (2015) Fabrication and characterization of conductive chitosan/gelatin-based scaffolds for nerve tissue engineering. Int J Biol Macromol 74:360–366

Bashari A, Rouhani Shirvan A, Shakeri M (2018) Cellulose-based hydrogels for personal care products. Polym Adv Technol 29:2853–2867

Batista CPM, Caetano AA, Rossi MA, Gonçalves MA (2019) Chitosan-iron oxide hybrid composite: mechanism of hexavalent chromium removal by central composite design and theoretical calculations. Environ Sci Pollut Res 26:15973–15988

Berger J, Reist M, Mayer JM, Felt O, Peppas NA, Gurny R (2004) Structure and interactions in covalently and ionically crosslinked chitosan hydrogels for biomedical applications. Eur J Pharm Biopharm 57:19–34

Bezerra RM, Neto DMA, Galvão WS, Rios NS, de Carvalho ACLM, Correa MA, Bohn F, Fernandez-Lafuente R, Fechine PBA, de Mattos MC, dos Santos JCS, Gonçalves LRB (2017) Design of a lipase-nano particle biocatalysts and its use in the kinetic resolution of medicament precursors. Biochem Eng J 125:104–115

Bonazza HL, Manzo RM, dos Santos JCS, Mammarella EJ (2018) Operational and thermal stability analysis of thermomyces lanuginosus lipase covalently immobilized onto modified chitosan supports. Appl Biochem Biotechnol 184:182–196

Brar V, Kaur G (2018) Preparation and characterization of polyelectrolyte complexes of hibiscus esculentus (Okra) gum and chitosan. Int J Biomater 7

Brigger I, Dubernet C, Couvreur P (2012) Nanoparticles in cancer therapy and diagnosis. Adv Drug Deliv Rev 64:24–36

Cadet J, Douki T, Ravanat J-L (2015) Oxidatively generated damage to cellular DNA by UVB and UVA radiation. Photochem Photobiol 91:140–155

Călinoiu L-F, Ştefănescu B, Pop I, Muntean L, Vodnar D (2019) Chitosan coating applications in probiotic microencapsulation. Coatings 9:194

Callewaert C, De Maeseneire E, Kerckhof FM, Verliefde A, Van de Wiele T, Boon N (2014) Microbial odor profile of polyester and cotton clothes after a fitness session. Appl Environ Microbiol 80:6611–6619

Calvo P, Alonso MJ, Vila A, Sanchez A, Tobıo M (2002) Design of biodegradable particles for protein delivery. J Control Release 78:15–24

Casadidio C, Peregrina DV, Gigliobianco MR, Deng S, Censi R, Di Martino P (2019) Chitin and chitosans: characteristics, eco-friendly processes, and applications in cosmetic science. Mar Drugs 17:369

Cazón P, Velazquez G, Ramírez JA, Vázquez M (2017) Polysaccharide-based films and coatings for food packaging: a review. Food Hydrocoll 68:136–148

Chaichi MJ, Ehsani M (2016) A novel glucose sensor based on immobilization of glucose oxidase on the chitosan-coated Fe_3O_4 nanoparticles and the luminol–H_2O_2–gold nanoparticle chemiluminescence detection system. Sens Actuators B Chem 223:713–722
Chandra Hembram K, Prabha S, Chandra R, Ahmed B, Nimesh S (2016) Advances in preparation and characterization of chitosan nanoparticles for therapeutics. Artif Cells Nanomedicine Biotechnol 44:305–314
Chandrasekar S, Vijayakumar S, Rajendran R (2014) Application of chitosan and herbal nanocomposites to develop antibacterial medical textile. Biomed Aging Pathol 4:59–64
Chellapandian M, Krishnan MRV (1998) Chitosan-poly (glycidyl methacrylate) copolymer for immobilization of urease. Process Biochem 33:595–600
Chen S-C, Duan K-J (2015) Production of galactooligosaccharides using β-galactosidase immobilized on chitosan-coated magnetic nanoparticles with tris(hydroxymethyl)phosphine as an optional coupling agent. Int J Mol Sci 16:12499–12512
Chen Q, Thouas GA (2015) Metallic implant biomaterials. Mater Sci Eng R Reports 87:1–57
Chen L, Zhu J, Li Y, Lu J, Gao L, Xu H, Fan M, Yang X (2013) Enhanced nasal mucosal delivery and immunogenicity of anti-caries DNA vaccine through incorporation of anionic liposomes in chitosan/DNA complexes. PLoS ONE 8:e71953
Chiari-Andréo BG, de Almeida-Cincotto MGJ, Oshiro JA, Taniguchi CYY, Chiavacci LA, Isaac VLB (2019) Nanoparticles for cosmetic use and its application. In: Nanoparticles in pharmacotherapy. Elsevier, pp 113–146
Choudhary RC, Kumari S, Amy RK, Sharma G, Kumar A, Budhwar S, Pal A, Raliya R, Pratim B, Saharan V (2019) Chitosan nanomaterials for smart delivery of bioactive compounds in agriculture. In: Raliya R (ed) Nanoscale engineering in agricultural management. CRC Press, Boca Raton, p 214
Chuan D, Jin T, Fan R, Zhou L, Guo G (2019) Chitosan for gene delivery: methods for improvement and applications. Adv Colloid Interface Sci 268:25–38
Cipolatti EP, Valério A, Henriques RO, Moritz DE, Ninow JL, Freire DMG, Manoel EA, Fernandez-Lafuente R, de Oliveira D (2016) Nanomaterials for biocatalyst immobilization-state of the art and future trends. RSC Adv 6:104675–104692
Corsi K, Chellat F, Yahia L, Fernandes JC (2003) Mesenchymal stem cells, MG63 and HEK293 transfection using chitosan-DNA nanoparticles. Biomaterials 24:1255–1264
Cota-Arriola O, Cortez-Rocha MO, Rosas-Burgos EC, Burgos-Hernández A, López-Franco YL, Plascencia-Jatomea M (2013) Antifungal effect of chitosan on the growth of *Aspergillus parasiticus* and production of aflatoxin B1. Polym Int 60:937–944
Crini NM, Lichtfouse E, Torri G, Crini G (2019) Fundamentals and applications of chitosan. In: Crini G, Lichtfouse E (eds) Sustainable agriculture reviews. Springer, Cham
Dabiri G, Damstetter E, Phillips T (2016) Choosing a wound dressing based on common wound characteristics. Adv Wound Care 5:32–41
Darwesh OM, Sultan YY, Seif MM, Marrez DA (2018) Bio-evaluation of crustacean and fungal nano-chitosan for applying as food ingredient. Toxicol Reports 5:348–356
de Masi A, Tonazzini I (2019) Chitosan films for regenerative medicine: fabrication methods and mechanical characterization of nanostructured chitosan films 1–9
de Oliveira UMF, Lima de Matos LJB, de Souza MCM, Pinheiro BB, dos Santos JCS, Gonçalves LRB (2018) Effect of the presence of surfactants and immobilization conditions on catalysts' properties of *Rhizomucor miehei* lipase onto chitosan. Appl Biochem Biotechnol 184:1263–1285
de Souza TC, de Fonseca TS, da Costa JA, Rocha MVP, de Mattos MC, Fernandez-Lafuente R, Gonçalves LRB, dos Santos JCS (2016) Cashew apple bagasse as a support for the immobilization of lipase B from Candida antarctica: application to the chemoenzymatic production of (R)-Indanol. J Mol Catal B Enzym 130:58–69
Demirci C, Marras S, Prato M, Pasquale L, Manna L, Colombo M (2019) Design of catalytically active porous gold structures from a bottom-up method: the role of metal traces in CO oxidation and oxidative coupling of methanol. J Catal 375:279–286

Devlieghere F, Vermeulen A, Debevere J (2004) Chitosan: antimicrobial activity, interactions with food components and applicability as a coating on fruit and vegetables. Food Microbiol 21:703–714
Dhandayuthapani B, Yoshida Y, Maekawa T, Kumar DS (2011) Polymeric scaffolds in tissue engineering application: a review. Int J Polym Sci 2011:1–19
Diab M, Curtil D, El-shinnawy N, Hassan ML, Zeid IF, Mauret E (2015) Biobased polymers and cationic microfibrillated cellulose as retention and drainage aids in papermaking: comparison between softwood and bagasse pulps. Ind Crops Prod 72:34–45
Dincer I (2018) Comprehensive energy systems (1st ed). Elsevier, Amsterdam
Divya K, Jisha MS (2018) Chitosan nanoparticles preparation and applications. Environ Chem Lett 16:101–112
Doan CT, Tran TN, Nguyen VB, Vo TPKV, Nguyen AD, Wang S (2019) Chitin extraction from shrimp waste by liquid fermentation using an alkaline protease-producing strain, *Brevibacillus parabrevis*. Int J Biol Macromol 131:706–715
Dongre RS (2018) Chitosan-derived synthetic ion exchangers: characteristics and applications
Dongre RS (2019) Chitosan formulations: chemistry, characteristics and contextual adsorption in unambiguous modernization of S&T. Hystersis Compos
dos Santos Rodrigues B, Lakkadwala S, Sharma D, Singh J (2019) Chitosan for gene, DNA vaccines, and drug delivery. In: Materials for biomedical engineering. Elsevier, Amsterdam, pp 515–550
dos Santos JCS, Garcia-Galan C, Rodrigues RC, de Sant' Ana HB, Gonçalves LRB, Fernandez-Lafuente R (2014) Improving the catalytic properties of immobilized Lecitase via physical coating with ionic polymers. Enzyme Microb Technol 60:1–8
dos Santos JCS, Rueda N, Gonçalves LRB, Fernandez-Lafuente R (2015a) Tuning the catalytic properties of lipases immobilized on divinylsulfone activated agarose by altering its nanoenvironment. Enzyme Microb Technol 77:1–7
dos Santos JCS, Rueda N, Sanchez A, Villalonga R, Gonçalves LRB, Fernandez-Lafuente R (2015b) Versatility of divinylsulfone supports permits the tuning of CALB properties during its immobilization. RSC Adv 5:35801–35810
dos Santos JCS, Bonazza HL, de Matos LJBL, Carneiro EA, Barbosa O, Fernandez-Lafuente R, Gonçalves LRB, de Sant' Ana HB, Santiago-Aguiar RS (2017) Immobilization of CALB on activated chitosan: application to enzymatic synthesis in supercritical and near-critical carbon dioxide. Biotechnol Reports 14:16–26
dos Silva MS, Cocenza DS, Grillo R, de Melo NFS, Tonello PS, de Oliveira LC, Cassimiro DL, Rosa AH, Fraceto LF (2011) Paraquat-loaded alginate/chitosan nanoparticles: preparation, characterization and soil sorption studies. J Hazard Mater 190:366–374
Dowling M (2019) Hydrophobically—modified chitosan for use in cosmetics and personal care applications
Duhan JS, Kumar R, Kumar N, Kaur P, Nehra K, Duhan S (2017) Nanotechnology: the new perspective in precision agriculture. Biotechnol Reports 15:11–23
Durán N, Marcato PD, de Souza GIH, Alves OL, Esposito E (2007) Antibacterial effect of silver nanoparticles produced by fungal process on textile fabrics and their effluent treatment. J Biomed Nanotechnol 3:203–208
El Knidri H, Belaabed R, Addaou A, Laajeb A, Lahsini A (2018) Extraction, chemical modification and characterization of chitin and chitosan. Int J Biol Macromol 120:1181–1189
El-Hefian EA, Nasef MM, Yahaya AH (2014) Chitosan-based polymer blends: current status and applications. Chem Soc Pak 36:11–27
Esfandiarpour-Boroujeni S, Bagheri-Khoulenjani S, Mirzadeh H, Amanpour S (2017) Fabrication and study of curcumin loaded nanoparticles based on folate-chitosan for breast cancer therapy application. Carbohydr Polym 168:14–21
Eshghi S, Hashemi M (2014) Effect of nanochitosan-based coating with and without copper loaded on physicochemical and bioactive components of fresh strawberry fruit (*Fragaria x ananassa Duchesne*) during storage. Food Bioprocess Techonoly 7:2397–2409

Fang H, Huang J, Ding L, Li M, Chen Z (2009) Preparation of magnetic chitosan nanoparticles and immobilization of laccase. J Wuhan Univ Technol Sci Ed 24:42–47

Farias CBB, Silva AF, Rufino RD, Luna JM, Gomes Souza JE, Sarubbo LA (2014) Synthesis of silver nanoparticles using a biosurfactant produced in low-costmediumas stabilizing agent. Electron J Biotechnol 17:122–125

Fernandez-Lopez L, Rueda N, Bartolome-Cabrero R, Rodriguez MD, Albuquerque TL, dos Santos JCS, Barbosa O, Fernandez-Lafuente R (2016) Improved immobilization and stabilization of lipase from *Rhizomucor miehei* on octyl-glyoxyl agarose beads by using $CaCl_2$. Process Biochem 51:48–52

Fernández-Urrusuno R, Calvo P, Remuñán-López C, Vila-Jato JL, Alonso MJ (1999) Enhancement of nasal absorption of insulin using chitosan nanoparticles. Pharm Res 16:1576–1581

Fisher MB, Mauck RL, Qu F, Lin J-MG, Esterhai JL (2013) Biomaterial-mediated delivery of degradative enzymes to improve meniscus integration and repair. Acta Biomater 9:6393–6402

Fu L, Wang Z, Dhankher OP, Xing B (2019a) Nanotechnology as a new sustainable approach for controlling crop diseases and increasing agricultural production. J Exp Bot

Fu S, Sun Z, Huang P, Li Y, Hu N (2019b) Some basic aspects of polymer nanocomposites: a critical review. Nano Mater Sci 1:2–30

Galvão WS, Pinheiro BB, Golçalves LRB, de Mattos MC, Fonseca TS, Regis T, Zampieri D, dos Santos JCS, Costa LS, Correa MA, Bohn F, Fechine PBA (2018) Novel nanohybrid biocatalyst: application in the kinetic resolution of secondary alcohols. J Mater Sci 53:14121–14137

Gan Q, Wang T (2007) Chitosan nanoparticle as protein delivery carrier-systematic examination of fabrication conditions for efficient loading and release. Colloids Surf B 59:24–34

Garg U, Chauhan S, Nagaich U, Jain N (2019) Current advances in chitosan nanoparticles based drug delivery and targeting. Adv Pharm Bull 19:195–204

Goh PS, Ong CS, Ng BC, Ismail AF (2019) Applications of emerging nanomaterials for oily wastewater treatment. Nanotechnol Water Wastewater Treat 101–113

Gopakumar DA, Pai AR, Pasquini D, Ben LS-Y, Khalil AHPS, Thomas S (2019) Nanomaterials—state of art, new challenges, and opportunities. In: Nanoscale materials in water purification. Elsevier, Amsterdam, pp 1–24

Grenha A (2012) Chitosan nanoparticles: a survey of preparation methods. J Drug Target

Habibie S, Hamzah M, Anggaravidya M, Kalembang E (2016) The effect of chitosan on physical and mechanical properties of paper. J Chem Eng Mater Sci 7:1–10

Hamed I, Özogul F, Regenstein JM (2016) Industrial applications of crustacean by-products (chitin, chitosan, and chitooligosaccharides): a review. Trends Food Sci Technol 48:40–50

Harkin C, Mehlmer N, Woortman DV, Brück TB, Brück WM (2019) Nutritional and additive uses of chitin and chitosan in the food industry. pp 1–43

Harris R, Lecumberri E, Mateos-Aparicio I, Mengíbar M, Heras A (2011) Chitosan nanoparticles and microspheres for the encapsulation of natural antioxidants extracted from Ilex paraguariensis. Carbohydr Polym 84:803–806

Hashem AM, Gamal AA, Hassan ME, Hassanein NM, Esawy MA (2016) Covalent immobilization of *Enterococcus faecalis* Esawy dextransucrase and dextran synthesis. Int J Biol Macromol 82:905–912

Hasnain MS, Nayak AK (eds) (2019) Natural polysaccharides in drug delivery and biomedical application. Academic Press, Cambridge

Hebeish A, Sharaf S, Farouk A (2013) Utilization of chitosan nanoparticles as a green finish in multifunctionalization of cotton textile. Int J Biol Macromol 60:10–17

Henderson MA (2011) A surface science perspective on TiO_2 photocatalysis. Surf Sci Rep 66:185–297

Huang K-S, Wu W-J, Chen J-B, Lian H-S (2008) Application of low-molecular-weight chitosan in durable press finishing. Carbohydr Polym 73:254–260

Huang J, Yin Y, Pan Z, Zhang G, Zhu A, Liu X, Jiao X (2010) Intranasal immunization with chitosan/pCAGGS-flaA nanoparticles inhibits *Campylobacter jejuni* in a White Leghorn Model

Ibrahim MM, Abd-Elgawad A-EH, Soliman OA-E, Jabłonski MM (2015) Natural bioadhesive biodegradable nanoparticle-based topical ophthalmic formulations for management of glaucoma. Transl Vis Sci Technol 4:12
Islam S, Butola BS, Roy A (2018) Chitosan polysaccharide as a renewable functional agent to develop antibacterial, antioxidant activity and colourful shades on wool dyed with tea extract polyphenols. Int J Biol Macromol 120:1999–2006
Islam N, Dmour I, Taha MO (2019) Degradability of chitosan micro/nanoparticles for pulmonary drug delivery. Heliyon 5:e01684
Itodo HU (2019) Controlled release of herbicides using nano-formulation: a review. J Chem Rev 1:130–138
Jadhav K, Dhamecha D, Bhattacharya D, Patil M (2016) Green and ecofriendly synthesis of silver nanoparticles: characterization, biocompatibility studies and gel formulation for treatment of infections in burns. Photochem Photobiol B Biol 155:109–115
Janes KA, Calvo P, Alonso MJ (2001) Polysaccharide colloidal particles as delivery systems for macromolecules. Adv Drug Deliv Rev 47:83–97
Jasour MS, Ehsani A, Samaneh S (2014) Chitosan coating incorporated with the lactoperoxidase system: an active edible coating for fish preservation. J Sci Food Agric 95:1373–1378
Jayakumar R, Prabaharan M, Sudheesh Kumar PT, Nair SV, Tamura H (2011) Biomaterials based on chitin and chitosan in wound dressing applications. Biotechnol Adv 29:322–337
Jiang D, Ni D, Rosenkrans ZT, Huang P, Yan X, Cai W (2019) Nanozyme: new horizons for responsive biomedical applications. Chem Soc Rev 48:3683–3704
Jimtaisong A, Saewan N (2014) Utilization of carboxymethyl chitosan in cosmetics. Int J Cosmet Sci 36:12–21
Joshi H, Somdutt, Choudhary P, Mundra S (2019) Future prospects of nanotechnology in agriculture. Int J Chem Stud 7:957–963
Ju H-Y, Kuo C-H, Too J-R, Huang H-Y, Twu Y-K, Chang C-MJ, Liu Y-C, Shieh C-J (2012) Optimal covalent immobilization of α-chymotrypsin on Fe_3O_4-chitosan nanoparticles. J Mol Catal B Enzym 78:9–15
Kalkan NA, Aksoy S, Aksoy EA, Hasirci N (2012) Preparation of chitosan-coated magnetite nanoparticles and application for immobilization of laccase. J Appl Polym Sci 123:707–716
Kashyap PL, Xiang X, Heiden P (2015) Chitosan nanoparticle based delivery systems for sustainable agriculture. Int J Biol Macromol 77:36–51
Kaushik A, Solanki PR, Ansari AA, Sumana G, Ahmad S, Malhotra BD (2009) Iron oxide-chitosan nanobiocomposite for urea sensor. Sens Actuators B Chem 138:572–580
Kermanizadeh A, Powell LG, Stone V, Møller P (2018) Nanodelivery systems and stabilized solid-drug nanoparticles for orally administered medicine: current landscape. Int J Nanomedicine 13:7575–7605
Khan N, Mukhtar H (2019) Tea polyphenols in promotion of human health. Nutrients 11:39
Khan I, Saeed K, Khan I (2017) Nanoparticles: properties, applications and toxicities. Arab J Chem
Khataee AR, Kasiri MB (2010) Photocatalytic degradation of organic dyes in the presence of nanostructured titanium dioxide: influence of the chemical structure of dyes. J Mol Catal A: Chem 328:8–26
Khattak S, Wahid F, Liu L-P, Jia S-R, Chu L-Q, Xie Y-Y, Li Z-X, Zhong C (2019) Applications of cellulose and chitin/chitosan derivatives and composites as antibacterial materials: current state and perspectives. Appl Microbiol Biotechnol 103:1989–2006
Klein MP, Nunes MR, Rodrigues RC, Benvenutti EV, Costa TMH, Hertz PF, Ninow JL (2012) Effect of the support size on the properties of β-galactosidase immobilized on chitosan: advantages and disadvantages of macro and nanoparticles. Biomacromol 13:2456–2464
Kolahalam LA, Kasi Viswanath IV, Diwakar BS, Govindh B, Reddy V, Murthy YLN (2019) Review on nanomaterials: synthesis and applications. Mater Today Proc
Kosheleva R, Mitropoulos AC, Kyzas GZ (2019) Effect of grafting on chitosan adsorbents, composite nanoadsorbents. Elsevier Inc, Amsterdam

Kosseva MR, Panesar PS, Kaur G, Kennedy JF (2009) Use of immobilised biocatalysts in the processing of cheese whey. Int J Biol Macromol 45:437–447

Koyani RD, Vazquez-Duhalt R (2016) Laccase encapsulation in chitosan nanoparticles enhances the protein stability against microbial degradation. Environ Sci Pollut Res 23:18850–18857

Krajewska B (2004) Application of chitin- and chitosan-based materials for enzyme immobilizations: a review. Enzyme Microb Technol 35:126–139

Kumar MNVR, Muzzarelli RAA, Muzzarelli C, Sashiwa H, Domb AJ (2004) Chitosan chemistry and pharmaceutical perspectives. Chem Rev 104:6017–6084

Kumar V, Dandapat S, Kumar A, Kumar N (2014) Preparation and characterization of chitosan nanoparticles "Alternatively, carrying potential" for cellular and humoral immune responses. Adv Anim Vet Sci 2:414–417

Kumar S, Bhushan P, Bhattacharya S (2017) Fabrication of nanostructures with bottom-up approach and their utility in diagnostics, therapeutics, and others. Environmental, chemical and medical sensors. Springer, Singapore, pp 167–198

Kumar S, Ye F, Dobretsov S, Dutta J (2019) Chitosan nanocomposite coatings for food, paints, and water treatment applications. Appl Sci 9:2409

Kuo C-H, Liu Y-C, Chang C-MJ, Chen J-H, Chang C, Shieh C-J (2012) Optimum conditions for lipase immobilization on chitosan-coated Fe_3O_4 nanoparticles. Carbohydr Polym 87:2538–2545

Lee K, Meng X, Kang T-Y, Ko S (2015) A dye-incorporated chitosan-based CO_2 indicator for monitoring of food quality focusing on makgeolli quality during storage. Food Sci Biotechnol 24:905–912

Leonida M, Belbekhouche S, Adams F, Bijja UK, Choudhary D-A, Kumar I (2019) Enzyme nanovehicles: histaminase and catalase delivered in nanoparticulate chitosan. Int J Pharm 557:145–153

Li Y, Yu J, Hu S, Chen Z, Sacchetti M, Sun CC, Yu L (2019) Polymer nanocoating of amorphous drugs for improving stability, dissolution, powder flow, and tabletability: the case of chitosan-coated indomethacin. Mol Pharm 16:1305–1311

Liang J, Yan H, Puligundla P, Gao X, Zhou Y (2017) Applications of chitosan nanoparticles to enhance absorption and bioavailability of tea polyphenols: a review. Food Hydrocoll 67:286–292

Lima GV, da Silva MR, de Sousa Fonseca T, de Lima LB, de da Oliveira MCF, de Lemos TLG, Zampieri D, dos Santos JCS, Rios NS, Gonçalves LRB, Molinari F, de Mattos MC (2017) Chemoenzymatic synthesis of (S)-Pindolol using lipases. Appl Catal A Gen 546:7–14

Liu Y, Wang M, Zhao F, Xu Z, Dong S (2005) The direct electron transfer of glucose oxidase and glucose biosensor based on carbon nanotubes/chitosan matrix. Biosens Bioelectron 21:984–988

Liu Y, Sun Y, Li Y, Xu S, Tang J, Ding J, Xu Y (2011) Preparation and characterization of α-galactosidase-loaded chitosan nanoparticles for use in foods. Carbohydr Polym 83:1162–1168

Liu M, Dai X, Guan R, Xu X (2014) Immobilization of *Aspergillus niger* xylanase A on Fe_3O_4-coated chitosan magnetic nanoparticles for xylooligosaccharide preparation. Catal Commun 55:6–10

Liu D, Yang F, Xiong F, Gu N (2016) The smart drug delivery system and its clinical potential. Theranostics 6:1306–1323

Lombardo D, Kiselev MA, Caccamo MT (2019) Smart nanoparticles for drug delivery application: development of versatile nanocarrier platforms in biotechnology and nanomedicine. J Nanomater 2019:26

Lone IH, Radwan NRE, Aslam J, Akhter A (2018) Concept of reverse micelle method for the synthesis of nano-structured materials. Curr Nanosci 15:129–136

Lu K-Y, Lin Y-C, Lu H-T, Ho Y-C, Weng S-C, Tsai M-L, Mi F-L (2019) A novel injectable in situ forming gel based on carboxymethyl hexanoyl chitosan/hyaluronic acid polymer blending for sustained release of berberine. Carbohydr Polym 206:664–673

Luangtana-anan M, Nunthanid J, Limmatvapirat S (2019) Potential of different salt forming agents on the formation of chitosan nanoparticles as carriers for protein drug delivery systems. J Pharm Investig 49:37–44

Ma Y, Liu P, Si C, Liu Z (2010) Chitosan nanoparticles: preparation and application in antibacterial paper. J Macromol Sci Part B Phys 49:994–1001

Ma F, Wang Y, Yang G (2019) The modulation of chitosan-DNA interaction by concentration and pH in solution. Polymers (Basel). 11

Magalhães J (1998) Urea potentiometric biosensor based on urease immobilized on chitosan membranes. Talanta 47:183–191

Manoel EA, dos Santos JCS, Freire DMG, Rueda N, Fernandez-Lafuente R (2015a) Immobilization of lipases on hydrophobic supports involves the open form of the enzyme. Enzyme Microb Technol 71:53–57

Manoel EA, Ribeiro MFP, dos Santos JCS, Coelho MAZ, Simas ABC, Fernandez-Lafuente R, Freire DMG (2015b) Accurel MP 1000 as a support for the immobilization of lipase from *Burkholderia cepacia*: application to the kinetic resolution of myo-inositol derivatives. Process Biochem 50:1557–1564

Manoel EA, Pinto M, dos Santos JCS, Tacias-Pascacio VG, Freire DMG, Pinto JC, Fernandez-Lafuente R (2016) Design of a core-shell support to improve lipase features by immobilization. RSC Adv 6:62814–62824

Manrich A, Komesu A, Adriano WS, Tardioli PW, Giordano RLC (2010) Immobilization and stabilization of xylanase by multipoint covalent attachment on agarose and on chitosan supports. Appl Biochem Biotechnol 161:455–467

Martínez-Ruvalcaba A, Sánchez-Díaz JC, Becerra F, Cruz-Barba LE, González-Álvarez A (2009) Swelling characterization and drug delivery kinetics of polyacrylamide-co-itaconic acid/chitosan hydrogels. Express Polym Lett 3:25–32

Matsuda LA, Lolait SJ, Brownstein MJ, Young AC, Bonner TI (1990) Structure of a cannabinoid receptor and functional expression of the cloned cDNA. Nature 346:561–564

Mauricio MD, Guerra-Ojeda S, Marchio P, Valles SL, Aldasoro M, Escribano-Lopez I, Herance JR, Rocha M, Vila JM, Victor VM (2018) Nanoparticles in medicine: a focus on vascular oxidative stress. Oxid Med Cell Longev 2018:20

Melo ADQ, Silva FFM, dos Santos JCS, Fernández-Lafuente R, Lemos TLG, Dias Filho FA (2017) Synthesis of benzyl acetate catalyzed by lipase immobilized in nontoxic chitosan-polyphosphate beads. Molecules 22

Meramo-Hurtado S, Herrera-Barros A, González-Delgado Á (2019) Evaluation of large-scale production of chitosan microbeads modified with nanoparticles based on exergy analysis. Energies 12:1200

Meryam Sardar RA (2015) Enzyme immobilization: an overview on nanoparticles as immobilization matrix. Biochem Anal Biochem 04

Mitra S, Gaur U, Ghosh PC, Maitra AN (2001) Tumour targeted delivery of encapsulated dextran-doxorubicin conjugate using chitosan nanoparticles as carrier. J Controlled Release 317–323

Mohammadpour Dounighi N, Damavandi M, Zolfagharian H, Moradi S (2012) Preparing and characterizing chitosan nanoparticles containing hemiscorpius lepturus scorpion venom as an antigen delivery system. Arch Razi Inst 67:145–153

Mohammed MA, Syeda JTM, Wasan KM, Wasan EK (2017) An overview of chitosan nanoparticles and its application in non-parenteral drug delivery. Pharmaceutics 9:53

Monteiro RRC, Lima PJM, Pinheiro BB, Freire TM, Dutra LMU, Fechine PBA, Gonçalves LRB, de Souza MCM, Santos JCS, Fernandez-Lafuente R (2019) Immobilization of lipase A from Candida antarctica onto chitosan-coated magnetic nanoparticles. Int J Mocelular Sci 20

Morganti P, Fabrizi G, Palombo P, Palombo M, Ruocco E, Cardillo A, Morganti G (2008) Chitin-nanofibrils: a new active cosmetic carrier. J Appl Cosmetol

Mouryaa VK, Inamdara NN, Tiwari A (2010) Carboxymethyl chitosan and its applications. Adv Mater Lett 1:11–33

Müller R, Zhou M, Liebert T, Landers J, Salamon S, Webers S, Dellith A, Borin D, Heinze T, Wende H (2017) Mobility investigations of magnetic nanoparticles in biocomposites. Mater Chem Phys 193:364–370

Mutalik S, Shetty PK, Venuvanka V, Jagani HV, Gejjalagere CH, Nayak UY, Musmade PB, Reddy MS, Kalthur G, Udupa N, Ligade VS, Rao CM (2015) Development and evaluation of sunscreen creams containing morin-encapsulated nanoparticles for enhanced UV radiation protection and antioxidant activity. Int J Nanomedicine 6477

Muxika A, Etxabide A, Uranga J, Guerrero P, de la Caba K (2017) Chitosan as a bioactive polymer: processing, properties and applications. Int J Biol Macromol 105:1358–1368

Muzzarelli RAA (1988) Carboxymethylated chitins and chitosans. Carbohydr Polym

Nagpal K, Singh SK, Mishra DN (2010) Chitosan nanoparticles: a promising system in novel drug delivery. Chem Pharm Bull 58:1423–1430

Neto CGT, Giacometti JA, Job AE, Ferreira FC, Fonseca JLC, Pereira MR (2005) Thermal analysis of chitosan based networks. Carbohydr Polym 62:97–103

Newman MD, Stotland M, Ellis JI (2009) The safety of nanosized particles in titanium dioxide– and zinc oxide–based sunscreens. J Am Acad Dermatol 61:685–692

Ng WL, Yeong WY, Naing MW (2016) Polyelectrolyte gelatin-chitosan hydrogel optimized for 3D bioprinting in skin tissue engineering. Int J Bioprinting 2:10

Nguyen LT, Yang KL (2017) Enzyme Microb. Technol. Elsevier Inc. 100

Nguyen VD, Styevkó G, Madaras E, Haktanirlar G, Tran ATM, Bujna E, Dam MS, Nguyen QD (2019) Immobilization of β-galactosidase on chitosan-coated magnetic nanoparticles and its application for synthesis of lactulose-based galactooligosaccharides. Process Biochem 0–1

Noori R, Perwez M, Sardar M (2019) Cross-linked enzyme aggregates: current developments and applications. In: Husain Q, Ullah M (eds) Biocatalysis. Springer, Cham, pp 83–112

Norton JE, Espinosa YG, Watson RL, Spyropoulos F, Norton IT (2014) Functional food microstructures for macronutrient release and delivery. Food Funct 6:663–678

Ntohogian S, Gavriliadou V, Christodoulou E, Nanaki S, Lykidou S, Naidis P, Mischopoulou L, Barmpalexis P, Nikolaidis N, Bikiaris D (2018a) Chitosan nanoparticles with encapsulated natural and UF-purified annatto and saffron for the preparation of UV protective cosmetic emulsions. Molecules 23:2107

Ntohogian S, Gavriliadou V, Christodoulou E, Nanaki S, Lykidou S, Naidis P, Mischopoulou L, Barmpalexis P, Nikolaidis N, Bikiaris DN (2018b) Chitosan nanoparticles with encapsulated natural and UF-purified annatto and saffron for the preparation of UV protective cosmetic emulsions. Molecules

O'Brien FJ (2011) Biomaterials and scaffolds for tissue engineering. Mater Today 14:88–95

O'Brien S, Brus L, Murray CB (2001) Synthesis of monodisperse nanoparticles of barium titanate: toward a generalized strategy of oxide nanoparticle synthesis. J Am Chem Soc 123:12085–12086

Oliveira HC, Gomes BCR, Pelegrino MT, Seabra AB (2016) Nitric oxide-releasing chitosan nanoparticles alleviate the effects of salt stress in maize plants. Nitric Oxide—Biol Chem 61:10–19

Ortega N, Kumar A, Scott JF, Katiyar RS (2015) Multifunctional magnetoelectric materials for device applications. Phys Condens Matter 27:504002

Ortega F, García MA, Arce VB (2019) Nanocomposite films with silver nanoparticles synthesized in situ: effect of corn starch content. Food Hydrocoll 97:105200

Ortiz C, Ferreira ML, Barbosa O, dos Santos JCS, Rodrigues RC, Berenguer-Murcia Á, Briand LE, Fernandez-Lafuente R (2019) Novozym 435: the "perfect" lipase immobilized biocatalyst? Catal Sci Technol 9:2380–2420

Oryan A, Sahvieh S (2017) Effectiveness of chitosan scaffold in skin, bone and cartilage healing. Int J Biol Macromol 104:1003–1011

Othman SH (2014) Bio-nanocomposite materials for food packaging applications: types of biopolymer and nano-sized filler. Agric Agric Sci Procedia 2:296–303

Pan C, Hu B, Li W, Sun Y, Ye H, Zeng X (2009) Novel and efficient method for immobilization and stabilization of β-d-galactosidase by covalent attachment onto magnetic Fe_3O_4–chitosan nanoparticles. J Mol Catal B Enzym 61:208–215

Pan C, Qian J, Fan J, Guo H, Gou L, Yang H, Liang C (2019) Preparation nanoparticle by ionic cross-linked emulsified chitosan and its antibacterial activity. Colloids Surf A Physicochem Eng Asp 568:362–370

Panão Costa J, Carvalho S, Jesus S, Soares E, Marques AP, Borges O (2019) Optimization of chitosan-α-casein nanoparticles for improved gene delivery: characterization, stability, and transfection efficiency. AAPS PharmSciTech 20:132

Pant A, Singh J (2018) Novel controlled ionic gelation strategy for chitosan nanoparticles preparation using TPP- β -CD inclusion complex. Eur J Pharm Sci 112:180–185

Parisi C, Vigani M, Rodríguez-Cerezo E (2015) Agricultural nanotechnologies: what are the current possibilities? Nano Today 10:124–127

Parreidt TS, Müller K, Schmid M (2018) Alginate-based edible films and coatings for food packaging applications. Foods 7:1–38

Pinheiro MP, Rios NS, de Fonseca TS, de Bezerra FA, Rodríguez-Castellón E, Fernandez-Lafuente R, de Mattos MC, dos Santos JCS, Gonçalves LRB (2018) Kinetic resolution of drug intermediates catalyzed by lipase B from Candida antarctica immobilized on Immobead-350. Biotechnol Progress 2–48. 10.1002/btpr.2630

Pinheiro BB, Rios NS, Rodríguez Aguado E, Fernandez-Lafuente R, Freire TM, Fechine PBA, dos Santos JCS, Gonçalves LRB (2019) Chitosan activated with divinyl sulfone: a new heterofunctional support for enzyme immobilization. Application in the immobilization of lipase B from Candida antarctica. Int J Biol Macromol 130:798–809

Prabaharan M, Mano JF (2004) Chitosan-based particles as controlled drug delivery systems. Drug Deliv 12:41–57

Qi L, Xu Z, Jiang X, Hu C, Zou X (2004) Preparation and antibacterial activity of chitosan nanoparticles. Carbohydr Res 339:2693–2700

Quiñones JP, Peniche H, Peniche C (2018) Chitosan based self-assembled nanoparticles in drug delivery. Polymers (Basel) 10:1–32

Radt S (2018) Perspective of recent progress in immobilization of enzymes. Mnemosyne 71:41–57

Ragab TIM, Nada AA, Ali EA, Shalaby ASG, Soliman AAF, Emam M, El Raey MA (2019) Soft hydrogel based on modified chitosan containing P. granatum peel extract and its nano-forms: multiparticulate study on chronic wounds treatment. Int J Biol Macromol 135:407–421

Raghavendra GM, Jung J, Kim D, Seo J (2017) Effect of chitosan silver nanoparticle coating on functional properties of Korean traditional paper. Prog Org Coatings 110:16–23

Rastegari B, Karbalaei-Heidari HR, Zeinali S, Sheardown H (2017) The enzyme-sensitive release of prodigiosin grafted β-cyclodextrin and chitosan magnetic nanoparticles as an anticancer drug delivery system: synthesis, characterization and cytotoxicity studies. Colloids Surf B Biointerfaces 158:589–601

Reis C, Sousa E, Serpa J, Oliveira R, Oliveira R, Santos J (2019) Design of immobilized enzyme biocatalysts: drawbacks and opportunities. Quim. Nova X:1–16

Rios NS, Pinheiro BB, Pinheiro MP, Bezerra RM, dos Santos JCS, Barros Gonçalves LR (2018) Biotechnological potential of lipases from Pseudomonas: sources, properties and applications. Process Biochem 75:99–120

Roberfroid MB (2000) Prebiotics and probiotics: are they functional foods? Am J Clin Nutr 71:1682S–1687S

Rodrigues RC, Virgen-Ortíz JJ, dos Santos JCS, Berenguer-Murcia Á, Alcantara AR, Barbosa O, Ortiz C, Fernandez-Lafuente R (2019) Immobilization of lipases on hydrophobic supports: immobilization mechanism, advantages, problems, and solutions. Biotechnol Adv 37:746–770

Rueda N, dos Santos JCS, Torres R, Ortiz C, Barbosa O, Fernandez-Lafuente R (2015) Improved performance of lipases immobilized on heterofunctional octyl-glyoxyl agarose beads. RSC Adv 5:11212–11222

Rueda N, dos Santos JCS, Ortiz C, Torres R, Barbosa O, Rodrigues RC, Berenguer-Murcia Á, Fernandez-Lafuente R (2016) Chemical modification in the design of immobilized enzyme biocatalysts: drawbacks and opportunities. Chem Rec 16:1436–1455

Sadighi A, Faramarzi MA (2013) Congo red decolorization by immobilized laccase through chitosan nanoparticles on the glass beads. J Taiwan Inst Chem Eng 44:156–162
Saewan N, Jimtaisong A (2015) Natural products as photoprotection. J Cosmet Dermatol 14:47–63
Saharan V, Pal A (2016) Current and future prospects of chitosan-based nanomaterials in plant protection and growth 43–48
Saharan V, Mehrotra A, Khatik R (2013) Synthesis of chitosan based nanoparticles and their in vitro evaluation against phytopathogenic fungi. Int J Biol Macromol 62:677–683
Saleh TA (2016) Nanomaterials for pharmaceuticals determination. Bioenergetics 5:226
Sánchez-Machado D, López-Cervantes J, Correa-Murrieta M, Sánchez-Duarte R, Cruz-Flores P, de la Mora-López G (2019) Chitosan. In: Nonvitamin and nonmineral nutritional supplements. Academic Press, pp 485–493
Santos JCSD, Barbosa O, Ortiz C, Berenguer-Murcia A, Rodrigues RC, Fernandez-Lafuente R (2015) Importance of the support properties for immobilization or purification of enzymes. ChemCatChem 7:2413–2432
Sarojini KS, Indumathi MP, Rajarajeswari GR (2019) Mahua oil-based polyurethane/chitosan/nano ZnO composite films for biodegradable food packaging applications. Int J Biol Macromol 124:163–174
Schoukens G (2019) Bioactive dressings to promote wound healing. In: Advanced textiles for wound care. Elsevier pp 135–167
Sedghi R, Shaabani A, Mohammadi Z, Samadi FY, Isaei E (2017) Biocompatible electrospinning chitosan nanofibers: a novel delivery system with superior local cancer therapy. Carbohydr Polym 159:1–10
Serpone N, Dondi D, Albini A (2007) Inorganic and organic UV filters: their role and efficacy in sunscreens and suncare products. Inorganica Chim Acta 360:794–802
Shagholani H, Ghoreishi SM, Mousazadeh M (2015) Improvement of interaction between PVA and chitosan via magnetite nanoparticles for drug delivery application. Int J Biol Macromol 78:130–136
Shahid-ul-Islam, Shahid M, Mohammad F (2013) Green chemistry approaches to develop antimicrobial textiles based on sustainable biopolymers—a review. Ind Eng Chem Res 52:5245–5260
Shahid-ul-Islam, Butola BS, Verma D (2019) Facile synthesis of chitosan-silver nanoparticles onto linen for antibacterial activity and free-radical scavenging textiles. Int J Biol Macromol 133:1134–1141
Shang Y, Hasan MK, Ahammed GJ, Li M, Yin H, Zhou J (2019) Applications of nanotechnology in plant growth and crop protection: a review. Molecules 24:2558
Sheldon RA, van Pelt S (2013) Enzyme immobilisation in biocatalysis: why, what and how. Chem Soc Rev 42:6223–6235
Shen X, Shamshina JL, Berton P, Gurau G, Rogers RD (2016) Hydrogels based on cellulose and chitin: fabrication, properties, and applications. Green Chem 18:53–75
Shoaebargh S, Karimi A (2014) RSM modeling and optimization of glucose oxidase immobilization on TiO2/polyurethane: feasibility study of AO7 decolorization. J Environ Chem Eng 2:1741–1747
Shukla S, Mishra A, Arotiba O, Mamba BB (2013) Chitosan-based nanomaterials: a state-of-the-art review. Int J Biol Macromol 59:46–58
Sivashankari PR, Prabaharan M (2019) Bioactive nanomaterials/chitosan composites as scaffolds for tissue regeneration. In: Polysaccharide carriers for drug delivery. Elsevier, pp 559–584
Song R, Murphy M, Li C, Ting KW, Soo CB, Zheng Z (2018) Current development of biodegradable polymeric materials for biomedical applications. Drug Des Devel Ther 2018:3117–3145
Souza MP, Vaz AFM, Vicente AA, Carneiro-da-cunha MG (2014) Quercetin-loaded lecithin/chitosan nanoparticles for functional food applications 1149–1159
Suescun A, Rueda N, Dos Santos JCS, Castillo JJ, Ortiz C, Torres R, Barbosa O, Fernandez-Lafuente R (2015) Immobilization of lipases on glyoxyl-octyl supports: improved stability and reactivation strategies. Process Biochem 50:1211–1217
Suh S, Meng X, Ko S (2016) Proof of concept study for different-sized chitosan nanoparticles as carbon dioxide (CO_2) indicators in food quality monitoring. Talanta 161:265–270

Sukhova AA, Gofman IV, Skorik YA (2019) Preparation and properties of chitosan–nanodiamond dispersions and composite films. Diam Relat Mater 98
Sun B, Zhang L, Yang L, Zhang F, Norse D, Zhu Z (2012) Agricultural non-point source pollution in China: causes and mitigation measures. Ambio 41:370–379
Suzuki S (2000) Biological effects of chitin, chitosan and their oligosaccharides. Biotherapy 14:965–971
Sweidan K, Jaberb A-M, Al-jbourc N, Obaidatd R, Al-Remawi M, Badwan A (2011) Original paper further investigation on the degree of deacetylation of chitosandetermined by potentiometric titration. Excipients Food Chem. 2
Tamura A, Hiramoto K, Ino K, Taira N, Nashimoto Y, Shiku H (2019) Genipin crosslinking of electrodeposited chitosan/gelatin hydrogels for cell culture. Chem Lett 48:1178–1180
Threepopnatkul P, Wongnarat C, Intolo W, Suato S, Kulsetthanchalee C (2014) Effect of TiO_2 and ZnO on thin film properties of PET/PBS blend for food packaging applications. Energy Procedia 56:102–111
Tokumitsu H, Ichikawa H, Fukumori Y, Block LH (1999) Preparation of gadopentetic acid-loaded chitosan microparticles for gadolinium neutron-capture therapy of cancer by a novel emulsion-droplet coalescence technique. Chem Pharm Bull (Tokyo) 47:838–842
Unsoy G, Khodadust R, Yalcin S, Mutlu P, Gunduz U (2014) Synthesis of Doxorubicin loaded magnetic chitosan nanoparticles for pH responsive targeted drug delivery. Eur J Pharm Sci 62:243–250
Urimi D, Agrawal AK, Kushwah V, Jain S (2019) Polyglutamic acid functionalization of chitosan nanoparticles enhances the therapeutic efficacy of insulin following oral administration. AAPS PharmSciTech 20:131
Varelis P, Melton L, Shahidi F (eds) (2019) Encyclopedia of food chemistry. Elsevier, Amsterdam
Vásconez MB, Flores SK, Campos CA, Alvarado J, Gerschenson LN (2009) Antimicrobial activity and physical properties of chitosan–tapioca starch based edible films and coatings. Food Res Int 42:762–769
Villalba M, Verdasco-Martín CM, dos Santos JCS, Fernandez-Lafuente R, Otero C (2016) Operational stabilities of different chemical derivatives of Novozym 435 in an alcoholysis reaction. Enzyme Microb Technol 90:35–44
Virgen-Ortíz JJ, dos Santos JCS, Ortiz C, Berenguer-Murcia Á, Barbosa O, Rodrigues RC, Fernandez-Lafuente R (2019) Lecitase ultra: a phospholipase with great potential in biocatalysis. Mol Catal 473:110405
Wang G, Xu J-J, Ye L-H, Zhu J-J, Chen H-Y (2002) Highly sensitive sensors based on the immobilization of tyrosinase in chitosan. Bioelectrochemistry 57:33–38
Wang X, Liu X, Zhao C, Ding Y, Xu P (2011) Biodiesel production in packed-bed reactors using lipase-nanoparticle biocomposite. Bioresour Technol 102:6352–6355
Wang S, Zhang C, Qi B, Sui X, Jiang L, Li Y, Wang Z, Feng H, Wang R, Zhang Q (2014) Immobilized alcalase alkaline protease on the magnetic chitosan nanoparticles used for soy protein isolate hydrolysis. Eur Food Res Technol 239:1051–1059
Weber M, Bechelany M (2019) Combining nanoparticles grown by ALD and MOFs for gas separation and catalysis applications. Pure Appl Chem 1–10
Xing K, Zhu X, Peng X, Qin S (2015) Chitosan antimicrobial and eliciting properties for pest control in agriculture: a review. Agron Sustain Dev 35:569–588
Xu Y, Du Y (2003) Effect of molecular structure of chitosan on protein delivery properties of chitosan nanoparticles. Int J Pharm 250:215–226
Xu J, Dai W, Wang Z, Chen B, Li Z, Fan X (2011) Intranasal vaccination with chitosan-DNA nanoparticles expressing pneumococcal surface antigen A protects mice against nasopharyngeal colonization by *Streptococcus pneumoniae*. Clin Vaccine Immunol 18:75–81
Yadav SK, Khan G, Bansal M, Vardhan H, Mishra B (2017) Screening of ionically crosslinked chitosan-tripolyphosphate microspheres using Plackett-Burman factorial design for the treatment of intrapocket infections. Drug Dev Ind Pharm 43:1801–1816

Yadav M, Goswami P, Paritosh K, Kumar M, Pareek N, Vivekanand V (2019) Seafood waste: a source for preparation of commercially employable chitin/chitosan materials. Bioresour Bioprocess 6

Ye J, Wang S, Lan W, Qin W, Liu Y (2018) Preparation and properties of polylactic acid-tea polyphenol-chitosan composite membranes. Int J Biol Macromol 117:632–639

Youssef AM, El-Sayed SM (2018) Bionanocomposites materials for food packaging applications: concepts and future outlook. Carbohydr Polym 193:19–27

Yu D, Tian L, Wu H, Wang S, Wang Y, Ma D, Fang X (2010) Ultrasonic irradiation with vibration for biodiesel production from soybean oil by Novozym 435. Process Biochem 45:519–525

Yu C, Guo X, Muzzio M, Seto CT, Sun S (2019) Self-assembly of nanoparticles into two-dimensional arrays for catalytic applications. ChemPhysChem 20:23–30

Zeng J, Du G, Shao X, Feng K-N, Zeng Y (2019) Recombinant polyphenol oxidases for production of theaflavins from tea polyphenols. Int J Biol Macromol 134:139–145

Zhao L-M, Shi L-E, Zhang Z-L, Chen J-M, Shi D-D, Yang J, Tang Z-X (2011) Preparation and application of chitosan nanoparticles and nanofibers. Brazilian J Chem Eng 28:353–362

Zhou C-E, Kan C-W, Sun C, Du J, Xu C (2019) A review of chitosan textile applications. AATCC J Res 6:8–14

Zottel A, Paska AV, Jovčevska I (2019) Nanotechnology meets oncology: nanomaterials in brain cancer research, diagnosis and therapy. Materials (Basel) 12:1588

Feng BH, Peng LF (2012) Synthesis and characterization of carboxymethyl chitosan carrying ricinoleic functions as an emulsifier for azadirachtin. Carbohydr Polym 88:576–582

Chapter 5
Nanotechnology as a Tool for Contaminants Detection in Milk or Milk Products

Pooja Singh, Smriti Singh, and Seema Nara

Milk is ladle out as complete food for adults and infants worldwide by remarkably fulfilling daily nutritional requirements. Milk or milk-based products form a significant part of human diet globally and hence its contamination or adulteration is a natural concern. Some of the known contaminants of milk or milk products include biological contaminants such as microbes or their toxins, antibiotics, chemicals like urea, melamine, natural or synthetic hormones, pesticides, etc. Poor hygienic conditions of the cattles, inappropriate processing, storage or packaging conditions could be a reason for natural microbial infections whereas increase in demand and less supply, decease nature of milk and unavailability of specific and rapid detection techniques can be realizable reasons for adulteration of milk. These contaminants may cause adverse health issues in infants and adults. These issues may be reflected instantly and with a time course too. Health problems due to intake of this adulterants milk are headache, nausea, vomiting, diarrhoea, eye sight problem, gastrointestinal complications, kidney problem, heart problem, cancer, and even death. Existing techniques for contaminants detection in milk and milk products are inadequate to detect low levels of these contaminants efficiently. Secondly, these techniques are sophisticated and centralized, and require skilled persons or are less specific. The emphasis is on developing point of use diagnostic assays of milk contaminant detection so that the end user can perform the test easily. Nanotechnology has shown a great potential in meeting this demand. Nanomaterials have been used in various ways like, as reporter, catalyst, quencher, or separator to develop point of use diagnostic assays. This chapter intend to present an overview of the nanomaterial-based diagnostic platforms reported so far for contaminant detection in milk or milk products.

P. Singh · S. Singh · S. Nara (✉)
Department of Biotechnology, Motilal Nehru National Institute of Technology Allahabad, Prayagraj, Uttar Pradesh 211004, India
e-mail: seemanara@mnnit.ac.in

R. F. do. Nascimento et al. (eds.), *Nanomaterials and Nanotechnology*,
Materials Horizons: From Nature to Nanomaterials,
https://doi.org/10.1007/978-981-33-6056-3_5

The status quo of the nanosensing methods for milk or milk products analysis would be presented here.

1 Introduction

Milk is a rich source of quickly available nutrients like protein, fat, calcium, minerals, etc., and is considered as natural, healthy staple food for many people throughout the world. Dairy industry is rapidly expanding with consumers demand for fresh, authentic, and flavored items. Hence, monitoring quality of milk or milk-based products is indispensible. Raw milk could get contaminated before or after mulching from cattle feed. This contamination could be natural or artificial leading to spoilage if not treated properly. Unfortunately, milk is being very easily adulterated throughout the world for cost cutting purpose or increasing the life of dairy product. Presence of such adulterants could have serious health issues and therefore proper quality assessment of milk before and after packaging and other milk products is highly indispensable. Most of the time, quality of milk is assessed on the basis of its nitrogen or fat content only. Presently, only small sample sizes are assessed for the presence of adulterants through HPLC-MS or GC-MS in centralized laboratories which is a costly affair in terms of time and resources. Therefore, rapid on-site methods of detection are immensely needed for rapid detection of milk adulterants or contaminants.

Nanotechnology has demonstrated its applicability in designing novel detection systems with enhanced sensitivity. Nanomaterials are preferred choice for labels/reporter molecules because of their easy synthesis, facile surface chemistry, and stability. Their increased surface area also enables them to be used as an agent for immobilization of capture reagents on solid surfaces. Their unique optical, electrical, or magnetic properties make them immensely versatile to develop different assays employing these properties. This chapter briefs about various milk contaminants, sources of contamination, and their permissible limits by different regulatory agencies. It also details the ways by which nanotechnology is contributing in detecting these milk contaminants. Chapter intends to acquaint the readers with recent approaches to use nanostructures and their inherent properties like fluorescence, surface plasmon resonance, enzyme mimetics, fluorescence resonance energy transfer, electrical conductivity, or magnetism for milk contaminant detection.

2 Major Contaminants of Milk or Milk Products

Milk and dairy products are an important part of daily diet for many people throughout the world. The quality of the milk or milk products can get deteriorated with the presence of various contaminants like pesticides, heavy metals, pathogens, mycotoxins, drugs or antibiotics, hormones, melamine, urea, and other nitrogenous compounds.

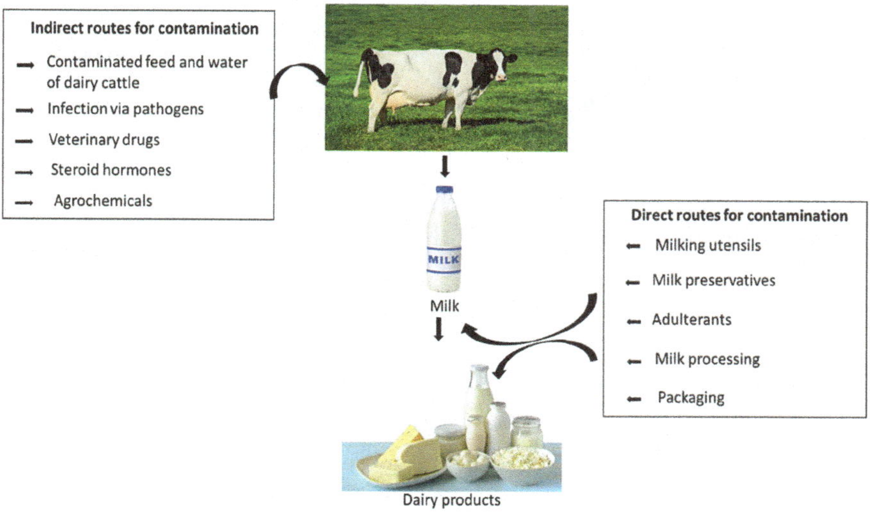

Fig. 1 Routes of contamination of milk or milk products

These contaminants may find entry in milk or milk products through various direct and indirect routes (Fig. 1).

Indirect route includes transfer of contaminants to milk via food chain, when the milk producing animals feed upon contaminated food, water, or gets infected with pathogens or are artificially injected with veterinary drugs, hormones, etc. Direct route includes direct contamination of milk/milk products during handling and processing, or by deliberate addition of contaminants, i.e., "adulteration" of these products. Some examples of adulteration of milk include addition of water, whey, vegetable oil, starch, sugar, preservatives, detergents, urea, melamine, etc. These adulterants falsely impact milk quality in various ways as depicted in Fig. 2.

Possible reasons behind the adulteration of milk may include demand and supply gap, perishable nature of milk, low purchasing capability of customer, and lack of suitable detection tests (Kamthania et al. 2014). This section details about some of the major milk contaminants along with their maximum residual limits as listed in Table 1.

Pesticides

Pesticides are widely used in the agriculture to protect the crops, vegetables, fruits, etc., from pests and insects. It includes insecticides, herbicides, rodenticides, fungicides, etc., that are applied during pre-harvest, post-harvest stage or during storage of crops. The pesticide residues present in these crops may get entry to the next level of food chain when cattle stocks feed on such crops. Pesticide residues could find entry in animal body through inhalation route during direct spray in the animal accommodation to invade the pest. Different classes of pesticides such as organo phosphorus (OPPs), organo chlorine pesticide (OCPs), pyrethroids, and carbamate (CB)

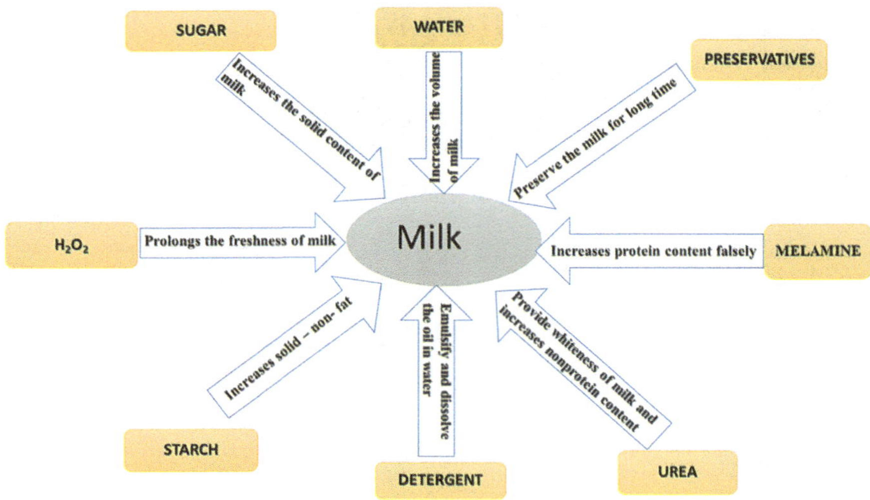

Fig. 2 Impact of various adulterants on milk

are reported to be present in milk and milk products. These pesticides possess wide range of toxicity and cause severe health effects such as carcinogenicity, neurotoxicity genotoxicity, and hormonal disturbances (Akhtar and Ahad 2017). According to the present regulations, the use of OCPs or mixtures of OCPs is no longer allowed in the European Union. However, traces of OCPs have been observed in milk and milk products because OCPs have been used for a long period of time and they persist in the environment due to their slow decomposition rate, long half-life, and high stability in the environment (Rusu et al. 2016). Pesticides residue can be accumulated in fatty tissues of animal body due to liposoluble nature and transfer into milk and dairy products.

Mycotoxins

Mycotoxins are secondary metabolites which are produced by certain types of filamentous fungi under favourable growth conditions. Out of various types of identified mycotoxins, few are very toxic for human health like aflatoxins, ochratoxin, patulin, fumonisin, Zearalenone, etc. Aflatoxins are mainly produced by *Aspergillus flavus* and *Aspergillus parasiticus*. When animals ingest contaminated feedstuffs, mycotoxins are metabolized, biotransformed and secreted in the milk, thus becoming a risk to human health.

Aflatoxins are known to contaminate milk of animals that fed on feed contaminated with Aflatoxin M1 (AFM1). AFM1 is an aflatoxin B1(AFB1) metabolite produced in the animal rumen and secreted in milk (Algeri et al. 2016). This AFM1 is a very mutagenic and carcinogenic compound and therefore since 2002, it has been categorized by the International Agency for Research on Cancer as group I carcinogen to humans (IARC, Lyon, France, 2002). Presence of aflatoxin above the

Table 1 Maximum residue limit of milk contaminants from different guidelines

Contaminants	Maximum allowable limit in milk		
	E.U. Regulation	U.S. FDA	FSSAI
Aflatoxin MI (μg/kg)	0.050	0.5	0.5
Dioxins and polychlorinated biphenyls (PCB) (pg/gm)	2.5	–	–
Urea	–	–	70 mg/100 ml
Chloramphenicol (mg/kg)	–	–	0.0003
Melamine (mg/kg)	0.5	0.063	1
Heavy metals			
Arsenic	–	–	0.05 ppm (infant milk) 0.1 ppm (milk)
Lead	0.02 mg/kg	–	0.2 ppm (for infant milk) 2.5 ppm (other milk products)
Mercury	–	–	1.0 ppm (all milk products)
Cadmium	–	–	0.1 ppm (infant milk) 1.5 ppm (other milk products)
Zinc	–	–	50.0 ppm(but) not less than 25.0)
Antibiotics			
Tetracycline	100 ppb	100 ppb	0.1 mg/kg
Oxytetracycline	–	100 ppb	0.1 mg/kg
Trimethoprim	50 ppb	50 ppb	0.05 mg/kg
Oxolinic acid	–	–	0.3 mg/kg
Streptomycin	200 ppb	200 ppb	–

Source U.S. FOOD & DRUG administration. Guidance for Industry: Action Levels for Poisonous or Deleterious Substances in Human Food and Animal Feed; WHO Technical Report Series. Evaluation of certain contaminants in Food. Joint FAO/WHO, 2007; Ministry of Health and Family Welfare. Food Safety and Standards Authority of India Notification New Delhi, dated the 1st August, 2011

allowable limit in milk and milk products causes acute poisoning, known as aflatoxicosis, as it usually damages the liver and can be life threatening. Due to its high toxicity, the levels of aflatoxin in milk are strictly regulated. Thus, in European Union, the maximum allowable limit for AFM1 in infant milk is 0.025 ng/mL and in milk for adult consumption is 0.05 ng/mL (European Commission (EC) Regulation 2003), whereas in USA, this limit is 0.5 ng/mL for all milk types (Commission Regulation 2009).

Ochratoxin A is produced by several species of *Aspergillus* (*A. ochraceus, A. melleus, A. sulphureus, Aspergillus section Nigri, A. carbonarius, A. awamori*) and *Penicillium* (*P. verrucosum, P. crysogenum and P. nordicum*) and is one of the well-known mycotoxin contaminants present in milk and milk products (Bayman and Baker 2006; Magan 2006; Zheng et al. 2005). Ochratoxin A is formed during the storage of crops and it causes a number of toxic effects in animal species. Several studies have reported the effects of ochratoxin A as hepatotoxic, nephrotoxic, teratogenic; thus, International Agency of Research in Cancer (IARC) has grouped OTA as a carcinogenetic of 2B class (Muscarella et al. 2004). This OTA is also responsible for causing Balkan endemic nephropathy (BEN) and chronic interstitial nephropathy (CIN), as well as other renal diseases in human (Bayman and Baker 2006; Pattono et al. 2011).

Melamine and Other Nitrogenous Compounds

Melamine (1,3,5-triazine-2,4,6-triamine, $C_3H_6N_6$) is a chemical compound mainly used for the synthesis of melamine resins and fertilizers. Recently, melamine has been illegally added into the milk products to falsely increase the level of apparent protein because of its high nitrogen content. Melamine is not carcinogenic compound and it has low oral acute toxicity. However, the intake of melamine above the safety limit results in the formation of insoluble melamine cyanurate crystal in kidney and finally cause renal failure (Maryam 2017).

The US Food and Drug Administration (FDA) declared the allowed limit of melamine as 1 mg/kg body weight in infant formula and 2.5 mg/kg for other dairy products (Kelong et al. 2009; Mauer et al. 2009). Whereas, the World Health Organization (WHO) also fixed the daily intake limit of melamine to 0.2 mg/kg body weight per day (Fischer 2015; Bates et al. 2017). Melamine has subsequently been detected in a variety of milk containing products and the level of melamine reported in dairy products ranged from 0.09 to 6200 mg/kg. Besides melamine, nitorhen content of milk is increased using several other chemical substances as adulterants such as urea, amidinourea, ammeline, cyanamide, dicyandiamide, 3-aminotriazole, 4-aminotriazole, guanidine, biuret, triuret, cyromazine, etc. (Fischer et al. 2015).

Heavy Metals

Milk and milk products contain a large number of toxic and nontoxic metals which are present in a variety of different chemical species. The toxicity of heavy metals to humans and animals is the result of exposure to long-term contamination in our environment, including the air we breathe, water, food, and so on. Heavy metals refer to any metal element that has a relatively high density and low toxicity. It includes lead (Pb), mercury (Hg), copper (Cu), cadmium (Cd), zinc (Zn), arsenic (As), chromium (Cr), and iron (Ziarati et al. 2012). These heavy metals enter in milk if cattle stock graze on land either naturally containing high level of certain metals or contaminated by industrial or other human activities (Heeschen et al. 1997). In addition, raw milk may get contaminated during its production (Ziarati et al. 2018).

Lead and cadmium residues in milk and dairy products are of particular concerns because of high consumption of milk and milk products by newborns and children. Heavy metals produce toxic effects by replacing essential metal ions existing as chelates in body. Excess intake of heavy metals causes serious health problems like nervous system disorders, renal failure, genetic mutations, types of cancers, neurological disorders, respiratory disorders, and cardiovascular, immune system weakening, and infertility (Ziarati et al. 2018). Food Safety and Standards Authority of India (FSSAI), 2011 proposed the maximum residue limit for arsenic to be 0.05 ppm in infant milk and 0.1 ppm in other dairy milk products, 2.5 ppm for lead, 1.0 ppm for mercury, and 0.1 ppm for cadmium.

Antimicrobial Drugs

Antimicrobials drugs are mainly administrated in dairy cattle primarily to treat or prevent disease and infections. The trace residues of almost all drugs administrated intentionally or unintentionally to animals remain in milk and milk products (Tsegaye et al. 2015). Most of the veterinary treatment of dairy cattle involves intramammary infusion of antibiotics to control mastitis. Mastitis is an inflammation of the mammary glands which is characterized by an increase in somatic cell count in the milk and pathological change in mammary tissue. The mastitis causing pathogens include mainly mycoplasma and pathogenic bacteria. The antimicrobial drugs commonly used for veterinary treatment are divided in five major classes, namely beta-lactams (e.g., penicillins and cephalosporins), tetracyclines (e.g., Oxytetracycline, tetracycline and chlortetracycline), amino glycosides (e.g., streptomycin, neomycin and gentamycin), macrolides (e.g., erythromycin), and sulfonamides (e.g., sulfamethazine) (Mitchell et al. 1998; Sundlof et al., 1995). These drugs are administrated to animals through different routes such as via injection, oral, topical dermal, or intramammary.

Problems associated with antimicrobial residues in milk include the risk of allergic reactions, toxicity, carcinogenicity, and teratogenic effects (Asredie and Engdaw 2015). In many countries, government authorities have established monitoring programs to determine the antibiotic levels in milk and set a maximum residue level (MRL) for these drugs. The European Union declared the maximum residue level as 4 μg/kg for penicillins, 25 μg/kg for sulfonamides, 100 μg/kg for tetracyclines, and 200 μg/kg for streptomycin.

Steroid Hormones

Steroid hormones have been widely used in animal husbandry because they can promote the growth of animals and improve feed conversion efficiency, i.e., the efficiency with which the bodies of livestock convert animal feed into the desired output. Steroid hormones include androgens, estrogens, glucocorticoids, and progestogens (Qu et al. 2017). Hormones in milk originate from the blood flow and are secreted into the milk through the mammary gland. Some hormones can also be synthesized by the mammary gland and excreted into the milk and can be classified in two groups: natural hormones (endoestrogens) and synthetic hormones (exoestrogens) (Xu et al. 2013). A number of studies have identified prolactin and steroids in dairy products, including estrogens, progesterone, glucocorticoids, and androgens and it is found that these compounds may have significant biological effects even at very low doses (Malekinejad and Rezabakhsh 2015).

The steroid hormones in milk and milk products have the potential to be toxic and carcinogenic, and their residues may be relevant to a variety of diseases including cancer of the breasts, ovaries, and prostate (Jouan et al. 2006; Xu et al. 2013; Qu et al. 2017). Yang et al. 2009, have reported the presence of 17α-hydroxyprogesterone, testosterone, progesterone, 4-androstene-3, 17-dione, and estradiol in the range of

0.11–6.10 μg/kg in whole milk. However, Chen et al. 2014 have reported the presence of natural steroid hormones estrone, 17α-estradiol, 17β-estradiol, and estriol in commercial milk samples in the range of 0–146.12 ng/L, 0–70.12 ng/L, 0–31.85 ng/L and 0–2.18 ng/L, respectively.

Other Contaminants

Except these key contaminants, there are several other compounds which contaminates the milk and milk products and make it unsuitable for consumption. This includes tap water, whey, hydrogen peroxide, synthetic urea, detergent, ammonium sulphate, formalin, glucose, starch, sugar, vegetable oil, etc. These contaminants can be used deliberately to increase the milk quality in a dishonest way by falsely increasing its fat percentage, SNF (Solid-not-Fat) percentage, protein content, and freezing point. Few of these adulterants may adversely affect human health. The presence of peroxides and detergents in milk can cause gastro-intestinal complications, which leads to gastritis and inflammation of the intestine. Excessive starch in the milk can cause diarrhoea due to the effects of undigested starch in colon. Carbonate and bicarbonates might cause disruption in hormone signalling that regulate development and reproduction.

3 Nanotechnology in Milk Contaminant Detection

Nanotechnology is an attractive technology that offers plethora of opportunities for developing novel nanomaterials and systems for applications in food and diagnostic industry. Nanomaterials, owing to their very small size (any one dimension in the range of 1 to 100 nm), possess unique chemical, optical, magnetic, and electronic properties, which distinguishes them from their bulk counterparts. These properties have been well utilized in developing sensitive and rapid methods of contaminant or adulterant detection milk or milk products. This section details new and recent approaches of using nanomaterials in milk contaminant detection employing their unique properties.

3.1 Employing Surface Enhanced Raman Scattering Property of Nanostructures

Surface Enhanced Raman Scattering (SERS) is an advanced technique to amplify weak Raman signals. This technique uses colloidal solution of metallic nanoparticles alone or nanoparticles with surface tagged/encapsulated Raman active molecules as probes. Use of nanoparticles as Raman probes greatly enhances the Raman signal intensity generally in the order of 10^{10}–10^{11}. Metallic nanoparticles significant amplifies the Raman signals and thus popularly used for increasing the lower limit of detection (LOD) of the detection assay. Laser excitation of the nanoparticles resonantly drives their surface charge, thereby, creating a highly localized plasmonic light field. When a target molecule is absorbed or lies at the surface closer to this enhanced field, a large enhancement in the Raman spectra is observed. Hence, SERS property of nanoparticles has been popularly used for developing sensitive detection methods for milk or milk products contaminants.

Lou et al. (2011) have developed a novel indirect SERS sensing method for detection of melamine in milk powder using 4-mercaptopyridine (MPY) modified gold nanoparticles as raman reporter molecules. Here, 13 nm AuNPs with three different Raman reporters (MPY, FITC, and rhodamine B) attached onto their surface via Au–S or Au–N bonds are used as SERS probes. SERS spectra of melamine-spiked milk powder with different AuNP reporters, MPY-based Raman probes are found to produce noticeable spectrum with strong raman-enhancing signals. The peak at 1015 cm^{-1} with MPY reporter is very prominent and used as an instructive peak for quantitative detection of melamine with a limit of detection of 0.1 ppb. Except enhanced signal melamine also induces the aggregation of AuNPs after addition of melamine. Selectivity of SERS method has been checked with metal ions such as K^+, Ca^{2+}, Na^+, Al^{3+}, Mg^{2+}, Zn^{2+}, Fe^{3+}, and Mn^{2+} as well as other adulterants like urea, threonine, phenylalanine, lactose, glucose, glycerine, and sucrose. Only melamine is found to greatly increase the Raman intensity even at 0.5 μM concentration whereas others have negligible effect even in the presence of 50 μM concentration.

Use of two different nanostructures, namely spherical magnetic-core gold-shell nanoparticles modified with 11-mercaptoundecanoic acid (11-MUA) and rod-shaped gold nanoparticles (nanorods) with surface tagged Raman-active compound 5,5′-Dithiobis (2-nitrobenzoic acid) [DTNB] has been demonstrated for melamine detection in milk (Yazgan et al. 2012). Here, the activated free carboxyl groups of 11-MUA on the surface of magnetic-core gold-shell nanoparticles interact with free amine groups of melamine whereas the DTNB raman label on gold nanorods surface links with the amino group of melamine through six double (NH…H) hydrogen bonds. These two nanoparticles crosslink through melamine and this complex is analyzed by Raman spectroscopy after magnetic sorting. As the size of complex increases the number of Raman-labeled nanoparticles also increases in the complex and the most intense Raman band of DTNB was observed at 1330 cm^{-1}. The intensity of SERS spectra at 1330 cm^{-1} is directly proportional to increasing melamine concentrations (0.2–20.0 mg L^{-1}). Melamine was extracted from skimmed milk samples using trichloroacetic acid (TCA) which precipitates proteins and extracts melamine simultaneously. The limit of detection (LOD) of this assay is 0.39 mg/L and the assay time is approximately 15 min (Fig. 3). To enhance the specificity and sensitivity for melamine detection, novel SERS substrates are fabricated. An aptamer-modified SERS nanosensor is used for the qualitative determination of melamine in milk (Dong et al. 2016). Aptamers are single-strand oligonucleotides that bind to their target in a 3-D conformation with high affinity and specificity. This study employs the stable complex formation strategy between thymine and melamine via NH–O and NH–N bond which indicates that thymine-rich DNA can be used for melamine recognition in aqueous medium with high specificity. Here, a Raman label DP(4,4′-dipyridyl) is

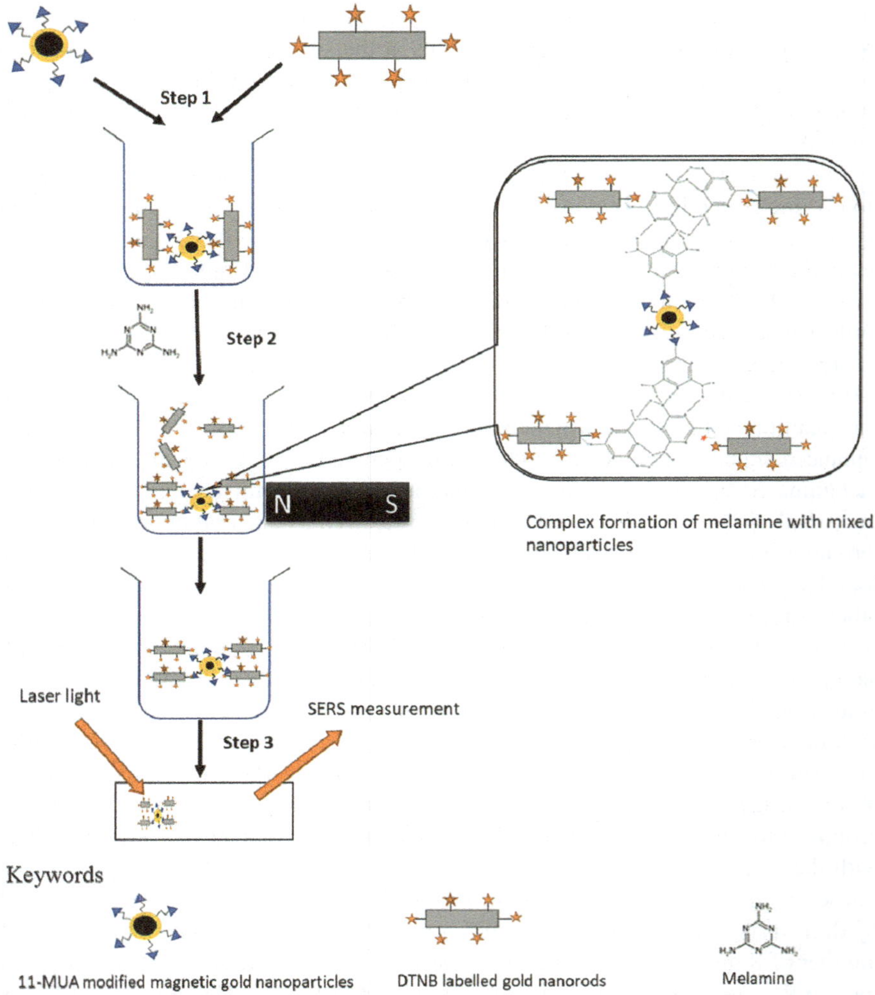

Fig. 3 Schematic diagram of melamine detection system. Step 1: mixing of 11-MUA modified gold nanoparticles and DTNB modified gold nanorods. Step 2: addition of melamine makes complex with mixed nanoparticles and separation from unbound nanoparticles magnetically. Step 3: measurement of SERS intensity (Yazgan et al. 2012)

tagged on the surface of gold nanoparticles. This is followed by introducing thiol-modified DNA (thymine rich aptamers) on the surface of these raman tagged gold nanoparticles to be finally used as SERS probe. In parallel, an oligonucleotide chip was fabricated by immobilizing aminated (5′-NH_2) DNA (thymine rich). Addition of SERS nanoprobes with melamine onto the chip leads to formation of "T–melamine–T" structure. SERS signal on the chip shows most intense Raman band of DP at

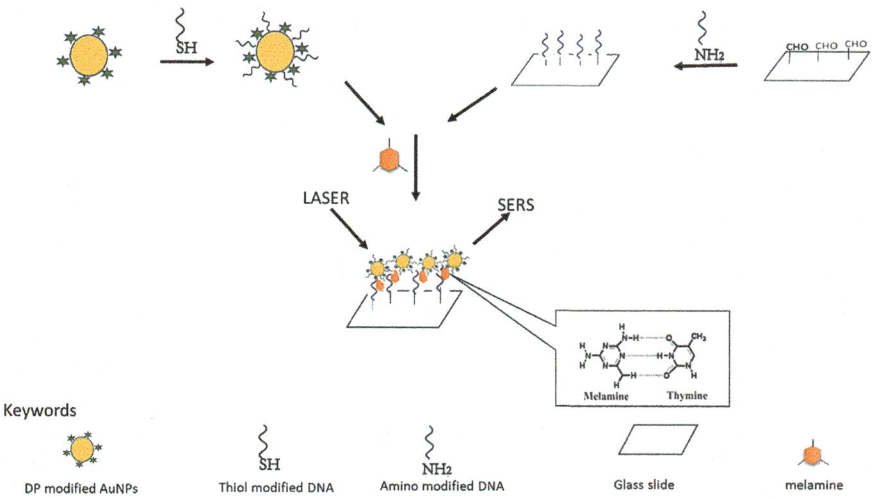

Fig. 4 Illustration of SERS detection using DNA-AuNP-Raman tag probes and oligonucleotide chip (Dong et al. 2016)

1612 cm^{-1} which increases with increasing melamine concentrations. This method enables to detect melamine with 1.0 pg/ml limit of detection (Fig. 4).

Aggregation of AuNPs in the presence of melamine enhances the intensity of SERS spectral peak at 715 cm^{-1}, characteristic of melamine (Giovannozzi et al. 2014). This enhancement forms the basis of melamine detection using 40 nm and 80 nm AuNPs. The raman intensity of 715 cm^{-1} peak is directly proportional to the melamine concentration which is used to plot a calibration curve. The lower limit of detection for melamine with this method is reported to be 0.17 mg/L. Chitosan modified silver nanoparticles (AgNPs) have been used for detection of Sodium sulfocyanate (SS), melamine (MEL), dicyandiamide (DCD) in whole milk powder (Li et al. 2016). This study combines the use of chromatography separation with surface-enhanced Raman scattering detection (SERS). Here, milk samples spiked with contaminants are pre-treated, filtered and then deposited directly on to the chitosan modified AgNPs paper (Ch/AgNPs/paper) at three points to detect three samples simultaneously (10 mm from each other). After drying, the substrate is placed upright inside a sealed jar pre-saturated with the mobile phase for chromatographic separation. After separation, the solvent front is marked and the chromatogram detected by the Raman spectrometer. The limit of detection for melamine, sulfocyanate, and dicyandiamide is reported to be 1 mg/L, 10 mg/L, and 100 mg/L, respectively.

Dhakal et al. (2018) have reported a simple surface-enhanced Raman spectroscopic method for on-site screening of tetracycline residue in whole milk. In this study, cyclodextrin capped silver colloid nanoparticles are used to enhance the SERS peak intensity specific to tetracycline (1322 cm^{-1}, 1621 cm^{-1}) in spiked milk. These cyclodextrin-capped Ag colloids nanoparticles produce low background signal and does not interfere with tetracycline detection. The characteristic spectral peaks are absent in unenhanced spectra, i.e., without silver colloids. The intensity of peak at 1322 cm^{-1} is observed at different concentrations of tetracycline (1000–0.01 ppm) in the presence of silver colloids and used to draw a calibration curve for determining the concentration of tetracycline in test sample. The peak at 1322 cm^{-1} can detect tetracycline in milk-tetracycline solution at concentration as low as 0.01 ppm. Recently, a SERS detection method uses AuNPs alone and Au@AgNPs as Raman-active substrates to produce a coffee ring effect for detecting urea and ammonium sulphate (AmS) adulterants in milk (Hussain et al. 2019). Pre-treated milk samplesspiked with urea and AmS) are mixed with AuNPs (1:1) and Au@AgNPs (2:1) and then dropped on gold-coated slide. This drop is evaporated under controlled condition at 25 °C and 50% humidity. This causes the formation of small coffee rings with regular circular shape of maximum 1.9 mm diameter as analyzed by microhyperspectral imaging system. The nanoparticles in the coffee rings enhance the Raman spectra of urea and ammonium sulphate at 1001 cm^{-1} and 980 cm^{-1}, respectively. The investigated limit of detection is found to be 5 mg/dL for both urea and ammonium sulphate. As evident from above studies, most of the SERS methods employ AuNPs or their composites as reporter probes probably because of their easy synthesis, facile chemistry, and distinct interaction with prime milk contaminants like melamine. The SERS methods are although sensitive but depend on the use of specialized and costly raman spectrometer. This restricts the translation of such methods to on-site detection.

3.2 Nanoparticles Based Colorimetric Detection

Localized surface plasmon resonance (LSPR) is another unique property of nanoparticles for which they are being popularly used in designing rapid and cost-effective detection methods. Such methods rely on the change in color of nanoparticles solution occurring due to the inter-particle coupling effect. According to this, analyte brings a change in inter-particle distance, thereby, causing nanoparticles aggregation to form large-sized particles and a shift in position and intensity of their LSPR peak (Fig. 5). This principle has been used to develop various nanoparticle-based colorimetric methods for detecting contaminants in milk or milk products.

Recently, metallic nanoparticle-based colorimetric detection has drawn more and more attention. Among all the metal nanoparticles explored in the colorimetric detection, gold nanoparticles have received much more consideration due to its characteristic properties such as biocompatibility, stability, easy preparation, and high extinction coefficients. Kelong et al. (2009), have reported visual detection of melamine using MTT(1-(2-mercaptoethyl)-1,3,5-triazinane-2,4,6-trione) stabilized gold nanoparticles. A color change induced by the triple hydrogen bonding recognition between melamine and gold nanoparticle forms the basis of its detection. MTT-stabilized gold nanoparticles (12 nm) synthesized by ligand-exchange reaction using MTT and citrate-stabilized Au nanoparticles are wine red in color with a LSPR

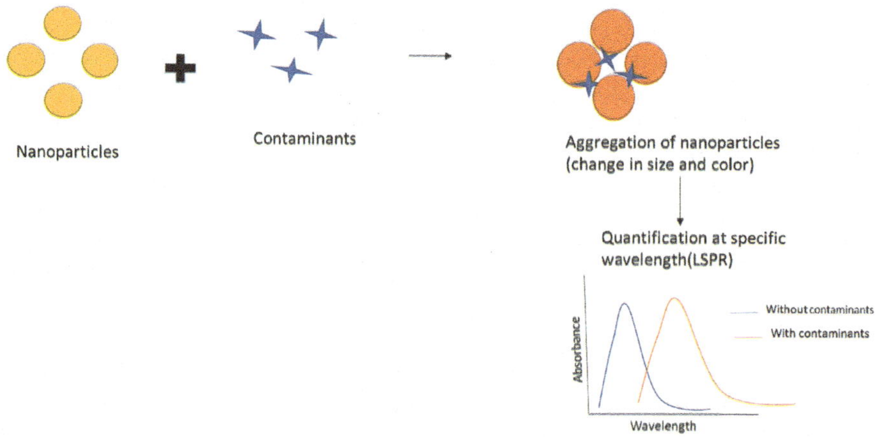

Fig. 5 Principle of nanoparticle based colorimetric methods

at 519 nm. Addition of melamine resulted in the aggregation of gold nanoparticles leading to change in wine red color of AuNPs to purple and to violet blue within a minute. The minimum concentration of melamine that brings a visual change in color of gold colloid is 20 nM or 2.5 ppb. The method is found to be specific when tested with milk samples spiked with melamine and other cross-reactants. It does not require any expensive instruments. Modification of gold nanoparticles with MTT could probably make this method cumbersome. To overcome this complexity, unmodified AuNPs have also been used as colorimetric probe. Fang et al. (2010), have used the interaction of melamine with 35 nm AuNPs through its amine groups because exocyclic amines are known to interact with AuNPs; however, other amines such as adenine may also show cross-reaction. To make the assay selective, cyanuric acid is added to melamine-AuNP solution. Cyanuric acid causes precipitation of melamine and brings a shift in absorbance peak of AuNPs-melamine solution; however, this shift is absent from AuNP-adenine solution. This dual signal readout makes melamine detection a very specific and rapid method with sensitivity of 40 ppb or 0.04 μg/ml. It is also demonstrated that readjusting salt concentration could govern the assay sensitivity and its dynamic range. Thus, by preconditioning AuNPs with specific NaCl concentration, assay detection range and sensitivity can be changed such that it falls in concentration range desired to be detected. The removal of casein is a pH dependent process and works efficiently at pH less than 6. Similar approach has been used by Li et al. (2010), by using AuNPs of two different sizes, i.e., 2.6 nm and 13 nm diameter. This study finds 2.6 nm AuNPs more sensitive to melamine probably due to their increased surface area. Change in the color of AuNPs solution from red to blue is noted spectrometric ally to achieve a limit of detection of 0.4 μg/ml. However, cross-reactivity with other amines like adenine was not investigated in this study. On the contrary, Hong et al. (2010), have used 13 nm unmodified AuNPs to sensitively (limit of detection of 0.025 μg/ml) detect melamine in infant milk powder with a melamine. It is emphasized that melamine interacts with AuNPs not only through its three exocyclic amino groups but also shows interaction of three nitrogen hybrid rings with weekly surface bound citrate ions and crosslinks the gold nanoparticles via coordinating interactions. Further, it is established that AuNPs does not show any color change in the presence of cross-reactants possessing single amino group but a change in color is increasingly noticed with cyclic compounds possessing two and three amine groups at a concentration of 0.1 μM and 10 μM, respectively. However, 1 μM melamine is sufficient to induce the color change in AuNPs. This study also proves that low concentration of $NaHSO_4$ (up to 0.8 mM) promotes the ligand exchange between citrate and melamine at AuNP surface. Hence, $NaHSO_4$ could significantly improve the aggregating sensitivity of AuNPs to trace melamine by enhancing the extinct ratio by ~2–4 fold.

Hu et al. (2017), have used 2,6-dimercaptopurine (DMP) modified AuNPs for colorimetric detection of $Cd^{2+}$and demonstrated its application in spiked milk samples. Here, DMP-AuNPs gets aggregated in the presence of cadmium due to chelation between the metal ion and the mercapto as well as amino groups of DMP. 4.0 μM DMP is found appropriate for highly sensitive response to Cd^{2+}. The limit of detection of the assay is 32.7 nM and the response is specific to Cd^{2+} when tested with other ions. The aggregation of modified or unmodified AuNPs in above cited literature is uncontrollable and non-directive resulting into formation of large aggregates. These large aggregates are unstable in solution as the resulting change in color fades away over time which leads to reduction in accuracy of quantitative determination and decreased long term stability. To address such issue, Chen et al. (2018), have used PEGylated AuNPs for sensitive colorimetric detection of melamine. Use of PEG with water-soluble inert groups and relatively long chain for asymmetrical modification of AuNPs increases their size and stabilizes these large size AuNPs in aqueous solution. This strategy of introducing a dense PEG layer on the restricted AuNP surface efficiently improves the long-term stability of 42-nm AuNPs and enhances the lower detection limit from 1 μmol L^{-1} to 1 nmol L^{-1}.

In addition to AuNPs, Han and Li (2010), have developed a sensitive and low-cost colorimetric method for detection of melamine using *p*-nitroaniline-modified(*p*NA) silver nanoparticles (Ag NPs). Here, *p*NA is used as a stabilizer and attached directly to AgNPs via linker carbon disulphide (CS2). When infant formula powder spiked with different concentrations of melamine is mixed with *p*NA modified AgNPs, the color of the resulting solution clearly changes from yellow to dark green within one minute. This color change indicates the aggregation of AgNPs and monitored by UV–vis spectroscopy. The absorption spectra of AgNPs exhibited a huge shift from 395 to 645 nm after addition of melamine with detection limit 0.1 ppm. Selectivity is also evaluated with different interferents such as 0.1 mM each of Na^+, $H_2PO_4^-$, HPO_4^{2-}, Cl^-, Ca^{2+}, Zn^{2+}, Fe^{3+}, glucose, 2-aminoethanesulfonic acid, phenylalanine, leucine, threonine and valine but no color change has been observed. This method showed relatively high sensitivity for melamine, however, the complex modification of nanoparticles limits its potential application. Kumar et al. (2014), and Ping et al. (2012), have reported the use of unmodified silver nanoparticles for detection of melamine in milk on the same principle, i.e., aggregation based color change. Addition of melamine causes aggregation of AgNPs bringing a change in color from yellow to red and a shift of absorption maxima. This aggregation is dependent upon the concentration of melamine which is used to achieve a limit of detection of ~0.04 mg/L and 2.32 μM, respectively. Ramalingam et al. (2017), have used three different silver nanoparticles (AgNPs): Borohydride-reduced AgNPs, citrate-capped AgNPs and polyvinyl pyrrolidone (PVP)-capped AgNPs for melamine detection. PVP-capped AgNPs are not found to be suitable for melamine detection. Whereas addition of Borohydride-reduced AgNPs and citrate-capped AgNPs in melamine spiked milk induces a color change from pale yellow to red. Besides visible color change based analysis, melamine is also quantified using RGB color sensor. The sensor operation is based on a color light to frequency conversion. It has ability to detect red, green and blue light and convert it into R, G, B color data. Here, a standard white light is allowed to incident on a cuvette with sample. The transmitted light is measured and correlated to melamine concentration for both AgNPs. All the methods discussed here require pre-treatment of milk sample to minimize the interference during detection.

3.3 Nanoparticle Based Fluorescence Assays

Nanoparticles have been used to develop fluorescent assays for detection of contaminants in milk samples broadly in two ways

1. **NPs as direct fluorescent probes**
   The inherent fluorescent property of some nanostructures makes them attractive alternative to organic fluorescent dyes to be used as direct fluorescent probes. Such probes have long decay time and enhanced extinction coefficients. Liposomes loaded with CdSe/3CdS/2ZnS core/shell/shell quantum dots (LQD) have been used as labels to develop a competitive fluorescent immunoassay for aflatoxin M1 detection in milk products (Beloglazova et al. 2013). Two carrier proteins viz cationic BSA and Ovalbumin are used to prepare AFM1 conjugates which are labelled with LQD. The IC50 value of this assay is 0.014 $\mu g\ kg^{-1}$ and could detect AFM1 at its minimum permissible limit set up by European commission (0.05 $\mu g\ kg^{-1}$). A smartphone-based fluorescence visualization device for visual detection of melamine in milk using gold nanoparticle@Carbon quantum dots nanocomposites (Au@CQDs) has been reported (Hu et al. 2018). This Au@CQDs is nonfluorescent complexes due to the fluorescence quenching of CQDs by AuNPs. Addition of melamine spiked milk samples of different concentrations results in AuNPs aggregation and CQDs liberation from the AuNPs surface with gain of fluorescence. The fluorescence of Au@CQDs enhanced gradually with an increasing amount of melamine achieving 3.6 nM limit of detection. The fluorescent images of milk samples are recorded by the smartphone under 365 nm UV light and the brightness of images is visually compared to determine melamine concentration. Song et al. (2015), have reported detection of antibiotic residues in milk by combining multi-color quantum dot-based

fluorescence immunoassay and visual detection by array simultaneously. Here, multicolor CdTe QDs with different wavelength (QD_{520nm}, QD_{565nm} and QD_{610n}) are conjugated with mAbs of Streptomycin (SM), tetracycline (TC) and penicillin G (PC-G). The sample of three antigens BSA-SM, BSA-TC, and BSA-PC-G are coated in each well of microtiter plate. The fluorescence intensity of each microwell in the plates is automatically recorded by a fluorescence spectrophotometer with the linear range of 0.01–25, 0.01–25, and 0.01–10 ng/mL, respectively. It is investigated that limit of detection for each of the three antibiotics is 0.005 pg/mL. This study also encompasses visual detection of QD_{520nm}, QD_{565nm}, and QD_{610nm} as hunter green, chartreuse, and vermillion fluorescence colors, respectively, under excitation with blue light. The quantity of each antibiotic is determined by different shades of each color image (Fig. 6).

2. **NPs for immobilization of capture reagents (antibodies/aptamers)**
High surface area and facile surface chemistry of nanoparticles is magnificently employed to immobilize capture reagents like antibodies and aptamers onto solid surfaces at high concentrations. Antibody-coated magnetic nanoparticles are used to develop a fluorescent assay for detection of aflatoxin M1 in raw milk (Atanasova et al. 2017). The AFM1 coupled with fluoresceinamine competes with unlabelled AFM1 in test sample to bind the antibodies on the surface of magnetic nanoparticles. The developed immunoassay has a linear range of 3.0 to 100 pg/mL in bovine milk with the detection limit 2.9 pg/mL. This study has also analyzed the influence of fat, pH and other milk species and observed that high fat concentration reduces the sensitivity of AFM1assay. Zhang et al. 2019, have reported a fluorescence aptasensor based on an autocatalytic Exonuclease III (Exo III)-assisted signal amplification strategy to detect AFM1 in milk.

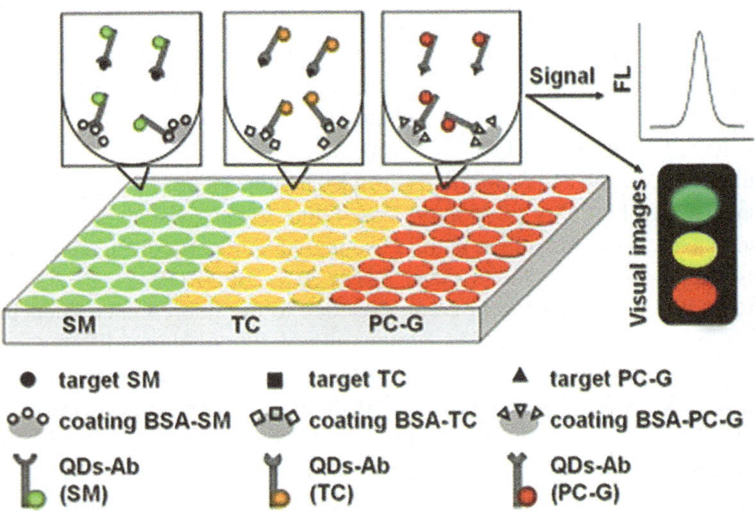

Fig. 6 A schematic illustration of the multi-analyte assay for three kinds of antibiotics by mQD-cFIA (Song et al. 2015)

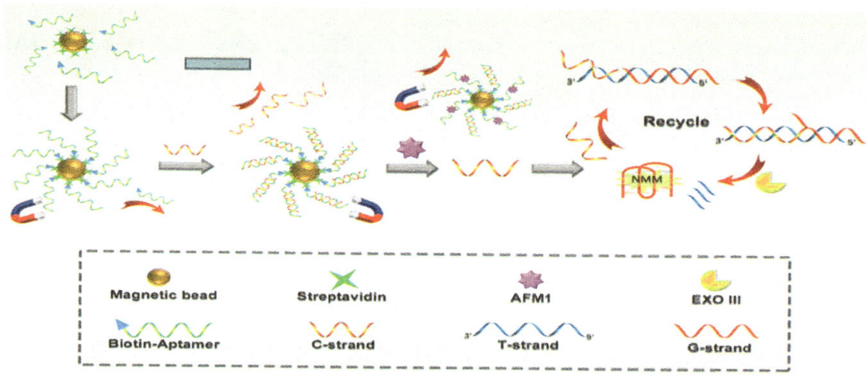

Fig. 7 Schematic representation of sensitive fluorescent detection of AFM1 based on Exo III-assistantsignal amplification recycles (adapted from Zhang et al. 2019 under the Creative Commons Attribution License)

Magnetic nanobeads loaded with aptamer/C-strand duplex DNA are used in the study. In the presence of AFM1, aptamer binds with AFM1; thereby, releasing C-strand. This released C-strand can trigger a simple fluorescence signal amplification recycle. Exo III is used as an amplifying biocatalyst in this amplification procedure to selectively digest duplex DNAs from blunt or recessed 3'-termini to produce a guanine-rich DNA strand with a G-quadruplex structure(Fig. 7). Here, the limit of detection is 0.01 ng/ml for all milk samples (whole milk, skimmed milk, and juice milk).

3.4 *Nanoparticle Based on-off Apta-Assays*

ssDNA or RNA oligonucleotides which adopts a three-dimensional conformation to recognize their targets specifically are known as aptamers. Aptamers have been used popularly as an alternative biorecognition element in developing assays for milk contaminant detection. Nanoparticle-based apta-assays are mostly on-off assays wherein, absorption of a single strand aptamer on the surface of nanoparticles alters (on/off) any of their unique property. This alteration in the property of nanoparticle is regained or reverted to original in the presence of analyte or contaminant. Because aptamers desorb from the nanoparticle surface in the presence of its specific target analyte owing to its higher affinity for the target (Fig. 8).

Song et al. (2012), have developed an aptasensor using ampicillin-specific AMP17 aptamer and gold nanoparticles. When FAM (fluorescein)-labelled AMP17 aptamer is adsorbed on AuNP surface, the fluorescence of FAM is quenched. In the presence of Ampicillin positive test or standard sample, the Fam-AFM17 aptamers desorbs from AuNP surface and hence the fluorescence of FAM is resumed. Ampicillin is detected with this principle upto a level of 2 ng/mL in milk samples. Zhao et al. (2017), have

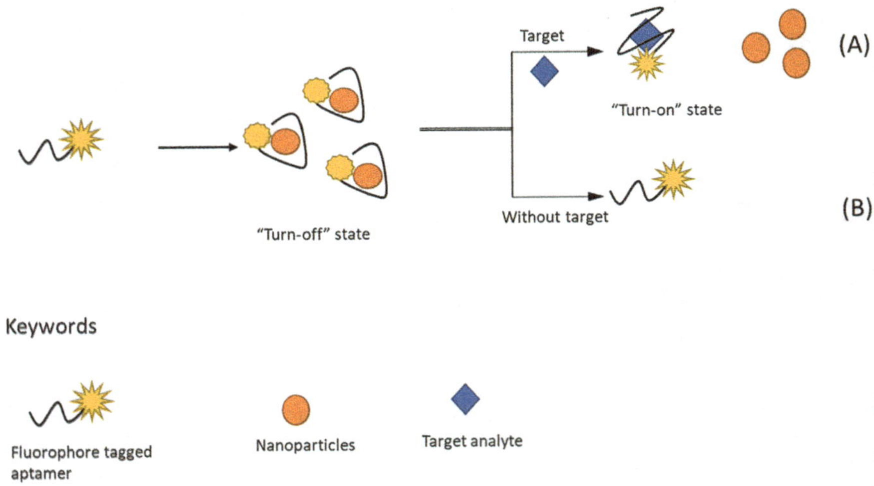

Fig. 8 Schematic illustration of nanoparticles quenching based sensing for target molecule detection

developed an on/off aptasensor using intrinsic peroxidase activity of gold nanoparticles and STR1 aptamer for specifically detecting streptomycin (STR) in milk. STR1 aptamers upon absorption on the surface of AuNPs inhibits their intrinsic peroxidase activity. In the presence of STR, STR1 aptamers leaches out of the AuNP surface and a regain in the peroxidase activity of AuNP is seen. The reaction is monitored calorimetrically by addition of colorimetric substrate to determine the peroxidase activity of AuNPs. The AuNPs catalyses oxidation of ABTS (peroxidase substrate) by hydrogen peroxide, which produces the ABTS radical cation ($ABTS^+$) with green color and displays a characteristic absorption peak at 733 nm. This citrate capped AuNPs are proven to be appropriate catalyst to fabricate this highly sensitive colorimetric aptasensor with LOD of 86 nM. Ren et al. (2019), have used combination of aptamer-coated magnetic nanoparticles (Apt-MNPs) and CdTe QD-labeled complementary DNA (QDs-ssDNA2) for detection of *S. Typhimurium* in milk. Here, QD-ssDNA2 is incubated with Apt-MNPs to form an aptamer-complementary DNA duplex as a detection probe. Upon addition of *S. Typhimurium*, QD-ssDNA2 is replaced by the bacteria and released from the Fe_3O_4 MNPs. Release of QD-ssDNA2 causes significant emission peak shifts from approximately 599 to 612 nm. The difference in fluorescence intensity is used to sensitively detect *S. Typhimurium*, with a low detection limit of 1 cfu/mL. The reported method is highly specific as no remarkable change in fluorescence intensity is observed for other pathogens such as *Staphylococcus aureus*, *Escherichia coli* O157:H7, *Listeria monocytogenes*, *Bacillus cereus*, *Salmonella enteritidis*, and *Pseudomonas aeruginosa*.

3.5 Nanotechnology Based Lateral Flow Immunoassay

The underlying principle of lateral flow assay (LFI) is based on the interaction of an analyte with the antibody immobilized on the membrane while the former moves across the strip laterally through capillary action. Along its path, the analyte in the sample first interacts with an antibody-probe forming a complex with it and subsequently interacting with the immobilized antibody. LFIA can be developed in competitive or sandwich formats (Fig. 9). A more detail information on LFIA principle and its formats can be found in various reviews (Tripathi 2018). LFIA commonly uses gold nanoparticles as probe or label because of easy visual detection, their facile surface chemistries, easy synthesis, and stability. This section presents LFIAs designed for milk contaminants detection by employing probes other than AuNPs or their composites.

Berlina et al. (2013), have developed quantum dots based lateral flow test for chloramphenicol (CAP) detection in milk. The quantum dots conjugated with mAb against CAP are used as a label. CAP-soybean trypsin inhibitor (STI) conjugate is immobilized onto the nitrocellulose membrane as the test line and goat antimouse antibody as the control line. The test strip is run with pre-treated CAP-spiked milk sample and the fluorescence intensity of the test and control lines is measured with a portable power-dependent photometer. The limit of visual detection is 1 ng/mL and with photometer is 0.2 ng/mL.

Haiyan et al. (2016), have developed sandwich immunochromatographic assay (ICA) for the screening of *Escherichia coli* O157:H7 (*E. coli* O157:H7) in bovine milk using IgG as the capture antibodies and modified IgY conjugated with a core/shell-structured super-paramagnetic Fe_3O_4/Au composite nanoparticle (GoldMag) as the detecting antibody. Gold allows visual detection and magnetic nanoparticles allow quantitative measurements by measuring the intensity of

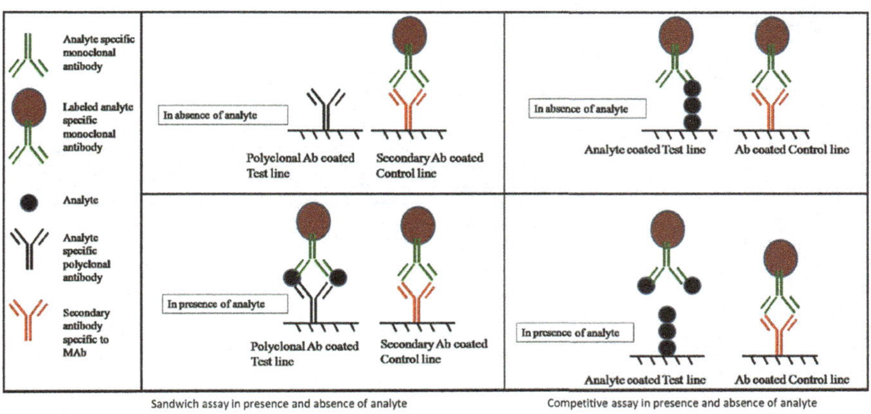

Fig. 9 Representation of sandwich and competitive formats in LFIA in absence and presence of analytes

magnetic signal (MS) using Magnetic Assay Reader (MAR). LOD of the assay is 32 CFU/mL and 10^2 CFU/mL in PBS and pre-treated free milk samples respectively. The specificity of this method is also tested with four kinds of *E. coli* O157:H7 strains and other related bacteria resulting in no cross reactivity. Wu et al. (2017), have reported background fluorescence quenching immunochromatographic assay (bFQICA) to detect aflatoxin M1 (AFM1) and chloramphenicol (CAP) in milk. Here, the nitrocellulose membrane is completely pre-coated with fluorescein dye to give background fluorescence (F_0) which is quenched in the presence of gold nanoparticles (F_1). AFM1and CAP antibodies are labelled with gold nanoparticles (AuNP). AFM1-BSA or CAP-BSA is coated as test line on nitrocellulose membrane. Test samples are incubated with AuNPs-labeled antibody and then dispensed on the sample pad. The unbound AuNPs-labeled antibody binds with AFM1 or CAP on the test line and quenches the fluorescence on the test line. Ratio of fluorescence on the test line (F_1/F_0) is inversely related to the concentration of AFM1 and CAP in the test sample. The detection limit for AFM1 is 0.0009 ng/mL, and for the CAP is 0.0008 ng/mL measured by bFQICA system (Fig. 10). Hu et al. (2019), have also employed similar principle of background fluorescence quenching using magnetic beads with colloidal gold nanoparticles (AuMB) for detection of sulphamethazine (SM_2) in raw milk. It differs from the method of Wu et al., in having background fluorescence only at the test line rather than on the complete NC membrane. For this, mixture of SM2–BSA with FM–BSA (fluorescein-BSA) is coated as the test line and anti-SM2 mAb labelled with AuMB is used as quencher of fluorescence at test line. The Fluorescence signal at the T line is measured by the FM-LFA reader. Qualitative detection is accomplished on the basis of naked-eye visualization of colloidal gold signals, and quantitative detection by the fluorescence quenching method. The

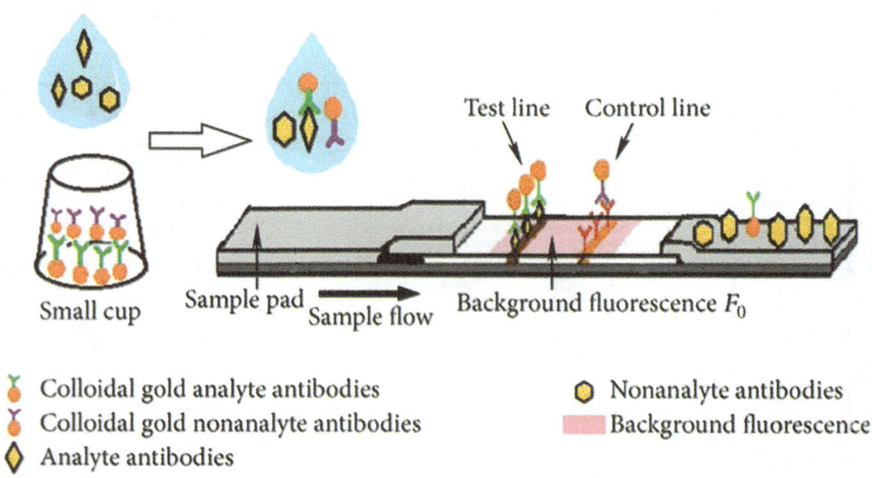

Fig. 10 Schematics of test strip used in the background fluorescence quenching immunochromatographic assay (adapted from Wu et al. 2017 under the Creative Commons Attribution License)

limit of detection and linear range achieved is 0.39 and 0.5–200 ng/mL, respectively. The specificity of the AuMB–LFA is evaluated by testing 10 sulphonamide drugs and 2 non-sulphonamide antibiotic drugs (enrofloxacin and tetracycline) with the concentration of 25 ng/mL. The cross reactivity of AuMB–LFA for sulphamerazine, sulphathiazole, sulphadimethoxine, sulphamonomethoxine, and sulphamethoxazole is 72.0%, 60.0%, 57.9%, 48.6%, and 29.4%, respectively, whereas the cross reactivity of the four other sulphonamide drugs and two non-sulphonamide antibiotics drugs is less than 10%.

3.6 Nanotechnology Based Electrochemical Biosensor

Nanoparticles or their hybrids have been used for electrochemical analysis of contaminants in milk samples either as reporter, or to immobilize capture reagents on working electrodes or to modify working electrode to enhance their conductivity. A typical three-electrode electrochemical cell consists of a working electrode that comprises chemically stable and conductive materials, such as platinum, gold, or carbon (e.g., graphene oxide, graphene nanoplatelets, carbon nanotubes), a reference electrode (e.g., Ag/AgCl), and a counter electrode (e.g., graphite, platinum, gold). The reactions between medium and working electrode generates a measurable potential or charge accumulation which can be analyzed by cyclic voltammetry, impedimetric, or amperometry.

Owino et al. (2008), have developed an electrochemical immunosensor based on Polythionine/Gold Nanoparticles for the detection of aflatoxin B_1. In this study, aflatoxin B1- BSA conjugate is immobilized on the polythionine (PTH)/gold nanoparticles (AuNP)-modified glassy carbon electrode (GCE) followed by addition of horseradish peroxidase (HRP) is used to prevent the non-specific binding. DPV measurements of GCE|PTH|AuNP|AFB1-BSA-HRP are performed in 3.2 μM H_2O_2. Formation of antigen-antibody complex prevents HRP from catalysing the breakdown of H_2O_2 thereby decreasing the signal intensity. Anti-AFB1 antibody is pre-incubated with various concentration of AFB1 and then coated on to the working electrode. Only free or unbound antibody forms a complex with AFB1 immobilized on working electrode. Hence, the concentration of AFM1 in test sample is inversely related to the signal intensity. The lower detection limit achieved is 0.07 ng/mL for aflatoxin B_1 (Fig. 11).

Ezhilan et al. (2016), have developed a highly sensitive acetylcholinesterase (AChE) cyclic voltammetric biosensor using zinc oxide (ZnO) nanospheres modified Pt electrode for the detection of melamine and urea in adulterated milk samples. Here, ZnO nanoparticles are used as a nano-interface between AChE enzyme and platinum (Pt) electrode due to their high conductivity and enhanced electron transfer capability. In this assay, AChE hydrolyses ATChCl (acetylthiocholine chloride) to thiocholine which is then converted to thiocholine (ox) by Pt/ZnO/AChE/Chitosan resulting in an effective electron transfer. Upon addition of melamine, urea or binary

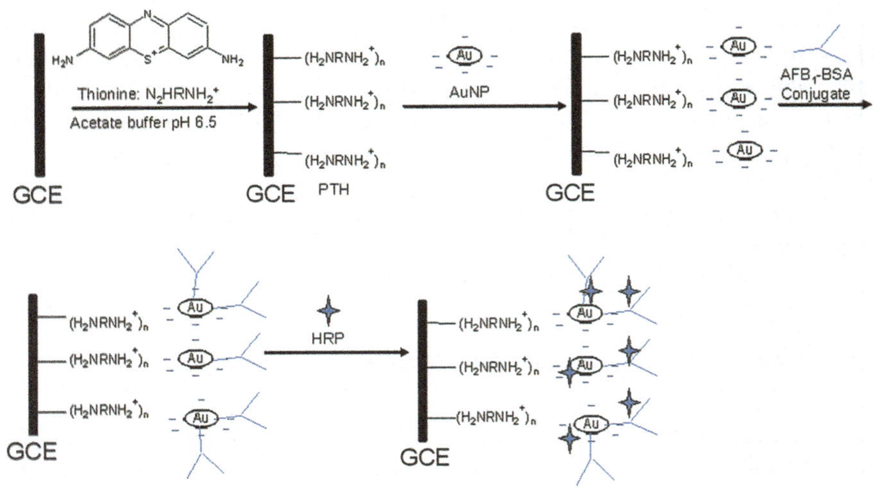

Fig. 11 Preparation of aflatoxin B1 immunosensor. Step 1: Polymerization of Thionine on Graphene carbon working electrode. Step 2: formation of AuNP layer. Step 3: loading of AFB1-BSA conjugate. Step 4: blocking of AFB1-BSA conjugate layer with HRP (adapted from Owino et al. 2008 under Creative Commons Attribution License)

mixture of melamine and urea inhibited AChE activity by forming AChE-melamine-ATChCl and AChE-urea-ATChCl complexes. This reduces the yield of thiocholine and a subsequent lowering down of electrochemical response. The lower detection limit for melamine and urea is 3 pM and 1 pM, respectively. Shadjou et al. (2018), have reported an electrochemical sensing method using silver nanoparticles dispersed on α-cyclodextrin/GQDs (graphene quantum dots) modified glassy carbon electrode (GCE) for the detection of aflatoxin M_1 (AFM1) in milk samples. Cyclic voltammetry measurements show increasing oxidation peak current with increasing concentration of AFM1 due to its electrostatic interaction with AgNPs; whereas, no peak appears in the absence of AFM1. AgNPs species is the active moiety that efficiently speeds up the oxidation of AFM1. The calibration curve for AFM1 concentration is linear in the range of 0.015 mM to 25 mM with a low limit of quantification of 2 μM. Kumar et al. (2017), have reported a mediator-free electrochemical sensor using graphitized nanodiamonds (GND) with polyaniline (PANI) for detection of urea in milk. The sensor is developed by immobilizing urease-conjugated GND/PANI nanocomposites onto the carbon nanotube screen printed electrodes (CNT-SPE). Upon addition of milk samples spiked with various concentrations of urea, the immobilized urease converted urea into ions, generating both positive and negative ions. These generated ions adhered onto the surface of nanomaterials, resulting in change in current at 0 V. The linear range for urea detection by GND/PANI modified electrodes lies in the range of 0.1–0.9 mg mL^{-1}, with a detection limit of 0.05 mg mL^{-1}. This group further used graphene nanoplatelets and graphitized nanodiamonds (f-GNPlts/GNDs) for urea detection on the same principle (Kumar et al. 2019). The method enhanced the limit of detection to 5 μg mL^{-1}. Khoshfetrata et al. (2017), have reported a closed bipolar electrode electrochemiluminescence (BPE-ECL) aptasensor for aflatoxin M1 (AFM1) detection in milk using luminol-functionalized silver nanoparticle-decorated graphene oxide (GO-L-AgNPs) as a signalling probe and gold-coated, magnetic nanoparticles (GMNPs) asthe immobilization support for the aptamer. The aptamer-conjugated GMNPs (Apt-GMNPs) are interacts with the GO-L-AgNPs via π-π interactions between the unhybridized parts of the aptamer and GO (Apt-GMNPs-GO-L-AgNPs). Apt-GMNP-GO-L-AgNPs is dropped onto the anodic poles of the individual gold BPE array under a magnetic field below the surface electrode. Thionine is used as electron-transfer mediator and able to serve as an electrochemical indicator to enhance the ECL signal at the anode through its reduction at the cathode (Fig. 12). AFM1 spiked milk samples are added on Apt-GMNP-GO-L-AgNPs at anodic pole, the GO-L-AgNPs nanocomposite releases from Apt-GMNPs surface as a result of the strong interaction between the aptamer and AFM1. The ECL signal of the luminol/H_2O_2 is monitored as an analytical signal using a smartphone in a dark room. On increasing the concentration of the AFM1 target (5–150 ng/mL), the ECL intensities of the luminol/ H_2O_2 system decreases linearly achieving detection limit of 0.01 ng/mL. An electronic tongue system using gold nanoparticles stabilized with poly (allylamine hydrochloride) (Au@PAH NPs) to form nanostructured Layer-by-Layer (LbL) films with seven bilayers with metal tetrasulfonated phthalocyanines (MTsPc) has been reported for milk quality analysis (Luiza et al. 2015). Two e-tongue sensors are fabricated, namely Au@PAH/CuTsPc and Au@PAH/NiTsPc LbL to distinguish milk samples with different fat contents (fat, semi-skimmed and skimmed milk samples). Milk samples with different fat concentrations are mixed

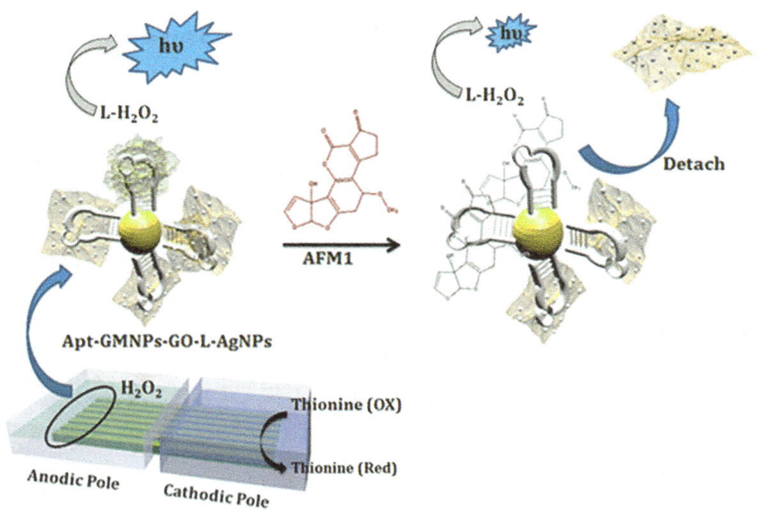

Fig. 12 Schematic illustration of the amplified visual electrochemiluminescence detection of AFM1 (Khoshfetrata et al. 2017)

with phosphate buffer and analyzed by the e-tongue system without any prior pre-treatment using electrical capacitance measurements. The data are collected in the range from 1 Hz up to 1 MHz, with an AC applied voltage of 50 mV.

3.7 *Other Nanoparticle Based Assays*

FRET Based Nanosensor

The upconversion nanoparticles (UCNPs) are nanoscale particles (diameter 1–100 nm) that exhibit photon upconversion in which two or more incident photons of relatively low energy are absorbed and converted into one emitted photon with higher energy in infrared region. They show unique properties like improved detection sensitivity, minimum photodamage, low toxicity, good chemical, and physical stability and no autofluorescence. On the basis of these advantage Wu et al. (2015), reported a nanosensor based on fluorescence resonance energy transfer between up conversion nanoparticles and gold nanoparticles for detection of melamine in raw milk. UCNPs of NaYF4:Yb3+, Er3+ show a strong fluorescent signal at 550 nm which gets quenched upon electrostatic interaction with AuNPs due to FRET. Maximum quenching is observed upon addition of 1.23 nM AuNPs into the UCNPs solutions. With melamine-spiked milk samples, fluorescence of UCNPs is "turn on" with a rapid color change from red to blue. This is due to interaction between AuNPs and melamine which releases gold nanoparticles from the UCNPs surface and their aggregation (Fig. 13). The assay achieved a limit of detection of 18 nM for melamine.

Nanoparticles for Contaminant Extraction from Milk Samples

Magnetic nanoparticles have been used as support to immobilize specific capture reagents like antibodies or aptamers on their surface. These modified nanoparticles are used to extract corresponding analyte from milk samples by using an external magnetic field. The nanoparticles bound analyte is analyzed either directly or after eluting from their surface. Aptamer-functionalized magnetic nanoparticles (AMNPs) have been used as magnetic solid-phase to extract AFM1 from treated milk samples (Khodadadi et al. 2018). This adsorbent is further washed and elution of AFM1 is

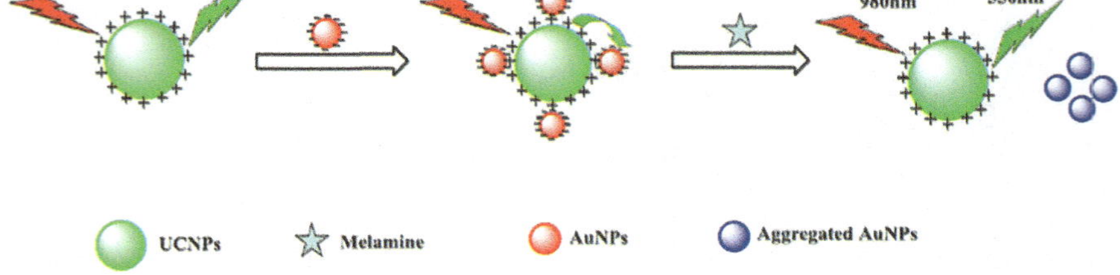

Fig. 13 Schematic illustration of the "turn-on" fluorescence assay for detection of melamine(Wu et al. 2015)

carried out by dichloromethane/methanol/acetic acid. The eluate containing AFM1 is dried and mixed with mobile phase and injected to HPLC for quantitative determination of AFM1. The calibration plot is linear over the 0.3 to 1 ng/L and 5 to 50 ng/L AFM1 concentration with a limit of detection of 0.2 ng/L. Fengying et al. (2015), have reported the use of hydrophilic polymer dextran coated and superhydrophilic polymer PMPC (PHEMA/PMPC IMB)-coated iron oxide immunomagnetic nanoparticles for detection of *Salmonella* in raw milk. Both nanoparticles have surface immobilized anti-*Salmonella* antibodies. The capture efficiency of PHEMA/PMPC IMB is higher (63–69%) than dextran-coated nanoparticle conjugates as measured by UV spectroscopy. Salmonella contaminated milk of various concentration is incubated with PHEMA/PMPC IMB and captured IMB-bacterial complex is further cultured with peptone. The precipitation of enrichment broth is directly added for PCR reaction resulting in a PCR product band even at the lowest concentrations of 10 cfu/mL. The detection limit achieved as low as 10^1 cfu/mL.

4 Conclusion

Milk is considered as a complete diet for infants, children and is a rich source of calcium for all. However, poor milk quality is a growing concern due to its adulteration with chemicals, antibiotics, etc., by human beings in order to satisfy their interests. This drives the need for developing some sensitive and rapid detection methods of widely used adulterants or contaminants of milk. Nanotechnology is contributing significantly in accurate and sensitive detection of contaminants in milk samples which is evidenced well in this chapter. Gold nanoparticles or their hybrids or composites are the most preferred nanoparticles for developing detection methodologies based on different principles. Besides gold nanoparticles, quantum dots, magnetic nanoparticles, and silver nanoparticles are also employed in devising detection systems for contaminant detection in milk. Most of the reported assays have established a proof of concept of some newer detection strategy and tested it on spiked milk samples. It is also observed that large number of scientific literatures uses nanoparticles for detecting melamine contamination in milk. Such studies have tracked selective changes in any inherent property of the nanostructures such as SPR, or SERS in the presence of varying contaminant concentrations. It could be concluded that the structure of the melamine (presence of amine groups) favours selective interaction with metallic nanostructures; in particular, gold-based nanostructures. However, for detection of other contaminants like aflatoxins, antibiotics, etc., nanoparticle surface needs to be modified with specific biorecognition elements including antibodies and aptamers. The emerging trend is toward developing a detection strategy independent of any bio-recognition element in order to reduce the cost and time of analysis. To achieve this, the contaminant in question must specifically change any unique property of the nanoparticle chosen. Hence, the interaction of different contaminants and their closely related structural analogues must be investigated with various nanostructures. The use of nanoparticles for extraction and enrichment of contaminants from milk samples also employs antibodies or aptamers. Alternate or cheaper extraction strategies must be explored to minimize the cost of assay.

References

Akhtarz S, Ahad K (2017) Pesticides residue in milk and milk products: mini review. Pak J Anal Environ Chem [S.l.] 18(1):37–45. ISSN 2221–5255

Algeri BTA, Castagnaro D, Bortoli K, Souza C, Drunkler DA, Badiale-Furlong E (2016) Mycotoxins in bovine milk and dairy products: a review. J Food Sci 81:R544–R552

Asredie T, Engdaw TA (2015) Antimicrobial residues in cow milk and its public health significance. World J Dairy Food Sci 10(2):147–153

Atanasova M, Vasileva N, Godjevargova T (2017) Determination of aflatoxin ml in milk by a magnetic nanoparticle-based fluorescent immunoassay. Anal Lett 50(3):452–469

Bates F, Busato M, Piletska E, Whitcombe MJ, Karim K, Guerreiro A, del Valle M, Giorgetti A, Piletsky S (2017) Computational design of molecularly imprinted polymer for direct detection of melamine in milk. Sep Sci Technol 52:1441–1453

Bayman P, Baker JL (2006) Ochratoxins: a global perspective. Mycopathologia 162:215–223

Beloglazova NV, Shmelin PS, Goryacheva IY, De Saeger S (2013) Liposomes loaded with quantumdots for ultrasensitive on-site determination of aflatoxin M1 in milk products. Anal Bioanal Chem 405:7795–7802

Berlina AN, Taranova NA, Zherdev AV, Vengerov YY, Dzantiev BB (2013) Quantum dot-based lateral flow immunoassay for detection of chloramphenicol in milk. Anal Bioanal Chem 405:4997–5000

Chen C, Mi X, Yuan Y, Chen G, Ren L, Wang K, Zhu D, Qian Y (2014) A preliminary risk assessment of potential exposure to naturally occurring estrogens from Beijing (China) market milk products. Food Chem Toxicol 71:74–80

Chen X-Y, Ha W, Shi YP (2018) Sensitive colorimetric detection of melamine in processed raw milk using asymmetrically PEGylated gold nanoparticles. J Talanta 10:070

Commission Regulation (EU) No 37/2010 of 22 December 2009 on pharmacologically active substances and their classification regarding maximum residue limits in foodstuffs of animal origin. J Eur Commun, No L 15

Dhakal S, Chao K, Huang Q, Kim M, Schmidt W, Qin J, Broadhurst CL (2018) A simple surface-enhanced raman spectroscopic method for on-site screening of tetracycline residue in whole milk. Sensors 18:424

Dong N, Hu Y, Yang K, Liu J (2016) Development of aptamer-modified SERS nanosensor and oligonucleotide chip to quantitatively detect melamine in milk with high sensitivity. Sensors and Actuators B 228:85–93

European Commission (EC) Regulation (2003) Commission Regulation (EC) No. 2174/2003 of 12 December 2003 amending Regulation (EC) No 466/2001 as regard aflatoxins. Off J Eur Union L326:12

Ezhilan M, Gumpu MB, Ramachandra BL, Nesakuma N, Babu KJ, Krishnan UM, Rayappan JB (2016) Design and development of electrochemical biosensor for the simultaneous detection of melamine and urea in adulterated milk samples. Sen Actuators 09:100

Fang W, Robert L, Stacy C, Steven L, Dean H, Na L (2010) Rapid detection of melamine in whole milk mediated by unmodified gold nanoparticles. Appl Phys Lett 96:133702

Fengying D, Miao Z, Bingbing H, Yongjun S, Qunwei T, Meihong D, Xin Z (2015) Immunomagnetic nanoparticles based on a hydrophilic polymer coating for sensitive detectionof Salmonella in raw milk by polymerase chain reaction. RSC Adv 5:3574–3580

Fischer WJ, Schilter B, Tritscher A, Stadler R (2015) Contaminants of milk and dairy products: contamination resulting from farm and dairy practices. Encyclopaedia Dairy Sci 887–897

Giovannozzi AM, Rolle F, Sega M, Maria CA, Daniela M, Andrea MR (2014) Rapid and sensitive detection of melamine in milk with gold nanoparticles by surface enhanced Raman Scattering. Food Chem 159:250–256

Haiyan X, Baoyan Z, Baoyuan H, Zhanyong W, Chao C (2016) Rapid immunochromatographic assay for *Escherichia coli* O157:H7 in Bovine milk using IgY labeled by Fe3O4/Au composite nanoparticles. Food Sci Technol Res 22(1):53–58

Han C, Li H (2010) Visual detection of melamine in infant formula at 0.1 ppm level based on silver nanoparticles. Analyst 135:583–588

Heeschen H, Bluthgen A, Burt R (1997) Residues and contamination in milk and milk products. International dairy federation

Hong C, Bianhua L, Guijian G, Zhongping Z, Ming-Yong H (2010) A simple, reliable and sensitive colorimetric visualization of melamine in milk by unmodified gold nanoparticles. Royal Soc Chem 135:1070–1075

Hu MH, Huang WH, Suo LL, Zhou LH, Ma LF, Zhu HF (2017) Gold nanoparticles functionalized with 2,6-dimercaptopurine for sensitive and selective colorimetric determination of cadmium (II) in food, biological and environmental samples. Anal Methods 9:5598–5603

Hu X, Shi J, Shi Y, Zou X, Arslan M, Zhang W, Huang X, Li Z, Xu Y (2018) Use of a smartphone for visual detection of melamine in milk based on Au@Carbon quantum dots nanocomposites. Food Chem

Hu S, Huang Z, Chen W-Y, Xing K-Y, Peng J, LaiW -H (2019) Dual signal insight: field-efficient qualitative/quantitative detection of sulphamethazine in raw milk. Food Agric Immunol 30(1):163–177

Hussain A, Sun DW, Pu H (2019) SERS detection of urea and ammonium sulphate adulterants in milk with coffee ring effect. Food Addit Contam Part A 36(6):851–862

IARC Monographs on the evaluation of carcinogenic risks to human. Some traditional herbal medicine, some mycotoxins, naphthalene and styrene IARC, Lyon, France, 2002, No. 82

Jouan PN, Pouliot Y, Gauthier SF, Laforest J-P (2006) Hormones in bovine milk and milk products: a survey. Int Dairy J 16:1408–1414

Kamthania M, Saxena J, Saxena K, Sharma DK (2014) Methods of detection & remedial measures. Int J Engg Tech Res 1:15–20 Trict Area, Romania

Kelong A, Yanlan L, Lehui L (2009) Hydrogen-bonding recognition-induced color change of gold nanoparticles for visual detection of melamine in raw milk and infant formula. J Am Chem Soc 131:9496–9497

Khodadadi M, Malekpour A, Mehrgardi MA (2018) Aptamerfunctionalized magnetic nanoparticles for effective extraction of ultratrace amounts ofaflatoxin M1 prior its determination by HPLC. J Chroma 06.022

Khoshfetrata SM, Bagherib H, Mehrgardia MA (2017) Visual electrochemiluminescence biosensing of aflatoxin M1 based on luminol-functionalized, silver nanoparticle-decorated graphene oxide. J Bios 09.035

Kumar N, Seth R, Kumar H (2014) Colorimetric detection of melamine in milk by citrate-stabilized gold nanoparticles. Anal Biochem 456:43–49

Kumar V, Mahajan R, Kaur I, Kim Ki-H (2017) A simple and mediator-free urea sensing based on engineered nanodiamonds with polyaniline nanofibers synthesized in situ. ACS Appl Mater Interfaces

Kumar V, Kaur I, Arora S, Mehla R, Vellingiri K, Kim K-H (2019) Graphene nanoplatelet/ graphitized nanodiamond-based nanocomposite for mediator-free electrochemical sensing of urea. Food Chem: X

Li L, Li B, Cheng D, Mao L (2010) Visual detection of melamine in raw milk using gold nanoparticles as colorimetric probe. J Food Chem 03.032

Li D, Lv DY, Zhu QX, Li H, Chen H, Wu MM, Chai YF, Lu F (2016) Chromatographic separation and detection of contaminants from whole milk powder using a Chitosan-modified silver nanoparticles surface-enhanced raman scattering device. Food Chem 12.040

Lou T, Wang Y, Li J, Peng H, Xiong H, Chen L (2011) Rapid detection of melamine with 4-mercaptopyridine-modified gold nanoparticles by surface-enhanced Raman scattering. Anal Bioanal Chem 401:333–338

Luiza AM, Vanessa PS, Adriana P, Rafaela CS, Luiz HCM, Daniel SC (2015) Electronic tongue based on nanostructured hybrid films of gold nanoparticles and phthalocyanines for milk analysis. J Nanomater 1–7

Magan N (2006) Mycotoxin contamination of food in Europe: early detection and prevention strategies. Mycopathologia 262:245–253

Malekinejad H, Rezabakhsh A (2015) Hormones in dairy foods and their impact on public health—a narrative review article. Iran J Public Health 44:742–758

Maryam J (2017) A review paper on melamine in milk and dairy products. Dairy Vet Sci J 1(4):555566

Mauer LJ, Chernyshova AA, Hiatt A, Deering A, Davis R (2009) Melamine detection in infant formula powder using near- and mid-infrared spectroscopy. J Agric Food Chem 57:3974–3980

Mitchell JM, Griffiths MW, McEwen SA, McNab WB, Yee AE (1998) Antimicrobial drug residues in milk and meat: causes, concerns, prevalence, regulations, tests and test performance. J Food Prot 61:742–756

Muscarella M, Palermo C, Rotunno T, Quaranta V, D'Antini P (2004) Survey of ochratoxin A in cereals from Puglia and Basilicata. Vet Res Commun 28:229–232

Owino JH, Arotiba OA, Hendricks N, Songa EA, Jahed N, Waryo TT, Ngece RF, Baker PG, Iwuoha EI (2008) Electrochemical immunosensor based on polythionine/gold nanoparticles for the determination of aflatoxin B1. Sensors 8:8262–8274

Pattono D, Gallo PF, Civera T (2011) Detection and quantification of Ochratoxin A in milk produced in organic farms. Food Chem 127:374–377

Ping H, Zhang M, Li H, Li S, Chen Q, Sun C, Zhang T (2012) Visual detection of melamine in raw milk by label-free silver nanoparticles. Food Control 23:191–197

Qu X, Su C, Zheng N, Li S, Meng L, Wang J (2017) A survey of naturally-occurring steroid hormones in raw milk and the associated health risks in Tangshan City, Hebei Province, China. Int J Environ Res Public Health 15(1):38

Ramalingam K, Devasena T, Senthil B, Ramakrishnan K, Ramasamy J (2017) Silver nanoparticles for melamine detection in milk based on transmitted light intensity. IET Sci Meas Technol 11(2):171–178

Ren J, Liang G, Man Y, Li A, Jin X, Liu Q, Pan L (2019) Aptamer-based fluorometric determination of Salmonella Typhimurium using Fe_3O_4 magnetic separation and CdTe quantum dots. PLoS ONE 14(6):e0218325

Rusu L, Harja M, Suteu D, Dabija A, Favier L (2016) Pesticide residues contamination of milk and dairy products. A case study: Bacau District Area, Romania. J Environ Prot Ecol 17(3):1229–1241

Shadjou R, Hasanzadeh M, Heidar-poor M, Shadjou N (2018) Electrochemical monitoring of aflatoxin M1 in milk samples using silver nanoparticles dispersed on α-cyclodextrin-GQDs nanocomposite. J MolRecognit e2699

Song K-M, Jeong E, Jeon W, Cho M, Ban C (2012) Aptasensor for ampicillin using gold nanoparticle based dual fluorescence–colorimetric methods. Anal Bioanal Chem 402:2153–2161

Song E, Yu M, Wang Y, Hu W, Cheng D, Swihart MT, Song Y (2015) Multi-color quantum DoT-based fluorescence immunoassay array for simultaneous visual detection of multiple antibiotic residues in milk. Biosens Bioelectron J Bios 05.018

Sundlof SF, Kaneene JB, Miller R (1995) National survey on veterinarian-initiated use in lactating dairy cows. J Am Vet Med Assoc 207:347–352

Tripathi P, Upadhyay N, Nara S (2018) Recent advancements in lateral flow immunoassays: a journey for toxin detection in food. Crit Rev Food Sci Nutr 58(10):1715–1734

Wu Q, Long Q, Li H, Zhang Y, Yao S (2015) An upconversion fluorescence resonance energy transfer nanosensor for one step detection of melamine in raw milk. J Talanta 01.005

Wu X, Tian X, Xu L, Li J, Li X, Wang Y (2017) Determination of aflatoxin M1 and chloramphenicol in milk based on background fluorescence quenching immunochromatographic assay. Biomed Res Int 8649314

Xu X, Liang F, Shi J, Zhao X, Liu Z, Wu L, Song Y, Zhang H, Wang Z (2013) Determination of hormones in milk by hollow fiber-based stirring extraction bar liquid-liquid microextraction gas chromatography mass spectrometry. Anal Chim Acta 790:39–46

Yang Y, Shao B, Zhang J, Wu Y, Duan H (2009) Determination of the residues of 50 anabolic hormones in muscle, milk and liver by very-high-pressure liquid chromatography-electrospray ionization tandem mass spectrometry. J Chromatogr B 877:489–496

Yazgan NN, Boyacı IH, Topcu A, Tamer U (2012) Detection of melamine in milk by surface-enhanced Raman spectroscopy coupled with magnetic and Raman-labeled nanoparticles. Anal Bioanal Chem 403:2009–2017

Zhang F, Liu L, Ni S, Deng J, Liu G-J, Middleton R, Inglis DW, Wang S, Liu G (2019) Turn-on fluorescence aptasensor on magnetic nanobeads for aflatoxin M1 detection based on an exonuclease III-assisted signal amplification strategy. Nanomaterials 9:104

Zhao J, Wu Y, Han T, Chen H, Yang W, Qiu S (2017) Colorimetric detection of streptomycin in milk based on peroxidase-mimicking catalytic activity of gold nanoparticles. RSC Adv 7:38471–38478

Zheng Z, Hanneken J, Houchins D, King RS, Lee P, Richard JL (2005) Validation of an ELISA test kit for the detection of ochratoxin A in several food commodities by comparison with HPLC. Mycopathologia 159:265–272

Ziarati P, Moghimi S, Sepideh A-B, Qomi M (2012) Risk assessment of heavy metal contents (lead and cadmium) in lipsticks in Iran. Int J Chem Eng Appl 3(6):450–452

Ziarati P, Shirkhan F, Mostafidi M, Zahedi MT (2018) An overview of the heavy metal contamination in milk and dairy products. Acta Sci Pharm Sci 2(7):08–21

Chapter 6
Natural Polymers in Pharmaceutical Nanotechnology

G. Leyva-Gómez, N. Mendoza-Muñoz, M. L. Del Prado-Audelo, S. A. Ojeda-Piedra, M. L. Zambrano-Zaragoza, and D. Quintanar-Guerrero

1 Definitions and Classification of Pharmaceutical Polymers

1.1 Introduction

The word polymer is derived from the Greek word "poly" meaning many and "mer," which means parts. Hence, polymers are large molecules of high relative molecular mass, ranging from a few thousand to as high as millions of grams/mole, composed of many repeated subunits molecules of low relative molecular mass, named monomers (Vert et al. 2012) (IUPAC 2019). Monomers are linked via polymerization, creating a covalently bonded structure; this large molecular mass compared to small molecule

G. Leyva-Gómez · M. L. Del Prado-Audelo
Departamento de Farmacia, Facultad de Química, Universidad Nacional Autónoma de México, 04510 Mexico City, Mexico

N. Mendoza-Muñoz
Laboratorio de Farmacia, Facultad de Ciencias Químicas, Universidad de Colima, 28400 Colima, Mexico

M. L. Del Prado-Audelo · D. Quintanar-Guerrero (✉)
Laboratorio de Posgrado en Tecnología Farmacéutica, Universidad Nacional Autónoma de México, FES-Cuautitlán, 54740 Cuautitlán Izcalli, Edo. de Mexico, Mexico
e-mail: quintana@unam.mx

S. A. Ojeda-Piedra · M. L. Zambrano-Zaragoza
Laboratorio de Procesos de Transformación y Tecnologías Emergentes En Alimentos, Universidad Nacional Autónoma de México, FES-Cuautitlán, 54714 Cuautitlán Izcalli, Edo. de México, Mexico

M. L. Del Prado-Audelo
Escuela de Ingeniería y Ciencias, Departamento de Bioingeniería, Tecnológico de Monterrey Campus Ciudad de México, 14380 Ciudad de México, Mexico

R. F. do. Nascimento et al. (eds.), *Nanomaterials and Nanotechnology*, Materials Horizons: From Nature to Nanomaterials,
https://doi.org/10.1007/978-981-33-6056-3_6

compounds produces unique physical properties such as toughness modification and form viscoelastic structures. Due to their broad range of properties, polymers play essential and ubiquitous roles in everyday life (Olatunji 2015).

1.2 Polymer Classification

Polymers can act like biomaterials, which are any substance or material introduced such as a medical device, intended to interact with biological systems in order to evaluate, treat, deliver, augment, or replace any tissue or function of the body. Polymeric biomaterials (biopolymers) are an attractive alternative to other substances like metals due to their unique properties. These include their tendency to be biocompatible, have a relatively lighter weight, and possess ease of chemical modification compared to metals. Biopolymers can be classified according to their origin.

Synthetic polymers are human-made while natural polymers are those that which refer to materials sourced from nature, mostly plants or animals with minimal or no alteration to their chemical structure. This classification also includes polymers produced by biological processes such as bacterial synthesis or fermentation.

Biopolymers can also be divided into two broad groups: biodegradable and non-biodegradable. At present, the polymers used in biological systems are intended to be biodegradable (Niaounakis 2015). Currently, natural polymers have gained attention in biomedicine and pharmaceutical applications, such as 3D printing and tissue engineering. Also, significant efforts have been employed in drug delivery systems for active molecules, proteins, and genes for targeted delivery; however, the oral route is the most common way of delivery into the body (Dutta et al. 2013). Tissue engineering includes the regeneration, as well as the creation of biological substitutes for defective or lost tissues and organs (Vo-Dinh 2005).

Biomaterials used in tissue engineering must exhibit adequate cell adhesion to serve as supportive structures or scaffolds, for the regeneration and restoring function, by acting as a temporary matrix for cell grown and must multiply into the desired tissue structure. Scaffolds are also employed for the controlled delivery of bioactive compounds such as drugs and DNA. Hydrogels, which are the three-dimensional polymeric structures supported by their physical or chemical cross-linking can act as tissue scaffolds (Hosseini and Hamdy Makhlouf 2016).

Some advantages in the usage of natural polymers in biomedicine and pharmaceutical application are their biosecurity. Their use for the release of drugs is attractive due to factors such as their being biodegradable, biostable, and biocompatible, exhibiting no adverse effects on the environment or the human body. Natural polymers undergo both enzymatic and hydrolytic degradation in the biological environment with body-friendly degradation by-products (Bhatia 2016). In addition to secondary properties such as high swelling capacity, bioadhesion, electric charges, they allow control of nanoparticle size, functional surface properties, and the ability of forming systems for targeted delivery or stimulus-sensitive factor systems (De Frates et al. 2018). Applied biopolymers must be inexpensive, with relative simplicity of isolation and the possibility for chemical modification, their processing and fabrication must be reproducible. They must permit sterilization and possess stability during packaging and secondary processes (Damodaran et al. 2016).

Some disadvantages of natural polymers could be found during their production: if they are exposed to external environmental conditions, there are opportunities for microbial contamination or chemical modifications. In addition, batch-to-batch variations are found: the manufacturing of synthetic polymers is a controlled procedure with fixed quantities of ingredients, while the production of natural polymers is dependent on environmental and physical factors, generating a slow rate of production (Bhatia 2016).

The most employed polymers include, mainly, polysaccharides, those most common in nature, and proteins, which exist in both plants and animals, providing mainly structural support (Olatunji 2015).

1.2.1 Polysaccharides

Polysaccharides are known to be, by far, the most abundant renewable resource in the world. They are often isolated and purified from renewable sources including plants, animals, and microorganisms. Their building blocks are the monosaccharides, which are joined together by o-glycosidic linkages (Fig. 1).

The structure of polysaccharides offers freely available hydroxyl functionalities, which render it possible to alter their physicochemical properties by modifying their structure (Veerapandian and Yun 2009). At present, different cross-linking techniques have been developed (via ionic or covalent bonding) to restrict the polysaccharide chain movement in order to control their water uptake, degradation, and to improve the mechanical properties.

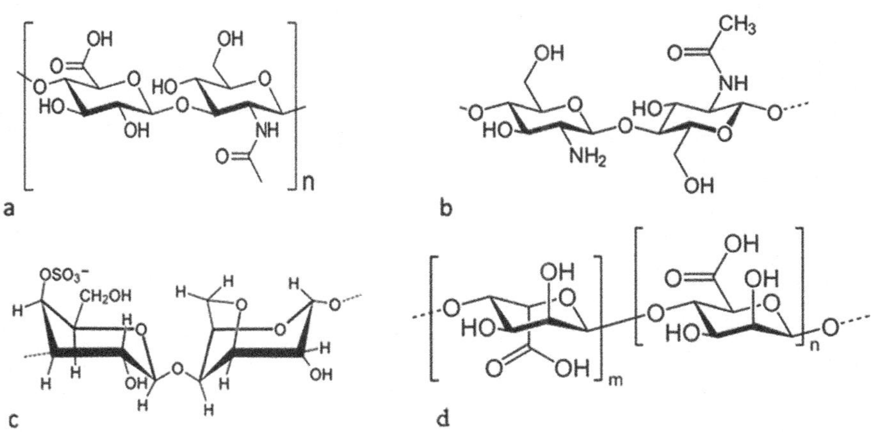

Fig. 1 Polysaccharide structures: **a** hyaluronic acid; **b** chitosan; **c** k-Carrageenan, and **d** alginic acid

Polysaccharide-based porous scaffolds, fiber matrices, hydrogels, and nanoparticles have been developed for a variety of tissue-regeneration and drug delivery applications (Bhatia 2016). Some of the most utilized polysaccharides are the following:

Hyaluronic Acid (HA)

Hyaluronic acid is a linear anionic polysaccharide consisting of alternating units of *N*-acetyl-D-glucosamine and glucuronic acid. It is widely distributed throughout connective, neural, and epithelial tissues; is water-soluble, and forms viscous solutions. Also, it is important in tissue repair by promoting mesenchymal—and epithelial—cell migration and differentiation; hence, it is extensively employed as a tissue engineering scaffold. The HA homopolymer can form hydrogels in solutions due to extensive intramolecular hydrogen bonding and can also be physically and covalently cross-linked to design biocompatible scaffolds in tissue engineering applications. Due to its unique viscoelastic properties, it is applied as a drug delivery vehicle, particularly in the form of nanoparticles. HA hydrogels have been utilized to deliver chemotherapeutics and antibiotics (Damodaran et al. 2016).

Chitosan

Chitosan is a cationic polysaccharide obtained from the deacetylation of chitin. Structurally, chitosan is very similar to Hyaluronic Acid, consisting of β-(1–4)-linked *N*-acetyl-glucosamine groups that form the exoskeleton of anthropoids. Due to the poor mechanical strength and high degree of hydrophilicity of chitosan, it has been cross-linked with other polymers to form stable chitosan-based scaffolds. Chitosan is non-toxic, biocompatible, biodegradable, and bioadhesive, all of the properties make it useful in regenerative medicine. It can be easily processed into various shapes ranging from sponges, fibers, microspheres, and hydrogels for tissue-engineering applications. Chitosan has been most widely used as a wound dressing due to its oxygen-permeable, bioactive, water-absorptive and hemostatic nature (Damodaran et al. 2016).

Carrageenan

Carrageenan is a naturally anionic sulfated polysaccharide. It is extracted by drying and washing seaweed in an alkaline environment. It is formed by alternate units of D-galactose and 3,6-anhydrogalactose that are joined by an α-1,3 and β-1,4 glycosidic linkage. On the basis of the type of bonding between galactose units and the position of the attachment of sulfate groups to the galactose unit, carrageenan is categorized as λ—(lambda), κ—(kappa), and ι—(iota), each having a different solubility and electric-charge density. It is employed for drug delivery due to its interesting characteristics such as its gelling, thickening, emulsifying, and stabilizing properties. It possesses the important property of the formation of water-retaining hydrogels, which is useful in the development of wound-coating materials. Some interesting features of carrageenan include its adhesiveness and that its positive surface charge provides an extra advantage in prolonging drug release in mucosal/epithelial tissues (George et al. 2019).

Sodium Alginate

Formally, the sodium salt of alginic acid is a hydrophilic polysaccharide included in all of the brown-algae species of the Phaeophyta family. It is a linear biopolymer constituted of two uronic acids, (1,4)-linked-β-D-mannuronic acid (M) and α-L-guluronic acid (G), making them anionic in nature by the presence of carboxylic groups (Olatunji 2015). This polymer exhibits a unique property of gel formation due to the ionic interaction between the guluronic acid residue and the presence of divalent cations such as calcium ions in aqueous medium at low pH. The drug delivery from alginate hydrogels is a pH-dependent phenomenon leading to a modified delivery; alginate shrinks in a low pH environment and the encapsulated drug is not released. The intrinsic properties of the calcium-alginate gel beads make it the most widely used carriers for controlled release (George et al. 2019).

Cellulose

Cellulose is the principal component of the plant cell wall, constituting about one half of the biomass of photosynthetic organisms. Therefore, cellulose is the most abundant molecule on earth. Cellulose cannot be digested by the human body; it is also water-insoluble due to the multiple hydrogen bonds on the same or on a neighbor chain, holding the chains firmly together, forming microfibrils with high tensile strength. Its primary structure had been established as a linear homopolymer of glucose residues with a D configuration linked by β-(1 $\rightarrow$ 4) glycosidic linkages (Olatunji 2015). Cellulosic nanocomposites are formed by adding cellulose nanoscale fillers to various polymer matrices, resulting in mechanical reinforcement and the alteration of other properties. Cellulose-based composites can be used as tissue-engineered cartilage scaffolds, wound dressings, artificial skin, dental implant, membranes for tissue-guided regeneration, or controlled-drug release carriers (Thomas et al. 2013).

Plant-Based Gums and Mucilages

The term "gum" refers to polysaccharides presents as hydrocolloids, which do not form a part of the cell wall, but that are exudates or slimes and are pathological products, while mucilages form part of the cell and physiological products. They are interesting polymers for the preparation of pharmaceutical formulations because of their high water-swelling ability and their non-toxicity. They are frequently employed as thickening, emulsifying, suspending, and stabilizing agents (Prajapati et al. 2013). Some of the most used gums are Arabic and Guar gum.

Arabic gum is the dried gummy exudation obtained from *Acacia arabica* (Leguminosae). This gum has been recognized as a branched molecule of 1, 3-linked β-D-galactopyranosyl units. It is an acidic polysaccharide containing D-galactose, L-arabinose, L-rhamnose, and D-glucuronic acid and is mainly utilized in oral and topical pharmaceutical formulations as a suspending and emulsifying agent. A previous study explored Arabic gum as a microencapsulating agent for enzymes. An increase in the amount of Arabic gum in the pellets decreased the rate of release (Bhatia 2016).

Guar gum (GG) is a high-molecular-weight, water-soluble, non-ionic biopolymer derived from the seeds of the *Cyamopsis tetragonolobus* family that has been widely applied in pharmaceutical products due to its unique ability to alter rheological properties. It consists of linear chains of $(1 \rightarrow 4)$-β-D-mannopyranosyl units with α-D-galactopyranosyl units attached by $(1 \rightarrow 6)$ linkages. Guar gum has gained much attention in oral drug delivery due to its propensity toward microbial degradation in the large intestine. This polymer can be employed as matrix tablets, hydrogel, nanoparticles, and as a coating for targeted drug delivery (Bhatia 2016).

Recently various mucilage gums have been investigated for their potential role in drug delivery and other pharmaceutical applications. Due to the high concentration of hydroxyl groups in polysaccharides, mucilages generally possess a high water-binding capacity (Prajapati et al. 2013). Some mucilages with pharmaceutical importance are presented in Table 1.

Proteins

Proteins are not only of animal origin; they also constitute an integral part of the plant cell wall, serving as structural and functional molecules. They are long chains consisting of a large number of amino acids, linked by amide bonds, also known as peptide bonds.

Typical polypeptide chains contain about 100–600 amino acid molecules and have a molecular weight of about 15,000–70,000 Da. Since amino acids have hydrophilic, hydrophobic, and amphiphilic groups, they tend to form locally ordered structures in an aqueous environment. They have a 3D structure, also called a secondary structure, which is characterized by a low-energy configuration with the hydrophilic groups outside of it and the hydrophobic groups inside. In general, simple proteins have a natural α-helix configuration. Another natural secondary configuration is a β-sheet. These two secondary configurations comprise the building blocks that assemble to form the final tertiary structure, which is held together by extensive secondary interactions, such as van der Waals bonding. Sometimes, several proteins are bound together to form supramolecular aggregates, also called quaternary structure (Vo-Dinh 2005).

Table 1 Pharmaceutical applications of some mucilages

Common name	Botanical name	Pharmaceutical application	References
Abelmoschus mucilage	*Abelmoschus esculentus*	Binder in tablets, sustained release	Kumar et al. (2009)
Aloe mucilage	*Aloe species*	Gelling and sustained release agent	Jani et al. (2007), Ahad et al. (2010)
Hibiscus mucilage	*Hibiscus esculentus*	Emulsifying, sustained release, and suspending agent	Kulkarni et al. (2002), Jani et al. (2007)
Ispagol mucilage	*Plantago psyllium/Plantago ovata*	Lubricant, demulcent, laxative, emulsifying, binder, and suspending agent, hydrogels, gastro retentive drug delivery system	Prajapati et al. (2013), Shidhaye et al. (2007), Singh et al. (2006)
Satavari mucilage	*Asparagus racemosus*	Binding and sustaining agent	Kulkarni et al. (2002)
Cactus mucilage	*Opuntia ficus-indica*	Gelling agent in sustained drug delivery	Cárdenas et al. (1998)

Proteins can be classified by their shape. Globular proteins are water-soluble types that are rather fragile in nature. Antibodies, enzymes, and hormones are typical examples of globular proteins. Fibrous proteins are tougher water-insoluble proteins. These are usually proteins found in structural tissues such as hair, nails, and skin (Olatunji 2015). Some proteins employed for medical proposes include the following:

Collagen

Collagen is the most abundant extracellular matrix protein present in different tissues of the human body, such as skin, tendon, bone, and cartilage. Collagen is a rod-shaped, fibrillar polymer (300 nm in length with a molecular weight of 300 kDa), is the most abundant protein in the human body, and is the most widely studied. It is a triple-stranded helix constituted of three polypeptide-alpha chains. The simple repetitive sequence of collagen is described as $[\text{Glycine-X-Y}]_n$ where X is proline and Y is hydroxyproline. This repeating structure sequence is responsible for the helical structure and for the mechanical strength of collagen. Due to its ease of degradation and unique biochemical and mechanical properties, collagen is extensively utilized in different forms (hydrogels, foams, membranes, and injectable viscous solutions). Collagen is even preferred for tissue-engineering scaffolds and devices, but its fragile mechanical strength could be a disadvantage. However, due to the high reactivity of collagen, it can be cross-linked by a variety of excipients, principally synthetic polymers (De Frates et al. 2018).

Fibrin

Fibrin is a fibrous, globular protein that is derived from fibrinogen and is involved in the natural blood-clotting process. Fibrinogen is a 340-kDa protein composed of three pairs of polypeptide chains bounded by disulfide linkages. Fibrin is biocompatible, biodegradable, promotes cell attachment, proliferation, angiogenesis, and is highly injectable. All of these properties make fibrin a desirable biopolymer for tissue-engineering applications. Fibrin-based scaffolds have also played a key role in bone, skin, in cardiovascular and neural-tissue engineering, and in drug delivery. Fibrin undergoes shrinkage and is mechanically weak. To improve the mechanical stiffness of fibrin-based scaffolds, collagen is applied in combination with other polymers such as HA. Fibrin-based scaffolds also support cell infiltration, the exchange of nutrients, and vascularization (Damodaran et al. 2016).

Albumin

Albumin-based nanoparticles have received attention due to their biological origin, easy accessibility, water solubility, non-immunogenicity, and their ability to accumulate in the tumor sites. Albumins are obtained from different sources, i.e., from egg (ovalbumin), and from bovine (BSA) and human serum (HSA). BSA and HSA are extensively utilized in targeted drug delivery applications due to the increase in the solubility of hydrophobic molecules in the blood, because of the large number of reactive functional groups on their surface, such as the thiol, amino, and carboxylic groups.

When the nanoparticles are administered to the bloodstream, thousands of plasma proteins surround them and form a "protein corona" that masks their actual function, leading to a successful in vitro result but a nearly unpredictable outcome in vivo. Coating these nanoparticles with a preformed albumin corona causes a camoflaging effect. This has turned out to be a potential method to improve nanoparticle circulation (George et al. 2019).

2 Applications of Natural Polymers for the Design of Drug Delivery Systems

Currently, the design of drug delivery systems (DDS) has encountered numerous challenges due to the novel and improved therapies that include substances of high molecular weight, such as peptides, proteins, and oligonucleotides. The latter are sensitive to chemical, physical, and enzymatic degradation; therefore, alternatives must be sought to trap these molecules in order to protect them from environmental conditions (Mao et al. 2010). The design of new DDS considers the improvement in the release-control, absorption, distribution, and release profile, considering that the new DDS are designed to respond to external conditions, with systems with diffusion control that are solvent-activated (George and Suchithra 2019).

Natural polymers represent an alternative with functions including polycations, polyanions, or neutral charge with the ability to develop more compatible systems that are less toxic, biodegradable, with high chemical reactivity, and with high drug entrapment and encapsulation efficiency for the controlled release of different molecules that are more compatible with different cells, tissues, and specific targets at the site-of-action (Bernkop-Schnürch 2018).

Natural polymers possess distinctive physicochemical characteristics for developing DDS, as follows: (1) diffusion-based release systems; (2) chemically controlled systems related with biodegradability; (3) systems externally activated by the effect of pH, ions, or temperature; (4) systems activated with solvents such as hydrogels, and (5) the size of systems, such as microparticles, nanoparticles, and fibers (Nep et al. 2018). It is considered that natural polymers are more capable of interacting with the cellular membrane, organelles, and tissues. They are able to be chemically controlled due to their biodegradation, and they can also be activated by environmental conditions such as pH, ions, and temperature, also giving rise to a modification in the presence of water or a solvent. Natural polymers can be activated by their responding to the medium conditions, allowing better absorption (Sirisha and Campus 2015).

Natural polymers can form matrices, with efficient entrapment of the drug, permitting the formation of small macromolecules with controlled release at a rate dependent on the diffusion coefficient. One of the most explored methods in the development of new drug delivery systems is that of hydrogels. The latter are soft biomaterials with a three-dimensional structure prepared with hydrophilic polymers. These allow their dispersion in water and in biological fluids, permitting the release of drugs, controlling the release speed associated with changes in their structure in response to environmental conditions, such as changes in temperature, pH, the electric and magnetic fields, light, or ions (Dragan and Dinu 2019; Pellá et al. 2018).

An important property of natural polymers is their adhesiveness to mucous membranes, permitting a controlled release through different routes. It is important to take advantage of the electrostatic attraction on the backbone chain with negatively or positively charged, which allows them to cross different barriers and proceed to the specific site-of-action. Depending on the surface charges, it interacts with tissues, cells, or organelles (Gopinath et al. 2018). The ability of natural polymers to target a site is directly influenced by particle size, since this depends on the drug crossing the biological barrier to reach the target site. The superficial charge distribution determines the systemic flow and interaction with cell membranes in such a way that it has specificity for the action site, determined by its hydrophobicity and/or hydrophilicity (George et al. 2019).

2.1 *Natural Polymers in the Design of Drug Delivery Systems*

Currently, there are many studies on the application of natural polymers in the design and development of release systems. In particular, polysaccharides and proteins have exhibited important properties in relation to their biocompatibility, high adhesion, disintegration, encapsulation efficiency, stability, and release. These polymers are extracted from different plant or animal sources, and their study will afford greater possibilities in the future for the development of new release systems.

These polymers are able to adhere, distribute, and retain in the types of administration forms considered for different administration routes: oral, vaginal, cutaneous, nasal, etc. Among the main natural polymers, we find zein, casein, whey protein, chitosan, alginate, mucilage, and gums that are extracted from different plant, animal, marine, and microbial sources, capable of releasing the drug. Moreover, controlled-release can be dependent on pH; for example, they could be expanded in the stomach as a function of polymer density, tending to float in the stomach and preventing them from crossing the pyloric sphincter. They can also improve transdermal absorption in order to obtain compatibility and adhesion vaginally, nasally, orally, etc. (Asnani et al. 2018; Irimia et al. 2018; Notario-Pérez et al. 2017). There are many considerations to be taken into account in the DDS, such as drug type, absorption site, system properties, release site, and changes associated with the administration route (Delcassian et al. 2019). Figure 2 shows the use of natural polymers on different administration routes and functions. Another important point to consider in DDS is it whether this comprises an alternative for unconscious patients or disabled patients who have difficulty in swallowing (dysphagia) (Notario-Pérez et al. 2017; Saidin et al. 2018).

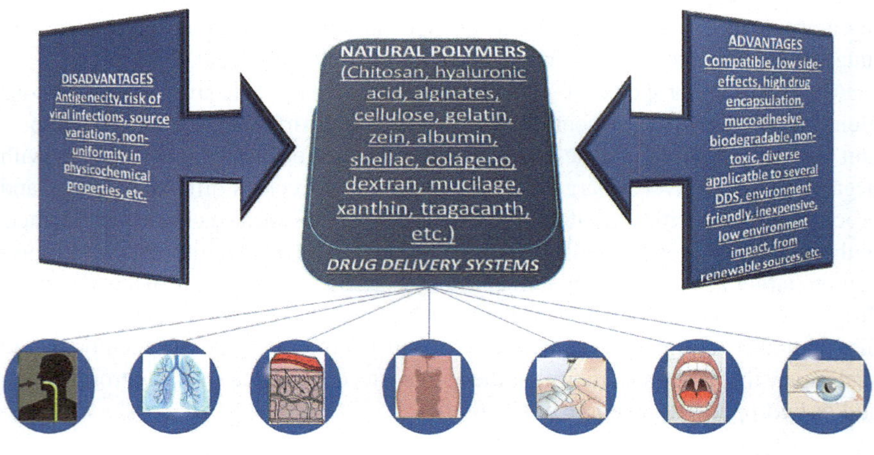

Fig. 2 Natural polymers employed for drug release: advantages and disadvantages

2.1.1 Natural Polysaccharides in DDS

Cationic polysaccharides have been shown to enhance the intensity of interaction with different drugs such as antibiotics like, gentamicin, amoxicillin, vancomycin and ciprofloxacin, and with analgesic/anti-inflammatory drugs such as ibuprofen, diclofenac sodium, indomethacin, curcumin, and emodin (Ghiorghita et al. 2019). In addition, among the polysaccharides utilized in the development of new drug delivery systems, including the gum and mucilage extract from plants, they are economical (Kaleemullah et al. 2017). Table 2 lists some examples of polysaccharide applications including, in DDS, those that employ gums or mucilage.

One of the most utilized administrations of drugs is the oral route. Here, the polysaccharides are employed in the preparation of capsules, tablets, suspensions, etc., since these contain approximately 30% of polymers, which exert a considerable influence on the release of the drug in the gastrointestinal tract. Polysaccharides are also applied as matrices or coatings in order to control swelling rate, controlled release, and bioavailability (Adrover et al. 2019; Assaf et al. 2019; Biswas et al. 2015). It is important to highlight that a large part of polysaccharides immersed in DDS is focused on the application of polycations, which have the capacity to respond to different pH. These two characteristics render them more suitable for use as a drug matrix (Mahto and Mishra 2019). Below are the main polysaccharides used in DDS in general, not including those of nanometric size; however, these are directly related to the current development of new systems of drug delivery.

Chitosan

Chitosan is undoubtedly one of most studied polysaccharides in the pharmaceutical area, in that it has shown affinity in the formulation of DDS for cancer treatment. Chitosan is employed alone or in combination with other polysaccharides, such as alginates, with anionic behavior, and which are capable of resisting the pH or approaching the ability to form gel in presence of divalent ions such as Ca^{+2}, Zn^{+2} or Ba^{+2} (Wang et al. 2016). Table 2 presents the natural polysaccharides utilized in DDS as a function of the site-of-action and specificity. Chitosan has also been employed in DDS for the release of highly water-soluble substances such as DNAzyme through the skin, taking advantage of the formation of polycations that in turn form ionic complexes, achieving DNA/chitosan polyplexes. Thus, the electrostatic interactions of chitosan on DNAzymes negatively charged lead to enhanced cell transfection (Eicher et al. 2019).

Hyaluronic Acid

Hyaluronic acid is another polysaccharide that has been utilized in DDS, because it has shown that it significantly prolongs the residence time of drugs at the target site, reducing the number of doses to be administered, increasing bioavailability,

Table 2 Natural polymers involved in drug delivery systems

Natural polymer	Administration route	System type	Drug	Relevant results
Aloe vera	Bucal cavity for oral squamous cell carcinoma Administration: oral	Hydrogel	Barbaloin	The film formulated as new delivery system demonstrates capacity to administer Aloin closely to the oral lesions, prolonging the retention of drug time on the targeted area with high drug concentration into the Tissue (Di Prima et al. 2019)
Chitosan	Pulmonary drug delivery system Administration: intranasal	Microspheres	Cyclosporine A	Microspheres of chitosan modified to form porous quaternized contribute to decrease the particle size, are biocompatible due to the low lactate dehydrogenase activity, less inflammatory cell and IL-4 and TNF- a cytokines (Yang et al. 2018)

(continued)

Table 2 (continued)

Natural polymer	Administration route	System type	Drug	Relevant results
Alginate-chitosan	Cancer therapy Administration: Oral, gastrointestinal tract	Hydrogel Smart pH-sensible	Resveratrol	The formulation exhibited high encapsulation efficiency (74%) whit hybrid systems with alginate-chitosan at mass ratio of 1:2 when 80 mg de resveratrol was loaded. It was observed that 98.1–99.8% the drug loaded at all pH was released into the medium with the first three hours. There was interaction between alginate-chitosan slowed down the diffusion of the drug intro the release medium controlling for longer periods (Nazlı and Açıkel 2019)
Chitosan	DNA therapy Administration: Dermal	Hydrogel	DNAenzymes	The hydrogels hold promise as a versatile dermal drug delivery system with multiple beneficial attributes. Chitosan hydrogel employed to deliver therapeutic oligonucleotides dermally prevents the API from enzymatic degradation (Eicher et al. 2019)

(continued)

Table 2 (continued)

Natural polymer	Administration route	System type	Drug	Relevant results
Hibiscus rosa-sinensis leaves mucilage-HPMC	Anti-inflammatory, analgesic, antipyretic Administration: Oral	Tablet Direct compression	Ketoprofen	Natural mucilage (40%) showed comparable dissolution profile to that of reference drug. The release kinetics were non-Fickian which involved both diffusion and erosion mechanism (Kaleemullah et al. 2017)
Portulaca oleracea mucilage-alginate	Cancer therapy Administration: Oral	Tablets with hydrogel pH-responsive	5-fluorouracil	Hydrogel was prepared by cross-link using borax, the polysaccharide used was alginate. Results showed that the proper selection of polymer ratio and cross-linker's concentration and the best encapsulation efficiency and to delay the release rate of the anticancer drug 5-flurouracil (Asnani et al. 2018)

(continued)

Table 2 (continued)

Natural polymer	Administration route	System type	Drug	Relevant results
Chitosan	Candidiasis treatment Administration: Vaginal	Tablet bio-adhesive	Garlic and ketoconazole	The use of chitosan increases the retention inside the vaginal tract, increasing the vaginal contact time, drug release with high mucoadhesion, and with the increase in garlic concentration. There was an increase in the anti-fungal activity by combination of garlic with ketoconazole and can be approach for drug release and effective treatment of vaginal candidiasis (Powar et al. 2017)
Elastin	Cancer therapy Administration: Intravenous	Thermosensitive liposomes	Doxorubicin	Thermosensitive liposomes formed spherical vesicles and had target-specific endocytosis. The stability of was maintained up to 12 h without the loss of their thermosensitive function for drug release between 42 to 45 °C (Nigatu et al. 2018)

(continued)

Table 2 (continued)

Natural polymer	Administration route	System type	Drug	Relevant results
Human serum albumin (HSA)	Rheumatoid arthritis (RA) Administration: Oral		Tumor necrosis factor-related apoptosis-inducing ligand (TRAIL)	Tumor necrosis factor related to apoptosis-inducing ligand with human serum albumin (HAS-TRAIL) and polyethylene glycol was studied for treatment of rheumatoid arthritis The enhanced therapeutic efficacy of HSA-TRAIL was demonstrated in collagen-induced arthritis mice, which presented a clear advantage of targeting RA and long systemic circulation by HSA and unique anti-inflammatory efficacy by TRAIL

and considerably reducing adverse reactions, in addition to being able to recognize specific binding sites with receptors overexpressed on cancer-cell membranes through receptor-mediated endocytosis (Huang and Huang 2018). It has also been shown that hyaluronic acid possesses good serum stability and active targeting capacity mediated by CD44, which is overexpressed in many cancer cells, and also employed in CD44-targeting drug delivery (Luo et al. 2019). However, hyaluronic acid possesses rapid clearance from the blood circulation due to HA receptors of the liver according to a report on nucleic acid in cancer therapy (Dosio et al. 2016).

Cellulose

Cellulose is a natural polymer that has been modified to produce cellulose esters such as cellulose acetate, cellulose acetate butyrate and cellulose acetate phthalate, which have been employed to better exploit its properties in tablet-coating systems in order to achieve better release control due to swelling properties, and the formation of a superficial gel that achieves greater retention at the gastrointestinal level. The polymer blend may be useful in modulating the dissolution of the drug as diclofenac sodium and prednisone (Ali et al. 2019). Furthermore, modified cellulose has been used as a pH-responsive system with the purpose of protecting peptic substances from gastrointestinal tract (Sarkar et al. 2017).

2.1.2 Protein in DDS

Gelatin

Conceivably, gelatin is the most studied and exploited protein in DDS. It possesses the ability to form cross-links using divalent cations, allowing modifications in release systems in response to pH changes. Gelatin has been shown to be versatile due to its intrinsic features that enable the design of different carrier systems, such as microparticles, fibers, hydrogels, and nanosystems. It serves as vehicles for cell proliferation and for the delivery of large molecules such as peptides and oligonucleotides. Hydrogels can trap molecules between the polymer's cross-link gaps, permitting these molecules to diffuse into the bloodstream. Another interesting area is the combination of bioadhesive gelatin with controlled drug release for pain management and wound healing (Foox and Zilberman 2015).

Zein

Zein is a natural polymer insoluble in water at $pH < 11$, it is utilized in the DDS with particular interest in oral administration systems, particularly in drugs absorbed at the colon level, since it has been shown that this protein contributes to the protection of drugs in gastrointestinal systems, limiting degradation and increasing the bioavailability of the drug. It has been shown that zein is digested by colonic microbiota, thereby controlling the release of the drug however, in general, this protein must be combined with other excipients or applied as coating to avoid the risk of premature drug release (Bisharat et al. 2019; Wang et al. 2018). In this respect, chitosan-zein mixtures have been studied for DNA therapies, establishing that the application of another component contributes to increasing the load capacity, high encapsulation efficiency, and protection of DNA under gastric conditions, with evidence of the induction of a primary immune response against the encoded model antigen (Farris et al. 2017).

Casein and Whey Milk

Caseinates and whey milk are employed mainly as coatings in tables in order to prolong the release of drugs in the digestive tract, protecting the drug, improving drug targeting, and absorption. At the molecular level, whey protein subfractions display appealing anticancer effects. In this regard, the most investigated whey protein subfractions include α-lactalbumin, human and bovine α-lactalbumin, bovine serum albumin, and lactoferrin (Teixeira et al. 2019).

Albumin

Albumin is emerging as a versatile protein carrier for drug targeting. It improves the pharmacokinetic profile of peptide-or protein-based drugs. Albumin possesses the ability to protract the circulation of short-life drugs and drug targeting. It is capable of accumulating in the tumor or locating inflamed tissues to subsequently carry out the release of an active substance (Wei et al. 2019). Human serum albumin exhibits an average half-life of 19 days. The functions and binding properties of human serum albumin contribute to solubilize long-chain agents and are essential in lipid metabolism (Kratz 2008). Studies that do not include nanosystems are mainly designed for the development of tissue treatments. Ovoalbumin is much used in the preparation of DDS as tablets, hydrogels, etc.

Elastin

Elastin is composed of valine, glycine, and leucine. Given its characteristics and solubility at 35 °C, it has been utilized in the development of DDS that are sensitive to different stimuli such as temperature, pH, and ionic strength. These conditions induce the phenomenon of reversible tropoelastin, which is exploited in the development of cancer therapy by the effect of stimuli-responsive controlled-release systems. Thermal triggering has advantages in terms of the controllability of the temperature at the desired location.

In general, the greatest application of natural polymers in DDS is their ability to have a pH response due to polycation activity. Moreover, one may consider their ability to form hydrogels due to their high water-absorption capacity, in that the mechanisms of release are associated with swelling and erosion. Therefore, there are many studies yet to be carried out to exploit their potential.

3 Natural Polymers Used in Pharmaceutical Nanotechnology

Today, nanotechnology, an emerging science, is already impacting the way in which the products we utilize every day are developed. In the pharmaceutical field, for example, the use of nanoscience has allowed the development of sophisticated drug delivery systems in which it is possible to deliver the drug in a specific site increasing bioavailability at the site-of-action and reducing the toxicity. In general, pharmaceutical nanotechnology seeks the development of particulate systems with sizes between 1 and 100 nm; however, systems sizes between 100 and 1000 nm are currently also considered pharmaceutical nanosystems. To complement the definition, nanosystems are therapeutic agents where the drug(s) are dispersed, adsorbed, or covalently bound in encapsulated vesicles, capsules, or polymer matrices (Babu et al. 2014).

To date, different platforms have been developed based on the manipulation of materials at the nanometric scale. The list is long; however, liposomes, micro- and nanoemulsions, nanocrystals, polymeric micelles, polyplexes, dendrimers, and polymeric nanoparticles can be highlighted. In some of the systems mentioned previously, the use of polymers plays an important role. For example, in nanoemulsions, polymers can function as materials that permit the formation and stabilization of the system, while in other systems such as that of polymeric nanoparticles, the polymer forms part of its structure. Basically, two types of polymers are employed in the preparation of nanopharmaceutical platforms: synthetic or natural.

In the following paragraphs, we present a description of the application of natural polymers in pharmaceutical nanotechnology. In this regard, nanoparticles have been classified according to their composition into protein-based and polysaccharides-based nanoparticles. Nanoparticles composed of nucleic acid polymers are not considered in this section.

3.1 Nanoparticles Based on Proteins

Proteins are natural macromolecules composed of amino acid monomers. They are present in all living organisms and play a broad number of biological functions. As materials, proteins possess interesting characteristics, such as their amphiphilic properties, their biodegradability, they are easy to modify chemically, are non-toxic, and are metabolizable. Two types of proteins have been tested as materials for the synthesis of pharmaceutical nanoparticles: animal-derived proteins, and plant-derived proteins. Examples of animal-derived proteins include albumin, gelatin, and silk, examples of plant-derived proteins include the gliadins, glutenins, and zein. Table 3 presents examples of nanosystems based on proteins.

3.1.1 Application of Animal-Derived Proteins in the Preparation of Nanoparticles

Human and bovine albumin is a globular protein present in the blood that has been utilized effectively in the design of nanoparticles for a few decades. Albumin-based nanoparticles have several technological and biopharmaceutical advantages that make it an excellent carrier; among these advantages we find their low cost, that they are processable at the industrial scale, that they do nor exhibit immunogenicity (when human albumin is used), they have a relatively long lifetime (19 days), and they possess the ability to form easily redispersible powder via lyophilization. An additional advantage lays the possibility of drug targeting by their covalent derivatization due to the presence of sulfur, amino, and carboxylic groups. Abraxane® was the first USFDA-approved nanomedicine composed of albumin bound to the antineoplastic agent paclitaxel, which indicated for the treatment of breast cancer, non-small-cell lung cancer, and pancreatic cancer. Methods to prepare albumin nanoparticles include desolvation, emulsification, thermal gelation, nano-spray drying, nab-technology, and self-assembly (Elzoghby et al. 2012). Albumin can play roles different from that of nanocarrier for constructing other nanosystems; for example, albumin can be employed as a template of inorganic nanoparticles, polymeric conjugate in polyplexes, as scaffold, or as a stabilizer in nanoemulsions (An and Zhang 2017).

Gelatin-based nanoparticles are another example of an inexpensive system. One technological aspect to consider in the manufacturing of gelatin-based nanoparticles is the existence of two types of gelatin: type A with isoelectric point (IEP) 7–9 and type B with IEP 4–5. At a neutral pH, type A gelatin has a positive net charge, while type B gelatin is negatively charged, with the electric characteristic affording different drug-release behavior. It is considered that type B gelatin nanoparticles demonstrate better potential in terms of drug delivery than type A (Yasmin et al. 2017). As other unmodified nanoparticles, gelatin-based nanoparticles possess the disadvantage of being phagocytosed by the mononuclear phagocyte system; despite that surface modification is easily achieved. There are several methods that are applied for the production of gelatin-based nanoparticles: the coacervation method; the two step desolvation method; the complex coacervation method; the solvent extraction or emulsion process; the nanoprecipitation method; the microemulsion method; the inverse miniemulsion technique, the nano-spray drying method, and the self-assembly method (Gabriel et al. 2018; Harsha et al. 2015; Yasmin et al. 2017).

Table 3 Examples of protein-based nanosystems intended for drug delivery

Source	Natural polymer	Type of nanosystem	Method of preparation	Drug loaded	Application	References
Animal protein	Albumin (bovine)	Nanospheres	Coacervation	Resveratrol	Antioxidant	Fonseca et al. (2017)
	Albumin (human)	Nanospheres	nab™ technology	Tracolimus	Anti-arthritic	Thao et al. (2016)
	Gelatin	Nanospheres	Nanoprecipitation	FITC-Dextran	Not determined	Khan and Schneider (2013)
	Gelatin	Nanospheres	Desolvation	Doxorubicin	Cancer	Long et al. (2014)
	Silk fibroin	Nanospheres	Desolvation	Paclitaxel	Cancer	Wu et al. (2013a)
	Sericin	Nanospheres	Desolvation	Atorvastatin	Antihyperlipidemic	Kanoujia et al. (2016)
Plant protein	Gliadin	Nanospheres	Desolvation	Clarithromycin and omeprazole	*H. pylori* infection	Ramteke and Jain (2008)
	Gliadin Gliadin/gelatin	Nanospheres	Electrospray deposition	Cyclophosphamide	Breast cancer	Kemp and Linhardt (2009)
	Zein	Nanospheres	Liquid–liquid dispersion method	5-fluorouracil	Cancer	Aswathy et al. (2012)
	Zein and propylene glycol alginate	Nanospheres	Solvent evaporation	None	Pickering stabilization	Dai et al. (2018)
	Soy protein	Nanospheres	Isoelectric precipitation	Curcumin	Cancer	Teng et al. (2013)

Silk is a natural fiber produced in the glands of arthropods; recently, silk fibroin nanoparticles have gain interest for biomedical applications due to their slow degradation and extraordinary mechanical properties. A broad spectrum of techniques could be utilized to develop silk fibroin nanoparticles, techniques such as milling, electrospray, microemulsion, the capillary microdot technique, and supercritical fluid technology (Crivelli et al. 2018). Tyrosine residues and RGD sequences could be excellent sites where functionalization can be performed for targeting capabilities.

3.1.2 Application of Plant-Derived Proteins in the Preparation of Nanoparticles

Zein is a proline-rich water-insoluble protein obtained from maize it is alcohol-soluble and generally recognized as safe (GRAS). Due to their hydrophobic nature, zein nanoparticles have been prepared as carriers to encapsulate several lipophilic drugs and bioactive compounds. In addition, zein nanoparticles possess the potential for use as Pickering-emulsion stabilizers (Dickinson 2017). Zein-based nanoparticulate systems can be synthetized by a large number of techniques, including phase separation coacervation, spray-drying, the supercritical anti-solvent approach, electrospinning, and self-assembly (Elzoghby et al. 2018).

Gliadin is a gluten protein found in wheat; it is biodegradable and biocompatible. Gliadin has attractive properties when is used as polymer for the preparation of polymeric nanoparticles. Gliadin exhibits bioadhesion phenomena due to its amine and disulfide groups, which are capable of developing bonds with mucin (Ramteke and Jain 2008). The application of gliadin nanoparticles can be developed for oral and topical drug delivery (Ramteke and Jain 2008; Umamaheshwari et al. 2004). Other plant proteins with the potential for being used in nanomedicine include glutenins and legumins.

3.2 Nanoparticles Based on Polysaccharides

Polysaccharides are one of the most abundant polymeric biomaterials in nature. They are constituted of monosaccharides and are biodegradable, safe, non-toxic, and can be processed at a low cost. Polysaccharides have several advantages when they are used to formulate nanoparticulated drug delivery systems. One of the latter advantages comprises the presence of hydrophilic groups in their structure, such as the hydroxyl, carboxyl, and amino groups. This allows the possibility or their being modified chemically and biochemically, with also the possibility to form hydrogen bonds with biological tissues. Polysaccharides could be classified according to the sources from which they can be obtained, from an algal origin (e.g., alginate), plant origin (e.g., pectin, guar gum), microbial origin (e.g., dextran, xanthan gum), and animal origin (chitosan, chondroitin) (Liu et al. 2008). Table 4 presents examples of nanosystems based on polysaccharides.

Table 4 Examples of polysaccharide-based nanosystems intended for drug delivery

Source	Natural polymer	Type of nanosystem	Method of preparation	Drug loaded	Application	References
Plant polysaccharide	Cellulose	Nanopowder	Milling	None	Pickering stabilization	Lu et al. (2018b)
	Starch	Nanopowder	Milling	None	Pickering stabilization	Lu et al. (2018a)
	Pectin	Nanopowder	Mechanical homogenization/freeze-drying	Itraconazole	Antifungal	Burapapadh et al. (2016)
	Guar gum	Nanospheres	Solvent precipitation	Ag85A	Antituberculosis	Kaur et al. (2015)
Microbial polysaccharide	Dextran	Nanospheres	Aqueous-aqueous freezing-induced phase separation	BSA/GM-CSF/β-galactosidase	Not specified	Wu et al. (2013b)
	Dextran	Nanospheres	Double emulsion	siRNA targeting the gene for the macrophage marker protein CD68	Hepatic tumor	Foerster et al. (2016)
	Xanthan gum	Nanospheres	Equilibrium	None	None	Xu et al. (2015)

(continued)

Table 4 (continued)

Source	Natural polymer	Type of nanosystem	Method of preparation	Drug loaded	Application	References
Animal polysaccharide	Chitosan/carrageenan	Nanospheres	Ionic complexation	Ovalbumin	None	Grenha et al. (2010)
	Chitosan	Nanospheres	Ionic gelation	*Ilex paraguariensis*	Antioxidant	Harris et al. (2011)
	Chitosan/alginate	Nanospheres	Ionotropic gelation	Amoxicillin	*H. pylori* infection	Arora et al. (2011)
	Chitosan/alginate	Nanocapsules	Solvent emulsion evaporation/Ionic gelation	Turmeric oil	Not specified	Lertsutthiwong et al. (2009)
	Hyaluronic acid/chitosan	Nanospheres	Ionotropic gelation	Doxorubicin	Cancer	Deng et al. (2014)
	Hyaluronic acid	Nanospheres	Self-assembling	Chlorin e6	Photodynamic therapy	Yoon et al. (2012)

3.2.1 Application of Algal-Derived Polysaccharides in the Preparation of Nanoparticles

Alginates (alginic acid, and sodium and potassium alginates) are polyanionic polysaccharides of β-D-mannuronic acid and α-L-guluronic acid that are widely distributed in the cell walls of brown algae. They are recognized as a GRAS material and have been well accepted in pharmaceutical applications due to their bioadhesive and mucoadhesive properties. Usually, alginate nanoparticles are obtained by the conventional ionotropic cross-linking method with calcium ions or by template approaches such as water-in-oil nanoemulsions (Venkatesan et al. 2017). Alginate-chitosan nanoparticles have been shown to be a promising carrier for the encapsulation of hydrophobic and hydrophilic drugs. These are prepared by the ionotropic pre-gelation of an alginate core, followed by chitosan polyelectrolyte complexation (Sosnik 2014).

3.2.2 Application of Vegetal-Derived Polysaccharides in the Preparation of Nanoparticles

Pectin is a natural hetero-polysaccharide found in the cell walls of many fruits and vegetables. Pectin and pectin derivatives are negatively charged at physiological pH. Pectin-based nanoparticles are promising drug delivery systems because they possess both mucoadhesive and targeting properties. The common methods of preparation for pectin-based nanoparticles are limited to ionotropic gelation, the nanoemulsion technique, and redox approaches (Zhao and Zhou 2016). A technological aspect to consider is the degree of methyl esterification of pectin. High methoxyl pectin, and low methoxyl pectin are the two types of pectins available. The degree of methyl esterification impacts in the gelation and other properties involved in the release mechanism of the drug.

Less common is the use of gums in the elaboration of nanoparticles, Soumya et al. (Sreenivasan et al. 2010) reported the preparation of guar gum nanoparticles with a size of 20–50 nm by nanoprecipitation and the cross-linking method. Pectin and guar gum nanoparticles could be a promising system for colon drug delivery, because both biomaterials are susceptible to microbial degradation in the large intestine, more specifically in the colon region.

3.2.3 Application of Microbial-Derived Polysaccharides in the Preparation of Nanoparticles

Dextran is a neutral polysaccharide structured of simple repeating glucose subunits. Dextran has been employed in the preparation of nanomedicines as a coating material; it is utilized as an alternative to PEGylation to avoid nanoparticle and opsonin interactions or to be biocompatible surface nanoparticles, for example, with iron-based nanoparticles or graphene nanoparticles (Banerjee and Bandopadhyay 2016;

Wasiak et al. 2016). Dextran micelles and vesicular nanoscaffolds are other two examples of dextran-based nanosytems. Depending on whether other molecules are included in the shell, hydrophilic and hydrophobic drugs can be loaded into the water core or polymeric shell, respectively.

Xanthan gum is an exo-polysaccharide that has been widely used in the food, cosmetic and pharmaceutical industries. However, in nanomedicine, xanthan gum possess attractive physicochemical properties for formulating tunable self-assembled nanogels for drug delivery applications (Xu et al. 2015). In the same manner, smart nanogels with pH/redox dual responsiveness were prepared by cross-linking for drug-controlled release of doxorubicin (Zhang et al. 2019b).

3.2.4 Application of Animal-Derived Polysaccharides in the Preparation of Nanoparticles

Chitosan is the most widely studied animal-derived polysaccharide in pharmaceutical nanotechnology. Due to its cationic characteristics, drug encapsulations are mediated by electrostatic interactions between the positively charged chitosan and the negatively charged drug. Chitosan also exhibits mucoadhesive properties and have been used to formulate mucosal drug delivery systems. The existence of one amino and two hydroxyl groups in the repeated glycosidic units of chitosan enables its chemical modification and the surface-covalent attaching of specific ligands. Chitosan has antimicrobial activity mediated by the amino-free groups; however, its antimicrobial properties are mostly limited to pH values below 6 (Sahariah and Másson 2017), despite that chitosan nanoparticles were developed for this purpose. Several methods have been proposed for preparing nanoparticles based on chitosan, including ionic gelation, emulsion cross-linking, spray-drying, the emulsion-droplet coalescence method, nanoprecipitation, the reverse micellar method, the desolvation method, modified ionic gelation with radial polymerization, and emulsion solvent diffusion (Naskar et al. 2019).

Finally, hyaluronic acid and chondroitin sulfate are two endogenous polysaccharides that exist in the extracellular matrix of cartilage and they have gained interest in recent years because they are able to bind specifically to CD44 receptors, which are overexpressed in several types of tumors (Lin and Lee 2018).

4 Vectorization of Nanoparticles Elaborated with Natural Polymers

4.1 Vectorization with Chitosan

Nanoparticles for drug delivery based on chitosan or coated with chitosan are widely investigated. Researchers bear in mind the different properties of chitosan depending on molecular weight, the deacetylation degree, or the pH medium. Size, morphology, and surface charge are factors that could exert an influence on the properties of the nanoparticles. In particular, the surface charge determines the cellular uptake, biodistribution, and interaction with biological environments of NP. The main applications regarding the vectorization of nanoparticles elaborated with natural polymers include cancer therapy, nasal drug delivery, pulmonary drug delivery, ocular drug delivery, vaccine delivery, and vaginal drug delivery. The advantages of chitosan nanoparticles, as previously mentioned, lie is their high mucoadhesion and gelation, increasing residence time, therefore permeation capacity in tissues. Interestingly, some studies have registered the absence of the burst effect attributed to the formation of a gel layer on the surface of nanoparticles, evidenced by TEM studies. Chitosan nanoparticles have been predominantly employed to encapsulate antiviral drugs, antibiotics, and intraocular-pressure regulators. The majority of these formulations has been obtained by the ionic gelation method and is characterized by not causing irritation. Nasal drug delivery has been classified as a non-invasive technique and one of the most appropriate strategies for practicality in transporting drugs to the central nervous system. Again, the mucoadhesive character improves permeation by increasing residence time. Sometimes, the formulation is accompanied by a hydrogel to facilitate administration. For oral drug delivery in situations of periodontal diseases, chitosan nanoparticles have been applied. Taking advantage of their mucoadhesion capacity, the application can be facilitated with the incorporation of the nanoparticles in mucoadhesive films. In particular, applications against cancer are prominent among the widespread uses of chitosan nanoparticles, especially nanoparticle formulations with additional ligands on the surface such as folate and glycol-modified-chitosan-NP, producing an increase in permeability and retention time.

4.2 *Vectorization with Other Natural Polymers*

Other polysaccharides have been explored to modify the surface of nanoparticles, for example, agarose, alginate, chondroitin, dextran, heparin, pullulan, hyaluronic acid, and starch. On describing some examples, dextran has been proposed for the coating of magnetic and polymeric nanoparticles, and dextran- and modified dextran-coated iron oxide nanoparticles are even available on the market as an MRI contrast agent. Vocelle et al. (2016) demonstrated that the dextran functionalization of silica nanoparticles enhanced the uptake and intracellular delivery of siRNA, the possible mechanism is being scavenger receptor-mediated endocytosis through a clathrin/caveolin-independent process. Heparin, a highly sulfated and linear natural polysaccharide, has been the subject of a few studies as a liver-specific surface coating for gold-nanoparticles. It has shown accumulation in Kupffer cells and in the endothelial-like cells lining the liver sinusoid in vivo (Kemp and Linhardt 2009; Sun et al. 2009). Heparin also has been effectively utilized for the surface functionalization of PLGA nanoparticles (Chung et al. 2010), providing an increase in the in vitro cellular uptake. In terms of hyaluronic acid, its high tumor specificity and biocompatibility makes it so that when it is used as a coating, it could significantly reduce cytotoxicity and increase the cellular uptake by cancer cells of nanoparticles. The latter was evidenced during the evaluation of hyaluronic acid as a coating biomaterial of nanoparticles of poly (butyl cyanoacrylate) (He et al. 2009). As mentioned in the previous section, chondroitin sulfate presents interesting high affinity by glycoprotein CD44, which is drastically upregulated in various cancers. Additionally, its high hydrophilicity makes it have a high hydrodynamic volume; therefore, chondroitin sulfate has been proposed as a ligand for tumor-targeted drug delivery. Zu et al. (2019) proved the capacity of chondroitin sulfate in colon cancer-targeted drug delivery. Because the affinity of some polysaccharides to CD 44 is mediated by the presence of β-1,3-linked *N*-acetyl galactosamine and β-1,4-linked D-glucuronic acid units, other polysaccharides such as hyaluronic acid also have been exploited for the functionalization of nanoparticles intended as a strategy for targeted cancer treatments (Xiao et al. 2015).

5 Gene Therapy Using Nanoparticles Prepared with Natural Polymers

Gene therapy is a procedure to improve the health condition of the patient by repairing, replacing, or regulating the patient's cells genetically by means of the administration of nucleic acids (Hardee et al. 2017). With the approximate number of 832 studies with the term "gene therapy" on the clinicaltrials.gov NIH website, by the NIH, the important progress of this new area of medicine is demonstrated. Early development is still recent, as approval of the first commercial products dates from 2016 by the European Medicines Agency and the U.S. Food and Drug Administration. As established by the new era of other medicines, approval by governmental entities, the regulatory establishment, the demonstration of industrial production, and, mainly, the approval of clinical trials with favorable results, will establish new research conduits for several novel medicines (Anguela and High 2019). The major contribution of gene therapy is conditioned by advances in the investigation of the type of mutation characterizing each disease and by the development of the technological development in new vectors that guarantee adequate administration of the new genetic material. The use of viral vectors generally guarantees a high transfection of the genetic material by the invasive nature of the peptides composing them (Ramamoorth and Narvekar 2015). During in vitro tests, the results are encouraging but, in the most complex biological models, risk associated with the degree of invasiveness increases significantly with respect to risk to life. For several years now, the option of non-viral vectors has represented an attractive alternative on seeking new gene-therapy treatments for humans. Non-viral vectors offer low immunogenicity, ease of large-scale production, and the option of repetitive administration, in addition to vectorization at the site required for the modification of the damaged gene and gradual release (Hardee et al. 2017). However, the main limitation in the majority of proposals may be poor transgene expression due to the lack of highly invasive tools for transfection. Examples cited in the literature consider cystic fibrosis, autoimmune diseases, and neural neuropathies as examples of applications of chitosan in non-viral gene delivery. The experimental proposal for the manufacture of non-viral vectors generally involves the use of cationic lipids and polymers of natural origin. The intention is to form complexes in the nanometric order through a cross-linking mechanism by opposite charges: positive from the polymer or lipid and negative from the genetic material (Barua et al. 2011). This type of plexes can be combined with ligands for vectorization, or with fusogenic agents for the crossing of cell membranes. Among cationic polymers, the use of polyethylenimine, polyamidoamine, poly-L-lysine, and chitosan can be mentioned. Chitosan is a biodegradable and biocompatible polysaccharide that is widely used in the pharmaceutical area (Saranya et al. 2011). The preceding compound, chitin, derives from a broad natural source in nature, usually the crustacean exoskeleton. The amino groups provide the chitosan with extensive functionality in a protonated state with pH values lower than 6.5, conferring a polycationic nature. A minor disadvantage of chitosan solutions lies in the insolubility of polysaccharides in neutral media. However, the polycationic state has been well used to counteract the presence of microorganisms. The preparation of non-viral chitosan vectors represents nanometric particle size with gentle manufacturing methods that maintain the integrity of the genetic material (Layek and Singh 2017). The required infrastructure can be simple if the ionic gelation method is the option chosen. In this way, the chitosan will form a shell or sponge to protect the DNA or RNA (Fig. 3). There are different methods in the manufacture of nanoparticles for the transport of genetic material, including the complex polyelectrolyte method. This method is based on the electrostatic interaction between anion and cation, followed by charge neutralization. The complexes obtained are called polyplexes (Luo and Wang 2014). One of the great advantages of this method is the mild conditions it requires so as not to damage the protein, peptide, or plasmid DNA. The other widely used method is the so-called ionotropic gelation method (Koukaras et al. 2012). The difference of this

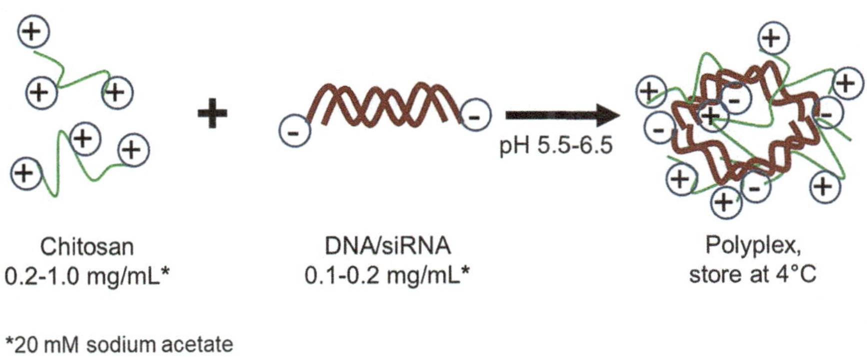

Fig. 3 Chitosan polyplex formation for gene delivery

method comprises the presence of a cross-linker, usually a polyanion (tripolyphosphate), to form the complex under the same principle of interaction by opposite electrostatic charges and followed by coacervation.

Other less frequent methods include microemulsion and the solvent evaporation method. In addition to the mild conditions of the previously mentioned methods, another advantage is the minimal infrastructure required to prepare the batches: basically a grill with stirring and heating (Hembram et al. 2016). The larger aspects to take into account in chitosan methods are the robustness of the process due to variations in particle size and the high polydispersity-index value associated with low control of the chitosan molecular weight. It is difficult to find a consensus on the size of the plexus according to factors such as the molecular weight of chitosan or the ratio of plasmid to polymer. There is more evidence for the latter relationship because it permits better condensation of the complex due to greater interaction of the opposite charges, although it will be necessary to ensure that greater condensation also allows the release of the plasmid at a certain place and time. Although some authors mention an increase in plexus size due to an increase in molecular weight, the preparation method appears to exert a greater influence. With regard to the surface charge, although it is always mentioned that one of the advantages of the chitosan plexus is the high residual positive electric charge, the truth is that it must be moderate to avoid alterations in the biological evaluation systems. Cationic polymers are, in principle, employed for gene therapy due to the affinity of the amino groups, generally, and their interaction with the negative charge of the plasmids. Greater affinity implies greater protection and transport, which translates into greater transfection (Barua et al. 2011). Nonetheless, as previously mentioned, an equilibrium point must be found in the polymer-plasmid relationship that permits plasmid release; otherwise, it will not be of clinical usefulness. An easily accessible tool for determining the degree of affinity of the plasmid is by agarose gel electrophoresis through the intercalation of ethidium bromide with the plasmid and its increase in emitted fluorescence. When chitosan is added to the same medium, it displaces ethidium bromide and decreases the light signal. Another characteristic-of-interest in the plexus is buffer capacity. For gene therapy, one of the limitations is the endosomal attack on admission to the cytoplasm. Cationic polymers, especially chitosan, attach protons to the amino groups acting like a sponge, which will then cause an increase in osmotic pressure and the rupture of the endosome membrane. Therefore, acid-base titrations offer an approximation of the buffer capacity. In general, chitosan is one of the natural cationic polymers that offer adequate sponge capacity. Finally, other characteristics-of-interest in plexus formation are its stability under different conditions such as temperature, pH, ionic strength, the capacity to protect the cargo, and the ability to transfection (Layek and Singh 2017). Usually, size, polydispersity index, and zeta potential are monitored. Some studies start development with known and stable plasmids, then later exchanging it for the plasmid-of-interest. Some products derived from chitosan have been proposed to improve transfection and gene delivery, including carboxymethylation, trimethyl chitosan, PEG, PEG-trimethyl chitosan, trimethyl chitosan-cysteine, *N*-methylene phosphonic chitosan, *N*,*N*-dimethylaminobenzyl derivatization, 4-pyridinylmethyl substitution, *N*-dodecylated chitosan, polyethyleneimine grafted with chitosan, *N*-dodecylated chitosan, folate-chitosan, PEG-folate chitosan, and hyaluronic acid-chitosan, among others (Liu et al. 2010b). These modifications have intended to improve the water solubility of chitosan and improve the condensation of DNA in polyplexes. As mentioned in this chapter, other polymers of natural origin and of interest for gene delivery include human serum albumin and gelatin.

6 Special Requirements in the Characterization of Excipients and Nanoparticles Elaborated with Natural Polymers

In nanopharmaceutical field, one of the major challenges is to achieve the reproducibility while ensuring the efficacy of new formulations. For this, nanoparticle characterization must be highly specialized and should be evaluated in order to analyze particle size, shape, and morphology, nanoparticle-surface stability, drug-polymer interactions, among others (Table 5). Some of the most representative characterization techniques would be described as follows. It should be noted that there are no standardized techniques, methodologies, or FDA protocols for nanoparticle characterization. However, we presented a brief description of the customary techniques applied to ensure the reproducibility and accuracy of the method applied.

Table 5 Different techniques usually applied in the physicochemical characterization of nanoparticles

Technique	Characterization parameters
Dynamic light scattering (DLS)	Hydrodynamic size distribution
Scanning electron microscopy (SEM)	Size and size distribution Shape Possible aggregation
Microscopy of atomic force (AFM)	Size and size distribution Shape Structure Possible aggregation
Transmission electron microscopy (TEM)	Size and size distribution Shape heterogeneity Possible aggregation
Zeta potential	Stability (surface charge)
Infrared spectroscopy (IR)	Functional group analysis Possible interactions of the drug and the polymer
Thermogravimetric analysis (TGA)	Thermal properties
Differential scanning calorimetry (DSC)	Thermal properties and possible interactions of the drug and the polymer

6.1 Size Analysis

Size analysis is one of the critical parameters when we speak of nanoparticles. The size and, more importantly, the reproducibility of it, are fundamental to the analysis of physicochemical properties and the possible biological interactions that the nanoparticles could present (Gaumet et al. 2008; He et al. 2010). Laser diffraction (LD) and dynamic light scattering (DLS) are the most used techniques to evaluate nanoparticle size; however, microscopy techniques, such as scanning electron microscopy (SEM) or microscopy of atomic force (AFM), could also be applied for this purpose.

6.1.1 DLS

The DLS technique is commonly applied to know the average particle size of a colloidal suspension. The colloidal sample is illuminated by a monochromatic laser light that is scattered into a photon detector (Xu 2008). The scattered light intensity fluctuates in time because the Brownian motion of the particles; these movements are related to particle size by means of an equation. The equation for spherical particles is as follows:

$$D_{\mathrm{h}} = \frac{k_{\mathrm{B}} T}{3\pi \eta D_{\mathrm{t}}}$$

where D_{h} is the hydrodynamic diameter, η is the relative viscosity of the solvent, k_{B} is the Boltzmann's constant, T is the temperature and D_{t} is the translational diffusion coefficient.

In addition to particle size, on using the technique it is also possible to determine the size distribution and the polydispersity index (PDI). Figure 4 presents two cases of particle dispersions, the blue line indicated a monodisperse solution, meaning a homogeneous sample with a low PDI, desirable results for nanodispersions. The red line represents a polydisperse population of particles, thus a heterogeneous solution.

The size, size distribution, and the PDI of nanoparticles for drug delivery are important characteristics due to that these are related to drug release, drug loading and also, may affect in vivo interactions such as cell uptake, the phagocytosis of immune cells such as macrophages, and crossing biological barriers (Azimi et al. 2013).

For example, the blood-brain barrier (BBB) comprises a very selective barrier that inhibits the free diffusion of foreign molecules from the body to the brain. Therefore, disorders related to the central nervous system remain untreatable, and natural polymeric nanoparticles are an attractive option as treatment as long as the nanoparticulate system maintains certain parameters, such as particle size and size distribution. In a study of Sahin et al. (2017), chitosan nanoparticles conjugated with antibodies were elaborated and characterized to evaluate the interaction between brain endothelial cells and the nanoparticles. The authors characterized the nanoparticles through DLS and observed that they presented an adequate average size in order to be capable of

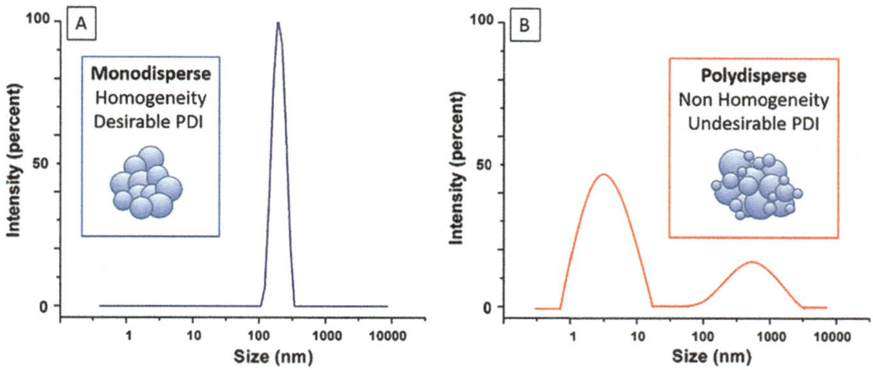

Fig. 4 Size-distribution example. **a** A monodisperse nanoparticle solution, nanoparticles are of about the same size particle, showing homogenous dispersion with a desirable PDI. **b** A nanoparticle solution with polydispersity, the size of the nanoparticle is not within the same range (heterogeneous solution), triggering an undesirable PDI

internalizing the cells. This observation was demonstrated by other techniques such as fluorescence microscopy, finding chitosan nanoparticles inside the cell cytoplasm. Similarly, in 2016, Zhao et al. (2016) developed a neuropeptide substance termed P-loaded gelatin nanoparticles in order to analyze their therapeutic effect in hemi-parkinsonian rats. The nanosystem was characterized by DLS, presenting an average particle size of less than 200 nm, which is suitable for brain-disorder applications due to that therapeutic biological particle with a size below 300 nm can bypass the BBB. As could be observed, the DLS technique is a useful and necessary tool in nanocarrier system characterization, especially with natural polymers. In addition, DLS and microscopy techniques have been applied as a size analysis approach; nevertheless, preparation of samples could be laborious and the integrity of the sample could be affected. These types of techniques are usually employed to evaluate nanoparticle morphology and shape.

6.2 Shape and Morphology

In order to complement the size analysis, the shape and the morphology of the nanoparticles should be evaluated. Additionally, nanoparticles often are not perfectly spherical, and shape has high importance in the behavior of NP with natural polymers and the biological barriers. The common microscopy techniques utilized for this purpose are SEM, TEM, and AFM.

6.2.1 SEM

In this surface imaging technique, the microscope scans the electron beams shooting from a burst of electrons on the sample surface and collects the scattered electrons for imaging. To build the three-dimensional images of the analyzed samples, this technique employs secondary electrons and backscattered electrons that have emanated from the sample. For this reason, electron beam energy does not need to be high and the sample does not require electron transparency, but requires the presence of their conductivity to avoid charging. The polymers utilized for nanoparticle elaboration are of an organic nature; therefore, they do not reflect an electron beam sufficiently. Consequently, the samples require a preparation based on metal thin-layer coating, creating a conductive layer on the sample surface and preventing surface charging. From the images obtained, one could infer the aggregation of the sample, in addition to the shape and morphology. On the other hand, this technique presents the disadvantage of a destructive sample preparation (Guo and Tan 2009; Samimi et al. 2019; Wang 2006).

6.2.2 TEM

This technique, applied to characterize the size and shape of nanoparticles, provides two-dimensional images formed from the electrons transmitted through the sample. TEM provides information on the inner-particle structures of the nanocarriers and is highly valuable for measuring the polymeric wall of nanoparticles. Also, this microscopy method could offer higher resolution than SEM, providing greater detail at the atomic scale (Crucho and Barros 2017; Hall and Mcneil 2007). However, due to the sample preparation method, the average particle size could be different compared with the data obtained through other techniques, such as DLS. For example, Motwani et al. (2008) elaborated nanoparticles of chitosan and sodium alginate as a novel vehicle for ophthalmic delivery. The apparent nanoparticle size observed by TEM was considerably smaller than the average size obtained by means of DLS. The authors suggest that the differences between these data could be related to the sample preparation based on dehydration, while the DLS measurements yield an ensemble average of particle size in suspension. Likewise, Rampino et al. (2013) observed chitosan nanoparticles by TEM, finding a different average size compared with the results obtained by this technique compared with DLS data. These authors indicated that this behavior was related to the aggregation of nanoparticles to form a homogeneous population. The micrographs obtained by the authors are presented in Fig. 5.

6.2.3 AFM

Another technique for analyzing the shape of nanoparticles is AFM. AFM uses a laser-beam deflection system in which a laser is reflected the form of the back of

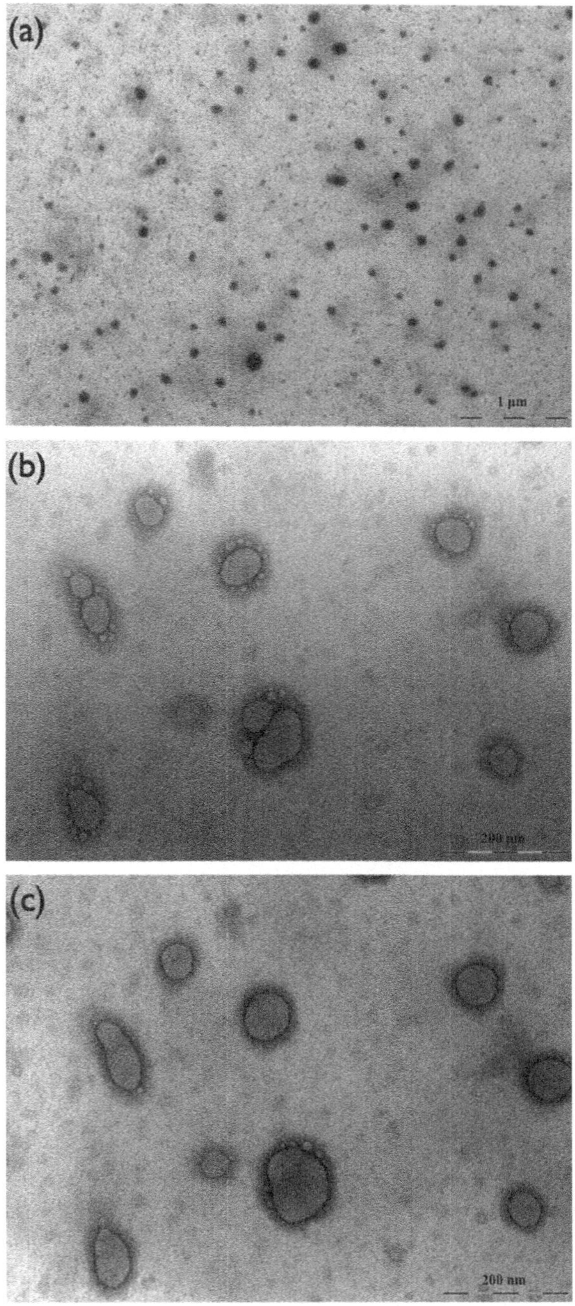

Fig. 5 TEM image of chitosan nanoparticles (**a** and **b**). Image **b** taken after a few minutes: the fusion process is clearly visible (**c**) (Rampino et al. 2013). Reprinted with permission from Elsevier 2019

the reflective AFM lever and onto a position-sensitive detector, and the AFM images provide topology information of the samples (Grobelny et al. 2011; Rao et al. 2007). For nanoparticles, the samples require being dispersed on flat surfaces. To ensure the obtaining of adequate images, the roughness of the surface should be much less than that of the nominal sizes of the nanoparticles. The most usual materials used as substrates for the AFM of nanoparticles are high-quality mica, atomically flat polycrystalline gold (deposited on mica), and single-crystal silicon.

6.3 *Surface Properties and Stability*

The stability of the NP suspension plays a key role in the elaboration and application of nanoparticles in order to analyze the quality control of the colloidal suspension.

6.3.1 Zeta Potential

The measurement of zeta potential is commonly employed to analyze the surface charge and stability of a nanoparticle suspension. In addition, the zeta potential can provide the basis for the analytical assessment of specific regulatory endpoints in relation to the safety implications of nanoparticles; in the nanotoxicological field, this parameter could be applied to predict its effects on human health (He et al. 2010). The analysis of zeta potential is generally obtained by the electrophoresis method (Kathe et al. 2014). The foundation of this method involves measuring the electrophoretic mobility of charged particles under an applied electrical potential. To calculate the zeta potential (z), the Henry equation is used:

$$U_{\mathrm{e}} = \frac{2\varepsilon z f(\kappa a)}{3\eta}$$

where U_{e} is the electrophoretic mobility, ε is the dielectric constant, η is the absolute zero-shear viscosity of the medium, $f(\kappa a)$ is the Henry function and κa is a measure of the ratio of the particle radio to the Debye length (Clogston and Patri 2011). Using the zeta potential, the physical stability of nanoparticles could be evaluated. Nanoparticles with a high absolute value of zeta potential are electrically stable, while those with a low absolute value of zeta potential tend to be less stable. Furthermore, the nanoparticle charge is associated with the mechanism of interactions with cells; thus, the attachment of nanoparticles to the cell membrane is highly affected by the surface charge (Froehlich and Fröhlich 2016; Fröhlich 2016; Jo et al. 2015). As revealed by Yue et al. (2011), changes on the surface charge of chitosan nanoparticles could modify cellular uptake profiles and intracellular trafficking. The authors prepared three types of chitosan nanoparticles, maintaining identical physicochemical properties except for surface charge. The nanoparticles were incubated with eight cell lines and the cellular uptake was analyzed. The results suggested that the surface charge considerably affected the cellular uptake, concluding that the positive charge promotes internalization; probably due to that positively charged nanoparticles prefer contact with the cell surface, which exhibits a negative charge. Depending on the nanoparticle application (slow drug delivery, anticancer therapy, or gene therapy, among others), the efficacy of nanoparticles could be associated with the internalization rate (slow or fast). In this respect, the zeta potential is an essential parameter to evaluate when natural polymer nanoparticles are elaborated.

6.4 *Interactions Between Polymers and Drugs*

Drug-loading efficiency and low drug release are parameters that exert a high impact on the field of medical nanotechnology. To obtain the best results possible, the type of polymer, used as well as drug/polymer crystallinity and the type of polymer-drug interactions must be taken into account. Different characterization techniques, such as Fourier transform Infrared (FT-IR), Raman spectroscopy, nuclear magnetic resonance (NMR) spectroscopy, thermogravimetric analysis (TGA), or differential scanning calorimetry (DSC), in combination with IR and RAMAN, have been applied to study these behaviors.

6.4.1 Fourier Transform Infrared (FT-IR) and Raman Spectroscopy

In addition to the zeta potential, FT-IR and Raman spectroscopy are techniques that could be applied in order to analyze the stability and structural properties of nanoparticles and, even more important, the molecular interactions between the drugs and the polymers. The basis of these techniques is the periodic change of dipole moments (FTIR) or polarizabilities (Raman) related to the characteristic molecular vibrations of molecules or groups of atoms. In particular, with Raman spectral information, analysis of real-time drug-polymer and polymer-stabilizer interaction could be elucidated. These techniques are non-destructive methods, require a small amount of sample, and are highly sensitive on analysis. The information obtained with these types of techniques could confirm the presence or purity of known compounds, the formation of new bonds among the excipients or between the drug and the polymers, and modifications in excipient structure due to the method applied in nanoparticle formation. For example, Ping Li et al. (Wang et al. 2015a) developed nanoparticles of chitosan-alginate loaded with nifedipine. These authors characterized the system by FT-IR, since this qualitative analysis is useful for estimating the extent of modification for functional groups. The FT-IR results implied that the nifedipine molecule was filled in the polymeric network; furthermore, on analyzing the obtained spectra, the authors suggested that the carboxylic groups of alginate were associated with ammonium groups of chitosan by electrostatic interactions. In the same manner, Bronze-Uhle et al. (2017) elaborated salicylic acid-loaded bovine serum-albumin nanoparticles to improve the drug's bioavailability and to avoid the decomposition of the salicylates. To evaluate the chemical and conformational changes present due to nanoparticle formation, the authors analyzed the system by FTIR. Comparison among the spectra of albumin, albumin nanoparticles, and salicylic acid-loaded albumin nanoparticles revealed differences in intensity and changes in amide bands, suggesting nanoparticle formation and the successful loading of salicylic acid in the nanoparticles.

6.4.2 NMR Spectroscopy

This technique is widely employed in the chemical field and is applied to investigate and analyze the structures and dynamic of molecules. The basis of this technique is the excitation of the spectral lines of different atomic nuclei triggered by the application of a strong magnetic field. It provides an advanced and convenient method for size analysis (diffusion-ordered (NMR) spectroscopy) and surface chemistry evaluations (Guo and Yarger 2018). This characterization technique allows evaluate the evaluation of changes in the chemical shifts of different elements in order to obtain nanoparticle information. Nanoparticle characterization could include the information of the stoichiometric formation affinity constant. For example, Campos et al. (2018) elaborated chitosan nanoparticles functionalized with β-cyclodextrin containing active agents. These authors evaluated, by ^{1}H NMR, the formation of inclusion complexes between β-cyclodextrin and carvacrol or linalool and the functionalized chitosan nanoparticles. Information on the changes in the chemical shifts of the hydrogens in the active agents permitted the authors to elucidate how the molecules interacted to form the complex. Similarly, Boonsongrit et al. (2008) elaborated chitosan nanoparticles and evaluated by NMR spectroscopy, the electrostatic interaction among the opposite charges of the drugs, that is, benzoic acid and insulin, and chitosan. However, these interactions were not detected, suggesting that drug entrapment is mainly physical due to the low ionic binding capacity for these molecules. Recently, Wang et al. (2015b) analyzed, by ^{1}H NMR spectroscopy, their nanoparticle system based on alginate, phytosterol, and folate, in order to confirm the conjugation. The authors compared the differences between NMR spectra of alginate and nanoparticles. They observed variations, suggesting that the phytosterols were effectively conjugated to carboxyl groups of alginate. Furthermore, changes in certain peaks implied the conjugation of phytosterols and folate, the authors concluding that the nanoparticulate system was successfully synthetized. In addition to NMR, thermal analysis techniques could be applied to elucidate the interactions among the excipients in nanoformulations.

6.4.3 TGA

This experimental technique is applied to measure the change in the mass of a sample as a function of temperature in a controlled atmosphere. It is very common technique in polymer research, and the analyzer consists of a thermobalance with high precision, which is connected to a pan/crucible holder inside a temperature-controlled furnace. The pan/crucible holder is located on a sensor that is supported by a thermocouple to measure the sample temperature. To control the sample environment, a purge gas is introduced into the furnace. The information rendered by TGA could provide a measure of the purity and composition of the sample. The possible changes in the oxidation or decomposition observed in thermograms of NP, in comparison with those of the excipients, could be related to variations on the NP surface and to chemical interactions.

In a recent work, Shi et al. (2018) developed chitosan-PLGA NP to improve the bioavailability of tolbutamide. The authors evaluated the nanoparticles through TGA. They observed changes in thermal profiles between the excipients and the nanoformulation, finding that the chitosan improved the thermodynamic stability of NP. Based on these results, the authors concluded that the chitosan coated the surface of the nanoparticles. Based on the degradation profile, thermograms could afford information on the remainder and the percentages of organic or inorganic contents. In addition, in order to obtain complementary information, the DSC technique could be useful in nanoparticle characterization and in the evaluation of different molecular interactions.

6.4.4 DSC

Similar to TGA, the DSC technique is useful to determine the polymorphism and crystallinity of the different samples. In nanotechnology, DSC is commonly used to analyze the physicochemical properties and possible interactions between the drug loaded and the polymeric nanoparticle (Crucho and Barros 2017). Whit the data collected by means of this analysis, one could obtain the glass transition temperature (Tg) of the sample (polymers, drugs, and polymeric NP with the drug loaded). Tg is defined as the temperature at or above which the molecular structure exhibits macromolecular mobility (Multur 2004). For this reason, Tg data could provide information on the molecular and physicochemical characteristics of NP. For example, the variation in Tg values could be related with the different stabilizers employed in the fabrication, exerting an influence on the physical properties of nanoparticles and affecting the biological characteristics as well. Other important parameters that could be elucidated based on the information obtained from DSC measurements comprise the crystallinity of the samples. Crystallinity is related to the physicochemical properties of the NP, and with this information, it could approximate the water-adsorbing capability, the biodegradation, the drug-release behavior, and the mechanical properties of the NP. The rigidity of the NP is an important factor that could influence the blood-resistance time; this rigidity is usually associated with the Tg value. It has been described that the high Tg of NP possesses a significantly higher blood-resistance time as compared with that of the low Tg of NP; however, the low Tg of NP causes high flexibility and an adequate level of surface interactions between the NP and the tissues or biological barriers. Sarmento et al. (2006) prepared and characterized alginate/chitosan nanoparticles and insulin-loaded alginate/chitosan nanoparticles. These authors evaluated the interactions among components by DSC. The thermograms showed endothermic and exothermic peaks for chitosan and for alginate. The physical mixture presented a wider endothermic event, which is probably due to the coalescence of the thermal profiles of both excipients. However, the nanoparticle thermogram exhibited peaks at different temperatures, which the authors related with the formation of new chemical bonds. Likewise, Mohammed et al. (2019) analyzed, by DSC, the sustained release system based on chitosan-PLGA with encapsulated mycophenolate mofetil. The thermogram of mycophenolate mofetil presents the phase transition peaks of the material, proving the crystalline form of the drug. Instead, in the thermogram of optimal formulation, this peak did not appear, suggesting that the mycophenolate mofetil present in the nanoparticles was in amorphous phase. There are many challenges involved in nanoparticle fabrication, with reproducibility and quality two of the most important of these. Furthermore, when natural polymers are the excipients for the nanoformulation, different properties could change due to the source of the materials. For example, in polymers such as chitosan, the molecular weight, the deacetylation degree, or the chain length could modify the characteristics of the material, obtaining nanoparticles with different characteristics. For these reasons, the physicochemical characterization of nanoparticles plays a key role in the nanotechnology field.

7 Nanotoxicology of Natural Polymers Used in the Pharmaceutical Area

For several decades, it has been well known that particles can affect human health as a function of their properties, such as chemical composition, size, shape and, more recently, aspects related to the interaction between the particles and the cells, which are capable of activating pathological mechanisms (Riediker et al. 2019). Thus, toxicological research on particles comprises a priority discipline and an inevitable aspect to satisfy the requirements of regulating agencies such as the FDA, as well as biosafety needs (Bouwmeester et al. 2011; Hunt et al. 2013; Kuempel et al. 2017). According to the International Standard ISO 10993-1:2018 (Biological Evaluation of Medical Devices), all materials utilized in humans are subjected to in vitro and in vivo biocompatibility tests to verify the response and behavior of the cells interacting with them. A material is assigned as biocompatible when it interacts with the body without inducing unacceptable toxic, immunogenic, thrombogenic, and carcinogenic responses, and any other adverse effects (Cambiaghi et al. 2018). In other words, a material is biocompatible if it possess the ability to perform medically without producing undesired local or systemic effects (Williams 2008). Biocompatibility includes both cytocompatibility and cytotoxicity. The first involves testing the particles with cells and evaluating the morphological changes. Cytotoxicity considers the substances that leach out of the particles (e.g., degradation products) on performing biochemical tests. It is noteworthy that biocompatibility is highly anatomically dependent; thus, particles can be affected in different ways (Goonoo et al. 2014). In general, biocompatibility tests include two levels: (i) biosafety testing, which can evaluate particle toxicity in cultured cells, hemolysis, allergic responses, and genetic alterations or tissue necrosis after animal implantation, and (ii) biofunctional testing, which focuses on the specific functions of particles being evaluated, preferably in vitro and in vivo, and on the responses of all of the cells and tissues in contact with the particles (Zhang et al. 2003).

In the last three lustrums, the global production of nanoplastics for all types of applications has increased, and the engineered nanomaterials with two or three dimensions have grown exponentially (Hernandez et al. 2017). This is explained by their better properties, such as ultra-small size, large surface area-to-mass ratio, and high reactivity. A great number of studies have shown new nanosystems with different natures, shapes, and structures, and that have potential applications in different fields, particularly in nanomedicine. This research has focused on developing drug delivery platforms and nano-engineered health-related products such as site-specific systems, theranostics, personalized medicine, tissue engineering, highly sensitive diagnostics (biosensors), and nanostructured lab-on-a-chip systems (Lamberti et al. 2014). This research activity is in contrast with a lesser number of papers evaluating the potential toxicity, which reveals an offset between nanotechnological development and likely adverse effects on health. Thus, a comprehensive understanding of nano/bio interactions is a key factor for developing a successful nanosystem and avoiding unexpected toxicities. Sanhai et al. (2008) established that, before nanomaterials can be routinely integrated into mainstream therapeutics; several issues such as their toxicology must be considered. Therefore, the material properties can exert different effects on human health, such as bulk material and nanoparticles (Choudhary and Kusum Devi 2015).

The term nanotoxicology has been employed in different research publications to explain the adverse effects of nanomaterials on the life of individuals. In particular, these studies have been directed toward nanoparticles (Nyström and Fadeel 2012; Sharifi et al. 2012). The potential toxic risk of nanoparticles has been recently documented and related to the production of reactive oxygen species (ROS) and, as a result, oxidative stress, which can give rise to mitochondrial damage, DNA damage, and cytotoxicity in human cells (AshaRani et al. 2009; Foldbjerg et al. 2009, 2011; Hackenberg et al. 2011; Hussain et al. 2005). These effects are more common in nanometal oxides such as titanium dioxide (TiO_2) and zinc oxide (ZnO) than polymeric nanoparticles but, independently of composition, the biocompatibility of nanoparticles should be confirmed. When nanoparticles are put in touch with tissues, different cellular responses can be obtained that are related to several factors, such as dose, exposure time, size, shape, surface chemistry, environmental properties, and cell type (Hunt et al. 2013; Zhang et al. 2014). Dose-response curves are necessary to determine the appropriate and effective number of nanoparticles to administer, in terms of median toxicity and the limits for human exposure, in order to prevent any adverse effects (Diana et al. 2013). The US-FDA and the Alliance for NanoHealth recognized from 1998 recommend six priority areas that remain as priorities to date. These include the following: (1) the development of imaging technologies; (2) determination of the distribution of nanovehicles in the body upon their systemic administration; (3) the biological affinity of nanodrugs, the pathways by which they are internalized, retention time, and their ability to translocate across barriers; (4) the development of new computational models for predicting the human health risks of exposure to nanomaterials, (5) the establishment of consensus toxicity-testing protocols, and (6) understanding the unexpected secondary adverse effects of NP (Halappanavar et al. 2018; Sanhai et al. 2008).

A common path followed by a formulator when designing nanoparticles is to employ natural or synthetic polymers due to their promising advantages, such as, ease of formulation and better characterization ability (Choudhary and Kusum Devi, 2015). As mentioned previously, natural polymers are more biosafety than synthetic polymers. Biopolymers can be biodegradable, biostable, and biocompatible and, in general, they have fewer adverse effects on the environment and humans and are considered friendly. Among natural polymers, the option most chosen for formulating nanoparticles, not only from the technological viewpoint but also due to its negligible toxicity, is chitosan. This polymer has been proposed for different administration routes (oral, pulmonary, nasal, vaginal, etc.) with several advantages, such as lower production cost, higher encapsulation capacity, and a better capability of commercialization due to its origin and minimal adverse effects. Recently, cellulose has attracted attention as a low toxic nanomaterial. Cellulose is the most abundant natural polymer, is readily available, renewable, biodegradable, inexpensive, and sustainable, with a low environmental impact. Other natural polymers employed with certain security are sodium alginate and gelatin (Cassano et al. 2012; Choudhary and Kusum Devi 2015; Natterodt et al. 2017; Saraogi et al. 2010).

Some reports (Lipovsky et al. 2011; Oberdörster et al. 2004) have documented that, when nanoparticles are sufficiently small, they are able to access skin, lungs, and brain (e.g., transcytosis). Some authors have (Gliga et al. 2014) evaluated the size and coating-dependence on the mechanism of toxicity of silver nanoparticles. These authors found that small nanoparticles (10 nm) are cytotoxic for human lung cells and that the toxicity observed could be associated with the rate of intracellular silver release, a "Trojan horse" effect. Liu et al. (2010a) investigated the toxic effect of silver nanoparticles with different sizes on four human cell models. The authors found that smaller nanoparticles enter cells more easily than larger ones, in agreement with higher toxic effects confirming the size-dependent effect. However, at present, there is insufficient scientific information to correlate the adverse effects (e.g., cytotoxic) of nanoparticles on human health, and particle size and polymer composition. Singh et al. (2019) reported that there is a limited understanding of the complex interaction among nanomaterials (e.g., nanoparticles), biological surfaces (at the protein, cellular, and whole-organism level), and the immediate environment (or ecosystem). Other authors (Bettinger et al. 2009; Petersen et al. 2012; Singh et al. 2019) highlight the importance of studying the following: (a) the surface interactions between nanoparticles and cells; (b) the influence of the chemical composition, morphology (nanoscale to microscale); (c) material softness; (d) biosafety; (e) targeting ligands for overexpressed receptors on tumor cells conjugated with NP, (f) the interplay between the target and nanoparticles; (g) the quantitative assessment of in vivo cellular uptake; (h) the interplay of protein adsorption on nanoparticles; and (i) techniques for the quantitative assessment of drug targeting.

Further, in order to consider the evaluation of the adverse effects of nanoparticles on the consumer, nanotoxicology needs to take into account the ecological impact exerted on organisms and the environment. In this context, the toxicological behavior of a tiny piece of plastic piece is totally different from that of a large one; nanoparticles have high surface curvature, a large surface area, a specific interfacial interaction with other materials, higher permeation, infiltration, and localization properties, which impact the environment in a different manner (Choudhary and Kusum Devi 2015; Zhang et al. 2019a). It is important to mention that nanoparticles from natural polymers can derive from the degradation of plastic structures used in pharmaceutical products that degrade into nanoplastics in different environments (e.g., aquatic, dumps). Thus, the use of natural polymers can possess certain advantages on consideration of their biosecurity properties; however, available data is scarce, difficult to interpret, and inconclusive (Boncel et al. 2015). Another aspect that can contribute to ecotoxicological effects with the use of nanoparticles is the preparation method. The use of huge amounts of organic solvents or pollutant stabilizers needs to be limited to achieve acceptable concentrations or to be definitively avoided, particularly if potential commercialization is planned. These aspects need to be considered during the R&D and design phases of nanoparticles. The term "safe-by-design" is a well-accepted approach that has been employed to address the global risk associated with industrial innovation processes and their safe application (Singh et al. 2019). This was designed to ensure three inter-related communities, such as the workplace, consumers, and the environment (Ahonen et al. 2017).

8 Conclusions and Future Trends

The present chapter has compiled a brief description from recently published research that addresses the design, preparation, characterization, application, and toxicological aspects of pharmaceutical nanosystems from natural polymers. Highlighted are the advantages and disadvantages of the use natural polymers to design drug delivery systems including nanosystems, in terms of their chemistry, biosecurity, biocompatibility, biodegradability, low ecotoxicity, and microbial behavior. Thus, natural polymers comprise an interesting alternative for formulating drug delivery systems, they can serve to develop systems that are based on diffusion and that are chemically controlled, solvent, and medium-activated (e.g., pH, temperature), etc. They can behave as polycations or polyanions, being able to be chemically modified, to adhere to mucosa, or to enable drug targeting. Polysaccharides (e.g., hyaluronic acid, chitosan, carrageenan, and sodium alginate) and proteins (collagen, fibrin, and gelatin) are those preferred for pharmaceutical use. Chitosan is the most studied natural polymer as a drug delivery platform, this due to its low toxicity, its capacity to interact electrostatically with drugs including DNA, and its ability to form nanogels with ions.

Several challenges concerning the safety and toxicology of polymeric nanoparticles, including natural polymers, must be of prime consideration and need to be seriously addressed, including the potential risk and toxicity of their preparation as well as their effects on human health. Studies need to be conducted to elucidate, at the molecular level, the clinical and mechanistic relationship between nanoparticles and biological entities in order to obtain sufficient evidence to exploit these systems.

Clarity continues to be lacking regarding dosimetry, ways to mitigate exposure, measures to be taken after untoward exposures, waste management, and regulatory aspects. Herein is discussed the need to have a comprehensive understanding of nano-/bio-interaction in order to prevent human toxicity and ecotoxic effects. These issues require extra effort with the application of optimization in every step of a research protocol; "safe-by-design" is an approach employed to address these global risks. However, massive research in human nanotoxicology and ecotoxicology studies is necessary to consolidate the benefits of polymeric nanoparticles.

Nanomedicine is moving in different directions to satisfy the technological and toxicological challenges, and the use of natural polymers could solve several of these. Future directions in nanotechnology using natural polymers will include the development of novel and better nanosystems, studies on toxicity and on safe human applications, and keeping the ecosystem healthy.

References

Adrover A, Paolicelli P, Petralito S, Di Muzio L, Trilli J, Cesa S, Tho I, Casadei MA (2019) Gellan gum/laponite beads for the modified release of drugs: Experimental and modeling study of gastrointestinal release. Pharmaceutics 11

Ahad HA, Reddy BKK, Ishaq BM, Kumar CH, Kumar CS (2010) Fabrication and in vitro evaluation of glibenclamide *Abelmoschus esculentus* fruit mucilage controlled release matrix tablets. J Pharm Res 3:943–946

Ahonen M, Kahru A, Ivask A, Kasemets K, Kõljalg S, Mantecca P, Vrček IV, Keinänen-Toivola MM, Crijns F (2017) Proactive approach for safe use of antimicrobial coatings in healthcare settings: opinion of the cost action network AMiCI. Int J Environ Res Public Health 14

Ali SFB, Afrooz H, Hampel R, Mohamed EM, Bhattacharya R, Cook P, Khan MA, Rahman Z (2019) Blend of cellulose ester and enteric polymers for delayed and enteric coating of core tablets of hydrophilic and hydrophobic drugs. Int J Pharm 567:118462

An FF, Zhang XH (2017) Strategies for preparing albumin-based nanoparticles for multifunctional bioimaging and drug delivery. Theranostics 7:3667–3689

Anguela XM, High KA (2019) Entering the modern era of gene therapy. Annu Rev Med 70:273–288

Arora S, Gupta S, Narang RK, Budhiraja RD (2011) Amoxicillin loaded chitosan-alginate polyelectrolyte complex nanoparticles as mucopenetrating delivery system for H. pylori. Sci Pharm 79:673–694

AshaRani PV, Hande MP, Valiyaveettil S (2009) Anti-proliferative activity of silver nanoparticles. BMC Cell Biol 10:65

Asnani GP, Bahekar J, Kokare CR (2018) Development of novel pH–responsive dual crosslinked hydrogel beads based on *Portulaca oleracea* polysaccharide-alginate-borax for colon specific delivery of 5-fluorouracil. J Drug Deliv Sci Technol 48:200–208

Assaf SM, Subhi Khanfar M, Bassam Farhan A, Said Rashid I, Badwan AA (2019) Preparation and characterization of co-processed starch/MCC/chitin hydrophilic polymers onto magnesium silicate. Pharm Dev Technol 24:761–774

Aswathy RG, Sivakumar B, Brahatheeswaran D, Fukuda T, Yoshida Y, Maekawa T, Kumar DS (2012) Biocompatible fluorescent zein nanoparticles for simultaneous bioimaging and drug delivery application. Adv Nat Sci Nanosci, Nanotechnol, p 3

Azimi B, Nourpanah P, Rabiee M, Arbab S (2013) Producing gelatin nanoparticles as delivery system for bovine serum albumin. Iran Biomed J 18:34–40

Babu A, Templeton AK, Munshi A, Ramesh R (2014) Nanodrug delivery systems: a promising technology for detection, diagnosis, and treatment of cancer. AAPS PharmSciTech 15:709–721

Banerjee A, Bandopadhyay R (2016) Use of dextran nanoparticle: a paradigm shift in bacterial exopolysaccharide based biomedical applications. Int J Biol Macromol 87:295–301

Barua S, Ramos J, Potta T, Taylor D, Huang H-C, Montanez G, Rege K (2011) Discovery of cationic polymers for non-viral gene delivery using combinatorial approaches. Comb Chem High Throughput Screen 14:908–924

Bernkop-Schnürch A (2018) Strategies to overcome the polycation dilemma in drug delivery. Adv Drug Deliv Rev 136–137:62–72

Bettinger CJ, Langer R, Borenstein JT (2009) Engineering substrate topography at the Micro- and nanoscale to control cell function. Angew Chemie Int Ed 48:5406–5415

Bhatia S (2016) Nanotechnology and its drug delivery applications. Natural polymer drug delivery systems. Springer International Publishing, Cham, pp 1–32

Bisharat L, Barker SA, Narbad A, Craig DQM (2019) In vitro drug release from acetylated high amylose starch-zein films for oral colon-specific drug delivery. Int J Pharm 556:311–319

Biswas S, Chattopadhyay M, Sen KK, Saha MK (2015) Development and characterization of alginate coated low molecular weight chitosan nanoparticles as new carriers for oral vaccine delivery in mice. Carbohydr Polym 121:403–410

Boncel S, Kyziol-Komosińska J, Krzyzewska I, Czupiol J (2015) Interactions of carbon nanotubes with aqueous/aquatic media containing organic/inorganic contaminants and selected organisms of aquatic ecosystems—a review. Chemosphere 136:211–221

Boonsongrit Y, Mueller BW, Mitrevej A (2008) Characterization of drug-chitosan interaction by 1H NMR, FTIR and isothermal titration calorimetry. Eur J Pharm Biopharm 69:388–395

Bouwmeester H, Poortman J, Peters RJ, Wijma E, Kramer E, Makama S, Puspitaninganindita K, Marvin HJP, Peijnenburg AACM, Hendriksen PJM (2011) Characterization of translocation of silver nanoparticles and effects on whole-genome gene expression using an in vitro intestinal epithelium coculture model. ACS Nano 5:4091–4103

Bronze-Uhle ES, Costa BC, Ximenes VF, Lisboa-Filho PN (2017) Synthetic nanoparticles of bovine serum albumin with entrapped salicylic acid. Nanotechnol Sci Appl 10:11–21

Burapapadh K, Takeuchi H, Sriamornsak P (2016) Development of pectin nanoparticles through mechanical homogenization for dissolution enhancement of itraconazole. Asian J Pharm Sci 11:365–375

Cambiaghi A, Medical E, Testing D (2018) Biological evaluation of medical devices as an essential part of the risk management process: updates and challenges of ISO 10993-1

Campos EVR, Proença PLF, Oliveira JL, Melville CC, Vechia JFD, De Andrade DJ, Fraceto LF (2018) Chitosan nanoparticles functionalized with β-cyclodextrin: a promising carrier for botanical pesticides. Sci Rep 8:1–15

Cárdenas A, Higuera-Ciapara I, Goycoolea FM (1998) Rheology and aggregation of cactus (Opuntia). In: Advances in natural polymers: comosites and nanocomposites, pp. 152–159

Cassano R, Trombino S, Ferrarelli T, Cavalcanti P, Giraldi C, Lai F, Loy G, Picci N (2012) Synthesis, characterization and in-vitro antitubercular activity of isoniazid-gelatin conjugate. J Pharm Pharmacol 64:712–718

Choudhary S, Kusum Devi V (2015) Potential of nanotechnology as a delivery platform against tuberculosis: current research review. J Control Release 202:65–75

Chung YI, Kim JC, Kim YH, Tae G, Lee SY, Kim K, Kwon IC (2010) The effect of surface functionalization of PLGA nanoparticles by heparin- or chitosan-conjugated Pluronic on tumor targeting. J Control Release 143:374–382
Clogston JD, Patri AK (2011) Zeta potential measurement. In: Mcneil SE (ed) Characterization of nanoparticles intended for drug delivery. Springer, PP 63–70
Crivelli B, Perteghella S, Bari E, Sorrenti M, Tripodo G, Chlapanidas T, Torre ML (2018) Silk nanoparticles: from inert supports to bioactive natural carriers for drug delivery. Soft Matter 14:546–557
Crucho CIC, Barros MT (2017) Polymeric nanoparticles: a study on the preparation variables and characterization methods. Mater Sci Eng, C 80:771–784
Dai L, Zhan X, Wei Y, Sun C, Mao L, McClements DJ, Gao Y (2018) Composite zein - propylene glycol alginate particles prepared using solvent evaporation: characterization and application as Pickering emulsion stabilizers. Food Hydrocoll 85:281–290
Damodaran VB, Bhatnagar D, Sanjeeva Murthy N (2016) Biomedical polymers synthesis and processing
De Frates K, Markiewicz T, Gallo P, Rack A, Weyhmiller A, Jarmusik B, Hu X (2018) Protein polymer-based nanoparticles: fabrication and medical applications. Int J Mol Sci 19:1–20
Delcassian D, Patel AK, Cortinas AB, Langer R (2019) Drug delivery across length scales. J Drug Target 27:229–243
Deng X, Cao M, Zhang J, Hu K, Yin Z, Zhou Z, Xiao X, Yang Y, Sheng W, Wu Y, Zeng Y (2014) Hyaluronic acid-chitosan nanoparticles for co-delivery of MiR-34a and doxorubicin in therapy against triple negative breast cancer. Biomaterials 35:4333–4344
Di Prima G, Conigliaro A, De Caro V (2019) Mucoadhesive polymeric films to enhance barbaloin penetration into buccal mucosa: a novel approach to chemoprevention. AAPS PharmSciTech 20:1–12
Diana V, Bossolasco P, Moscatelli D, Silani V, Cova L (2013) Dose dependent side effect of superparamagnetic iron oxide nanoparticle labeling on cell motility in two fetal stem cell populations. PLoS One 8
Dickinson E (2017) Biopolymer-based particles as stabilizing agents for emulsions and foams. Food Hydrocoll 68:219–231
Dosio F, Arpicco S, Stella B, Fattal E (2016) Hyaluronic acid for anticancer drug and nucleic acid delivery. Adv Drug Deliv Rev 97:204–236
Dragan ES, Dinu MV (2019) Polysaccharides constructed hydrogels as vehicles for proteins and peptides. A review. Carbohydr Polym 225:115210
Dutta PK, Dutta J, Anal A (2013) Multifaceted development and application of biopolymers for biology, biomedicine and nanotechnology. Springer
Eicher AC, Dobler D, Kiselmann C, Schmidts T, Runkel F (2019) Dermal delivery of therapeutic DNAzymes via chitosan hydrogels. Int J Pharm 563:208–216
Elzoghby AO, Samy WM, Elgindy NA (2012) Albumin-based nanoparticles as potential controlled release drug delivery systems. J Control Release 157:168–182
Elzoghby A, Freag M, Mamdouh H, Elkhodairy K (2018) Zein-based nanocarriers as potential natural alternatives for drug and gene delivery: focus on cancer therapy. Curr Pharm Des 23:5261–5271
Farris E, Brown DM, Ramer-Tait AE, Pannier AK (2017) Chitosan-zein nano-in-microparticles capable of mediating in vivo transgene expression following oral delivery. J Control Release 249:150–161
Foerster F, Bamberger D, Schupp J, Weilbächer M, Kaps L, Strobl S, Radi L, Diken M, Strand D, Tuettenberg A, Wich PR, Schuppan D (2016) Dextran-based therapeutic nanoparticles for hepatic drug delivery. Nanomedicine 11:2663–2677
Foldbjerg R, Olesen P, Hougaard M, Dang DA, Hoffmann HJ, Autrup H (2009) PVP-coated silver nanoparticles and silver ions induce reactive oxygen species, apoptosis and necrosis in THP-1 monocytes. Toxicol Lett 190:156–162

Foldbjerg R, Dang DA, Autrup H (2011) Cytotoxicity and genotoxicity of silver nanoparticles in the human lung cancer cell line, A549. Arch Toxicol 85:743–750

Fonseca DP, Khalil NM, Mainardes RM (2017) Bovine serum albumin-based nanoparticles containing resveratrol: characterization and antioxidant activity. J Drug Deliv Sci Technol 39:147–155

Foox M, Zilberman M (2015) Drug delivery from gelatin-based systems. Expert Opin Drug Deliv 12:1547–1563

Froehlich E, Fröhlich E (2016) The role of surface charge in cellular uptake and cytotoxicity of medical nanoparticles. Int J The role of surface charge in cellular uptake and cytotoxicity of medical nanoparticles. Int J Nanomed 5577–5591

Fröhlich E (2016) The role of surface charge in cellular uptake and cytotoxicity of medical nanoparticles. Int J Nanomed 7:5577–5591

Gabriel T, Brhane Y, Gabriel T (2018) Recent advances in preparation and modification of gelatin nanoparticles for pharmaceutical applications. Int J Pharm Sci Nanotechnol 11:1–8

Gaumet M, Vargas A, Gurny R, Delie F (2008) Nanoparticles for drug delivery: the need for precision in reporting particle size parameters. Eur J Pharm Biopharm 69:1–9

George B, Suchithra TV (2019) Plant-derived bioadhesives for wound dressing and drug delivery system. Fitoterapia 137

George A, Shah PA, Shrivastav PS (2019) Natural biodegradable polymers based nano-formulations for drug delivery: a review. Int J Pharm 561:244–264

Ghiorghita C-A, Bucatariu F, Dragan ES (2019) Influence of cross-linking in loading/release applications of polyelectrolyte multilayer assemblies. A review. Mater Sci Eng C 105:110050

Gliga AR, Skoglund S, Odnevall Wallinder I, Fadeel B, Karlsson HL (2014) Size-dependent cytotoxicity of silver nanoparticles in human lung cells: the role of cellular uptake, agglomeration and Ag release. Part Fibre Toxicol 11:1–17

Goonoo N, Bhaw-Luximon A, Jhurry D (2014) In vitro and in vivo cytocompatibility of electrospun nanofiber scaffolds for tissue engineering applications. RSC Adv 4:31618–31642

Gopinath V, Saravanan S, Al-Maleki AR, Ramesh M, Vadivelu J (2018) A review of natural polysaccharides for drug delivery applications: special focus on cellulose, starch and glycogen. Biomed Pharmacother 107:96–108

Grenha A, Gomes ME, Rodrigues M, Santo VE, Mano JF, Neves NM, Reis RL (2010) Development of new chitosan/carrageenan nanoparticles for drug delivery applications. J Biomed Mater Res Part A 92:1265–1272

Grobelny J, DelRio FW, Pradeep N, Kim D-I, Hackley V, Cppok R (2011) Size measurement of nanoparticles using atomic force microscopy. In: Mcneil SE (ed) Characterization of nanoparticles intended for drug delivery. Humana Press, pp 71–82

Guo Z, Tan L (2009) Nanomaterials characterization. In: Fundamentals and applications of nanomaterials. Artech House, pp 75–92

Guo C, Yarger JL (2018) Characterizing gold nanoparticles by NMR spectroscopy. Magn Reson Chem 56:1074–1082

Hackenberg S, Scherzed A, Kessler M, Hummel S, Technau A, Froelich K, Ginzkey C, Koehler C, Hagen R, Kleinsasser N (2011) Silver nanoparticles: evaluation of DNA damage, toxicity and functional impairment in human mesenchymal stem cells. Toxicol Lett 201:27–33

Halappanavar S, Vogel U, Wallin H, Yauk CL (2018) Promise and peril in nanomedicine: the challenges and needs for integrated systems biology approaches to define health risk. Wiley Interdiscip Rev Nanomed Nanobiotechnol 10:1–7

Hall JB, Mcneil SE (2007) Characterization of nanoparticles for therapeutics: physicochemical characterization 2:789–803

Hardee CL, Arévalo-Soliz LM, Hornstein BD, Zechiedrich L (2017.) Advances in non-viral DNA vectors for gene therapy. Genes (Basel) 8

Harris R, Lecumberri E, Mateos-Aparicio I, Mengíbar M, Heras A (2011) Chitosan nanoparticles and microspheres for the encapsulation of natural antioxidants extracted from Ilex paraguariensis. Carbohydr Polym 84:803–806

Harsha SN, Aldhubiab BE, Nair AB, Alhaider IA, Attimarad M, Venugopala KN, Srinivasan S, Gangadhar N, Asif AH (2015) Nanoparticle formulation by Büchi b-90 nano spray dryer for oral mucoadhesion. Drug Des Devel Ther 9:273–282
He M, Zhao Z, Yin L, Tang C, Yin C (2009) Hyaluronic acid coated poly(butyl cyanoacrylate) nanoparticles as anticancer drug carriers. Int J Pharm 373:165–173
He C, Hu Y, Yin L, Tang C, Yin C (2010) Effects of particle size and surface charge on cellular uptake and biodistribution of polymeric nanoparticles. Biomaterials 31:3657–3666
Hembram KC, Prabha S, Chandra R, Ahmed B, Nimesh S (2016) Advances in preparation and characterization of chitosan nanoparticles for therapeutics. Artif Cells Nanomed Biotechnol 44:305–314
Hernandez LM, Yousefi N, Tufenkji N (2017) Are there nanoplastics in your personal care products? Environ Sci Technol Lett 4:280–285
Hosseini M, Hamdy Makhlouf AS (2016) Industrial applications for intelligent polymers and coatings. Industrial applications for intelligent polymers and coatings
Huang G, Huang H (2018) Hyaluronic acid-based biopharmaceutical delivery and tumor-targeted drug delivery system. J Control Release 278:122–126
Hunt G, Lynch I, Cassee F, Handy RD, Fernandes TF, Berges M, Kuhlbusch TAJ, Dusinska M, Riediker M (2013) Towards a consensus view on understanding nanomaterials hazards and managing exposure: knowledge gaps and recommendations. Materials (Basel) 6:1090–1117
Hussain SM, Hess KL, Gearhart JM, Geiss KT, Schlager JJ (2005) In vitro toxicity of nanoparticles in BRL 3A rat liver cells. Toxicol Vitr 19:975–983
Irimia T, Dinu-Pîrvu CE, Ghica MV, Lupuleasa D, Muntean DL, Udeanu DI, Popa L (2018.) Chitosan-based in situ gels for ocular delivery of therapeutics: a state-of-the-art review. Mar Drugs 16
Jani G, Shah D, Jain V, Patel M, Vithalani D (2007) Evaluating mucilage from Aloe Barbadensis Miller as a pharmaceutical excipient for sustained-release matrix tablets
Jo DH, Kim Jin Hyoung, Lee TG, Kim Jeong Hun (2015) Size, surface charge, and shape determine therapeutic effects of nanoparticles on brain and retinal diseases. Nanomed Nanotechnol Biol Med 11:1603–1611
Kaleemullah M, Jiyauddin K, Thiban E, Rasha S, Al-Dhalli S, Budiasih S, Gamal OE, Fadli A, Eddy Y (2017) Development and evaluation of Ketoprofen sustained release matrix tablet using Hibiscus rosa-sinensis leaves mucilage. Saudi Pharm J 25:770–779
Kanoujia J, Singh M, Singh P, Saraf SA (2016) Novel genipin crosslinked atorvastatin loaded sericin nanoparticles for their enhanced antihyperlipidemic activity. Mater Sci Eng, C 69:967–976
Kathe N, Henriksen B, Chauhan H (2014) Physicochemical characterization techniques for solid lipid nanoparticles: principles and limitations. Drug Dev Ind Pharm 40:1565–1575
Kaur M, Malik B, Garg T, Rath G, Goyal AK (2015) Development and characterization of guar gum nanoparticles for oral immunization against tuberculosis. Drug Deliv 22:328–334
Kemp MM, Linhardt RJ (2009) Heparin-based nanoparticles. Wiley Interdiscip Rev Nanomed Nanobiotechnol 2:77–87
Khan SA, Schneider M (2013) Improvement of nanoprecipitation technique for preparation of gelatin nanoparticles and potential macromolecular drug loading. Macromol Biosci 13:455–463
Koukaras EN, Papadimitriou SA, Bikiaris DN, Froudakis GE (2012) Insight on the formation of chitosan nanoparticles through ionotropic gelation with tripolyphosphate. Mol Pharm 9:2856–2862
Kratz F (2008) Albumin as a drug carrier: design of prodrugs, drug conjugates and nanoparticles. J Control Release 132:171–183
Kuempel ED, Jaurand MC, Møller P, Morimoto Y, Kobayashi N, Pinkerton KE, Sargent LM, Vermeulen RCH, Fubini B, Kane AB (2017) Evaluating the mechanistic evidence and key data gaps in assessing the potential carcinogenicity of carbon nanotubes and nanofibers in humans. Crit Rev Toxicol 47:1–58
Kulkarni GT, Gowthamarajan K, Satish Kumar MN, Suresh B (2002) Vilages: therapeutic and pharmaceutical applications. Nat Prod Radiat 1:10–17

Kumar R, Patil MB, Patil SR, Paschapur MS (2009) Evaluation of *Abelmoschus esculentus* mucilage as suspending agent in paracetamol suspension. Int J PharmTech Res 1:658–665
Lamberti M, Zappavigna S, Sannolo N, Porto S, Caraglia M (2014) Advantages and risks of nanotechnologies in cancer patients and occupationally exposed workers. Expert Opin Drug Deliv 11:1087–1101
Layek B, Singh J (2017) Chitosan for DNA and gene therapy, chitosan based biomaterials. Elsevier Ltd.
Lertsutthiwong P, Rojsitthisak P, Nimmannit U (2009) Preparation of turmeric oil-loaded chitosan-alginate biopolymeric nanocapsules. Mater Sci Eng C 29:856–860
Lin WJ, Lee WC (2018) Polysaccharide-modified nanoparticles with intelligent CD44 receptor targeting ability for gene delivery. Int J Nanomed 13:3989–4002
Lipovsky A, Nitzan Y, Gedanken A, Lubart R (2011) Antifungal activity of ZnO nanoparticles-the role of ROS mediated cell injury. Nanotechnology 22
Liu Z, Jiao Y, Wang Y, Zhou C, Zhang Z (2008) Polysaccharides-based nanoparticles as drug delivery systems. Adv Drug Deliv Rev 60:1650–1662
Liu W, Wu Y, Wang C, Li HC, Wang T, Liao CY, Cui L, Zhou QF, Yan B, Jiang GB (2010a) Impact of silver nanoparticles on human cells: effect of particle size. Nanotoxicology 4:319–330
Liu Z, Zhang Z, Zhou C, Jiao Y (2010b) Hydrophobic modifications of cationic polymers for gene delivery. Prog Polym Sci 35:1144–1162
Long, JT, Cheang T, Zhuo SY, Zeng RF, Dai QS, Li HP, Fang S (2014) Anticancer drug-loaded multifunctional nanoparticles to enhance the chemotherapeutic efficacy in lung cancer metastasis. J Nanobiotechnol 12:1–11
Lu X, Xiao J, Huang Q (2018a) Pickering emulsions stabilized by media-milled starch particles. Food Res Int 105:140–149
Lu X, Zhang H, Li Y, Huang Q (2018b) Fabrication of milled cellulose particles-stabilized pickering emulsions. Food Hydrocoll 77:427–435
Luo Y, Wang Q (2014) Recent development of chitosan-based polyelectrolyte complexes with natural polysaccharides for drug delivery. Int J Biol Macromol 64:353–367
Luo Z, Dai Y, Gao H (2019) Development and application of hyaluronic acid in tumor targeting drug delivery. Acta Pharm Sin B
Mahto A, Mishra S (2019) Design, development and validation of guar gum based pH sensitive drug delivery carrier via graft copolymerization reaction using microwave irradiations. Int J Biol Macromol 138:278–291
Mao S, Sun W, Kissel T (2010) Chitosan-based formulations for delivery of DNA and siRNA. Adv Drug Deliv Rev 62:12–27
Mohammed M, Mansell H, Shoker A, Wasan KM, Wasan EK (2019) Development and in vitro characterization of chitosan-coated polymeric nanoparticles for oral delivery and sustained release of the immunosuppressant drug mycophenolate mofetil. Drug Dev Ind Pharm 45:76–87
Motwani SK, Chopra S, Talegaonkar S, Kohli K, Ahmad FJ, Khar RK (2008) Chitosan-sodium alginate nanoparticles as submicroscopic reservoirs for ocular delivery: formulation, optimisation and in vitro characterisation. Eur J Pharm Biopharm 68:513–525
Multur S (2004) Thermal analysis of composites using DSC. In: Kessler M (ed) Advanced topics in characterization of composites, pp 11–33
Naskar S, Koutsu K, Sharma S (2019) Chitosan-based nanoparticles as drug delivery systems: a review on two decades of research. J Drug Target 27:379–393
Natterodt JC, Petri-Fink A, Weder C, Zoppe JO (2017) Cellulose nanocrystals: surface modification, applications and opportunities at interfaces. Chimia (Aarau) 71:376–383
Nazlı AB, Açıkel YS (2019) Loading of cancer drug resveratrol to pH-Sensitive, smart, alginate-chitosan hydrogels and investigation of controlled release kinetics. J Drug Deliv Sci Technol 53:101199
Nep EI, Mahdi MH, Adebisi AO, Ngwuluka NC, Conway BR, Smith AM, Asare-Addo K (2018) Hydro-alcoholic media effects on theophylline release from sesamum polysaccharide gum matrices. Drug Dev Ind Pharm 44:251–260

Niaounakis M (2015) Biopolymers : processing and products
Nigatu AS, Ashar H, Sethuraman SN, Wardlow R, Maples D, Malayer J, Ranjan A (2018) Elastin-like polypeptide incorporated thermally sensitive liposome improve antibiotic therapy against musculoskeletal bacterial pathogens. Int J Hyperth 34:201–208
Notario-Pérez F, Martín-Illana A, Cazorla-Luna R, Ruiz-Caro R, Bedoya LM, Tamayo A, Rubio J, Veiga MD (2017) Influence of chitosan swelling behaviour on controlled release of tenofovir from mucoadhesive vaginal systems for prevention of sexual transmission of HIV. Mar Drugs 15:1–16
Nyström AM, Fadeel B (2012) Safety assessment of nanomaterials: implications for nanomedicine. J Control Release 161:403–408
Oberdörster G, Sharp Z, Atudorei V, Elder A, Gelein R, Kreyling W, Cox C (2004) Translocation of inhaled ultrafine particles to the brain. Inhal Toxicol 16:437–445
Olatunji O (2015) Natural polymers: industry techniques and applications. Natural polymers: industry techniques and applications
Pellá MCG, Lima-Tenório MK, Tenório-Neto ET, Guilherme MR, Muniz EC, Rubira AF (2018) Chitosan-based hydrogels: from preparation to biomedical applications. Carbohydr Polym 196:233–245
Petersen AL, Hansen AE, Gabizon A, Andresen TL (2012) Liposome imaging agents in personalized medicine. Adv Drug Deliv Rev 64:1417–1435
Powar TA, Hajare AA, Patil-Vibhute PB, Nadaf SJ, Jarag RJ (2017) Bioadhesive garlic and ketoconazole vaginal tablets for treatment of candidiasis. Indian J Pharm Educ Res 51:239–248
Prajapati VD, Jani GK, Moradiya NG, Randeria NP (2013) Pharmaceutical applications of various natural gums, mucilages and their modified forms. Carbohydr Polym 92:1685–1699
Ramamoorth M, Narvekar A (2015) Non viral vectors in gene therapy—an overview. J Clin Diagnostic Res 9:GE01–GE06
Rampino A, Borgogna M, Blasi P, Bellich B, Cesàro A (2013) Chitosan nanoparticles: preparation, size evolution and stability. Int J Pharm 455:219–228
Ramteke S, Jain N (2008) Clarithromycin- and omeprazole-containing gliadin nanoparticles for the treatment of *Helicobacter pylori*. J Drug Target 16:65–72
Rao A, Schoenenberger M, Gnecco E, Glatzel T, Meyer E, Brändlin D, Scandella L (2007) Characterization of nanoparticles using atomic force microscopy. J Phys Conf Ser 61:971–976
Riediker M, Zink D, Kreyling W, Oberdörster G, Elder A, Graham U, Lynch I, Duschl A, Ichihara G, Ichihara S, Kobayashi T, Hisanaga N, Umezawa M, Cheng TJ, Handy R, Gulumian M, Tinkle S, Cassee F (2019) Particle toxicology and health—where are we? Particle and fibre toxicology
Sahariah P, Másson M (2017) Antimicrobial chitosan and chitosan derivatives: a review of the structure-activity relationship. Biomacromol 18:3846–3868
Sahin A, Yoyen-Ermis D, Caban-Toktas S, Horzum U, Aktas Y, Couvreur P, Esendagli G, Capan Y (2017) Evaluation of brain-targeted chitosan nanoparticles through blood–brain barrier cerebral microvessel endothelial cells. J Microencapsul 34:659–666
Saidin NM, Anuar NK, Meor Mohd Affandi MMR (2018) Roles of polysaccharides in transdermal drug delivery system and future prospects. J Appl Pharm Sci 8:141–157
Samimi S, Maghsoudnia N, Eftekhari R, Dorkoosh F (2019) Lipid-based nanoparticles for drug delivery systems. In: Mohapatra S, Ranjan S, Dasgupta N, Mishra R, Thomas S (eds) Characterization and biology of nanomaterials for drug delivery. Elsevier
Sanhai WR, Sakamoto JH, Canady R, Ferrari M (2008) Seven challenges for nanomedicine. Nat Nanotechnol 3:242–244
Saranya N, Moorthi A, Saravanan S, Devi MP, Selvamurugan N (2011) Chitosan and its derivatives for gene delivery. Int J Biol Macromol 48:234–238
Saraogi GK, Gupta P, Gupta UD, Jain NK, Agrawal GP (2010) Gelatin nanocarriers as potential vectors for effective management of tuberculosis. Int J Pharm 385:143–149
Sarkar A, Zhang S, Murray B, Russell JA, Boxal S (2017) Modulating in vitro gastric digestion of emulsions using composite whey protein-cellulose nanocrystal interfaces. Colloids Surfaces B Biointerfaces 158:137–146

Sarmento B, Ferreira D, Veiga F, Ribeiro A (2006) Characterization of insulin-loaded alginate nanoparticles produced by ionotropic pre-gelation through DSC and FTIR studies. Carbohydr Polym 66:1–7

Sharifi H, Nayebi AM, Farajnia S (2012) The effect of chronic administration of buspirone on 6-hydroxydopamine-induced catalepsy in rats. Adv Pharm Bull 2:127–131

Shi Y, Xue J, Jia L, Du Q, Niu J, Zhang D (2018) Surface-modified PLGA nanoparticles with chitosan for oral delivery of tolbutamide. Colloids Surfaces B Biointerfaces 161:67–72

Shidhaye S, Kadam V, Desai A (2007) Possible use of psyllium husk as a release retardant. Indian J Pharm Sci 69:206

Singh B, Chauhan GS, Sharma DK, Kant A, Gupta I, Chauhan N (2006) The release dynamics of model drugs from the psyllium and N-hydroxymethylacrylamide based hydrogels. Int J Pharm 325:15–25

Singh S, Hussain A, Shakeel F, Ahsan MJ, Alshehri S, Webster TJ, Lal UR (2019) Recent insights on nanomedicine for augmented infection control. Int J Nanomed 14:2301–2325

Sirisha VL, Campus K (2015) Polysaccharide nanoparticles: preparation and their potential application as drug delivery systems. Int J Res Appl Nat Soc Sci 3:69–94

Sosnik A (2014) Alginate particles as platform for drug delivery by the oral route: state-of-the-art. ISRN Pharm. 2014:1–17

Sreenivasan R, Ghosh S, Abraham ET (2010) Preparation and characterization of guar gum nanoparticles. Int J Biol Macromol 46:267–269

Sun IC, Eun DK, Na JH, Lee S, Kim IJ, Youn IC, Ko CY, Kim HS, Lim D, Choi K, Messersmith PB, Park TG, Kim SY, Kwon IC, Kim K, Ahn CH (2009) Heparin-coated gold nanopartieles for liver-specific CT imaging. Chem A Eur J 15:13341–13347

Teixeira FJ, Santos HO, Howell SL, Pimentel GD (2019) Whey protein in cancer therapy: a narrative review. Pharmacol Res 144:245–256

Teng Z, Luo Y, Wang T, Zhang B, Wang Q (2013) Development and application of nanoparticles synthesized with folic acid conjugated soy protein. J Agric Food Chem 61:2556–2564

Thao LQ, Byeon HJ, Lee C, Lee S, Lee ES, Choi HG, Park ES, Youn YS (2016) Pharmaceutical potential of tacrolimus-loaded albumin nanoparticles having targetability to rheumatoid arthritis tissues. Int J Pharm 497:268–276

Thomas S, Vissakh P, Mathew AP (2013) Advances in natural polymers: composites and nanocomposites, advanced structured materials

Umamaheshwari RB, Ramteke S, Jain NK (2004) Anti-helicobacter pylori effect of mucoadhesive nanoparticles bearing amoxicillin in experimental gerbils model. AAPS PharmSciTech 5:60–68

Veerapandian, Yun K (2009) The state of the art in biomaterials as nanobiomaterials and nanopharmaceutics. Dig J Nanomater Biostructures 4:243–26

Venkatesan J, Anil S, Singh SK, Kim S (2017) Preparations and applications of alginate nanoparticles. In: Venkatesan J, Anil S, Se-Kwon K (eds) Seaweed polysaccharides. Elsevier Inc., pp 249–266

Vert M, Doi Y, Hellwich K-H, Hess M, Hodge P, Kubisa P, Rinaudo M, Schué F (2012) Terminology for biorelated polymers and applications (IUPAC Recommendations 2012). Pure Appl Chem 84:377–410

Vocelle D, Chesniak OM, Malefyt AP, Comiskey G, Adu-Berchie K, Smith MR, Chan C, Walton SP (2016) Dextran functionalization enhances nanoparticle-mediated siRNA delivery and silencing. Technology 04:42–54

Vo-Dinh T (2005) Protein nanotechnology: protocols, instrumentation, and applications. Human Press

Wang ZL (2006) Scanning microscopy for nanotechnology. In: Zhou W, Wang ZL (eds) Fundamentals of scanning electron microscopy (SEM). Springer, pp 1–41

Wang A, Li P, Dai Y, Zhang J, Wang Ai-qin, Wei Q (2015a) Chitosan-alginate nanoparticles as a novel drug delivery system for Nifedipine. Int J Biomed Sci 4:221–228

Wang J, Wang M, Zheng M, Guo Q, Wang Y, Wang H, Xie X, Huang F, Gong R (2015b) Folate mediated self-assembled phytosterol-alginate nanoparticles for targeted intracellular anticancer drug delivery. Colloids Surfaces B Biointerfaces 129:63–70
Wang QS, Wang GF, Zhou J, Gao LN, Cui YL (2016) Colon targeted oral drug delivery system based on alginate-chitosan microspheres loaded with icariin in the treatment of ulcerative colitis. Int J Pharm 515:176–185
Wang H, Zhang X, Zhu W, Jiang Y, Zhang Z (2018) Self-assembly of Zein-based microcarrier system for colon-targeted oral drug delivery. Ind Eng Chem Res 57:12689–12699
Wasiak I, Kulikowska A, Janczewska M, Michalak M, Cymerman IA, Nagalski A, Kallinger P, Szymanski WW, Ciach T (2016) Dextran nanoparticle synthesis and properties. PLoS ONE 11:1–17
Wei Y, Wang C, Jiang B, Sun CC, Middaugh CR (2019) Developing biologics tablets: the effects of compression on the structure and stability of bovine serum albumin and lysozyme. Mol Pharm 16:1119–1131
Williams DF (2008) On the mechanisms of biocompatibility. Biomaterials 29:2941–2953
Wu P, Liu Q, Li R, Wang J, Zhen X, Yue G, Wang H, Cui F, Wu F, Yang M, Qian X, Yu L, Jiang X, Liu B (2013a) Facile preparation of paclitaxel loaded silk fibroin nanoparticles for enhanced antitumor efficacy by locoregional drug delivery. ACS Appl Mater Interfaces 5:12638–12645
Wu F, Zhou Z, Su J, Wei L, Yuan W, Jin T (2013b) Development of dextran nanoparticles for stabilizing delicate proteins. Nanoscale Res Lett 8:1–8
Xiao B, Han MK, Viennois E, Wang L, Zhang M, Si X, Merlin D (2015) Hyaluronic acid-functionalized polymeric nanoparticles for colon cancer-targeted combination chemotherapy. Nanoscale 7:17745–17755
Xu R (2008) Progress in nanoparticles characterization: sizing and zeta potential measurement. Particuology 6:112–115
Xu W, Jin W, Li Z, Liang H, Wang Y, Shah BR, Li Y, Li B (2015) Synthesis and characterization of nanoparticles based on negatively charged xanthan gum and lysozyme. Food Res Int 71:83–90
Yang TT, Wen BF, Liu K, Qin M, Gao YY, Ding DJ, Li WT, Zhang YX, Zhang WF (2018) Cyclosporine A/porous quaternized chitosan microspheres as a novel pulmonary drug delivery system. Artif Cells Nanomed Biotechnol 46:552–564
Yasmin R, Shah M, Khan SA, Ali R (2017) Gelatin nanoparticles: a potential candidate for medical applications. Nanotechnol Rev 6:191–207
Yoon HY, Koo H, Choi KY, Lee SJ, Kim K, Kwon IC, Leary JF, Park K, Yuk SH, Park JH, Choi K (2012) Tumor-targeting hyaluronic acid nanoparticles for photodynamic imaging and therapy. Biomaterials 33:3980–3989
Yue ZG, Wei W, Lv PP, Yue H, Wang LY, Su ZG, Ma GH (2011) Surface charge affects cellular uptake and intracellular trafficking of chitosan-based nanoparticles. Biomacromol 12:2440–2446
Zhang W, Torabinejad M, Li Y (2003) Evaluation of cytotoxicity of MTAD using the MTT-tetrazolium method. J Endod 29:654–657
Zhang T, Wang L, Chen Q, Chen C (2014) Cytotoxic potential of silver nanoparticles. Yonsei Med J 55:283–291
Zhang CL, Jiang HS, Gu SP, Zhou XH, Lu ZW, Kang XH, Yin L, Huang J (2019a) Combination analysis of the physiology and transcriptome provides insights into the mechanism of silver nanoparticles phytotoxicity. Environ Pollut 252:1539–1549
Zhang L, Xu J, Wen Q, Ni C (2019b) Preparation of xanthan gum nanogels and their pH/redox responsiveness in controlled release. J Appl Polym Sci 136:6–11
Zhao XJ, Zhou ZQ (2016) Synthesis and applications of pectin-based nanomaterials. Curr Nanosci 12:103–109

Zhao YZ, Jin RR, Yang W, Xiang Q, Yu WZ, Lin Q, Tian FR, Mao KL, Lv CZ, Wáng YXJ, Lu CT (2016) Using gelatin nanoparticle mediated intranasal delivery of neuropeptide substance P to enhance neuro-recovery in hemiparkinsonian rats. PLoS ONE 11:1–18

Zu M, Ma L, Zhang X, Xie D, Kang Y, Xiao B (2019) Chondroitin sulfate-functionalized polymeric nanoparticles for colon cancer-targeted chemotherapy. Colloids Surfaces B Biointerfaces 177:399–406

Chapter 7
Progress and Challengers of Nanomaterials in Water Contamination

Vicente de Oliveira Sousa Neto, Antonio Joel Ramiro de Castro, Gilberto Dantas Saraiva, and Ronaldo Ferreira do Nascimento

1 Introduction

In recent years, there has been a greater concern and interest in environmental issues. This has motivated the development of new materials that can be applied in the protection and remediation of the environment. In this specific point, nanosized materials have been used in some cases with great success in the decontamination of water.

Nanomaterials have been increasingly explored due to their excellent performance and low cost for removing contaminants. This removal capacity is related to its high surface area (Yang et al. 2012; Suriyaraj et al. 2014; Khandare and Mukherjee 2019). Inorganic nanomaterials are basically represented by nanosized metals (Zero-valent iron, ZVI; zero-valent copper, ZVC) and metallic oxides such as ferric oxides, aluminum oxides, manganese oxides, titanium oxides, magnesium

V. de Oliveira Sousa Neto (✉)
Laboratory of Study and Research in Pollutants Removal by Adsorption—LERPAD, Department of Chemistry, State University of Ceará (UECE-FECLESC), Rua José de Queiroz Pessoa, N° 2554, University Plateau, Quixadá, Ceará 63.900-000, Brazil
e-mail: vicente.neto@uece.br

A. J. R. de Castro
Federal University of Ceará, Quixadá Campus, Av. José de Freitas Queiroz, 5003, Quixadá, Ceará 63902-580, Brazil

G. D. Saraiva
Laboratory of Synthesis and Characterization of Materials—LASCAM, Department of Physics, State University of Ceará (UECE-FECLESC), Rua José de Queiroz Pessoa, N° 2554, University Plateau, Quixadá, Ceará 63.900-000, Brazil

R. F. do Nascimento (✉)
Laboratory of Trace Analysis (LAT), Department of Analytical and Physical Chemistry, Federal University of Ceará (UFC), Campus do Pici, Bloco 940, S/N, Fortaleza 60455-970, Ceará, Brazil
e-mail: ronaldo@ufc.br

R. F. do. Nascimento et al. (eds.), *Nanomaterials and Nanotechnology*,
Materials Horizons: From Nature to Nanomaterials,
https://doi.org/10.1007/978-981-33-6056-3_7

oxides, and cerium oxides (Nabid et al. 2013,Taman et al. 2015). Extensive studies have shown that these classes of nanomaterials exhibit a sufficiently high and selective adsorption capacity to remove contaminants such as arsenic (Kanel et al. 2006), cadmium (Boparai et al. 2011), chromium (Hu et al. 2005), and other pollutants common, such as phosphate and organic (Khan and Malik 2019; Khalaj et al. 2018).

The performance of nanomaterials can be evaluated according to some parameters such as shape, size, porosity, and optical properties. Theses parameters can affect both the kinetics and the thermodynamics of chemical reactions. Therefore, the performance of removing environmental contaminants by any nanomaterial must consider such effects that will certainly have economic and environmental impacts.

(i) ***Shape***: The shape effect is associated with the direct relationship between shape and size. This ratio leads to a change in the surface area. The shape effect is also observed in the properties inherent to nanomaterials, such as bandgap energy.
(ii) ***Size***: Smaller particles have a huge contact surface area, which increases the chances of particle collisions in the reaction process, then it is noted that increasing the reaction rate.
(iii) ***Porosity***: In general, any solid has a degree of porosity, detectable or not, resulting from the presence of cavities, channels (or interstices). Porosity influences some physical properties of materials, including density, thermal conductivity, and mechanical resistance. Consequently, the control of the porous structure is of great importance, for example, in the design of catalysts, industrial adsorbents, membranes, and pottery.
(iv) ***Surface charge***: Generally, it is possible to say that there is a relationship between the surface charge of a solid and the pH of the medium in which it is dispersed. A convenient index of the tendency of a surface to become positively or negatively charged as a function of pH is the charge value required for the net charge of the solid to be zero, the so-called zero charge point (pH_{PZC}). For pH values lower than (pH_{PZC}), the surface charge is positive, and anion adsorption is favored; and for pH values higher than (pH_{PZC}), the surface charge is negative, and cation adsorption is favored.

In this chapter, an overview of the development of nanomaterials based on metals and metal oxide as well as their applications in water decontamination will be addressed.

2 Water Decontamination

Nanomaterials have provided innovative solutions for water treatment systems. As previously mentioned, nanomaterials are manufactured with physico-chemical properties that are important from an industrial point of view and are useful in technology designed to remove pollutants. Nanomaterials also are advantageous in almost treating wastewater since they eliminate (or removing) contaminants. In this case, they contribute to the recycling process to obtain treated water. In the following item, some steps of the water treatment processes will be addressed in which nanomaterials can be applied.

2.1 Adsorption and Separation

In the adsorption process, the species adsorbed to the solid surface is called adsorbate, and the solid with this property is called adsorbent. Adsorption is a mass transfer operation from a fluid phase to a continuous phase. Basically, it is a surface phenomenon and will be favored the greater the contact surface area of the adsorbent. The adsorption phenomenon is a very vast and complex subject. Therefore, the understanding of the processes involved in adsorption, such as the study of equilibrium, kinetics, and mechanisms is necessary for the development of new materials with good adsorption capacity. In the industrial development of adsorbents, some factors can determine the economic viability of the material such as low cost, availability, efficiency, and regenerative capacity.

2.2 Sensing

Currently, conventional sensing and monitoring methods are not able to detect the extremely low concentration of micro-pollutants in water. Rapid and in situ detection of pathogens and highly toxic pollutants are of great significance in environmental crisis management.

Carbon points (CD) are an emerging family of nanosystems that exhibit a variety of properties that are interesting for the development of nanosensors (Nguyen et al. 2016). These materials can generally be described as small functionalized carbonaceous nanoparticles on the surface, characterized by an intense and tunable fluorescence, a marked sensitivity to the environment, and a variety of interesting photochemical properties. Such properties have motivated more intense research about CDs with direct applications in several areas of knowledge such as bioimaging, the capture of solar energy, nanosensitivity, light-emitting devices, and photocatalysts (Pan et al. 2015; Cayuela et al. 2015; Rong et al. 2015; Sciortino et al. (2018a, b). This chapter aims to address the most recent advances in the development of CDs, with a focus on water treatment.

2.3 Disinfection

Currently the major challenge of disinfection processes is related to the production of undesirable by-products that are harmful to human health and the life of environmental ecosystems. In this way, increasing demand for efficient treatment systems demands new technologies for efficient disinfection and microbial control (Zang et al. 2019). According to Li et al. (2008), there are several engineered nanomaterials that have demonstrated strong antimicrobial properties. The mechanisms including photocatalytic production of reactive oxygen species that damage cell components and viruses (e.g., TiO_2, ZnO, and fullerol) compromising the bacterial cell envelope. According to Xu et al. (2004a, b), and Gogoi et al. 2006, most the case, particles of less than 10 nm are more toxic to bacteria such as *Escherichia coli* and *Pseudomonas aeruginosa.*

2.4 Catalysis

Nanoparticles have great potential as catalysts due to their large surface area and high chemical reactivity. Therefore, in the development of the design of the nanomaterial, its performance in the catalysis will depend on the shape and size. Common catalytic nanoparticles include nanosized semiconductor materials such as nano-TiO_2, CeO_2, ZnO, CdS, CuO, CuO_2, Fe^0, Zn^0, and Cu^0. They are generally applied in the degradation of organic pollutants via catalysis or redox reaction. They are generally applied in the degradation of organic pollutants via catalysis or redox reaction. Among the organic pollutants removed via photocatalytic degradation are polychlorinated biphenyls—PCBs (Liu and Zhang 2010), Bisphenol A (Tang et al. 2020), Phenol (Ahmad et al. 2020), industrial dyes (Ahmed et al. 2019), insecticide (Mudhoo et al. 2019), and nitroaromatics (Tong et al. 2011) among others. Table 1 summarizes some conventional methods used in water treatment plants with their advantages and limitations.

3 Nanomateriais

3.1 Metal and Metal Oxides Nanomaterials

Metal oxide-based nanomaterials are among the most produced compounds due to their very specific properties such as nonlinear optical properties, greater ductility at elevated temperatures, superparamagnetic behavior, unique catalytic, sensitive, and selective activity. In the nanosize dimension, physical property is modified. For example, the melting point of nanosized material is less than that of a bulk material with the same composition. These properties can be exploited in order to promote better performance during their application. However, it is necessary to be careful as the limitations imposed by operational conditions. Take nanoparticles as an example: NPs exhibit unusual adsorptive properties and fast diffusivities, but they are not stable in critical conditions.

Table 1 Advantages and limitations associated with conventional water treatment

Conventional methods	Advantage	Limitations	References
Distillation	Method of low operational complexity	Expensive method and ineffective to organic pollutants with boiling points lower than 100 °C. They cannot be removed efficiently and can actually become concentrated in the product water. Distillation requires large amounts of energy and water	Kunduru et al. (2017)
Coagulation and flocculation	Facilitates the removal of suspended solids and colloidal particles of the solution	This is a complex and less-efficient method and requires alkaline additives to achieve optimum pH	Kooijman et al. (2020)
Biological	The organic contaminants to be destroyed are used and transformed by bacteria or other organisms as a source of food	Microorganisms are sensitive to environmental factors and difficult to control.	Yildiz (2012)
Reverse osmosis (RO)	*Reverse osmosis (RO) is the most economical method of removing 95% to 99% of all contaminants*	RO membranes are very restrictive, they yield very slow flow rates. Storage tanks are required to produce an adequate volume in a reasonable amount of time	Kunduru et al. (2017)
Ultraviolet treatment	With special lamps, total organic carbon (TOC) levels in high purity water can be reduced to 5 ppb	Expensive method and inactivated by water cloudiness and turbidity. Ineffective for heavy metals and other nonliving contaminants removal	Zhang et al. (2019)

The potential technological applications of metal oxides, in nanosized form, have stimulated an interest in the many fields of science as chemistry (Zhang et al. 2015), physic (Ke et al. 2014), environment, energy (Ke et al. 2014; Chavali and Nikolova 2019), biology (Etefagh et al. 2013), and catalysis (Chavali and Nikolova 2019).

3.2 Copper-Based Nanomaterials (CuBNMs)

Copper is a metal of recognized technological importance due to its exceptional chemical, electrical, optical, and thermal properties. Copper oxides such as CuO and Cu_2O are p-type semiconductors that are well known and studied for their narrow bandgap (Isherwood 2017; Zhang et al. 2019). Figure 1 shows the unitary cells of CuO and Cu_2O.

Copper-based nanomaterials have been able to provide an efficient and cost-effective alternative via for the treatment of persistent effluents. The environmental application of CuNM in the removal of contaminates has been successfully explored in recent years and has gained more attention. Various method methods were developed to obtain CuNM such as chemical reduction (Ahmad et al. 2020), thermal decomposition (Hao et al. 2015), microwaves, and sonochemical reduction (Li et al. 2019). Some modifications of copper-based materials have been proposed with a specific focus on protecting the environment.

The application of copper-based nanomaterials has been explored in several ways with regard to the removal of pollutants from water. Lucchetti et al. (2017), applied zero-valent copper in the removal, in situ, of nitrate from surface waters. Copper has loaded on the titania (P25) nanoparticles. The system used glycerol as a sacrificial agent and UV-A/Vis radiation. The experimental results suggested that total nitrogen (i.e., nitrate, nitrite, and ammonia) removal performance over 93% was achieved for initial nitrate concentrations up to 150 mg/l. It was noted that reaction rates were dependent on the following parameters: pH, glycerol, and nitrate starting concentrations in the mixture. Lucchetti et al. (2017), demonstrated that there is a

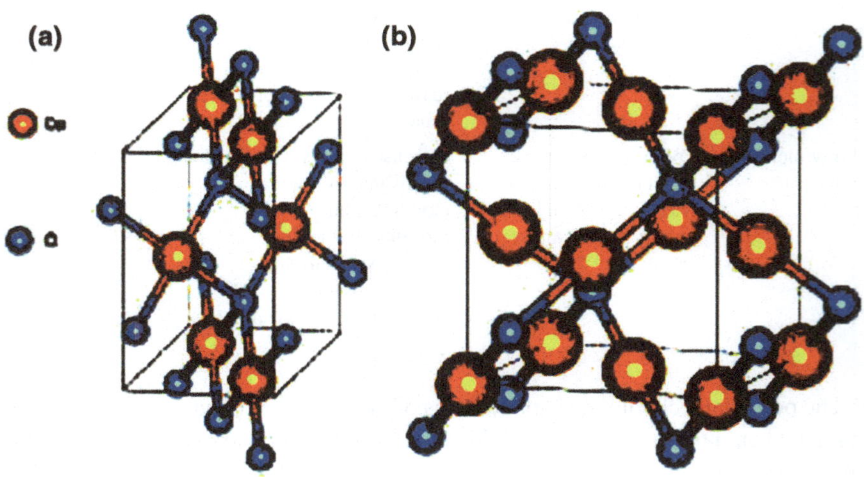

Fig. 1 Crystal structure of **a** CuO and **b** Cu_2O, (Zhang et al 2019). This is an open access article distributed under the Creative Commons Attribution License which permits unrestricted use, distribution, and reproduction in any medium, provided the original work is properly cited link: https://www.mdpi.com/2079-6412/9/2/137/htm#

connection between the concentration of hydrogen and the increase in the concentration of glycerol. The authors concluded that the removal of nitrate from aqueous media by photocatalytic processes employing zero-valence nanocopper/nano-P25 is efficient and feasible. However, it is necessary to consider that the reaction rates are strongly affected by the following factors: concentration of the hole scavenge, nitrate concentration, and pH of the mixture, mainly due to the concentration of protons required for photocatalytic reactions. The system was characterized by high photocatalyst activity, but in industrial-scale, additional parameters need to be investigated because of the corresponding value of ***electrical energy-per-order***, EE/O (62.8 kWh m^{-3} order $^{-1}$) was considered far from satisfactory from the economic point of view for pre-pilot scale applications.

Liú et al. (2016), applied zero-valent copper (ZVCu), obtained via the reduction method (Fig. 2), in the degradation of methyl orange (MO) present in water. According to these authors, ZVCu was synthesized using hydrazine hydrate as a reducing reagent to reduce copper (II) acetate, Cu $(CH_3COO)_2 \cdot H_2O$, in ethylene glycol. According to Liú et al. (2016), nanocopper is an alternative to remove MO by degradation due to its visible-light photocatalytic activity.

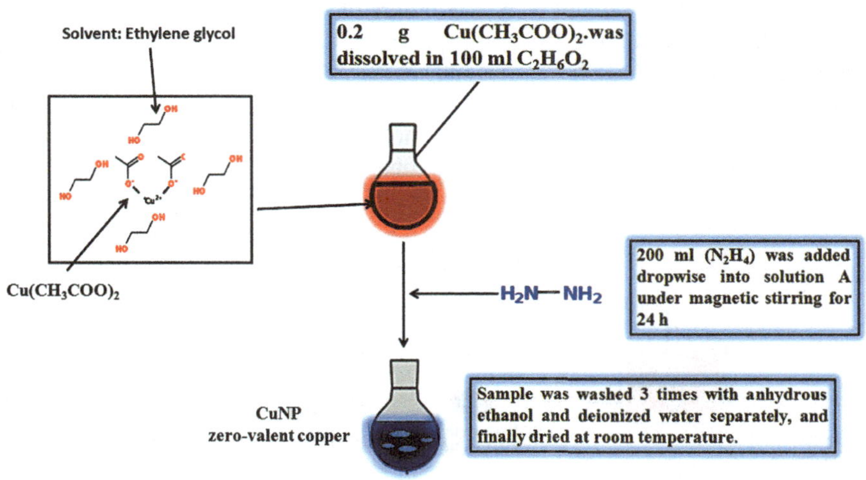

Fig. 2 Synthesis of zero-valent copper via reduction method using hydrazine hydrate as a reducing agent according to Liú et al. (2016). This figure by Vicente Neto is licensed under the Creative Commons Attribution 4.0 International license. Link: https://commons.wikimedia.org/wiki/File:Synthesis_of_zero-valent_copper_via_reduction_method_using_hydrazine_hydrate_as_a_red ucing_agent.tif

3.3 Iron-Based Nanomaterials for Water Decontamination

Zero-valence iron (ZVI) is a very effective material in the decontamination of water and wastewater. Its reactivity has been enhanced with the development of nanosized structure, (nZVI). The nZVI is highly reactive to contaminants due to its large specific area, tiny sizes that help both to promote effective surface dispersion and injection in aqueous systems for remediation of the contaminated area (Tosco et al. 2014; Sun et al. 2020).

There are several techniques for producing nZVI particles that are often divided into the bottom-up and top-bottom method. In the first method, the synthesis of nZVI from ions or smaller particles via nucleation, deposition, precipitation, agglomeration, etc. The top-bottom method obtains nZVI by grinding and ablation techniques. These techniques reduce the initial particle size to a nanosized scale.

In general, nZVI shows the capacity to remove contaminants through reduction, adsorption, precipitation/co-precipitation, and other methods due to its nanoscale effects. nZVI has been used successfully to reduce organic pollutants present in both water and soil, including PCBs, TCE, and DDT (El-Temsah and Joner 2013). Kanel's studies showed that the size of ZVI from micron to nanoscale affected positively the rate constant for As(V) removal by 1–3 orders of magnitude (Kanel et al. 2006). Shang et al. (2017) evaluated the adsorption performance of biochar (nZVI/BC) for the removal of Cr(VI) from aqueous solutions. According to the authors, the composite material consisting of nanoscale zero-valent iron particles supported on herb-residue biochar. Figure 3 shows a thermochemical conversion of the residue of an herb, Astragalus membranaceus, under controlled conditions

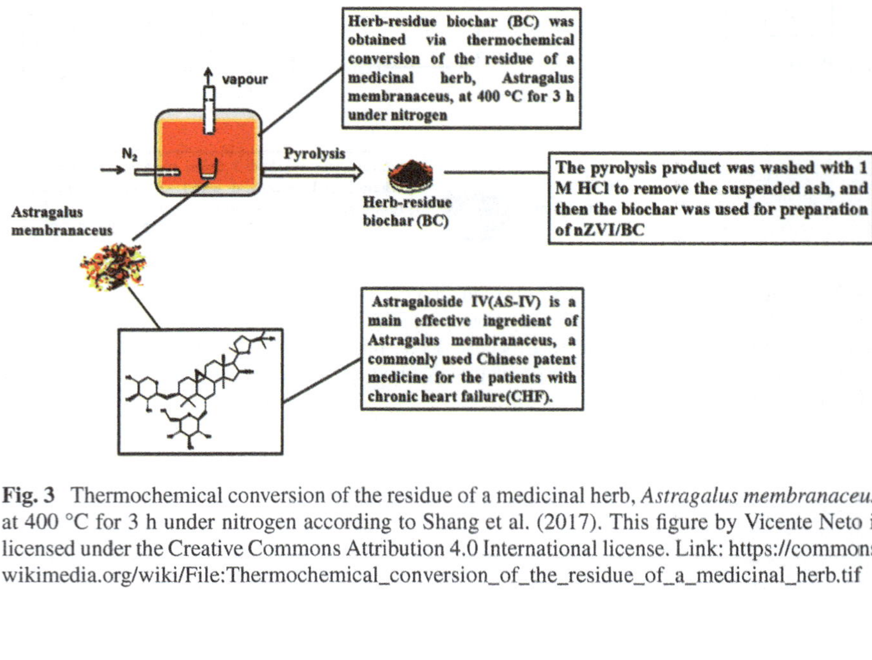

Fig. 3 Thermochemical conversion of the residue of a medicinal herb, *Astragalus membranaceus*, at 400 °C for 3 h under nitrogen according to Shang et al. (2017). This figure by Vicente Neto is licensed under the Creative Commons Attribution 4.0 International license. Link: https://commons.wikimedia.org/wiki/File:Thermochemical_conversion_of_the_residue_of_a_medicinal_herb.tif

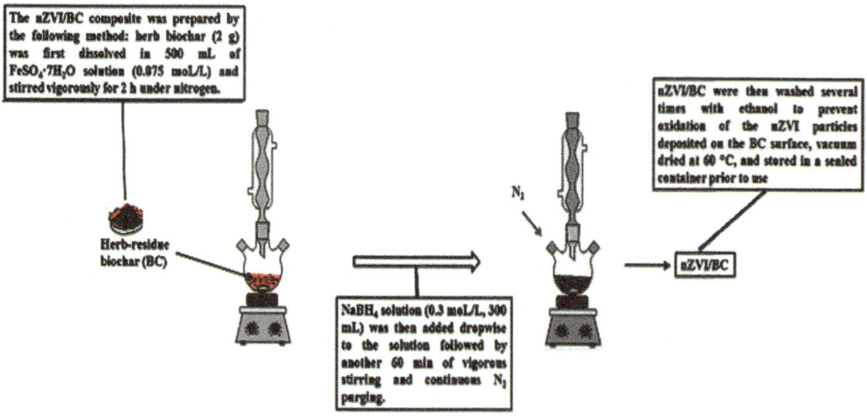

Fig. 4 Chemical reduction step for nZVI synthesis from Su et al. (2016). This figure by Vicente Neto is licensed under the Creative Commons Attribution 4.0 International license. https://commons.wikimedia.org/wiki/File:Chemical_reduction_step_for_nZVI_synthesis.tif

To investigate the adsorption mechanism of removal of Cr(VI) by nZVI/BC, Shang et al. (2017) applied XPS analysis to measure the chemical compositions and states of Cr and Fe on the surface of nZVI/BC. XPS spectra suggested that Cr(VI) was in fact reduced to Cr(III) on the surface of nZVI/BC.

Figure 4 shows the chemical reduction of the residue obtained from the thermochemical conversion of an herb, Astragalus membranaceus, under controlled conditions.

The data also showed that there was the formation of chromium hydroxide, via coprecipitation, during Cr(VI) adsorption and reduction chromium hydroxide. These results are according to previous researches that proposed that the removal of Cr(VI) by nZVI/ZVI involves the adsorption, coprecipitation, and subsequent reduction of Cr(VI) to Cr(III) with nZVI corrosion. Shang et al. (2017) concluded by adsorption study that nZVI/BC can be applied as an effective adsorbent for removing Cr(VI) from solutions.

3.4 *Nanometal Oxides and Polymers Supports*

Oxides have great natural abundance, low cost, and high pH_{PZC}. These characteristics make them suitable to be used as potential adsorbents to removing pollutants present in water. However, there is a limitation of the use of nanoparticles (nZVMs or nanooxides in water treatment that is associated with the difficulty of separating them from the treated water at the end of the process. An alternative to circumvent this limitation has stimulated the development/preparation of supports in which nanooxides can be impregnated. When the supports chose are polymers, the final result is a composite. Then, there are many optional techniques to be applied in the synthesis.

Generally, the preparation process involves two categories: direct compound and in situ synthesis. Direct compounding is a low-cost method, and it has easy operation which is suitability for massive production. This method involves two steps. Firstly, nanofillers and polymer supporters are prepared separately. After that, they are compounded by various forms such as solution, emulsion, fusion, or mechanical forces (Zhao et al. 2011). However, this method also has a limitation that is related to the difficulty of controlling the distribution of the nanoparticles in the most homogeneous way possible in the polymer matrix or on it. One explanation for this limitation is due to the fact that nanoparticles exhibit a tendency to aggregate into larger particles during mixing. The polymeric support also has limitations as they are sensitive to high temperatures and can severely degrade after melting. Degradation affects the phase separation process. In order to reduce these effects, some surface treatments for nanoparticles have been developed with good results. To obtain a good dispersion, parameters such as temperature and time, shear force, and reactor configuration should be adjusted. Then, they are able to apply to both in the synthesis stage and in the operational conditions in the composition stage.

Yew et al. (2006), applied the direct compounding method to develop a polyhydroxybutyrate (PHB) film containing nanosized TiO_2 disperse. The purpose of this study was to examine the decolorization efficiency of the TiO_2/PHB nanocomposite film through the photodegradation, in the aqueous medium, of the dyes methylene blue and cane pink. The authors also studied the durability of disinfection by photocatalytic sterilization using *E. coli* as the standard biological indicator of the efficiency. Figure 5 shows a flowchart summarizing the steps for preparing the PHB–TiO_2 film composites according to Yew et al. (2006).

Yew et al. (2006), concluded that the PHB–TiO_2 composite film was found to be effective in both the sterilization and decolorization of methylene blue and rose bengale. According to the authors, the use of the PHB film containing TiO_2 dispersed and immobilized photocatalysts was important also from an operational point of view due to not require the separation process after photocatalytic treatment. Another important aspect is that composite films are biodegradable and environmental-friendly. Table 2 shows the advantages of applying TiO_2 in photocatalysis as well as the advantages of PHB films as support. Table 2 shows the advantages of applying TiO_2 in photocatalysis as well as the advantages of PHB films as support.

The in situ synthesis is more appropriate for nanocomposites whose support base is of polymeric origin. In this method, nanoparticles are preloaded within the polymeric phase.

Wu et al. (2009), developed a synthesis of zero-valent copper-chitosan nanocomposites via in situ method. They applied this nanomaterial for the treatment of hexavalent chromium. According to the authors, the chitosan (Fig. 6) was used as support because of some advantages such as it is inexpensive, and its removal process for the recycled metal is simpler and cleaner. Added to that, the complete pyrolysis of chitosan under oxygen conditions permits to obtain metal or its oxide, and the amino and hydroxyl groups of chitosan are able to form metal complexes that can be used as a precursor of a heterogeneous molecular catalyst.

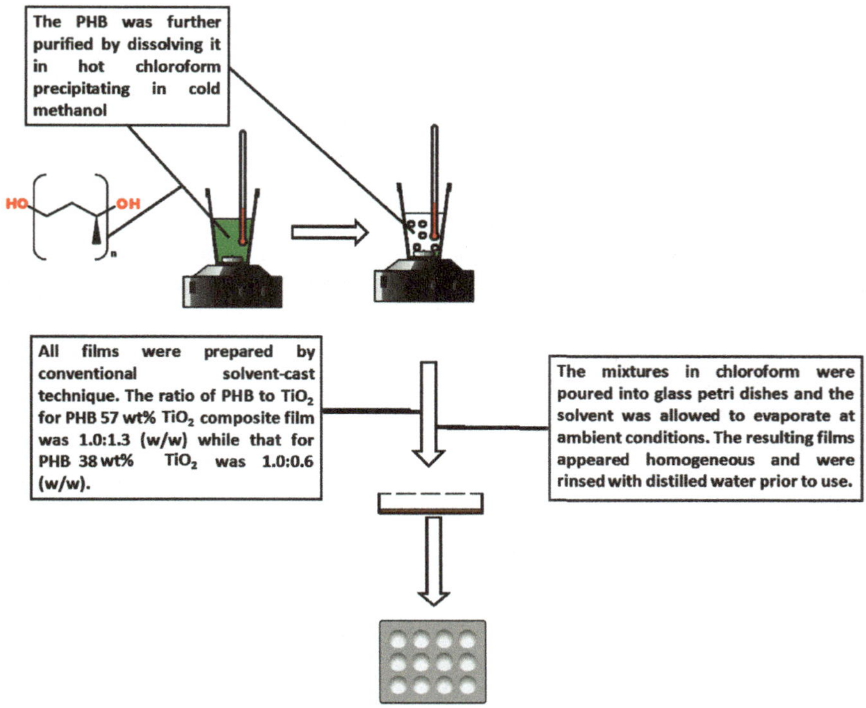

Fig. 5 Flowchart summarizing the steps for preparing the PHB–TiO_2 film composites according to Yew et al. (2006). This figure by Vicente Neto is licensed under the Creative Commons Attribution 4.0 International license. Link: https://commons.wikimedia.org/wiki/File:Flow_chart_summarizing_the_steps_for_preparing_the_PHB-TiO2_film_composites.tif

In the method used by Wu et al. (2009), KBH_4 is a strong reducing agent that is widely used in the synthesis of zero-valence metals because it is a stable reagent in alkaline solutions and under normal atmospheric conditions (Eq. 1); however, it requires care in handling because it has the risk of burning when exposed to acids or oxidants (Eq. 2). Figure 7 shows a chart flux summarizing the zero-valence copper method preparation.

$$4CuSO_4 + KBH_4 + 8KOH \rightarrow 4Cu + 4K_2SO_4 + KBO_2 + 6H_2O \quad (1)$$

$$NaBH_4 + 2H_2O \xrightarrow[\text{or } H^+]{\text{cat}} NaBO_2 + 4H_2 \uparrow \quad (2)$$

Wu et al. (2009), concluded that the zero-valent copper-chitosan nanocomposites showed a good removal capacity by adsorption of Cr(VI) ions.

Table 2 Advantages of using TiO_2 and PHB (as support) in photocatalysis

Component		Advantage
TiO_2	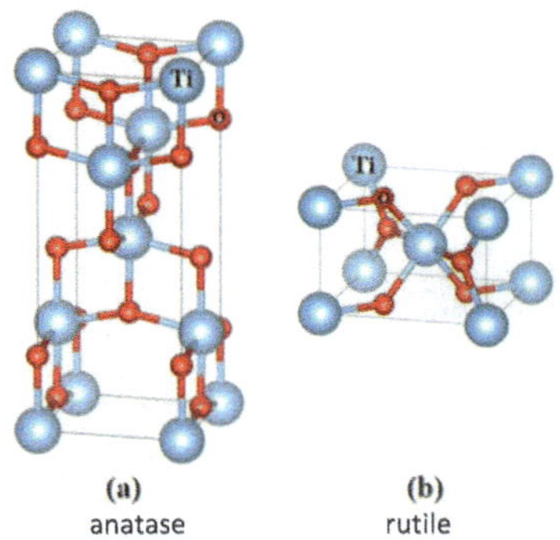	TiO_2 had been approved by the food testing laboratory of the United States Food and Drug Administration (FDA), and it is considered safe and harmless to human It is the most preferable material among the various photocatalysts available, and it is proven to be able to degrade organic carbon into CO_2 effectively
PHB	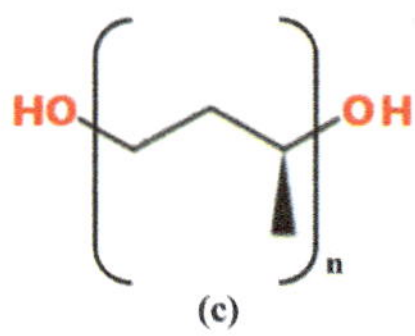	PHB is the most common polymer among the polyhydroxyalkanoates (PHAs). PHA is a bio-based thermoplastic synthesized by various microorganisms, and it can be decomposed and assimilated by many microbial species (biodegradable)

Figures **a** and **b** were extracted from Jia et al. (2016).

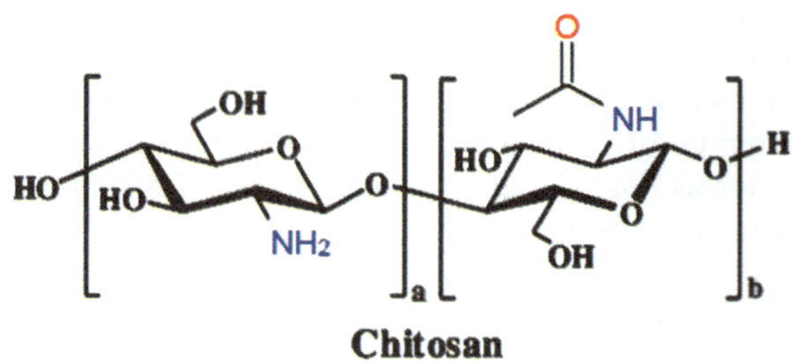

Fig. 6 Structure of chitosan. This figure by Vicente Neto is licensed under the Creative Commons Attribution 4.0 International license. Link: https://commons.wikimedia.org/wiki/File:Chitosan.tif

3.5 Fluorescent Carbon Nanoparticles

Fluorescent carbon nanoparticles (CNPs) or carbon nanodots (CDs) are one of the precursors of the development of the carbon nanoscience. CDs are nanoparticles smaller than ≈10 nm (Fig. 1) and basically composed of carbon, oxygen, nitrogen, and hydrogen (Lu et al. 2012). CDs are produced with low production costs, exhibit

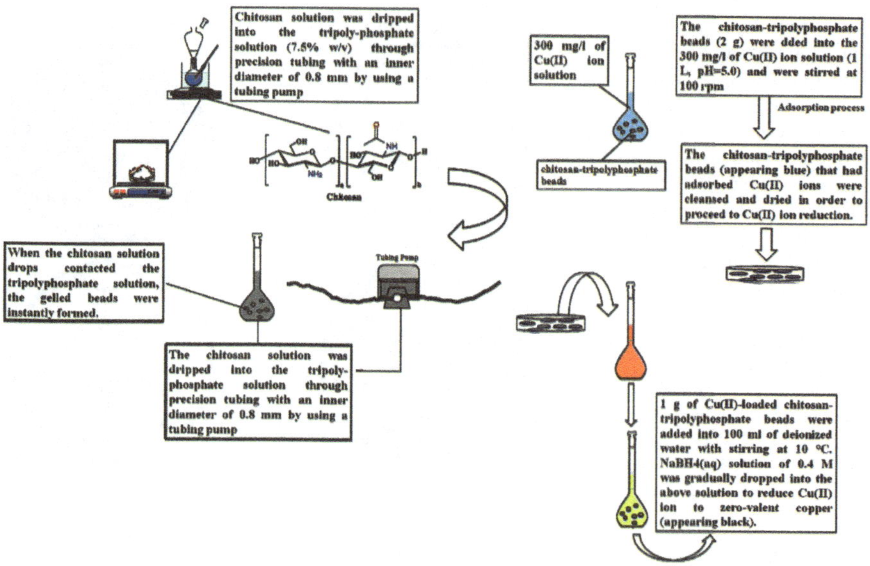

Fig. 7 Chart flux summarizing the zero valence copper method preparation. This figure by Vicente Neto is licensed under the Creative Commons Attribution 4.0 International license. Link https://commons.wikimedia.org/wiki/File:Chart_flux_sumarizing_the_zero_valence_copper_method_preparation.tif

chemical inertia to various substances, as well as non-intermittent fluorescence (Ray et al. 2009). According to Nguyen et al. (2016), CDs are potentials substitutes for the present fluorescent nanomaterials due essentially to the combination of their specific properties of intense photoluminescence (PL), low toxicity, and high aqueous solubility.

3.5.1 Synthesis of CDs

There are several synthesis routes to obtain CDs. They were grouped into two main techniques: top-down and bottom-up approaches. The following methods have been applied to the synthesis of the CDs: laser ablation (Sun et al. 2006), electrochemical synthesis (Liu et al. 2016), arc discharge (Xu et al. 2004a, b), and chemical oxidation (Sciortino et al. 2017).

In general, top-down methods involve the disruption of bulk carbon precursors or nanomaterials with higher dimensionality than CDs, while bottom-up syntheses usually exploit the carbonization (more or less complete) of different molecular precursors, as citric acid or sucrose (Sciortino et al. 2018a, b), Cong and Zhao (2018).

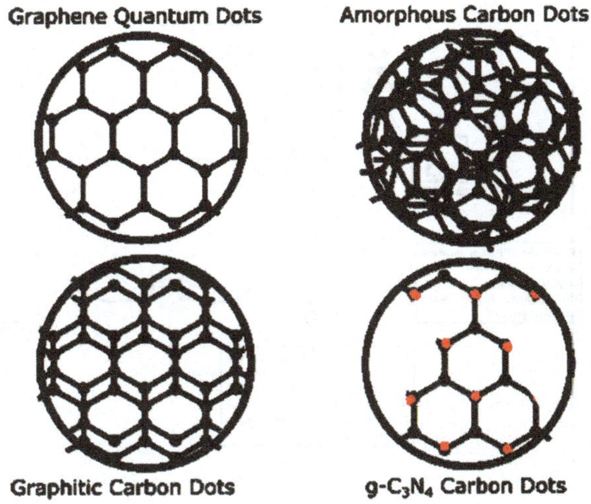

Fig. 8 Schematic representation of CD structures. Black and red dots represent carbon and nitrogen atoms, respectively. With permission from Sciortino et al. (2018a, b). This article is an open access article distributed under the terms and conditions of the Creative Commons Attribution (CC BY) license. Link https://www.mdpi.com/2311-5629/4/4/67

3.5.2 The Structure of CDs

From an experimental point of view, it is already known that the core and surface structures of CDs are highly dependent on synthesis. This behavior opens up several possibilities for designing different types of CDs. According to Sciortino et al. (2018a, b), no universal classification was currently accepted in the literature. Spherical-shaped graphitic CDs are probably the most common in the literature (Fig. 8).

Similarly to graphite, the core of these dots is composed of layers of sp^2-hybridized carbon stacked on top of each other. It is noted that the lateral dimension of the layers is limited by the overall dot diameter, typically of a few nanometers.

3.5.3 Application

Shahnazi et al. (2020) obtained new nanocomposite by surface molecularly imprinted technique. This way, a molecularly imprinted polymer layer coated on TiO_2/carbon nanodots (TiO2/CNDs/MIP) was successfully synthesized. Figure 9 shows a flowchart for obtaining the TiO_2/CNDs nanoparticle. From this one, the final composite was obtained (Fig. 9).

According to authors, O-phenylenediamine was used both as the functional monomer for molecularly-imprinted polymer (MIP) and as the precursor of carbon nanodots. Nanocomposite obtained was successfully applied in selective photocatalytic degradation of herbicides. Figure 10 shows a flowchart for obtaining the TiO_2/CNDs/MIP nanocomposite and photodegradation mechanism according to Shahnazi et al. (2020).

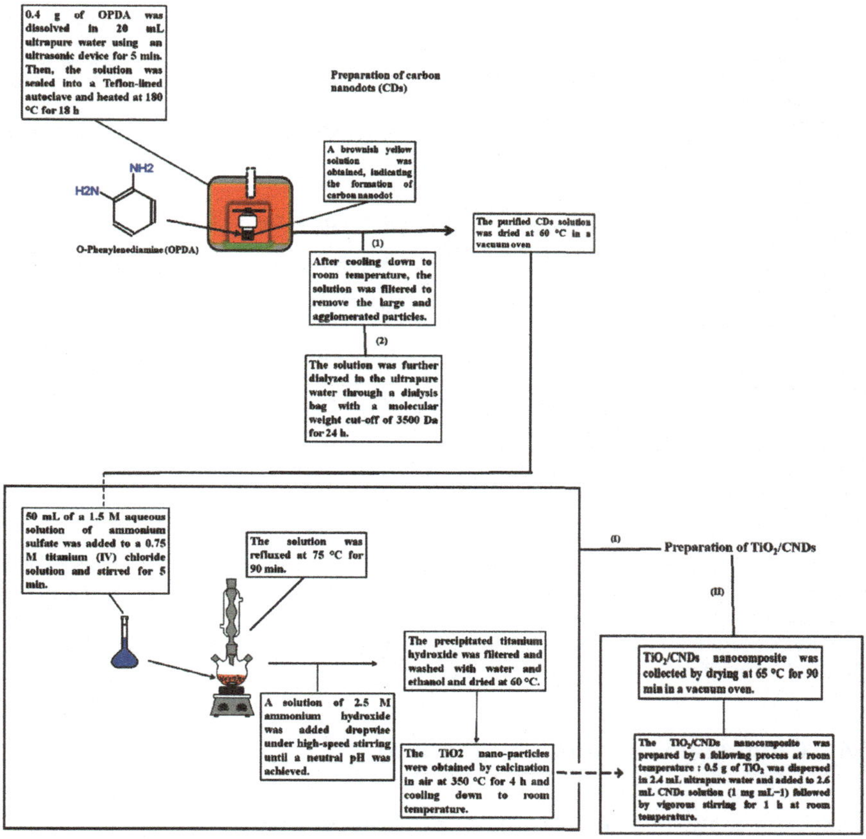

Fig. 9 Flowchart for obtaining the TiO_2/CNDs nanoparticle. This figure by Vicente Neto is licensed under the Creative Commons Attribution 4.0 International license. Link https://commons.wikimedia.org/wiki/File:Flowchart_for_obtaining_the_TiO2-CNDs_nanoparticle.tif

According to the authors, TiO_2/CNDs/MIP nanocomposite exhibited a higher adsorption capacity for the PM target than TiO2/CNDs/NIP nanocomposite. This resulted can be attributed to the imprinted cavities and the specific recognition sites on the surface of MIP. From experimental data, the authors concluded that TiO_2/CNDs/MIP nanocomposites increased its photocatalytic activity as well as better selectivity for PM target degradation compared to TiO_2, TiO_2/CNDs nanoparticles under visible light.

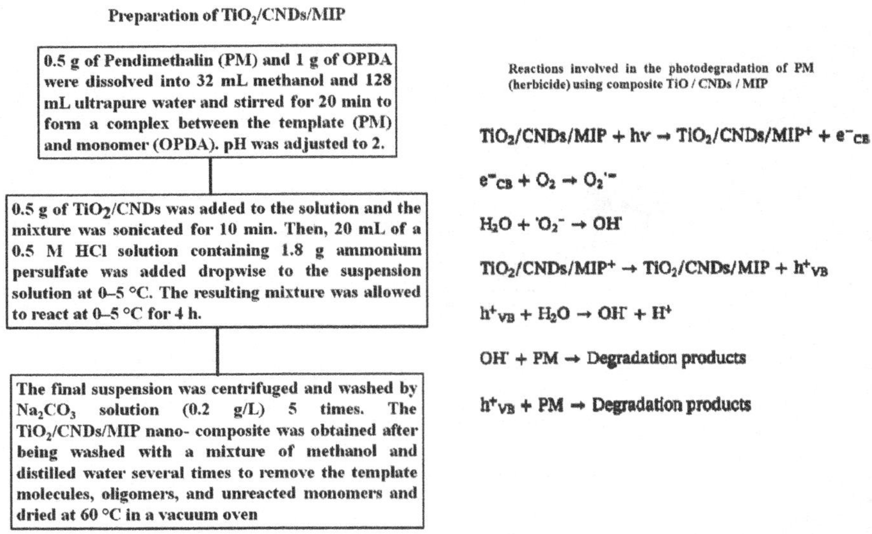

Fig. 10 Flowchart for obtaining the TiO_2/CNDs/MIP nanocomposite and photodegradation mechanism proposed by Shahnazi et al. (2020). This figure by Vicente Neto is licensed under the Creative Commons Attribution 4.0 International license. Link: https://commons.wikimedia.org/wiki/File:Flowchart_for_obtaining_the_TiO2-CNDs-MIP_nano-_composite_and_photodegradation_mechanism.tif

3.6 Challenges in Pollutant Removal

3.6.1 Concentration and Chemical Speciation of Contaminants

Environmental pollution is caused by a large kind of chemical species such as heavy metals, pesticides, medications, and others. Almost all these pollutants are associated with industrial activities, such as metal-mechanic, electroplating, pesticide, pigments, photographic materials, and pharmaceutical industry.

In the process of the pollutants removing, their chemical speciation in the environment in which it is should be considered. According to IUPAC, speciation of chemical specie is defined as the distribution of this element among defined chemical species in a system (Fig. 11).

In the case of heavy metals, the stability of the species in an environment is determined by the metal-ligand binding energy (M-L). The bond that occurs is characterized quantitatively by the constants that describe the balance of the complexes. For a general example.

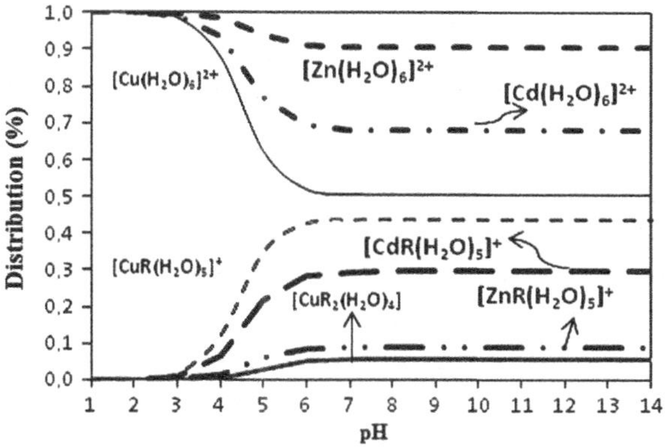

Fig. 11 Distribution of species in solution as a function of pH where R=CH3COO– acetate group. The concentration of acetic acid [ligand,L] is $\times 10^{-3}$M, $\log \beta_{1,Cd(II)} = 1.93$ and $\log \beta_{2,Cd(II)} = 3.15$, $\log \beta_{1,Cu(II)} = 2.23$ and $\log \beta_{2,Cu(II)} = 3.63$, $\log \beta_{1,Zn(II)} = 1.28$ and $\log \beta_{2Zv(II)} = 2.09$

3.6.2 Surfacial Charges

The theory of superficial charges in adsorbents surface suggests that a certain adsorbent in aqueous solution can adsorb OH^- or H^+ ions. Therefore, the surface charge of each particle will depend on the pH of the solution. Thus, the superficial groups of each active site of the adsorbent can dissociate or associate protons from the solution depending on the properties of the adsorbent and the pH of the solution. Consequently, the surface of the adsorbent becomes positively charged when associated with the protons from the solution, under acidic conditions, or negatively charged when the loss of protons to the solution occurs, under alkaline conditions. The point of zero charges (PZC) is the pH (pH_{pzc}) at which the surface of the adsorbent is globally neutral. Below the pH_{pzc}, the surface is positively charged; beyond this value, it is negatively charged.

This implies that in the study of pollutant removal, it is of fundamental importance to know what types of species are present in the aqueous medium. It is important to emphasize that the species distribution [as well as the surface of an adsorbent potential] is dependent on pH and that the pH variation will favor the formation of one species in relation to others, and this data must be analyzed with the already discussed early by the zero charge point (PZC). Nanomaterials with a negative surface charge will have a greater affinity for positive species. The same principle applies to nanomaterials with a positive surface charge.

3.6.3 Stability Constant

The occurrence of metal ions in aquatic environments is mainly due to industrial activities. From the point of view of public health and environment protection, contamination of environments by these metals represents a serious risk both for the human population and also for the survival of other animals, microorganisms, and plants.

Almost all heavy metals found in aqueous medium are in a form called complex. The stability of the complexes is measured by stability constant. In general, the stability of the complexes can be represented by Eq. (3):

$$M^{y+} + xL \leftrightarrow [ML_x]^{y+} \quad (3)$$

where M^{y+} represents the metallic element, neutral donor ligand specie (e.g., water molecules), and $[MLx]^{y+}$ is a complex formed. The stability constant is then defined by Eq. (4) according to the law of mass action:

$$K = \frac{[ML_x]^{+y}}{[M^{+y}][L]^x} \quad (4)$$

This constant defines the relationship between the concentration of the complex metal ion and the concentrations of metal ions and free binders in the solution. Consequently, the higher the K value, the greater the stability of the metal complex.

When it comes to the formation of complexes involving successive reaction steps, as in the following example:

$$M + L \leftrightarrow ML \quad k_1 = \frac{[M]}{[M][L]} \quad (5)$$

$$ML + L \leftrightarrow ML_2 \quad k_2 = \frac{[ML_2]}{[ML][L]} \quad (6)$$

$$ML_2 + L \leftrightarrow ML_3 \quad k_3 = \frac{[ML_3]}{[ML_2][L]} \quad (7)$$

$$\vdots$$

$$ML_{n-1} + L \leftrightarrow ML_n \quad k_n = \frac{[ML_n]}{[ML_{n-1}][L]} \quad (8)$$

the constants k_1, k_2, k_3, k_n, are called partial formation constants, and their product is the absolute or total formation constant. Combining all the above equations by multiplication, the stability constant or complex formation constant (β):

$$\beta_1 = k_1; \; \beta_2 = k_1 \cdot k_2; \; \beta_3 = k_1 \cdot k_2 \cdot k_3; \; \beta_n = k_1 \cdot k_2 \cdot k_3 \ldots k_n \quad (9)$$

In aqueous solutions, the distribution in the equilibrium of species M^{+2} in medium containing an L^- ligand (Fig. 12) can be represented by Reactions 10–13 where β_1–β_4 are the stability constants of the complexes.

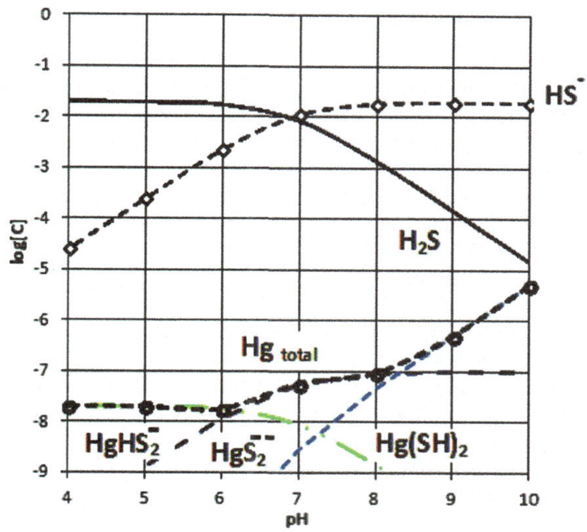

Fig. 12 Distribution (log [C]) of mercury in hydrogen sulfide (HS + HS⁻ = 0.019 M, $\mu = 1$) at 20C as a function of pH

$$M^{+2} + L^{-\leftrightarrow} \leftrightarrow [ML]^{+} \quad \beta_1 = \frac{[ML]^{+1}}{[M^{+2}][L^{-1}]} \tag{10}$$

$$M^{+2} + 2L^{-} \leftrightarrow [ML_2] \quad \beta_2 = \frac{[ML_2]}{[M^{+2}][L^{-1}]^2} \tag{11}$$

$$M^{+2} + 3L^{-} \leftrightarrow [ML_3]^{-} \quad \beta_3 = \frac{[ML_3]^{-1}}{[M^{+2}][L^{-1}]^3} \tag{12}$$

$$M^{+2} + 4L^{-} \leftrightarrow [ML_4]^{-2} \quad \beta_4 \frac{[ML_3]^{-2}}{[M^{+2}][L^{-1}]^4} \tag{13}$$

4 Final Considerations

Nanomaterials have the potential to be the key to the development of new materials with specificity properties to many fields of science as physic, chemistry, biology, medicine, and environmental science. Although replication of natural systems is one of the most promising areas of this technology, scientists are still trying to grasp their astonishing complexities. Overall, several nanomaterials such as CDS, nanometal oxides, and nZVM's, can be applied as a complementary step in the traditional water treatment plants.

References

Ahmad T, Iqbal J, Bustam MA, Zulfiqar M, Muhammad N, Al Hajeri BM, Irfan M, Anwaar Asghar HM, Ullah S (2020) Phytosynthesis of cerium oxide nanoparticles and investigation of their photocatalytic potential for degradation of phenol under visible light. J Mol Struct 1217:128292

Ahmed A, Usman M, Yu B, Ding X, Peng Q, Shen Y, Cong H (2019) Efficient photocatalytic degradation of toxic Alizarin yellow R dye from industrial wastewater using biosynthesized Fe nanoparticle and study of factors affecting the degradation rate. J Photochem Photobiol B 202:111682

Boparai HK, Joseph M, O'Carroll DM (2011) J Hazard Mater 186–1:458–465

Cayuela A, Soriano ML, Valcàrcel M (2015) Photoluminescent carbon dot sensor for carboxylated multiwalled carbon nanotube detection in river water. Sens Actuators 207:596–601

Chavali MS, Nikolova MP (2019) Metal oxide nanoparticles and their applications in nanotechnology. Springer International Publishing, SN Applied Sciences

Cong S, Zhao Z (2018) Carbon quantum dots: a component of efficient visible light photocatalysts. Visible-light photocatal. Carbon-based mater. https://doi.org/10.5772/intechopen.70801

El-Temsah YS, Joner EJ (2013) Effects of nano-sized zero-valent iron (nZVI) on DDT degradation in soil and its toxicity to collembola and ostracods. Chemosphere 92:131–137

Etefagh R, Azhir E, Shahtahmasebi N (2013) Synthesis of CuO nanoparticles and fabrication of nanostructural layer biosensors for detecting *Aspergillus niger* fungi. Sci Iran 20:1055–1058

Gogoi SK, Gopinath P, Paul A, Ramesh A, Ghosh SS, Chattopadhyay A (2006) Green fluorescent proteinexpressing *Escherichia coli* as a model system for investigating the antimicrobial activities of silver nanoparticles. Langmuir 22(22):9322–9328

Hao G, Liu J, Gao H, Xiao L, Ke X, Jiang W, Zhao F, Gao H (2015) Preparation of nano-sized Copper β-resorcylate (β-Cu) and its excellent catalytic activity for the thermal decomposition of ammonium perchlorate. Prop Explos Pyrotech 40(6):848–853

Hu J, Chen G, Lo IM (2005) Removal and recovery of Cr(VI) from wastewater by maghemite nanoparticles. Water Res 39(18):4528–4536

Isherwood PJM (2017) Copper zinc oxide: investigation into a p-type mixed metal oxide system. Vacuum 139:173–177

Jia J, Yamamoto H, Okajima T, Shigesato Y (2016) On the crystal structural control of sputtered TiO_2 thin films on the crystal structural control of sputtered TiO_2 thin films. Nanoscale Res Lett 11:324

Kanel SR, Greneche JM, Choi H (2006) Arsenic(V) removal from groundwater using nano scale zero-valent iron as a colloidal reactive barrier material. Environ Sci Technol 40:2045–2050

Ke FS, Jamison L, Huang L, Zhang B, Li JT, Zhou XD, Sun SG (2014) Negative electrode comprised of Fe_3O_4 nanoparticles and Cu nanowires for lithium ion batteries. Solid State Ionics 262:18–21

Khalaj M, Kamali M, Khodaparast Z, Jahanshahi A (2018) Copper-based nanomaterials for environmental decontamination—an overview on technical and toxicological aspects. Ecotoxicol Environ Saf 148:813–824

Khan ST, Malik A (2019) Engineered nanomaterials for water decontamination and purification: from lab to products. J Hazard Mater 363:295–308

Khandare D, Mukherjee S (2019) A review of metal oxide nanomaterials for fluoride decontamination from water environment. Mater Today Proc 18:1146–1155

Kooijman G, de Kreuk MK, Houtman C, van Lier JB (2020) Perspectives of coagulation/flocculation for the removal of pharmaceuticals from domestic wastewater: a critical view at experimental procedures. J Water Process Eng 34:101161

Kunduru KR, Nazarkovsky M, Farah S, Pawar RP, Basu A, Domb AJ (2017) Nanotechnology for water purification: applications of nanotechnology methods in wastewater treatment, Water Purification

Li Q, Mahendra S, Lyon DY, Brunet L, Liga MV, Li D, Alvarez PJJ (2008) Antimicrobial nanomaterials for water disinfection and microbial control: potential applications and implications. Water Res 42:4591–4602

Liu Z, Zhang FS (2010) Nano-zerovalent iron contained porous carbons developed from waste biomass for the adsorption and dechlorination of PCBs. Bioresour Technol 101:2562–2564

Li B, Zeng HC (2019) Synthetic chemistry and multifunctionality of an amorphous Ni-MOF-74 shell on a Ni/SiO2 hollow catalyst for efficient tandem reactions. Chem Mater 31(14):5320–5330

Liu M, Xu Y, Niu F, Gooding JJ, Liu J (2016) Carbon quantum dots directly generated from electrochemical oxidation of graphite electrodes in alkaline alcohols and the applications for specific ferric ion detection and cell imaging. Analyst 141:2657–2664

Liú D, Wang G, Liŭ D, Lin J, He Y, Li X, Li Z (2016) Photocatalysis using zero-valent nano-copper for degrading methyl orange under visible light irradiation. Opt Mater 53:155–159

Lu W, Qin X, Liu S, Chang G, Zhang Y, Luo Y, Asiri AM, Al-Youbi AO, Sun X (2012) Economical, green synthesis of fluorescent carbon nanoparticles and their use as probes for sensitive and selective detection of mercury(II) ions. Anal Chem 84:5351–5357

Lucchetti R, Onotri L, Clarizia L, Natale F Di, Somma I Di, Andreozzi R, Marotta R (2017) Removal of nitrate and simultaneous hydrogen generation through photocatalytic reforming of glycerol over "in situ" prepared zero-valent nano copper/P25. Appl Catal B Environ 202:539–549

Mudhoo A, Bhatnagar A, Rantalankila M, Srivastava V, Sillanpää M (2019) Endosulfan removal through bioremediation, photocatalytic degradation, adsorption and membrane separation processes: a review. Chem Eng J 360:912–928

Nabid MR, Sedghi R, Gholami S, Oskooie HA, Heravi MM (2013) Preparation of new magnetic nanocatalysts based on TiO_2 and ZnO and their application in improved photocatalytic degradation of dye pollutant under visible light. Photochem Photobiol 89:24–32

Nguyen V, Si J, Yan L, Hou X (2016) Direct demonstration of photoluminescence originated from surface functional groups in carbon nanodots. Carbon N Y 108:268–273

Pan L, Sun S, Zhang A, Jiang K, Zhang L, Dong C, Huang Q, Wu A, Lin H (2015) Truly fluorescent excitation-dependent carbon dots and their applications in multicolor cellular imaging and multidimensional sensing. Adv Mater 27:7782–7787

Ray SC, Saha A, Jana NR, Sarkar R (2009) Fluorescent carbon nanoparticles: synthesis, characterization, and bioimaging application. J Phys Chem C 113:18546–18551

Rong M, Song X, Zhao T, Yao Q, Wang Y, Chen X (2015) Synthesis of highly fluorescent P, O-g-C_3N_4 nanodots for the label-free detection of Cu^{2+} and acetylcholinesterase activity. J Mater Chem C 3:10916–10924

Sciortino A, Cayuela A, Soriano ML, Gelardi FM, Cannas M, Valcarcel M, Messina F (2017) Different natures of surface electronic transitions of carbon nanoparticles. Phys Chem Chem Phys 19:22670–22677

Sciortino A, Cannizzo A, Messina F (2018a) Carbon nanodots: a review—from the current understanding of the fundamental photophysics to the full control of the optical response. C 4, 67

Sciortino A, Mauro N, Buscarino G, Sciortino L, Popescu R, Schneider R, Giammona G, Gerthsen D, Cannas M, Messina F (2018b) β-C_3N_4 nanocrystals: carbon dots with extraordinary morphological, structural, and optical homogeneity. Chem Mater 30:1695–1700

Shahnazi A, Nabid MR, Sedghi R (2020) Synthesis of surface molecularly imprinted poly-o-phenylenediamine/TiO_2/carbon nanodots with a highly enhanced selective photocatalytic degradation of pendimethalin herbicide under visible light. React Funct Polym 151:104580

Shang J, Zong M, Yu Y, Kong X, Du Q, Liao Q (2017) Removal of chromium (VI) from water using nanoscale zerovalent iron particles supported on herb-residue biochar. J Environ Manag 197:331–337

Su H, Fang Z, Tsang PE, Fang J, Zhao D (2016) Stabilisation of nanoscale zero-valent iron with biochar for enhanced transport and in-situ remediation of hexavalent chromium in soil. Environ Pollut 214:94–100

Sun Y, Zhou B, Lin Y, Wang W, Fernando KAS, Pathak P, Meziani MJ, Harruff BA, Wang X, Wang H et al (2006) Quantum-sized carbon dots for bright and colorful photoluminescence. J Am Chem Soc 128:7756–7757

Sun Y, Zheng F, Wang W, Zhang S, Wang F (2020) Remediation of Cr(VI)-contaminated soil by nano-zero-valent iron in combination with biochar or humic acid and the consequences for plant performance. Toxics 8:26

Suriyaraj SP, Vijayaraghavan T, Biji P, Selvakumar R (2014) Adsorption of fluoride from aqueous solution using different phases of microbially synthesized TiO_2 nanoparticles. J Environ Chem Eng 2:444–454

Taman R, Ossman ME, Mansour MS, Farag HA (2015) Metal oxide nano-particles as an adsorbent for removal of heavy metals. J Adv Chem Eng 5:125. 4172/2090-4568

Tosco T, Petrangeli M, Papini M P, Viggi CC, Sethi R (2014) Nanoscale zerovalent iron particles for groundwater remediation: a review. J Cleaner Prod 77:10–21

Tang Y, Yin X, Mu M, Jiang Y, Li X, Zhang H, Ouyang T (2020) Anatase TiO_2@MIL-101(Cr) nanocomposite for photocatalytic degradation of bisphenol A. Colloids Surfaces A Physicochem Eng Asp 596:124745

Tong M, Yuan S, Long H, Zheng M, Wang L, Chen J (2011) Reduction of nitrobenzene in groundwater by iron nanoparticles immobilized in PEG/nylon membrane. J Contam Hydrol 122:16–25

Wu S-J, Liou T-H, Mi F-L (2009) Synthesis of zero-valent copper-chitosan nanocomposites and their application for treatment of hexavalent chromium. Bioresour Technol 100:4348–4353

Xu X, Ray R, Gu Y, Ploehn HJ, Gearheart L, Raker K, Scrivens WA (2004a) Electrophoretic analysis and purification of fluorescent single-walled carbon nanotube fragments. J Am Chem Soc 126:12736–12737

Xu XH, Brownlow WJ, Kyriacou SV, Wan Q, Viola JJ (2004b) Real-time probing of membrane transport in living microbial cells using single nanoparticle optics and living cell imaging. Biochemistry 43(32):10400–10413

Yang Y, Wang H, Li J, He B, Wang T (2012) Novel functionalized nano-TiO_2 loading electrocatalytic membrane for ole wastewater treatment Environ Sci Technol 46:6815−6821

Yew S-P, Tang H-Y, Sudesh K (2006) Photocatalytic activity and biodegradation of polyhydroxybutyrate films containing titanium dioxide Polym. Degrad Stab 91:1800–1807

Yildiz BS (2012) Water and wastewater treatment: biological processes. Metrop Sustain Underst Improv Urban Environ 406–428

Zhang G, Li Z, Zheng H, Fu T, Ju Y, Wang Y (2015) Influence of the surface oxygenated groups of activated carbon on preparation of a nano Cu/AC catalyst and heterogeneous catalysis in the oxidative carbonylation of methanol. Appl Catal B Environ 179:95–105

Zang YT, Bing S, Li YJ, Shu DQ (2019) Application of slightly acidic electrolyzed water and ultraviolet light for *Salmonella enteritidis* decontamination of cell suspensions and surfaces of artificially inoculated plastic poultry transport coops and other facility surfaces. Poult Sci 98:6445–6451

Zhang N, Sun J, Gong H (2019) Transparent p-type semiconductors: copper-based oxides and oxychalcogenides. Coatings 9:1–27

Zhao X, Lv L, Pan B, Zhang W, Zhang S, Zhang Q (2011) Polymer-supported nanocomposites for environmental application: a review. Chem Eng J 170:381–394

Chapter 8
New Advances of the Nanotechnology in Textile Engineering: Functional Finishing with Quantum Dots and Others Nanoparticles

J. H. O. Nascimento, B. H. S. Felipe, R. L. B. Cabral, Awais Ahmad, A. B. da Silva, N. F. A. Neto, A. P. S. Júnior, and A. L. C. Teófilo

The development of textile materials with functional nanoparticles has been driven by the advancement of materials science, the globalized market, competitiveness and the relentless pursuit of solutions that generate innovations in processes and products environmentally correct. Advancement in the studies with quantum dots and semiconductors nanoparticles applied to the surface modifications such as textile fibers and plastics with the purpose of adding specific properties has been one of the reasons for the growth of nanotechnology applied in the textile industry, mainly in the area of multifunctional finishing. The application of many of these inorganic nanocoatings on textiles allows functionalize so as to improve their performance in a wide variety

J. H. O. Nascimento (✉) · R. L. B. Cabral · A. B. da Silva
Graduate Program in Textile Engineering, Center of Technology, Federal University of Rio Grande do Norte—UFRN, Natal, Rio Grande do Norte, Brazil
e-mail: joseheriberto@ct.ufrn.br

J. H. O. Nascimento · B. H. S. Felipe · R. L. B. Cabral · A. B. da Silva · A. P. S. Júnior · A. L. C. Teófilo
IMNG—Innovation in Micro and Nanotechnologies Group, Center of Technology, Federal University of Rio Grande do Norte—UFRN, Natal, Rio Grande do Norte, Brazil

J. H. O. Nascimento · B. H. S. Felipe · A. L. C. Teófilo
Graduate Program in Mechanical Engineering, Center of Technology, Federal University of Rio Grande do Norte—UFRN, Natal, Rio Grande do Norte, Brazil

A. Ahmad
Department of Chemistry, University of Lahore, 54590 Lahore, Pakistan

Department of Applied Chemistry, Government College University Faisalabad, Faisalabad 38000, Pakistan

N. F. A. Neto
LSQM—Laboratory of Chemical Synthesis of Materials, Department of Materials Engineering, Federal University of Rio Grande do Norte—UFRN, Faisalabad 38000, Pakistan

R. F. do. Nascimento et al. (eds.), *Nanomaterials and Nanotechnology*,
Materials Horizons: From Nature to Nanomaterials,
https://doi.org/10.1007/978-981-33-6056-3_8

of uses ranging from technical textiles (geotextiles, medical, microelectronic, solar cells and many others) to the conventional textile, giving them new properties, such as the photoluminescence, antibacterial properties, fungicides, self-cleaning, UV protection, flame retardant, supercapacitors, sensors and controlled drugs release. In the years 50–70 have emerged many patents related to the inorganics material coating on textile fibers for technical applications, but it was only in the early 90 that appeared the first patents and publications with application of quantum dots and others inorganics nanoparticles in coating of optical fibers, glass fibers, cellulosic fibers, wool, silk, non-woven and paper. Various techniques of application of functional coatings have been studied: Chemical techniques (wet finishing) carried out mainly by reactions, depletion and chemicals dispersions: examples: sol-gel, electrodeposition, self-assembly and other; techniques carried out by physical and chemical methods of low environmental impact, examples, ALD, PVD, PECVD, CDV, PLD and others.This chapter aims to describe the evolution of research and publications related to the development and application of functional and intelligent textiles using quantum dots and other inorganic nanocoatings in recent years. And yet in this chapter intends to describe to new physical and chemical processes of nanocoatings with different semiconductors quantum dots, metals and ceramics nanoparticles (Au, Ag, AgCl, ZnO, TiO_2, SiO_2, Al_2O_3 and others), carbon nanotubes, graphenes, in order to obtain a smart textile material, as well as describe the properties that textiles may have showing their performance and applications.

1 Nanotechnology as Industrial Innovation

Nanotechnology had its concept first addressed by Richard Feynman in 1959. Understood as: the study, manipulation, the construction of materials, molecules, advanced devices, objects with nanometric dimensions ($1\ nm = 10^{-9}\ m$) and that have dependent properties of this scale. Technology in nanoscale or simply nanotechnology is a comprehensive term that has, in recent decades, aroused the interest of industries and the scientific community in the various areas such as: pharmaceutical, chemical, microelectronics, food and textile, due to the new techniques obtaining, characterization and numerous possibilities of application (Gogotsi 2006; Singh 2017).

The application of nanoscience and nanotechnology has been changing the entire industrial segment worldwide due to the potential of its different applications. Nanoscience and nanotechnology are based on two concepts that are called top-down and bottom-up. The first approach is to miniaturize to the nano-scale from the micro-scale. This type of approach requires technological mastery of all microelectronic technology, involving a large volume of resources and highly qualified labor. The second approach is simpler and requires far fewer resources than the first, with the scientist's creativity being the key ingredient. In this case, nanostructured systems (SNs) are formed from atoms and molecules and can be "assembled" through the manipulation of atoms or by self-assembly (self-assembly) processes. This is what happens with deposition techniques by mass transfer in gaseous medium (e.g., Physical Vapor Deposition) and in liquid medium (e.g., Sol-gel) (Porter and Youtie 2009).

According to ISO/TS 80004-2:2015, a nanoobject is a nanomaterial with one, two or three dimensions at the nanoscale (approximately, between 1 nm and 100 nm). Examples include:

- Nanoplates, nanocoating, nanofilm, quantum well present only one of the external dimensions in nanoscale;
- Nanofibers, nanowires, nanotubes, nanorods, quantum wire have two nanoscale dimensions and a significantly larger third dimension;

Nanoparticles with three dimensions—organic nanoparticles (dendrimers, nanocapsules, etc.) and inorganic (silver, oxides, zinc, aluminum, gold, etc.), dendrimes, fullerenes, nanoclusters, quantum dots.

There are specific areas where the manipulation, modification and control of matter at atomic and molecular levels have become important (Mansoori 2005; Busnaina 2018; Long live the NNI 2019):

I Quantum mechanics is the theory of physics responsible for studies in physical systems whose dimensions are close to or below the atomic scale (molecules, atoms, electrons, protons and other subatomic particles). When manipulated, they modify the micro- and macro-properties of materials such as melting temperature, magnetic, electrical and optical properties, without changing their chemical composition.

II The fundamental characteristic of biological entities (genes, proteins, viruses, among others) is their systematic organization on a nanoscale. The development in nanoscience and nanotechnology allows us to place artificial materials at the nanoscale inside living cells, controlled release of drugs, possibility of targeting the drug to specific targets, diagnosis of diseases using semiconductor nanocrystals, as well as to develop biomimetic materials that mimic the nature. This is certainly a powerful combination of biological sciences and materials science.

III Nanoscale materials are ideal for use in composites as well as in systems involving catalytic reactions, surface activation, storage, transformation and release of energy by fuel cells (hydrogen and natural gas), solar cells, dispersed nanoparticulate systems (nanoparticles dispersed in a emulsion, microemulsion or aqueous nanoemulsion), among others.

IV Macroscopic systems composed of nanostructures can have a much higher charge density than those composed of microstructures. With this, they can be better conductors of electricity and lead to new concepts of electronic devices, smaller, faster, with sophisticated functions and at the same time, reduce power consumption through interactions and control of its complex nanostructure.

Today, this relatively new field of science and technology has a potential impact on the development of all sectors of the market, both socially and economically and industrially, such as transportation, energy, telecommunications, food production, medicines, materials, the environment and engineering (Ramakrishna 2005; Tour 2007; Sasaki and Tour 2008). As well as the demand for nanotechnology in all industries in the world is evidenced by the growing volume of investment made by private investors (corporations and funds) and at the expense of state budgets.

1.1 R&D in Nanotechnology

The growth in the development and consumption of new products with applied nanotechnology in the USA, member countries of the European Union, Russia, China, Japan has increased every day and, more and more universities, research centers and companies are involved in this evolution. The USA remains a global leader in the volume of government nanotechnology investments through the end of 2018. The cumulative investment of the National Nanotechnology Initiative (NNI) since fiscal 2001, including orders placed in 2018, now exceeds $25 billion. In addition, more than $1.1 billion has been invested cumulatively since 2004 in support of nanotechnology for small businesses through profile programs of authorized federal agencies. In the last 20 years, the number of international patent applications related to nano-objects, nanotechnologies and nanoproducts has increased. According to NNI, the countries that applied for the most patents were China, USA, Japan and South Korea (National Science and Technology Council 2018). The nanostructured materials most applied in the global patent market are graphenes (oxidized and reduced), fullerenes, semiconductor nanoparticles, nanofills, nanocomposites, quantum dots and carbon nanotubes (Rusnano, no date). As for the use of nanotechnology in textile materials, the numbers that started to appear in the world economy as expressed since 2007 and with a continuous increase until today (StartNano, no date; WMWTA—World Markets for Woven Textiles and Apparel 2009).

A survey of our authorship, carried out using a patent search system from WIPO (World Intellectual Property Organization) in relation to patent registration agencies worldwide and using only a nanotechnology-related search based on the CPC (Cooperative Patent Classification) B82, found a large increase in the number of registered patents, going from 13 applications in 1990–9770 in 2018, causing an accumulation of more than 90 thousand patents accumulated in 20 years, involving nanotechnology (Fig. 1). The patent system for the protection of inventions, trademarks and copyrights has evolved since its inception. The advantages offered by patents, in addition to incentives for technological development, incentives for scientific research, dissemination of practical and economic knowledge, creation of new markets and satisfaction of consumers' latent needs.

Currently, the importance of patents and their link with technological and social development has a direct relationship with industry 4.0, since processes and products that will be used during the 4th industrial revolution need to meet the need for consumption in this period (Fauss 2008; Long live the NNI 2019).

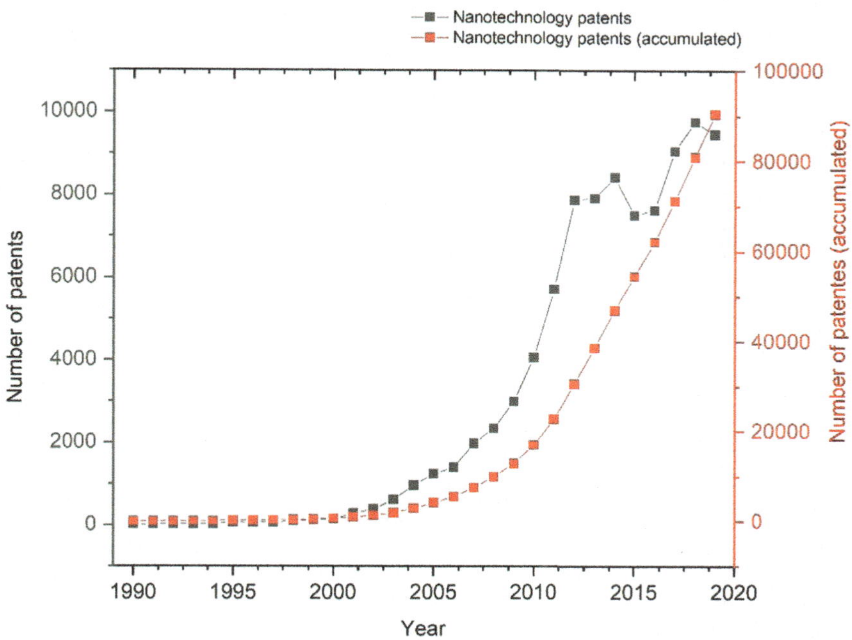

Fig. 1 Graphical data on the number of patents and patents accumulated between 1990 and 2019 on the basis of WIPO (World Intellectual Property Organization)

It is known that the forecast for the 3rd generation of nanotechnology will begin in this new revolution, even with the advente of the SARS-CoV-2 (COVID 19) pandemic that contaminated the population of several countries, causing several deaths in the population and also, generated health crises and the economy on all continents. The patent system for the protection of inventions, trademarks and copyrights has evolved since its inception.

This increase in patent applications in the after the 2008 crisis, as shown in Fig. 1, was caused due to a considerable amount of resources allocated to nanotechnology studies by government agencies, public agencies and private research centers, universities and industry. Taking into account that research involving nanotechnology has been growing and fast, mainly in areas such as microelectronics, new processes and materials have been produced quickly in various sectors such as cosmetics, textiles, electronics, civil construction, computing, making developed countries compete each other in a real technological "war." When the USA took the role of world leader in investment in R&D in nanotechnology, in 2016, the other countries started major investments in this area and thus, USA, China, South Korea, Japan and the European Union, appear as the countries with the largest number of patents involving nanotechnology in the world, as shown in Fig. 2.

The defense industry, energy, microelectronics, computer, communication and biomedical sectors for nanotechnology studies are entering into prominence in China,

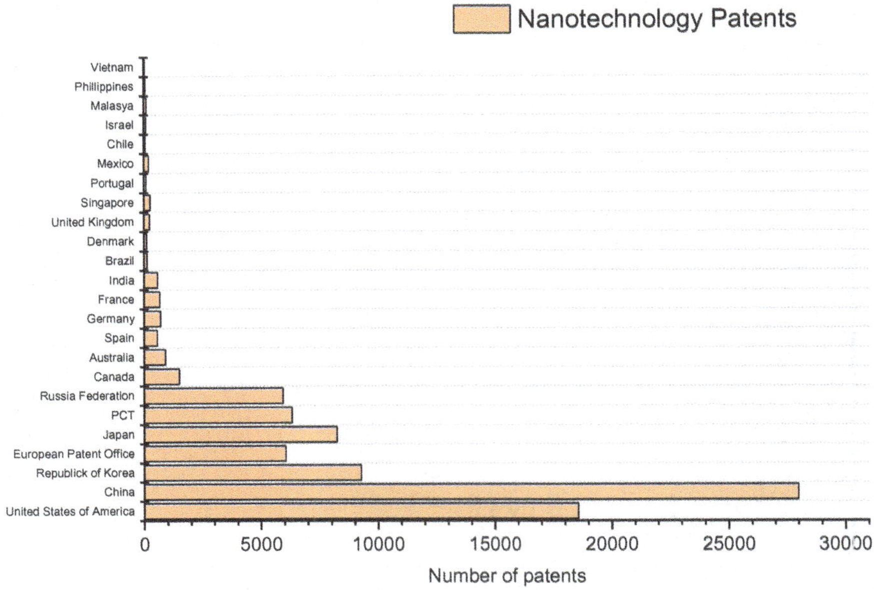

Fig. 2 Graphical data on the number of patents by countries registered between 1990 and 2019, based on WIPO (World Intellectual Property Organization)

South Korea and Japan that conducts R&D studies in recent years. China is the largest holder of the largest number of patents with nanotechnology in the world, followed by the USA, Republic of Korea, European Union, Japan and Russia. Latin America, shown in Fig. 2, by Mexico, Brazil and Chile, has 201, 126 and 58 patents involving nanotechnology, which proves the low investment in this area in Latin American countries, when compared to the number of patents in China and USA.

1.2 Nanotechnology in Textile Engineering

Despite being considered a traditional sector, the textile industry was one of the pioneers to introduce nanotechnology in its products. Textiles are not only fashion oriented, they have important applications in the aerospace, automotive, construction, health and sportswear industries. Since then, a new generation of products with new performances, functionality and other resources has been investigated, developed and placed on the market, increasing the competitiveness of innovative agents (research centers, large companies and universities), encouraging the opening of new markets and decreasing costs. This was due to the sensitivity of several researchers to see that nanoscience can contribute to the development of innovative textile products, with different functionalities and performances, for example, Dr. David Soane, who in

1998 founded Nano-Tex, one of the first companies to use nanotechnology in traditional textiles to give it new features (The New York Times, no date). Then came other companies such as the German NANO-X GmbH and Clariant, the Japanese Gunze Sangyo Inc., the American Ciba SPecialty Chemicals and Nanophase Technologies Corp., the Swedish Texcote Technology Ltd., and the Chinese Beijing Zhong-Shang Centennial Nanotech Co. Ltd., among others (Soane et al. 2005; Kaounides et al. 2007).

In 2006, the European Technology Platform for the Future of Textiles and Clothing presented an industrial vision around expectations to change the structure, activities and global competitiveness of the situation of this industry for the next 14 years. The research priorities proposed by a group of experts can be summarized in 9 areas, as follows (Euratex—The European Apparel and Textile 2006).

- New varieties of fibers and composites for innovative textile products;
- Functionalization of textile materials and related processes;
- Processing of bio-based materials (BBM), biotechnologies and ecological textile materials;
- New products of textile raw material for better human performance;
- New textile products for innovative technical applications;
- Textiles and smart clothes;
- Customization of innovative products;
- Implementation of new design and development of new products, integrated with quality and life cycle management, to achieve and personalize clothing and fashion products;
- Use of micro- and nanotechnologies for the development of new products and manufacturing processes.

Since then, a new generation of products with new performances, features and other resources has been investigated and patented, developed and placed on the market, increasing the competitiveness of innovative agents (research centers, large companies and universities), encouraging the opening of new ones decreasing markets and costs. Based on this information, use the same methodology previously used to survey patents involving nanotechnology. All these data (Table 1) refer to the amount of results in the patent search system of WIPO (World Intellectual Property Organization) in relation to patent filing agencies around the world. All of these data (Table 1) refer to the amount of results in the patent search system of WIPO (World Intellectual Property Organization) in relation to patent registration agencies worldwide. The research related to nanotechnology was based on CPC (Cooperative Patent Classification) B82. The search for topics with nanotechnology applied to the textile area was based on the search for the following keyword involving the CPC codes "B82 e (D01 or D02 or D03 or D04 or D05 or D06 or D07 or D08 or D09 or D10)" analyzed data were from 01/01/1990 to 01/01/2019. And all the terms searched for in relation to textiles are also explained in Table 1, all of these terms are usual keywords called processes and textile products.

After the survey on the platform, the graph was obtained in Fig. 3, which shows an increase in the number of patents accumulated during the years 1990 and 2019.

Table 1 The following table describes the cooperative patent classification (CPC) codes used in this research

CPC description	Classification symbol	Reference area
Nanotechnology	B82	Nanotechnology
Natural or man-made threads or fibers; spinning	D01	Textile
Yarns; mechanical finishing of yarns or ropes; warping or beaming	D02	Textile
Weaving	D03	Textile
Braiding; lace-making; knitting; trimmings; non-woven fabrics	D04	Textile
Sewing; embroidering; tufting	D05	Textile
Treatment of textiles or the like; laundering; flexible materials not otherwise provided for	D06	Textile
Ropes; cables other than electric	D07	Textile
Indexing scheme associated with subclasses of section d, relating to textiles	D10	Textile

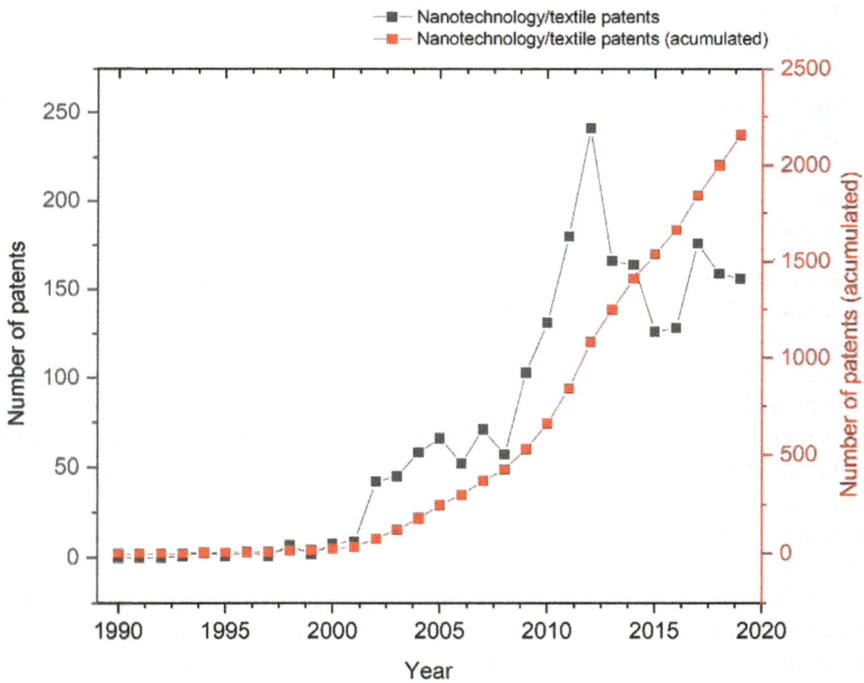

Fig. 3 Graphical data on the number of textile patents and patents accumulated between 1990 and 2019 on the basis of WIPO (World Intellectual Property Organization)

Analyzing the data obtained, it is observed that most patents have nanotechnology applied to the textile industry as a way to innovate and develop textile products with greater fabric durability, resistance to light and washing, comfort in technical clothes and clothing, materials with self-cleaning properties and with production cost reduced (Dolgin 2015). Another objective of the research is to develop innovative products using textile fibers as a support, due to their flexibility (Coyle et al. 2007), in particular smart and functional technical textiles, with innovative features and functions, applied to the area of electronics and optical-electronic, namely sensors, photovoltaic cells (Kaounides et al. 2007) and applications in biomedical materials (Abouraddy et al. 2007).

These dates prove the advancement of nanotechnology over the years. Even at the time of the world economic crisis that started in 2008, for example, many textile products based on nanotechnology were being developed or were in the final stages or being applied for by patents. Textile companies, research institutes and universities, in general, did not interrupt their investments in innovation, since the amount spent was large and interrupting the investment would mean losing everything. They continued to research and today, according to the data in Fig. 3, which show the number of patent applications per year and the accumulated value. This graph (Fig. 3) shows a considerable and notable increase. Since the first patent in 1993, involving textile materials with nanotechnology until 2019, there has been an increase in the accumulated X patents, with an average of patents in the last 10 years of 150 patents/year (Fig. 4).

The countries that hold the largest number of international patents involving materials and textile applications according to WIPO between 1990 and 2019 are the USA and China, both holding 682 and 330 patents, respectively. These numbers have increased considerably because it is a strategy of the textile segment to innovate and meet the essential demands for consumers of this new nanotechnological industry revolution, thus, the textile industry has moved from traditional textiles to advanced textile materials, with optimized processes and high production. But, applying nanotechnology requires intense research and very high cost, but these costs are expected to decrease in the not too long term due to the advancement of nanomaterials innovations. However, although it is already possible to find several of these textile material products with incorporated nanotechnology, they still represent a small part of the market when compared to traditional textile, so there is still a long way to go before nanotechnology is fully adopted by the textile industry.

Specific spinning processes, such as electrospinning, can be used to obtain nonwovens with new or improved characteristics and with multiple applications. Nanoscale surface treatments, using different physical-chemical processes, can bring great and important advantages in the finishing stage. All of this contributes to the development of multifunctional and high-performance textiles for the most diverse applications.

The flowchart of Fig. 5 shows how much nanotechnology can contribute to the different segments of the textile industry, from the production process of fibers and nonwovens, through finishing (step where these nanomaterials are most applied), to the wastewater treatment.

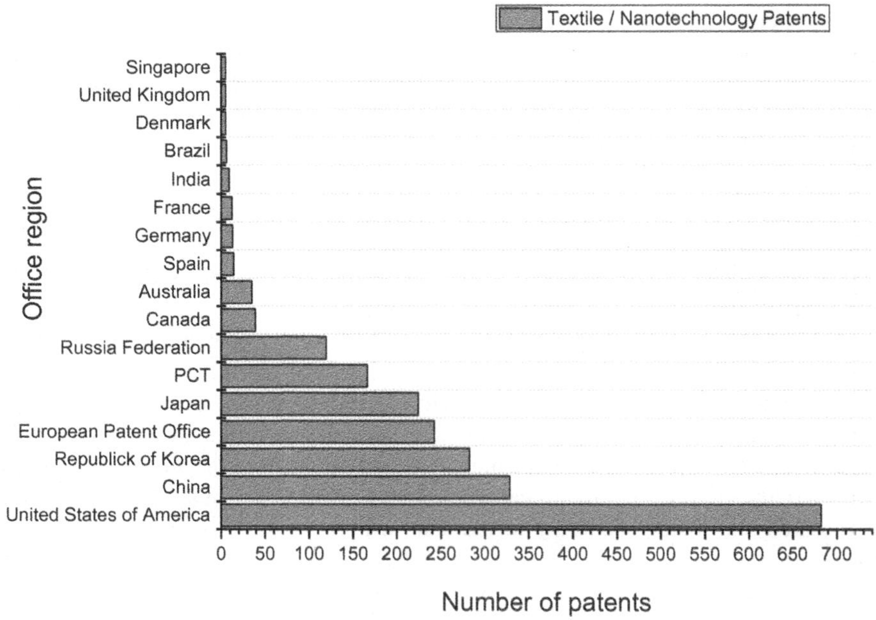

Fig. 4 Graphical data on the number of textile patents by countries registered between 1990 and 2019, based on WIPO (World Intellectual Property Organization)

2 Functional Finishing with Quantum Dots and Others Nanoparticles

Textile materials in the form of fibers, threads and fabrics have been used as an object of consumption since ancient times, when animal skins and fur were used as protective clothing from the weather, such as sun, rain, light, heat, cold. Over the centuries, clothing has become an object of desire for both technical applications such as home textiles and fashion design. With the first industrial revolution of the century, large-scale production of textile products was diversified, transforming textiles into products of high consumption, added value and essential for human life. With noble applications ranging from medical products to flexible roofs for airports, reinforcements in civil construction to ballistic protective clothing. And this diversity became more advanced with the revolution, with the emergence of new polymers, dyes, pigments and making the textile product cheaper and easier to consume for the world population (Rivero et al. 2015).

One of the ways to make the textile material more attractive, adding color, softness and other properties is the surface finish. Finishing on textile materials to promote this superficial modification, visible or not, but mandatorily, adding functionality to textiles has been the object of research since antiquity. In ancient China, functional fabrics were obtained with *Perilla frutescens* seed oil and were used in the process of

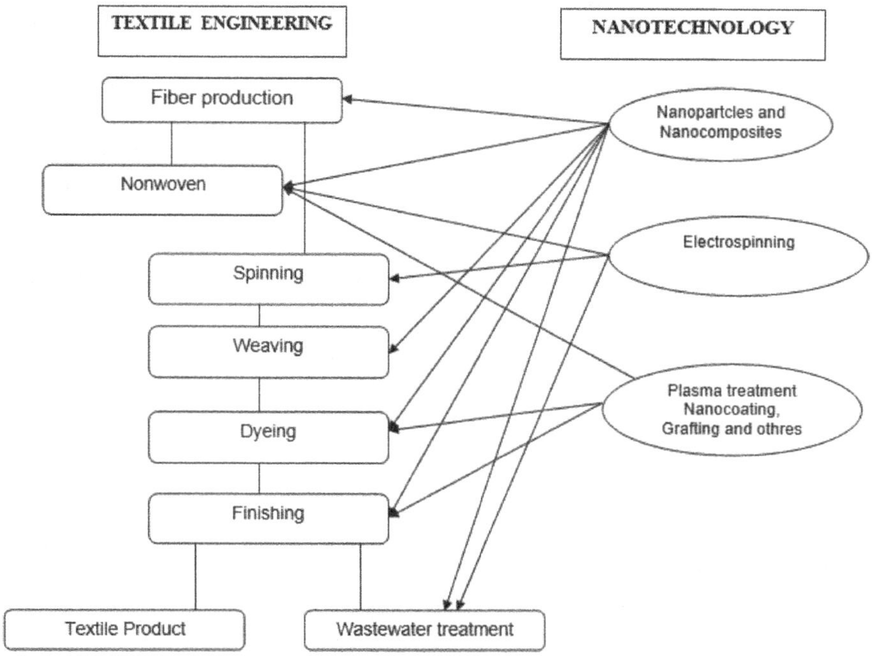

Fig. 5 Flowchart—contribution of nanotechnology in the textile engineering

dyeing, hydrophobization and antimicrobial action of flax, hemp and silk. In ancient Egypt, the linen fabrics used for mummification were finished with beeswax, resin from coniferous plants and castor oil (Tchapla et al. 2004; Haufe et al. 2008). In Central and South America, the finishing of fabrics with the objective of making them repellent and resistant to water was made with natural latex, natural waxes and pitch. In ancient Rome and the Byzantine empire, textiles were coated with metals such as gold and silver for decorative applications. And from 1930, textiles metallized by electrochemical processes began to be patented (Sen 2007; Hegemann et al. 2009). The advance in studies with nanofinishes applied to the modification of surfaces such as glass fibers, ceramic fibers and polymer fibers, with the purpose of adding specific properties, has been one of the reasons for the growth of nanotechnology applied in the textile industry, mainly in this area of functional finishes (Dahotre and Nayak 2005; Cavaleiro and de Hosson 2007; Liu et al. 2008; Gowri et al. 2010).

The application of many of these nanomaterials on textile materials allows them to be functionalized, in order to improve their performance in a wide variety of uses, from technical textiles (geotextiles, medical textiles, electronics and many others) to conventional textiles, giving them new properties, such as antibacterial, fungicidal, self-cleaning, conductive and controlled drug delivery systems (Mukhopadhyay et al. 2005; Gosh 2006; van Lente and van Til 2008). Between the 1950s and the 1970s, many patents related to the coating of ceramic material on fibers for technical applications emerged, but it was only in the early 1990s that the first patents and publications with nanotechnology applications appeared (Fig. 3), in the coating optical fibers, glass fibers, lignocellulosic fibers, wool, nonwovens and paper. Then, the nanofinishes were also applied to photodegradation of effluents (Pozzo et al. 1997; Pelton et al. 2006). According to Reijnders (2008), these materials, in addition to being flexible and inexpensive, can be used to immobilize the nanoparticles, avoiding the problem, in relation to the environment, of the use of these nanomaterials in the effluent due to the difficulty of separation, treatment and recovery after your application.

Several application techniques for these functional coatings have been studied:

- Chemical techniques (wet finish) performed mainly by reaction, depletion and dispersion of chemicals. Examples are: sol-gel, electrodeposition and self-assembly;
- Treatment techniques in the solid–steam–solid (dry finish) phases, where the process is carried out by physical-chemical, physical and mechanical methods, with low environmental impact. Examples are: PVD, CDV, PLD, etc.

The methods specified here for applying the nanoparticles are the process specified here for applying the nanoparticles are wet nanofinish methods. And the first nanoparticles to be applied in the textile sector for functional finishing had their first studies in the late 1990s, and many of them are already commercially available as nanoparticles of silver, copper, gold, palladium and platinum (Yetisen et al. 2016); and zinc oxide (Vigneshwaran et al. 2006), silica (Daoud et al. 2004) and titanium dioxide, carbon nanotubes (CNT's), polyaniline, polypyrrole and Teflon (Avila and Hinestroza 2008; Shim et al. 2008; Liu et al. 2009; Onar et al. 2009). These groups of nanomaterials have very interesting properties such as catalytic photoactivity of organic compounds, electrical conductivity, UV protection, superhydrophilicity or superhydrophobicity.

The vast majority of research on the applicability of these nanomaterials is concentrated on antimicrobial activity, especially silver, copper and gold nanoparticles. In addition to this functionality, metallic nanoparticles exhibit pigment behavior when applied to textile textiles in order to give them color. These have a good applicability, fixation and solidity and can be applied both by the continuous process (padding) and by the discontinuous process (exhaustion), stamping, among other unconventional processes in the industry (Gomes, Sampaio and Maia, no date; Gang and Dapeng 2006; Ahmed and El-Shishtawy 2010).

2.1 Antimicrobial and Colored Nanofinishing with Nanoparticles and Quantum Dots on Textile Materials

In recent years, antimicrobial finishes have become extremely important in the production of protective textile products applied to the most diverse areas such as the clothing, pharmaceutical, medical, agricultural, food and interior design industries. The antimicrobial finish on textiles protects users from pathogenic microorganisms or odor generators, which can cause health problems. In addition, they can protect the textile structure itself from undesirable aesthetic changes caused by natural decay. As a consequence of its importance, the number of different antimicrobial agents suitable for textile application on the market has increased dramatically. These antimicrobial agents differ in their chemical structure, effectiveness, method of application, mechanism of activity, influence on people and the environment, resistance to washing, as well as cost (Padmavathy and Vijayaraghavan 2008).

According to the action, antimicrobial nanofinishes may be associated with mortality or inhibition of microbial growth, according to Fig. 6. Textiles that have the ability to only inhibit microbial activity are also known as biostats, whereas those that have the property of killing pathogenic microorganisms are known as biocides, which could make a difference in antimicrobial functions. The biostatic finishes are suitable for the preservation of textiles, reducing the formation of odors and may not be able to provide biological protection functions with high efficiency. Biocidal nanofinishes are indicated mainly for medical applications, where biological agents must offer biocidal rather than biostatic functions since textiles are expected to completely and quickly eliminate pathogens by contact.

The use of textiles materials with biocidal action should be moderate, as they can cause the complete elimination of microorganisms on human skin, which can damage the human natural protection system (Brayner et al. 2006; Meruvu et al. 2011). In general, the agents most used in antimicrobial treatments in textile materials are metallic nanoparticles (e.g., Ag, Cu and Au) and metallic oxides (e.g., TiO_2,

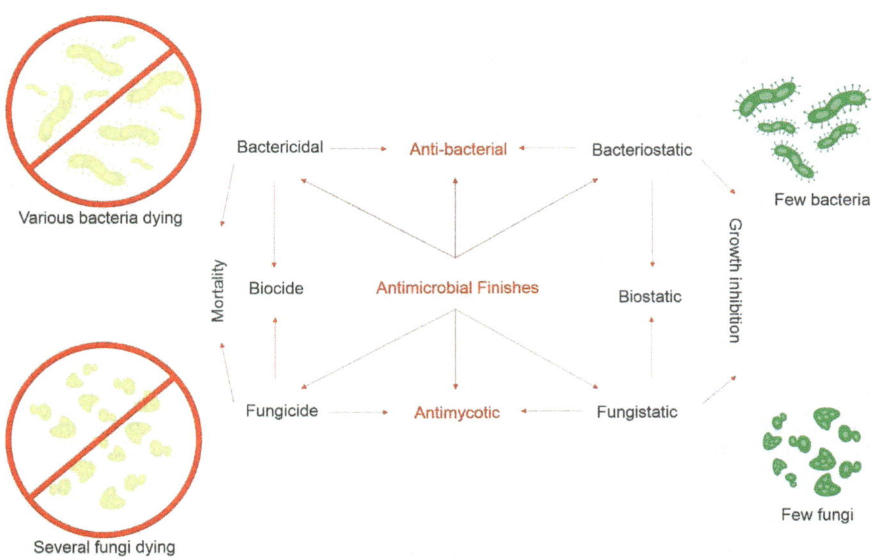

Fig. 6 Definitions of antimicrobial finishes applied to textiles

ZnO and CaO). Some studies point out bactericidal activity (*Escherichia coli* and *Staphylococus aureus*) of graphene quantum dots in the presence of ultraviolet light (Li et al. 2012; Reddy et al. 2014; Arakha et al. 2015).

Metallic nanoparticles have been extensively investigated on textile substrates and some, such as silver, have already been used on an industrial scale as an antimicrobial agent. With the advancement of research on these nanomaterials, new applications are emerging, for example, in the treatment of Toxic Epidermal Necrolysis (TEN) or Lyell's Syndrome, a mucocutaneous disease caused by reactions to medications or by the action of staphylococcal toxin. In addition to having the property of an antimicrobial agent, they are also used to color the textile fabric and according to the size and geometry of these nanoparticles, they can present different colors in the visible region (Asz et al. 2006; Nickoloff 2008; Engel et al. 2009; Sharma and Saravolatz 2009).

Gold nanoparticles are multifunctional, have several properties and applications in nanobiotechnology, both in cancer therapy and in biosensors for its diagnosis (Engel et al. 2009; Sharma and Saravolatz 2009; Silva et al. 2019). The reason for this emerges from the optical and electronic properties of these nanoparticles. Color is caused by a physical phenomenon known as plasma surface resonance. Some electrons in the gold particles are not anchored to individual atoms, but form a gas that floats freely on the surface. The light that falls on these electrons can cause them to oscillate. During this collective oscillation, the surface plasma (plasmons) acquires a certain wavelength, or color, different from the incident light, which also depends on the size of the nanoparticles (Liz-Marzán 2004; Huang et al. 2009; Rai et al. 2009).

When these nanoparticles have dimensions between 2 and 10 nm, they present an intense red color, due to the transition between the electronic bands that accommodate the surface electrons (bands of the surface plasma). However, in larger sizes, these nanoparticles begin to present a color tending to violet, due to the increase in the density of electronic states and, consequently, to the decrease in the transition energy between the bands. Thus, gold nanoparticles with average diameters equal to 9, 15, 22, 48 and 99 nm show maximum absorption at wavelengths 517, 520, 521, 533 and 575 nm, respectively, in aqueous media (Shalaev and Kawata 2006). Johnston et al. (Richardson and Johnston 2007; Johnston et al. 2009) taking advantage of the plasmonic properties of these nanostructures, gold (Au) nanoparticles were synthesized with stability in aqueous solutions, in different sizes and colors, and later dyed wool and cotton fibers with these nanoparticles. Subsequently, wool and cotton fibers were dyed. Other metallic nanoparticles have also been studied for potential application on textile fabrics such as the case of palladium and platinum nanoparticles that were applied by the self-assembly method to cotton fabrics and showed the ability to purify the air, acting as a bactericidal agent (Ashby et al. 2009).

Many other techniques have been applied to textile materials and their behavior has been evaluated and published in several scientific articles. As an example, we have the ultrasonic irradiation nanofinishing technique successfully tested by Abassi et al. (Abbasi and Morsali 2011) with silver nanoparticles on silk and by Perelshtein et al. (2008, 2009b). First, this group made nanocoatings with silver nanoparticles, by sonochemical reaction, on cotton, polyamide and polyester fabrics, and then tested the antibacterial activity of these substrates with *Escherichia coli*, obtaining excellent results (Perelshtein et al. 2008). In later works, using the same technique, nanocoating cotton bandages with zinc oxide nanoparticles (Perelshtein et al. 2009a), copper and MgO/Al_2O_3 (Perelshtein et al. 2009b), obtained almost 100% reduction in the bacteria *E. coli* and *S. aureus*. Due to these results, this group, together with Abramov (Abramov et al. 2009), developed a sonochemistry device in a pilot unit to nanocoated textile fibers with nanoparticles.

Semiconductor nanoparticles have also been the subject of studies in a large number of articles related to the antimicrobial properties of ZnO nanoparticles applied to fabrics (Brayner et al. 2006; Padmavathy and Vijayaraghavan 2008; Ahmed and El-Shishtawy 2010; Meruvu et al. 2011; Arakha et al. 2015), mainly because this semiconductor has structural characteristics that allow the generation of reactive oxygen species through UV radiation, the superoxide anion (O^{2-}) and the hydroxyl radical (HO^{2-}), released by the Zn^{+2} ions, which accumulate in the cell and damages the DNA of the bacteria (Premanathan et al. 2011; Wang et al. 2014; Zhao et al. 2018). Although there is still no complete and broad explanation of the antimicrobial capacity of ZnO nanoparticles, due to the different results that arise from the influence of different parameters of the synthesis conditions, sizes, shapes, defects, doping and zinc precursors (Zhao et al. 2018), the mechanism formation of zinc from reactive oxygen species (ROS) as a bacterial growth inhibitor is widely accepted. When the semiconductor is exposed to UV or visible radiation, the excitonic bonds promoted by quantum leaps are diffused to the outer face of the nanostructure, resulting in a series of redox reactions. These bonds composed of pairs of free electrons (e^-) in the layer of lower electronic density (conduction band) corresponding to a positive orifice (h^+) in the layer of higher electronic density (valence band) allow e^- with oxygen from the medium guarantees the obtaining of H_2O_2, while the h^+ cations ionize the water to generate free radicals OH* can easily rupture the plasma membrane and cause enormous oxidative stress in the cellular components (Seil and Webster 2012).

Therefore, the greater the amount of ZnO nanoparticles dispersed in the medium, the greater the cell inhibition. Dasari et al. (2013) reported that under light conditions there was a greater formation of ROS, raising the toxicity levels of ZnO nanoparticles to *E. coli* cells, unlike to lower levels in the absence of light. Likewise, Raghupathi et al. (2011) with UV light containing ZnO nanoparticles compared to bacteria irradiated without the nanoparticles. In the absence of radiation, it is assumed that the main role of ZnO antimicrobial activity is more focused on its absorption, where Joe et al. (2017) investigated a higher concentration of Zn^{2+} ions in the cytoplasm of *S. Aureus* and *Klebsiella pneumoniae* cells due to their dissolution of the ZnO nanoparticles bound to the walls of the plasma membrane.

Studies related to antimicrobial textile fibers functionalized by ZnO nanoparticles are more extensive in cellulosic fibers, mainly for cotton fabrics (Kamali and Talebian 2018; Tan et al. 2019), where it is the most commercialized textile material in the world. In addition, silk fabrics (Lumbreras-Aguayo et al. 2019), polyester (Verbič et al. 2019), polyurethane/PVA were also used as a functional textile. Table 2 shows

Table 2 Finishing method in textile materials with ZnO nanoparticles

Textile fabric	Nanofinishing method	References
Cotton	Dip coating	Tan et al. (2019)[a], Agrawal et al. (2019)[ab], Ran et al. (2018)[g]
	Atomic layer deposition	Popescu et al. (2019)[ab]
	Ex situ immersion	d'Água et al. (2018)[a,a*, c,d], Hassabo et al. (2019)[ab]
	sonichemical	Lumbreras-Aguayo et al. (2019)[bc], Salat et al. (2018)[ab], Kamali and Talebian (2018)[ab]
	pad-dry-cure	Aslam et al. (2018)[ab], Gao et al. (2019)[ab], Khan et al. (2018)[ab], El-Naggar et al. (2018)[ab]
Polyester	Dip coating	Fiedot-Toboła et al. (2018)[ab], Nourbakhsh et al. (2018)[ab]
	Sonochemical	Karimi and Ansari (2018)[ab]
Polystyrene	LBL	McGuffie et al. (2016)[ac]
Polyurethane	Dip coating	Kim et al. (2018)[b]
Polypropylene	Pulsed laser deposition	Ramamurthy et al. (2017)[ab]
Silk	In situ	Liao et al. (2013)[ab]
Bamboo	Dip coating	Zhang et al. (2013)[ab]
Modal	Dip coating	Nanjappan et al. (2018)[ab]
Viscose	Pad-dry-cure	Salama and El-Sayed (2014)[ab]
Cotton/Polyester	Pad-dry-cure	Farouk et al. (2012)[bm]

S. aureus[a]; *Methicillin Resistant Staphylococcus aureus* (MRSA)[a*]; *E. coli*[b]; *Staphylococcus epidermidis*[c]; *Propionibacterium acnes*[d]; *Gluconobacter cerinus*[g;]; *Micrococcus luteus*[m]

a sample of the development of the most recent work on textile materials modified with ZnO nanoparticles under different nanofinishing methods with biocidal action.

TiO_2 nanoparticles are classified as inorganic nanostructures, in which they stand out for being stable, non-toxic, wide band gap (3.2 eV in the anatase phase), etc. Through these characteristics it is possible to use TiO_2 in several applications, such as self-cleaning, UV protection and antibacterial properties (Dastjerdi and Montazer 2010). In antibacterial applications, this particle is generally used by means of thin films and can be deposited by sol-gel, deposition of pulsed laser, deposition by chemical solution, etc. The mechanism that occurs in this property combined with the photocatalytic effect is based on the fact that when receiving the energy greater than its band gap, electron/hole pairs are formed and through the reactions they form highly reactive species such as h^+ , O^{2-} and OH^-, contributing significantly to the death of microorganism cells.

Kangwansupamonkon et al. (2009) obtained a finish on cotton fabric with titanium dioxide (TiO_2) and apatite applied by the dry curing method. The nanofinished fabric was subjected to antimicrobial tests with four different bacteria: *S. aureus*, MRSA 20627, *E. coli* and *Micrococcus luteus* and for three different lighting situations, irradiation with black light, irradiation with visible light and in the dark. Where the antibacterial effect was greater for black light than for visible light and, finally, in the dark. Confirming the fact that through UV irradiation occurs an activation of TiO_2 showing a better antimicrobial activity.

Regarding the functional finishing of quantum dots on textile materials for antimicrobial application, there are few studies in the literature. Only Sun et al. (2014) obtained an antibacterial system in cotton fabric combining quantum dots of graphene (GQDs) with low dose of H_2O_2, where GQDs were obtained through the synthesis of graphene oxide (GO) from graphite by the hummers method modified. The presence of GQDs can convert H_2O_2 (low antibacterial activity) into $^{\cdot}OH$ radicals (high antibacterial activity), thus improving the antibacterial performance of H_2O_2, which makes it possible to avoid the use of a high concentration of H_2O_2 in the disinfection of wounds. The designed system showed antibacterial properties against Gram-negative (*E. coli*) and Gram-positive (*S. aureus*) bacteria in vitro. In addition, to assess the antibacterial efficacy of the system designed for wound disinfection, GQD-Band-Aids were prepared and showed excellent antibacterial properties in vivo with the aid of low H_2O_2 concentration. The results indicate that GQD-Band-Aids has the potential to disinfect wounds.

2.2 *Photocatalytic and Self-cleaning Nanofinishing with Nanoparticles and Quantum Dots on Textile Materials*

In a world where products must be increasingly suitable to meet high-quality requirements, adapted to different applications and, at the same time, have a minimal environmental impact, self-cleaning textile materials are gaining more and more importance. This is because they are associated with the inhibition of the formation of stains, increasing the durability of the material, less accumulation of organic matter that promotes microorganisms and odors, less need for washing and, consequently, less water consumption (Elmaaty et al. 2018).

The processes involving surface treatment of textile materials for self-cleaning properties are mainly related to the deposition of nanomaterials, which causes a sudden reduction in the surface energy of the fabric, leading to hydrophobicity, with which some liquids tend to form stains (such as coffee and red wine) are in contact with the fabric for a shorter time. In addition, some solid dirt particles present on the fabric surface tend to be carried more easily when the fabric is wet. The change in surface energy can also make the fabric oleophobic, with less tendency to impregnate oils or fats in the fibers. In general, factors such as the chemical structure of the surface, roughness and external factors, such as temperature, are mainly responsible for the level of wettability of the material (Adera et al. 2013).

The semiconductor nanofinish results in self-cleaning fabrics with photocatalytic properties, so that organic dirt deposited on the fabric surface tends to be degraded with greater intensity by means of a cleaning mechanism aided by sunlight or in the presence of artificial illuminants. However, metal oxides can provide other types of functionality to textile materials (Wijesena et al. 2015; Xu et al. 2015; Pavlidou and Paul 2018; Zahid et al. 2018).

Heterogeneous photocatalysis applied to textile materials has also been widely used in the development of self-cleaning fabrics (Pakdel et al. 2013; Ahmad and Kan 2017; Lu et al. 2017; Zhu et al. 2017; Luna et al. 2018; Rilda et al. 2018). Its principle, in contact with a textile surface treated with photocatalytic agents under the presence of light radiation in a specific wavelength range, the organic dirt particles are degraded. Many semiconductors have been used as photocatalysts, such as: TiO_2, ZnO, WO_3, Fe_2O_3, CuO, CdS and NiO. Each is characterized by an energy band gap and specific oxidizing power. TiO_2 has been the most studied and applied due to its high efficiency in the photocatalysis reaction of organic pollutants, chemical stability, non-toxicity and low cost. Some studies indicate an increase in photocatalytic efficiency when associated with TiO_2 to GQDs (Martins et al. 2016; Safardoust-Hojaghan and Salavati-Niasari 2017; Leong et al. 2018; Lim et al. 2018; Ou et al. 2019; Sun et al. 2019).

The principle of heterogeneous photocatalysis can be summed up in the electronic excitation of a semiconductor, where oxidative and reducing sites are generated on its surface. As shown in Fig. 7, a semiconductor is characterized by having in its electronic structure a valence band (BV) and a conduction band (BC), where the difference in energy that separates them is known as "band gap" energy (for example). When the absorption of a photon (hν) of energy greater than or equal to the energy of the "gap" by the semiconductor particle, an electron from the valence band is promoted to the conduction band, generating a gap in the valence band (h^+_{BV}) and excess electrons in the conduction band (e^-_{BC}) (Herrmann 1999).

The main processes that occur in a semiconductor particle when photoexcited are: recombination of electron orifices, reduction of an electron acceptor and oxidation of an electron donor by photogenerated orifices. The electrons and the photogenerated orifices can react with various organic compounds and also with the water molecule that generates radicals such as H^+, OH, $O_2^{\bullet -}$, OH_2, which are also responsible for the degradation of these compounds by oxidation and reduction with CO_2 and inorganic compounds (CARP 2004). In general, the efficiency of the processes involving heterogeneous photocatalysis is mainly influenced by factors such as: the species, distribution, concentration, doping and crystalline form of the semiconductor (anatase, rutile, brookite or a combination); the type and concentration of the organic contaminant; presence and concentration of auxiliary oxidants, such as H_2O_2, O_2 and O_3; form of use of the catalyst, which can be immobilized, supported or suspended; characteristics of the incident light source such as spectrum range and intensity; operation system, geometry and hydrodynamic parameters of the reactor; temperature, pH and presence of anions (Mills and Hunte 1997).

A TiO_2 semiconductor is activated by a mechanism involving reactive radical species, according to Eqs. 1 to 8 (Konstantinou and Albanis 2004):

$$TiO_2 + hv(UV) \rightarrow TiO_2\left(e^-_{CB} + h^+_{VB}\right) \quad (1)$$

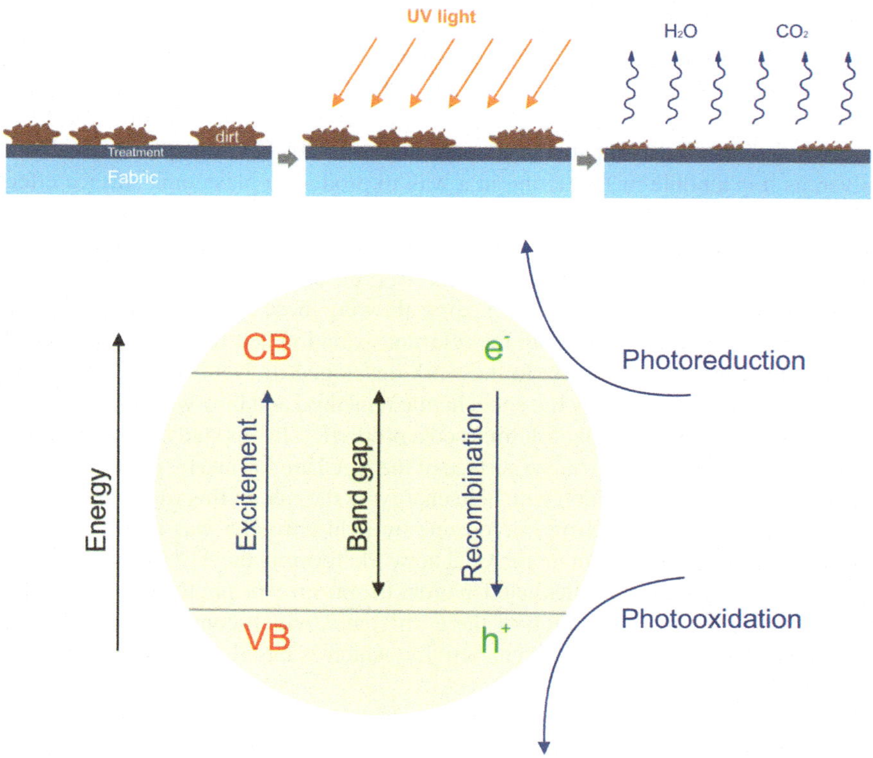

Fig. 7 Scheme of self-cleaning process on fabric by heterogeneous photocatalysis

$$TiO_2(h^+_{VB}) + H_2O \rightarrow TiO_2 + H^+ + OH^{\cdot} \quad (2)$$

$$TiO_2\left(h^+_{VB}\right) + OH^- \rightarrow TiO_2 + OH^{\cdot} \quad (3)$$

$$TiO_2\left(e^-_{CB}\right) + O_2 \rightarrow TiO_2 + O_2^{\cdot -} \quad (4)$$

$$O_2^{\cdot -} + H^+ \rightarrow HO_2^{\cdot} \quad (5)$$

$$\text{Organic compost} + OH^{\cdot} \rightarrow \text{reduced products} \quad (6)$$

$$\text{Organic compost} + h^+_{VB} \rightarrow \text{reduced products} \quad (7)$$

$$\text{Organic compost} + e^-_{CB} \rightarrow \text{reduced products} \quad (8)$$

The equations shown above describe the generation mechanisms of the main reactive oxygen species that act in the heterogeneous photocatalytic process governed by TiO_2, which is the most studied semiconductor material for such applications; however, they are similar for almost all semiconductor materials. In addition to TiO_2, silver-based materials gain attention in the study of photocatalytic properties due to silver, as it is a noble metal, acting in a way to produce a plasmonic surface effect (EPS). EPS occurs when a nanoparticulate noble metal is deposited on the surface of particles of semiconductor materials, where it will act in order to absorb a greater spectrum of incident radiation, in addition to preventing the recombination of the photogenerated electron/hole (e^-/h^+) pairs, allowing these charges to migrate to the surface of the material, producing the relevant oxidation and reduction reactions to degrade the organic compound. In the work developed by Neto et al. (2017) AgCl particles were synthesized by the sonochemical method and they were characterized by their photocatalytic and photoluminescent properties. In this study, it was observed that the silver nanoparticles on the surface of the AgCl are formed by photoreduction, where when focusing radiation with high energy (in the case of this work, the electron beam of the scanning electron microscope by field emission was used) there is a reduction in the silver present in the AgCl network, forming the Ag^0 nanoparticles on the surface. These nanoparticles acted in order to optimize the photocatalytic activity, by preventing the recombination of the e^-/h^+ pairs, which consequently provided a reduction in the photoluminescent activity, which is based on the recombination speed of these pairs.

The use of semiconductors to optimize the degradation process of organic contaminants present in the medium must also take into account, in addition to the intrinsic photocatalytic capacity of the material, properties such as the possibility of this material being used in consecutive cycles, without the need for specific treatments and the way of removing this material from the medium, so that the formation of a secondary solid residue does not occur. Thus, there needs to be a combination between the photocatalytic property, the maintenance of this activity during the performance of several photocatalytic cycles and the ease of removing this material from the medium so that it can be considered an ideal material for this application (Neto et al. 2019). As an example, one can mention the difference between the photocatalytic properties of loose particulate materials, immobilized particulates and thin films.

Particulate materials stand out in the photocatalytic activity due to their high surface area, compared to the others. The larger surface area allows more active sites to react with the organic molecules present in the medium, thus increasing the catalytic efficiency (Neto et al. 2018). On the other hand, particulate materials, mainly nanoparticles, make reuse and separation after the process difficult, providing even in low quantities, the generation of a secondary waste, making it impossible to use it on a large scale. The use of materials in the form of thin films eliminates the problems of material loss after photocatalytic processes; however, they have a very low surface area, in addition to having only an active plan to act as a catalyst, requiring much longer times for degradation of organic compounds (Garcia et al. 2018).

Thus, the functionalization of fabrics with nanoparticles appears as a better alternative for photocatalytic applications. The fabrics used normally have an organic chain, which is not harmful to the environment, in addition to contributing to the surface area of the system, being flexible and joining the three basic needs for photocatalytic processes. Guan et al. (2019) functionalized cotton fabric with AgCl/Ag, TiO_2 and ZIF particles through hydrothermal treatment in an autoclave and studied the effect of functionalization on photocatalytic activity under visible light. Through the micrographs presented in this study, functionalization is evident, where the particles are well dispersed in the fibers of the cotton fabric, allowing a high surface area, aggregating the fixation of the particles, allowing their total removal after the process. In addition, photocatalytic tests confirmed the maintenance of photocatalytic activity over the course of the cycles, with only a reduction in discoloration being observed due to saturation of the adsorptive effects from the fabric, as is already expected.

Hatamie et al. (2015) produced ZnO nanorods in situ in polyamide/silver fabric (conductor), showing that the catalyst-nanocoated fabric was necessary for the most efficient photodegradation under UV irradiation in methylene blue and Congo red dyes than in the absence of radiation. Lee et al. (2014) developed a cotton fabric with photocatalytic properties to degrade liquid and gaseous pollutants. The cotton was coated with CNTs/TiO_2 and acrylate copolymer (AC) using dip-pad-cure technique. In general, cotton fabrics with a porous structure favored the absorption of contaminants (methylene blue dye), promoted sufficient contact between pollutants and catalysts and prolonged the contact time, which is beneficial for the photocatalytic process. As a result, the coating samples with excellent photocatalytic capacity are mainly attributed to the intense absorption capacity of the fabrics combined with the efficient photocatalytic property of the TiO_2/AC—CNTs nanofinishing.

Nanofinishes with a self-cleaning properties are also desirable and can be used in military clothing, medical textiles, sports textiles and for external use. This property is accomplished through hydrophobicity, where water droplets roll through the fabric, or hydrophilicity where water penetrates the fabric leading to its dirt. Associated with either of these two approaches is the use of semiconductor materials, which have photocatalytic properties and thus it is possible to chemically degrade the dirt that was on the fabric. Among the main semiconductor materials used are TiO_2, ZnO, WO_3, Fe_2O_3, among others (Saad et al. 2016).

In view of the applications described in the works above, there are researches that describe that, varying the size of the semiconductor particles, different properties can be obtained, because besides the change in the surface area, there is also the change in the behavior of the particle, mainly in optical matters, biological and electronic. Among the categories used in nanostructures, the one that has shown promise is that of quantum dots, considered to be zero dimension when related to their volume, in which they are a set of particles, also called semiconductor nanocrystals, where they have a controlled size on a scale of 1 at 10 nm and which, due to this small size, undergoes quantum confinement (Movilla et al. 2005). Its property is influenced by several factors that will form the intrinsic properties of this material, such as size, crystallinity, composition, defects, impurities, among others. Regarding the size factor, it is a relevant property in quantum dots, it will influence its quantum confinement, which can be explained as the increase in the gap when the particle size is below a specific size, which will depend on the type of semiconductor (Bera et al. 2010).

The new properties exhibited by quantum dots, such as photoluminescence and electroluminescence, are interesting and occur in particles in general, due to this small dimension. One of the main advantages of using quantum dots with this property is its high stability to external environments, as well as, its prolonged useful life. This property is commonly used for solar cells, biosensors, bioimages, which are still being investigated and need optimization. This property can be combined with textiles to replace, for example, organic dyes, or be incorporated into fibers, obtaining hybrid materials for other applications, such as in the fashion industry for the authentication of textile products (Zeller and Johnston, no date).

2.3 *UV Protection Nanofinishing with Nanoparticles and Quantum Dots on Textile Materials*

Prolonged exposure to ultraviolet (UV) radiation from the sun can result in a number of skin problems, such as accelerating aging, burns, blemishes or even cancer. In this context, the use of textile products with anti-UV properties has become a practical and efficient form of personal protection (Dubrovski 2010). The solar radiation that passes through the earth's atmosphere has a spectrum in the range of 290–3000 nm, as can be seen in Fig. 8. The region corresponding to the wavelengths between 290

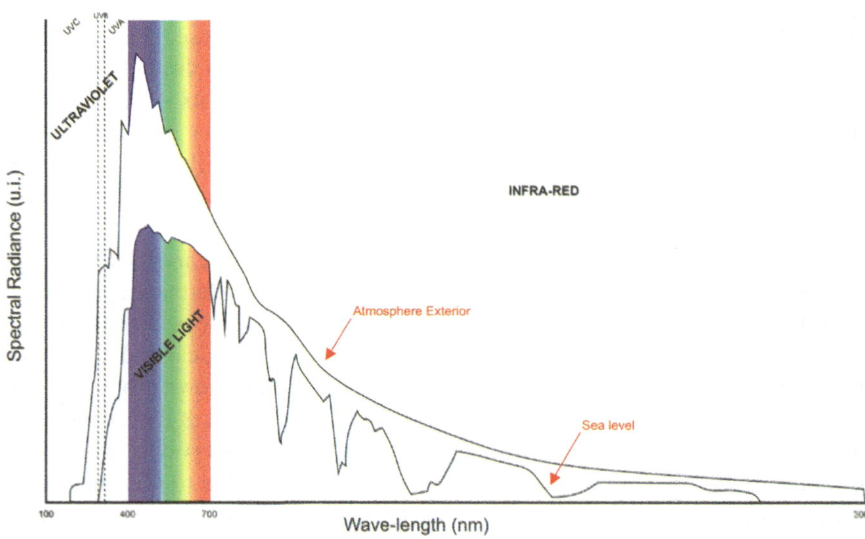

Fig. 8 Sun radiance spectrum

Table 3 Blocked UV index for UPF

Category	Factor	Blocked UV index (%)
Excellent	40–50/50+	97.5>
Very good	25–39	95.9 at 97.4
Good	15–24	93.3 at 95.8
Low	0–14	<93.3

and 400 nm refers to the range of ultraviolet radiation. The ultraviolet radiation from the sun can be divided from the point of view of biologists into: UVA (400–320 nm), UVB (320–290 nm), UVC (290–190 nm). UVA radiation causes little visible reaction on the skin, but has been shown to decrease the immune response of epithelial cells. UVB is primarily responsible for the development of skin cancer. UVC, although it can cause serious damage to human health, is completely absorbed by the earth's atmosphere (Hatch and Osterwalder 2006).

The ultraviolet protection factor (UPF) is the scientific term used to quantify the level of protection against UV radiation that a fabric can provide to human skin. While the sun protection factor (SPF) is similar to UPF, with the difference that it quantifies the level of UV protection of sunscreens. Thus, UPF tests are performed in vitro, while FPS tests are performed in vivo. Theoretically, the values of equal UPF and SPF should indicate the same level of sun protection under identical environmental conditions. However, some studies have shown that the results of the values of both indicators when compared do not have a high statistical correlation (Alebeid and Zhao 2017). Table 3 shows the categories of UPF based on the AS/NZS 4399: 1996 standard and the respective percentage of UV radiation that is blocked by the textile material.

The ultraviolet protection factor (UPF) is the scientific term used to quantify the level of protection against UV radiation that a fabric can provide to human skin. While the sun protection factor (SPF) is similar to UPF, with the difference that it quantifies the level of UV protection of sunscreens. Thus, UPF tests are performed in vitro, while FPS tests are performed in vivo. Theoretically, the values of equal UPF and SPF should indicate the same level of sun protection under identical environmental conditions. However, some studies have shown that the results of the values of both indicators when compared do not have a high statistical correlation (Alebeid and Zhao 2017). Table 3 shows the categories of UPF based on the AS/NZS 4399: 1996 standard and the respective percentage of UV radiation that is blocked by the textile material.

The value of UPF in a fabric is associated with its ability to absorb and/or reflect ultraviolet radiation to the outside, preventing light from passing through the material and reaching human skin, as shown in Fig. 9. Even some textiles that visibly appear not to transmit light can transmit significant amounts of erythema-inducing UV irradiation. The UPF of a textile depends on the synergy of all the parameters selected for its manufacture and the conditions of use. Thus, the factors that most influence the UV protection of a fabric are (Dubrovski 2010; Alebeid and Zhao 2017):

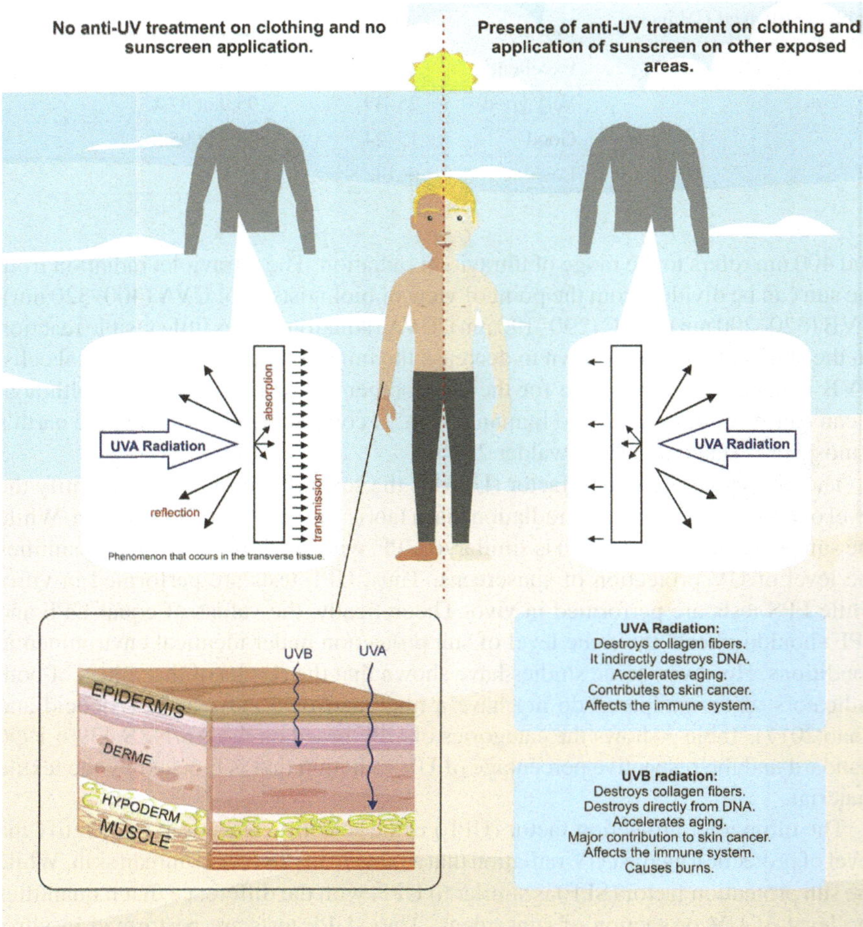

Fig. 9 Scheme of the UV protection principle in textile material

- The type of fiber: Natural fibers such as cotton usually have a lower UPF than synthetic fibers.
- Mixing fibers: Mixing cotton with polyester, for example, provides greater protection than pure cotton.
- Chemical processes: Cotton bleaching is responsible for a high reduction in UPF. On the other hand, a later application of optical brightener can increase protection.
- The structure: Satin, for example, tends to form less porosity than a twill or canvas, as a result, it presents greater resistance to the passage of UV radiation.
- The coverage factor.
- The color: Usually dyed fabrics in dark tones have a higher UPF than when dyed in light tones.
- The type of dye: Photoluminescent dyes can increase UPF of the fabric.

- Finishes: Some that are associated with products that absorb UV radiation can also significantly increase the protection of the material.
- Fabric moisture: The increase reduces the protection of the fabric.
- Wear: The wash resistance varies a lot depending on the application technique and the products used.

The textile industry normally uses AS/NZS 4399: 1996—Sun protection clothing—Evaluation and classification, AATCC Test Method 183-2004, European standard EN 13758-1: 2002 or EN 13758-2: 2003 as parameters for determining the UPF of a fabric. They differ in terms of scanning intervals, fabric positioning on the instrument, designation, classification and marking of the erythematous action spectrum. However, for all UPF it is calculated according to Eq. (9) (Dubrovski 2010):

$$\mathrm{UPF} = \frac{\sum_{290}^{400} E(\lambda) \cdot S(\lambda)}{\sum_{290}^{400} E(\lambda) \cdot S(\lambda) \cdot T(\lambda)} \times 100 \tag{9}$$

where $E(\lambda)$ is the CIE reference erythema dose spectrum, $S(\lambda)$ is the radiation intensity distribution of sunlight, and $T(\lambda)$ is the diffuse transmittance spectrum (%).

To calculate the UPF rating, the transmittance spectrum at more than four different wavelengths is measured for the sample

$$\begin{aligned} \mathrm{UPE\ rating} &= \mathrm{UPF}_{\mathrm{ave}} - E \\ \mathrm{UPF}_{\mathrm{ave}} &= \frac{\mathrm{UPF}_1 + \mathrm{UPF}_2 + \cdots + \mathrm{UPF}_n}{n} \\ E &= {}^{t_{k,a}}/\sqrt{n} \cdot \mathrm{SD} \\ \mathrm{SD} &= \sqrt{\frac{\sum_{i=1}^{n} (\mathrm{UPF} - \mathrm{UPF}_{\mathrm{ave}})^2}{n-1}} \end{aligned} \tag{10}$$

where $t_{k,a}$ is the value which provides 0.5% of the border value of probability of one side in the t distribution, a is the probability of one side (0.005), and k is the degree of freedom $(n - 1)$. If the UPF rating is smaller than the minimum of each UPF, the calculated value is rounded down by 5. If the UPF rating is more than 50, the UPF rating is defined at 50+.

UVA transmittance is calculated by the equation below using the average transmittance from 315 to 410 nm:

$$\mathrm{UVA} = \frac{T_{315} + T_{320} + T_{325} + \cdots + T_{400}}{18} \tag{11}$$

UVB transmittance is calculated by the equation below using the average transmittance from 290 to 315 nm:

$$UVA = \frac{T_{290} + T_{295} + T_{300} + \cdots + T_{315}}{6} \quad (12)$$

Anti-UV agents are organic or inorganic compounds characterized by strong absorption in the UV region of the electromagnetic spectrum. These compounds, when incorporated into textile fibers, are electronically excited by ultraviolet radiation and can convert it into thermal energy and/or radiation of greater wavelength, to function as radicais scavengers and singlet oxygen suppressors. Alternatively, isomerization can occur and the UV absorber can then fragment into non-absorbent isomers (Holme 2003). Anti-UV treatments on fabrics can be carried out using methods such as dry-pad-cure (El-Hady et al. 2013), sol-gel (Pakdel et al. 2017), hydrothermal (Pandiyarasan et al. 2017), printing (Pakdel et al. 2017), dip-coating (Ergindemir et al. 2016), layer-by-layer (Yang et al. 2019) and, etc.

An efficient anti-UV agent must cover the entire UVA/UVB spectrum and be stable under the conditions in which they will be applied. In some cases, more than one compound is used for better use of the spectrum. The main organic anti-UV agents are derived from O-hydroxybenzophenones, O-hydroxyphenothiazines and O-hydroxyphenylhydrazines (Hatch and Osterwalder 2006), while inorganic ones are basically titanium dioxide and zinc oxide nanoparticles. In addition, recent studies demonstrate the great potential for applying graphene-derived nanoparticles as blocking agents for ultraviolet radiation in fibrous materials (Holme 2003; Alebeid and Zhao 2017; Pandiyarasan et al. 2017) [147–149].

The high UV protection capacity of ZnO nanoparticles is linked to the optical property of absorbing UV radiation, mainly in the spectral range that comprises UV-A. Its mechanism is based on its high refractive index, allowing a good part of the UV rays that reach the fabric surface to be reflected/dispersed to the medium and another portion to be absorbed, preventing radiation from being transmitted through the fabric structure until the skin, therefore, the high index of refraction causes white dispersion in the powder of particle size <50 nm. Factors such as particle size and shape, concentration, synthesis method, nanofinishing technique have great influence on UV protection, as it is verified that the smaller the size of the ZnO nanoparticles, the specific superficial area is high, which fills a large part of the surface of the improving the UV protection factor. Fabric nanocoated with ZnO nanoparticles to improve anti-UV activity have been achieved using different deposition techniques. Li et al. (2014) synthesized quantum dots of ZnO by the sol-gel method. From these quantum dots, ZnO nanobonds were obtained on the cotton fabric, through the following steps: immersion in the ZnO quantum dot solution; dissolution-recrystallization process; hydrothermal growth process of nanobonds. The fabric coated with the ZnO nanobars showed excellent anti-UV properties with an UPF value of 118.12.

Carneiro et al. (2011) performed a TiO_2 nanocoating on PLA (lactic polyacid) fabric using the DC magnetron sputtering method, where they used an asymmetric bipolar pulsation, with titanium target 60 mm away from the substrate, at room temperature, with a reverse phase 40%. During deposition, argon (Ar) and oxygen (O_2) were used under controlled pressure and current. The fabric with the nanocoating was classified with excellent protection with UPF of 88.8 and even after 10 washing cycles of the treated fabric, it continued with this same classification, decreasing its UFP value to 81.3. Sójka-Ledakowicz et al. (2009) functionalized TiO_2 nanoparticles with aminosiloxane and applied them on cotton and non-PES fabrics. They obtained excellent UV protection results for the cotton fabric and for the PES non-woven fabric they achieved a uniform TiO_2 nanocoating also with good UV protection.

Researches using quantum dots for this purpose, Zuo (2019) applied a finish on cotton fabrics with carbon quantum dots co-doped with boron and nitrogen (BN-CDs) which were prepared using a hydrothermal carbonization method. The particles were immobilized on fabric by a dip-coating technique that involved immersing the fabric in a solution of BN-CDs and PVA, followed by drying at 100 °C. The treated fabric showed UPF values considered very good even after washing, with 38.6 being the initial value, which dropped to 30.6 after 20 washes.

2.4 Luminescents Nanofinishing with Nanoparticles and Quantum Dots on Textile Materials

Luminescence is the phenomenon of radioactive emission of ultraviolet, visible or infrared photons of a species that was electronically excited and then returned to its fundamentals state. Luminescence can be classified according to the response time (τ^c) after the excitation action:

- Fluorescence: $\tau c < 10^{-8}$ s
- Phosphorescence $\tau c > 10^{-8}$ s

That is, fluorescence is a phenomenon that occurs almost instantly, while phosphorescence has a more considerable latency and with notable temperature dependence (Barton and Davidson 2008).

Luminescence can also be classified according to the mechanism that causes electronic excitation such as chemical reactions (chemiluminescence), heating (thermoluminescence), electric current (electroluminescence) or absorption of visible or UV radiation (photoluminescence). The latter being the most used by the textile industry as it is associated with applications involving luminescence of fabrics exposed to UV radiation, white enhancers and UV filter.

Optical bleaching agents are often used to improve the appearance of fabric color, promoting a "purer" white appearance. They are a group of molecules that absorb electromagnetic radiation predominantly in the ultraviolet band and re-emit in a lower energy band (usually light in the blue-violet band), thus causing a lightening of the substrate by emitting more than 100% of the visible light incident on the surface, in addition to the possibility of neutralizing the yellowish color of the textile substrate (Barton and Davidson 2008). The anti-UV finishes on textiles, on the other hand, are associated with the application of products with properties to absorb the ultraviolet region of radiation that affects the fabric, such as some dyes, pigments or oxides. Optical brighteners commonly used in textile processes can also add a high UV

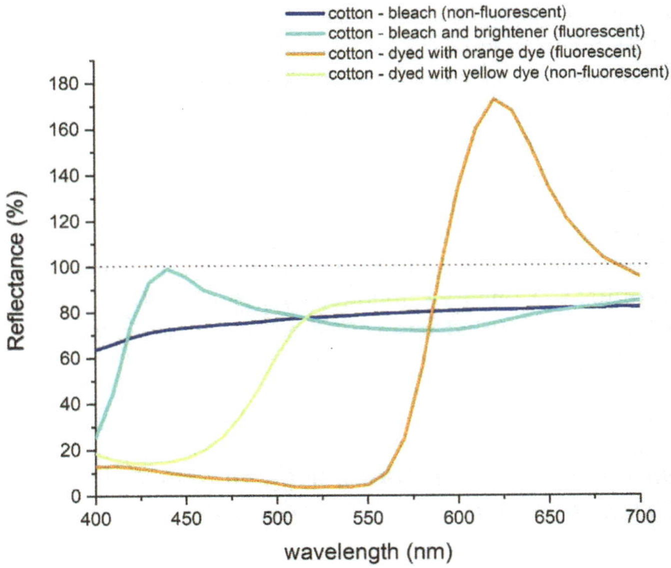

Fig. 10 Comparison between the typical reflectance behavior of cotton fabric under different treatment conditions

protection factor to the fabric under certain conditions, as they start from a similar mechanism of action (Algaba et al. 2007).

Fluorescent dyes, on the other hand, are defined as compounds that absorb UV or visible light and strongly re-emit in a longer wavelength region of the visible electromagnetic spectrum, and that owe their potential application to their properties of intense fluorescence (Clark 2011). It can be seen in Fig. 10 a comparison of a typical spectrum of a cotton fabric with and without bleaching agent, dyed with fluorescent dye and dyed with dye without fluorescent behavior.

The general principles of luminescence involve mechanisms of excitation and emission. Usually, the release of photons (quantum of electromagnetic radiation) and phonons (quasiparticle that designates a quantum of vibration in elastic structures of interacting particles), as shown in Fig. 11. Where E0 represents the ground state; while E1 and E5 are the excited states. After excitation, the electron that was initially at the lower energy level E0 passes to level E5. It is observed that the energy intervals between the adjacent levels of E2 to E5 are small, while the interval between E1 and E2 is large. If the interval between an excited level and the nearest adjacent one is small, the excited material tends to show a non-radioactive decay due to the phonon emission, releasing energy in the form of heat. Otherwise, that is, if the electron passes from the E5 energy level directly to E1 or E0, there is the effect of radioactive decay and, with that, the photon release. It is a fact that electromagnetic radiation, resulting from a radioactive decay from a higher electronic level to the ground state, only occurs when the interval to the lowest adjacent level is above a

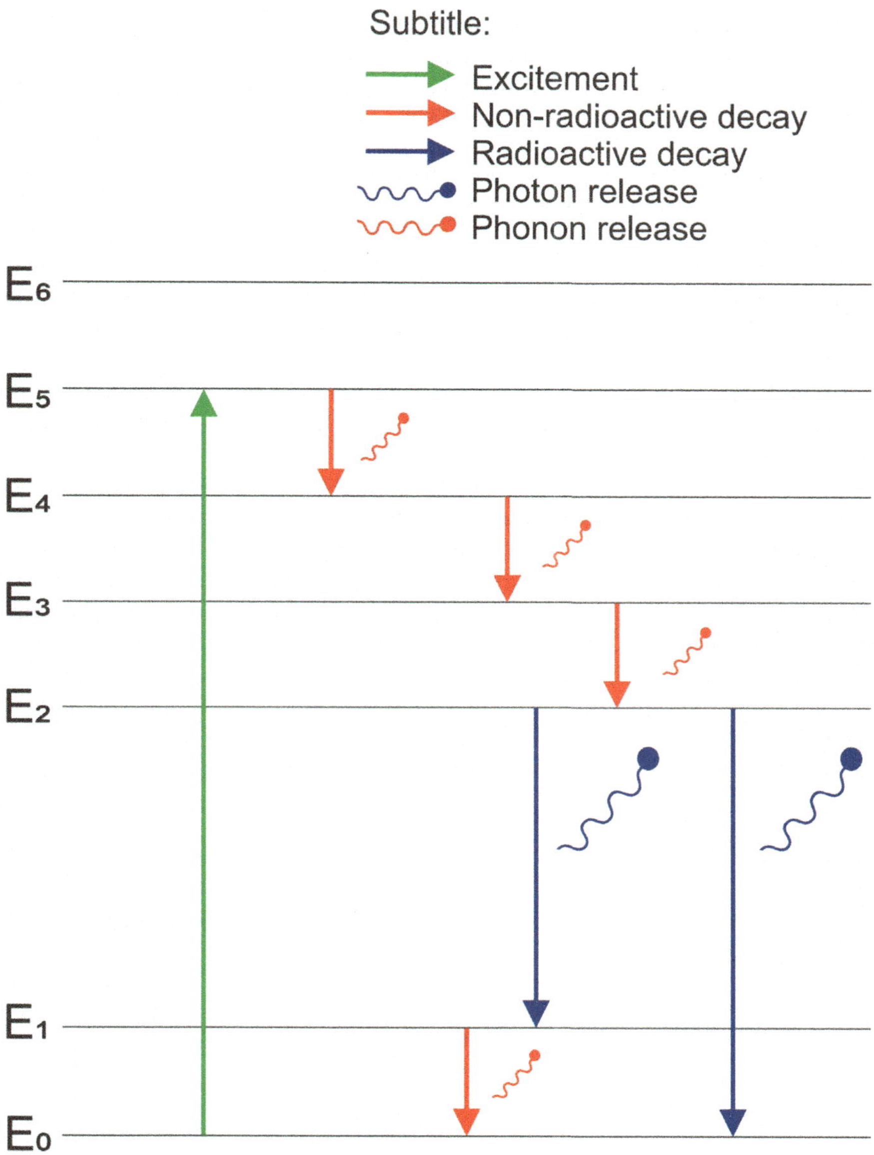

Fig. 11 Excitation and emission processes for a luminescent material (down-conversion)

critical value (Schawabl 2008). The excited molecule can be of an organic, inorganic or organometallic nature.

The processes for converting energy into luminescent materials can be classified into:

- Down-conversion: When the emission energy is lower than the energy that caused the excitation, the Stokes effect occurs.
- Up-conversion: When the emission energy is greater than the energy that caused the excitation, the anti-Stokes effect occurs.

The photoluminescent down-conversion effect can be considered as a transition from the lower molecular orbital (LUMO) to the higher molecular orbital (HOMO). This is the most common effect in studies involving QDs (quantum dots). In an up-conversion process a group of low-energy photons excites the electrons in the π orbital, the π electrons change to a high-energy state, such as LUMO, and then the electrons move to a state of less than the initial one, according to Fig. 12b. Thus, the emission of energy in the form of photons and phonons occurs when the electrons make the transition back to the σ orbital. Although the electrons of the σ orbital can also be excited, the energy of the emitted photons will be less than the excitation energy (Shen et al. 2011).

The luminescent effects of QDs are also strongly associated with the so-called quantum confinement effect, which occurs when the particle size is around dimensions less than 10 nm. The small region confines the movement of electrons, making the energy levels discrete and this increases or expands up to the band interval and, consequently, the band gap energy (Eda et al. 2010; Tuerhong et al. 2017). The luminescent properties of QDs have been widely investigated for applications such as sensors, biomarkers and filters for lamps; however, few are aimed at the textile area.

The substitution of dyes or organic pigments with QDs based, for example, carbon and ZnO can bring different benefits, such as chemical inertness, low cytotoxicity and high biocompatibility (Sun et al. 2012; Vaseem et al. 2012).

Fig. 12 Scheme of luminescent down-conversion and up-conversion processes

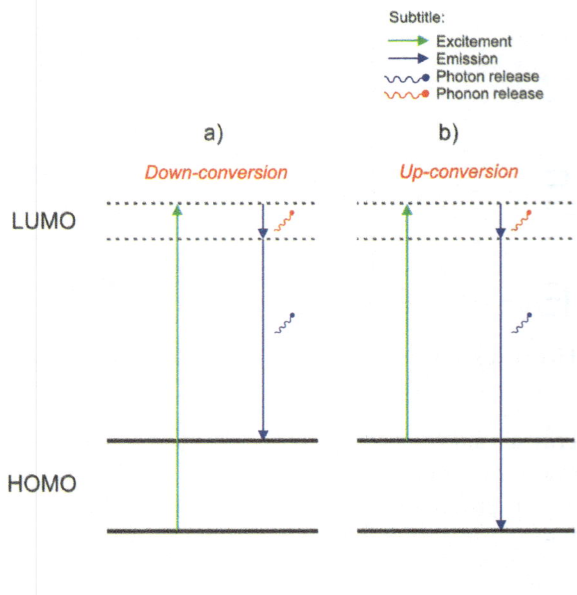

Yu et al. (2017) performed a synthesis of carbon QDs within the mesopores and the lumen of cotton fibers using a hydrothermal method using citric acid and urea as the main precursors. Briefly, the cotton fibers were mixed with a solution with the precursors and reaction aids, the mixture was placed in an autoclave for 3 h at 150 °C, then washed and dried. The treated fibers started to have fluorescence dependent on the excitation wavelength. That is, under the excitation of green, UV and blue light, the fibers emitted the colors red, blue and green, respectively. Unlike conventional techniques for immobilizing nanoparticles in textile fibers, in this case the immobilization of quantum carbon dots occurred simultaneously with the synthesis process.

Kuo et al. (2017) obtained an easy and scalable route to synthesize a high-quality fluorescent carbon (CD), through the reaction of citric acid (CA) with ethylenediamine (EDA) at 150 °C under atmospheric pressure, thus obtaining carbon points fluorescent (FCDs). The FCDs were successfully mixed with commercial inks and maintained their fluorescence in the solid state, showing that the FCDs are promising for use in applications involving wearable optoelectronics.

Pang et al. (2018) used N-GQDs (quantum dots of nitrogen doped graphene) to prepare a quick-drying fluorescent paint for anti-counterfeiting applications. The technique was summarized in the simple mixture of N-GQDs with glycerol in ethanol medium. The ink showed a high yield of quantum fluorescence, satisfactory photostability and good solubility in ethanol. When a substrate has the paint applied to its surface, a photoluminescent emission of bright blue light occurs in the paint area when exposed to ultraviolet light and this effect is responsible for proving that the material is not counterfeit, but an authentic material containing dots ink. quantum. associated with the ink.

When studying materials with high photoluminescent properties, one finds the use of lanthanide materials (rare earths) acting as dopants in the semiconductor material network. The rare earth compounds are characterized by the transitions $4f$-$4f$ or $5d$-$4f$, responsible for the electromagnetic emission in the ultraviolet regions until the region close to the infrared (Pinatti et al. 2019).

Such compounds are used as doping materials because, in addition to their high cost, they allow to obtain considerable properties in very low amounts in the network, generally in amounts less than 1 mol%, in relation to the main network cation. Rare earths are also characterized by their emission in specific regions of the spectrum, with very well-defined emission ranges. For example, praseodymium and europium normally emit in the red region, as samarium emits in the orange region.

The terbium, in turn, emits shorter wavelengths, in the region of green and thulium, in blue. Thus, because they have well-defined emission bands, encode, mix the quantities and the type of rare earths used, it allows to obtain emissions in any color of the visible spectrum, as well as in white, a very desired property for the manufacture of diodes and LEDs (Zhai et al. 2016). Erdman et al. (2016) studied the functionalization of cellulose fibers using nanoparticles of Sr_2CeO_4, CeF_3 doped with terbium and $Gd_4O_3F_6$ doped with europium in the photoluminescent properties. The photoluminescent results indicated that, by focusing the ultraviolet radiation on the functionalized cellulose fibers, they emit the specific colors of the nanoparticles used, being blue, green and red, respectively, proving the efficiency of rare earths in modifying the emission color of the cellulose fibers.

2.5 *Flame Retardants Nanofinishing with Nanoparticles and Quantum Dots on Textile Materials*

Serious accidents involving fire have been frequently reported all over the world since the beginning of the manipulation of this element of nature by mankind to the present day, often causing irreparable damage to material goods, human beings and nature. Materials with flammable characteristics are primarily responsible for the rapid spread of flames in fires and, consequently, can potentiate such disasters. Textiles made from natural (e.g., cotton, linen and silk) or synthetic fibers (e.g., polyesters, polyamides, acetates) are widely used for their excellent fiber-forming properties, but they share a common problem, i.e., most of them spread fire easily. Therefore, flame retardant treatments on textiles are a highly efficient way to prevent or minimize disasters involving fire.

Flame retardant fabrics are applied in several areas, such as in the manufacture of uniforms (for firefighters, military personnel, police and industrial workers); high-performance sports applications; transport (e.g., seats and covers for cars and airplanes); decoration (e.g., carpets, curtains); home textiles (mattress covers, table cloths); Pajamas for children and the elderly; tents, etc. Flame retardants are generally organic and inorganic compounds based on phosphorus (P), nitrogen (N), sulfur (S), bromine (Br), chlorine (Cl), fluorine (F), silicon (Si), aluminum (Al), magnesium (Mg), antimony (Sb), tin (Sn), boron (B), zinc (Zn), carbon (C), zirconium (Zr), titanium (Ti) and calcium (Ca). After exposure to fire, these compounds become active in the condensed or gaseous phase and can form a residual layer on the surface of the fabric responsible for inhibiting contact between the oxygen-rich atmosphere and the interior of the fabric; sometimes they can release inert or reactive gases with oxygen from the air, reducing the presence of active oxygen on the surface of the fabric exposed to fire. Today's focus is on developing finishes that do not harm the environment or human health; therefore, halogen-based compounds are being replaced by alternatives such as phosphorus or nitrogen-based compounds, highly effective, economical and advantageous. The phosphorus contained is known for the formation of coal residues, which act as a barrier to extinguish the fire. In recent studies, graphene-based compounds are also being used for applications as a flame retardant. In general, the large surface area of graphene-based compounds can function as a molecular barrier.

Khose et al. (2018) developed a method to synthesize a water-dispersible flame retardant, effective, based on GQDs and that helps to keep the color of the fabric intact. The flame retardant of the quantum dots of phosphorus-functionalized graphene (P-GQD) was synthesized using sources of graphene oxide and phosphorus (monosodium phosphate and polyphosphoric acid) through a hydrothermal treatment and then applied to the cotton fabric. It was observed in one of the experiments carried out on the developed material that the treated fabric, when in contact with the flame, emitted little smoke and the color was changed to black, but afterward it did not catch fire for more than 5 min and maintained its primary shape with a slight tilt, while the untreated fabric caught fire in 5 s and was completely burned in 15 s, producing a thin black mass. Due to the characteristics of the materials used in the treatment, the material did not have a harmful effect on mankind or the environment and, therefore, may be applicable to daily use.

Another alternative that helps to reduce the spread of flame fabric is the deposition of ZnO nanoparticles. It is proposed that, due to its high thermal stability, the inorganic structure coated on the polymer surface creates an insulating layer, reducing the heat transfer between the flame and the fiber, reducing the combustion rate. Although not much work has been done on the use of ZnO nanoparticles as a flame retardant, it appears that their implementation in the finish has resulted in significant effects in reducing the flame propagation rate on cotton fibers, cotton/polyester (Algaba et al. 2007), jute and sisal (Samanta et al. 2017; Sheshama et al. 2017).

2.6 Textile Fabric Nanofinished with Carbon Nanotubes to Smart Batteries and Sensor Applications

The development of alternative and renewable sources to generate and store energy has aroused great worldwide interest, especially with regard to the possible depletion of fossil fuels and the issue of environmental sustainability. It is in this field that the so-called supercapacitors, which are energy storage devices in the double electrical layer, formed at the interface between an electrolytic solution and an electronic conductor, are favorable to play a very important role in the near future, with significant advantages, which include since the great power density and the long life cycle (Winter and Brodd 2004; Hu et al. 2015).

The characteristics of traditional supercapacitors are rigid, bulky and heavy, therefore, reducing their weight and volume is of fundamental importance to increase the range of applications. To meet the recent technological demands of modern society, the invention of energy storage devices that are lighter, more flexible and portable, also called flexible supercapacitors, are being studied. For this reason, there has been a great interest in the production of these new supercapacitors, as they have high power density over traditional batteries, high-energy density, longer life cycle, greater stability in movement when compared to conventional capacitors (Huang et al. 2016). Which means that they can charge and discharge stored energy in a minimum time. Therefore, they can be used not only in power generation processes, but also for medical applications, use in the military area, in addition to use in hybrid electric vehicles, such as trains and buses, for example, where a large amount of cargo is verified quickly to break the state of inertia and start the movement (Paladini et al. 2007).

The types of electrochemical storage devices that can be used as textile systems are: double layer electrochemical capacitor (double layer electrochemical capacitor—DLEC), pseudocapacitors and batteries. EDLS and pseudocapacitors are also called supercapacitors (Jost et al. 2014).

The use of carbon as an electrode in EDLC devices generally increases the double layer capacitance, obtaining a specific area that reaches hundreds of square meters per strip, providing an increase in the stored charge. At EDLS there is no chemical reaction, which makes the life cycle of the supercapacitor superior to the battery. The development of the first flexible textile batteries and supercapacitors occurred around 2011, through the process of immersing cotton fabrics in an ink containing single layer carbon nanotubes (SWCNT) (Hu et al. 2010). Carbon-derived materials and their allotropic forms are the most suitable for energy storage, as they have a large surface area, high conductivity and controllable porosity, also contributing to the power and volume ratio (Kopczynski et al. 2016).

On average 1/3 of CNT's are metallic and 2/3 are semiconductors. In general, we can say that CNT's behave as one-dimensional conductors with high electrical conductivity. Its metallic properties originate from the multilayer structure consisting of tubes with different electrical properties, where electronic coupling occurs between layers (Iijima 1991). The structure of the CNT is determined depending on the procedure adopted in the synthesis, and it is possible to obtain three different types of CNT's: single wall (SWCNT) consisting of a single graphene cylinder; double wall (DWCNT) consist of two concentric cylinders; multiple walls (MWCNT) formed by a concentric arrangement of multiple CNT, like a coaxial cable, which may have conductive or semiconductor characteristics depending on the type obtained (Iijima 1991; Ci et al. 2003).

The best known synthesis methods are: arc discharge, where an alternating current plasma arc is generated between two electrodes in an inert atmosphere, at high temperature (3000–4000 °C); laser ablation consists of a graphite target being vaporized by radiation at high temperatures (1200 °C), under inert gas flow; chemical vapor deposition (DCV) in which a carbon-rich gas is decomposed into a substrate, in the presence of particles of metallic catalyst at a temperature of approximately 600 °C; conversion of high pressure carbon monoxide (HiPco) to produce CNTs through the continuous flow of gas (CO) and the precursor catalyst Fe(CO) 5 in a heated reactor; and pyrolysis, which consists of the thermal degradation of the organic material in the partial or total absence of an oxidizing agent, or even in an environment with an oxygen concentration capable of preventing the intensive gasification of the organic material, at a temperature ranging from 400 °C to the start of gasification. In all existing methods, the presence of the catalyst is essential to obtain CNTs, which in the process are nucleated and grown by the carbon atoms resulting from the precursor decomposition (Iijima 1991; Ci et al. 2003). In Table 4, we can compare the methods mentioned above.

The functionality of the materials can be done through various ways of impregnating these DWCNT's to fabrics, such as: spray coating; exhaustion or conventional dyeing; incorporation in Inherently Conductive Polymers (ICPs) such as polypyrrole, poly (aniline) and PDOT, among others; in addition, through processes such

Table 4 Methods of synthesis of CNT's and their comparisons

Synthesis method	Description of material obtained
Electric arc discharge	CNT's produced in good quality and with little defects
Laser ablation	CNT's produced with uniform diameters. Metals such as nickel, cobalt and iron are used
Chemical vapor deposition (CVD)	Lower cost compared to other methods due to the use of low temperatures, however the CNT's produced have a large number of defects
High pressure carbon monoxide conversion	Quality CNT's are produced, but production is very low
Pyrolysis	Pure CNT's are produced due to the absence of catalyst substances in the simplified synthesis process

as extrusion spinning, surface grafting with PBA (poly-butylacrylate), airgel, ink-jet printing, etc. (Shahidi and Moazzenchi 2018). With DWCNT's properly incorporated into the textile fabric, and ready to develop clothes and accessories, seeking to promote high performance of these pieces through their multifunctionalities, as illustrated in Fig. 13. Producing, then, functional textiles, which are distinguished from traditional textiles by presenting technical functions that go beyond a simple application as in clothing, aesthetics, fashion and decoration.

Studies in this field of nanotechnology in the textile industry are constantly growing, in order to produce new sensors, supercapacitors and flexible portable electronic devices, from these textile substrates. Thus, these technologies allow a wide applicability of these clothes that can be used for fashion clothing or technical textiles.

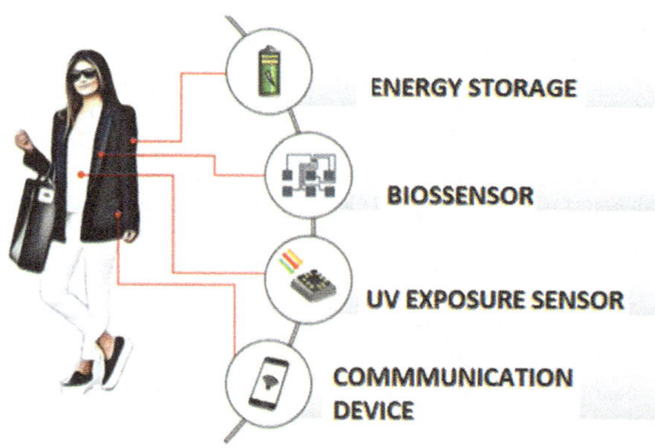

Fig. 13 Applications of carbon nanotubes on textiles fabric

3 Conclusion

With the technological advancement, the application of the development of textile products with nanotechnology has been a trend for the textile industry to advance to the 4th industrial revolution with the capacity to revolutionize clothing and textiles processes, as was observed with the growing increase in international patents. Applied in the clothing and fabric industry segment adding multifunctionalities such as UV protection, antimicrobial protection, self-cleaning, luminescence and intelligence. And also, nanotechnology with its nanoparticles and quantum dots is capable of developing new fibers and fabrics conductive, supercapacitors, with the purpose of storing energy and acting as a battery and body sensors. Both carbon nanotubes, as well as nanoparticles and quantum dots have the capacity to act as flame retardants, that is, one more property added to clothing. In conclusion, functional and intelligent textile materials developed with the tools of nanotechnology and functionalized by means of industrial finishes is a reality and is part of the 2nd generation of nanotechnology and which in the near future can be easily found in the consumer market.

References

Abbasi AR, Morsali A (2011) Synthesis and properties of silk yarn containing Ag nanoparticles under ultrasound irradiation. Ultrason Sonochem 18(1):282–287

Abouraddy AF et al (2007) Towards multimaterial multifunctional fibres that see, hear, sense and communicate. Nat Mater 6(5):336–347

Abramov OV et al (2009) Pilot scale sonochemical coating of nanoparticles onto textiles to produce biocidal fabrics. Surf Coat Technol 204(5):718–722

Adera S et al (2013) Non-wetting droplets on hot superhydrophilic surfaces. Nat Commun. 4:1–7. Available at: http://dx.doi.org/10.1038/ncomms3518

Agrawal N et al (2019) Green synthesis of robust superhydrophobic antibacterial and UV-blocking cotton fabrics by a dual-stage silanization approach. Adv Mater Interfaces 6(11):1900032

Ahmad I, Kan C (2017) Visible-light-driven, dye-sensitized TiO_2 photo-catalyst for self-cleaning cotton fabrics. Coatings 7(11):192. https://doi.org/10.3390/coatings7110192

Ahmed NSE, El-Shishtawy RM (2010) The use of new technologies in coloration of textile fibers. J Mater Sci 45(5):1143–1153

Alebeid OK, Zhao T (2017) Review on: developing UV protection for cotton fabric. J Text Inst 108(12):2027–2039. https://doi.org/10.1080/00405000.2017.1311201

Algaba† IM, Pepió‡ M, Riva*† A (2007) Modelization of the influence of the treatment with two optical brighteners on the ultraviolet protection factor of cellulosic fabrics. Am Chem Soc. https://doi.org/10.1021/ie060723c

Arakha M et al (2015) The effects of interfacial potential on antimicrobial propensity of ZnO nanoparticle. Sci Reports 5:9578

Ashby MF, Ferreira PJ, Schodek DL (2009) Nanomaterial product forms and functions. In: Ashby MF, Ferreira PJ, Schodek DL (eds) Nanomaterials, nanotechnologies and design: an introduction for engineers and architects. Elsevier, New York, pp 403–465

Aslam S et al (2018) Multifunctional finishing of cotton fabric. Autex Res J. Sciendo, 1(ahead-of-print)

Asz J et al (2006) Treatment of toxic epidermal necrolysis in a pediatric patient with a nanocrystalline silver dressing. J Pediatr Surg 41(12):e9–e12

Avila AG, Hinestroza JP (2008) Smart textiles: tough cotton. Nat Nanotechnol 3(8):458

Barton D, Davidson H (2008) Fluorescent brighteners. Rev Prog Color Relat Top 5(1):3–11. https://doi.org/10.1111/j.1478-4408.1974.tb03786.x

Bera D et al (2010) Quantum dots and their multimodal applications: a review. Materials 3(4):2260–2345

Brayner R et al (2006) Toxicological impact studies based on *Escherichia coli* bacteria in ultrafine ZnO nanoparticles colloidal medium. Nano Lett 6(4):866–870

Busnaina A (2018) Introduction to nanomanufacturing. In: Handbook of nanoscience, engineering, and technology. CRC Press, pp 374–389

Carneiro JO et al (2011) Photocatalytic activity and UV-protection of TiO2 nanocoatings on poly (lactic acid) fibres deposited by pulsed magnetron sputtering. J Nanosci Nanotechnol 11(10):8979–8985

Carp O (2004) Photoinduced reactivity of titanium dioxide. Prog Solid State Chem 32(1–2):33–177. https://doi.org/10.1016/j.progsolidstchem.2004.08.001

Cavaleiro A, de Hosson JT (2007) Nanostructured coatings. Springer Science & Business Media

Ci L et al (2003) Double wall carbon nanotubes with an inner diameter of 0.4 nm. Chem Vap Deposition 9(3):119–121. https://doi.org/10.1002/cvde.200304142

Clark M (2011) Principles, processes and types of dyes. Woodhead Pub. Available at: https://www.sciencedirect.com/book/9781845696955/handbook-of-textile-and-industrial-dyeing. Accessed 25 Mar 2019

Coyle S et al (2007) Smart nanotextiles: a review of materials and applications. MRS Bull 32(5):434–442

d'Água RB et al (2018) Efficient coverage of ZnO nanoparticles on cotton fibres for antibacterial finishing using a rapid and low cost in situ synthesis. New J Chem 42(2):1052–1060

Dahotre NB, Nayak S (2005) Nanocoatings for engine application. Surf Coat Technol 194(1):58–67

Daoud WA, Xin JH, Tao X (2004) Superhydrophobic silica nanocomposite coating by a low-temperature process. J Am Ceram Soc 87(9):1782–1784

Dasari TP, Pathakoti K, Hwang H-M (2013) Determination of the mechanism of photoinduced toxicity of selected metal oxide nanoparticles (ZnO, CuO, Co_3O_4 and TiO_2) to *E. coli* bacteria. J Environ Sci 25(5):882–888

Dastjerdi R, Montazer M (2010) A review on the application of inorganic nano-structured materials in the modification of textiles: focus on anti-microbial properties. Colloids Surf B 79(1):5–18

Dolgin E (2015) Textiles: fabrics of life. Nature 519

Dubrovski PD (2010) Woven fabrics and ultraviolet protection. In: Dubrovski PD (ed) Woven fabrics and ultraviolet protection. Sciyo

Eda G et al (2010) Blue photoluminescence from chemically derived graphene oxide. Adv Mater 22(4):505–509

El-Hady MMA, Farouk A, Sharaf S (2013) Flame retardancy and UV protection of cotton based fabrics using nano ZnO and polycarboxylic acids. Carbohyd Polym 92(1):400–406. https://doi.org/10.1016/j.carbpol.2012.08.085

Elmaaty TA et al (2018) One-step green approach for functional printing and finishing of textiles using silver and gold NPs. RSC Adv 8(45):25546–25557

El-Naggar ME, Shaarawy S, Hebeish AA (2018) Multifunctional properties of cotton fabrics coated with in situ synthesis of zinc oxide nanoparticles capped with date seed extract. Carbohyd Polym 181:307–316

Engel LS, Sanders CV, Lopez FA (2009) Fever and rash in critical care. In: Infectious diseases in critical care medicine. CRC Press, pp 39–68

Erdman A et al (2016) Preparation of multicolor luminescent cellulose fibers containing lanthanide doped inorganic nanomaterials. J Lumin 169:520–527

Ergindemir H et al (2016) Synthesis of novel UV absorbers bisindolylmethanes and investigation of their applications on cotton-based textile materials. Molecules 21(6):718. https://doi.org/10.3390/molecules21060718

Euratex—The European Apparel and Textile (2006) The future is…textiles! strategic research agenda of the European technology platform for the future of textile and clothing

Farouk A et al (2012) ZnO nanoparticles-chitosan composite as antibacterial finish for textiles. Int J Carbohydr Chem

Fauss E (2008) The silver nanotechnology commercial inventory. University of Virginia

Fiedot-Toboła M et al (2018) Deposition of zinc oxide on different polymer textiles and their antibacterial properties. Materials 11(5):707

Gang S, Dapeng L (2006) Dyeing textiles using nanoparticles. Available at: https://patents.google.com/patent/US7048771B2/en?oq=7048771

Gao D et al (2019) Construction of durable antibacterial and anti-mildew cotton fabric based on P (DMDAAC-AGE)/Ag/ZnO composites. Carbohyd Polym 204:161–169

Garcia LMP et al (2018) Photocatalytic activity and photoluminescence properties of TiO_2, In 2 O_3, TiO_2/In 2 O 3 thin films multilayer. J Mater Sci Mater Electron 29(8):6530–6542

Gogotsi Y (2006) Nanomaterials Handbook. CRC Press, Boca Raton

Gomes J, Sampaio S, Maia F (no date) Colored nanoparticles: composition and application to protein fibers and hair

Gosh SK (2006) Functional coatings. Weimheim. Wiley-VCH, Germany

Gowri VS et al (2010) Functional finishing of polyamide fabrics using ZnO–PMMA nanocomposites. J Mater Sci 45(9):2427–2435

Guan X et al (2019) Fabrication of Ag/AgCl/ZIF-8/TiO_2 decorated cotton fabric as a highly efficient photocatalyst for degradation of organic dyes under visible light. Cellulose 26(12):7437–7450

Hassabo AG et al (2019) Development of multifunctional modified cotton fabric with tri-component nanoparticles of silver, copper and zinc oxide. Carbohyd Polym 210:144–156

Hatamie A et al (2015) Zinc oxide nanostructure-modified textile and its application to biosensing, photocatalysis, and as antibacterial material. Langmuir 31(39):10913–10921

Hatch KL, Osterwalder U (2006) Garments as solar ultraviolet radiation screening materials. Dermatol Clin 24(1):85–100

Haufe H et al (2008) Bioactive textiles by sol–gel immobilised natural active agents. J Sol-Gel Sci Technol 45(1):97–101

Hegemann D et al (2009) Recent developments in Ag metallised textiles using plasma sputtering. Mater Technol 24(1):41–45

Herrmann J-M (1999) Heterogeneous photocatalysis: fundamentals and applications to the removal of various types of aqueous pollutants. Catal Today 53(1):115–129

Holme I (2003) UV absorbers for protection and performance. Int Dyer 13:9–10

Hu L et al (2010) Stretchable, porous, and conductive energy textiles. Nano Lett 10(2):708–714. https://doi.org/10.1021/nl903949m

Hu JW et al (2015) Folding insensitive, high energy density lithium-ion battery featuring carbon nanotube current collectors. Carbon 87:292–298

Huang X, Neretina S, El-Sayed MA (2009) Gold nanorods: from synthesis and properties to biological and biomedical applications. Adv Mater 21(48):4880–4910

Huang S et al (2016) Electrodeposition of polypyrrole on carbon nanotube-coated cotton fabrics for all-solid flexible supercapacitor electrodes. RSC Adv 6(16):13359–13364. https://doi.org/10.1039/c5ra24214b

Iijima S (1991) Helical microtubules of graphitic carbon. Nature 354(6348):56–58. https://doi.org/10.1038/354056a0

Joe A et al (2017) Antibacterial mechanism of ZnO nanoparticles under dark conditions. J Ind Eng Chem 45:430–439

Johnston JH, Burridge KA, Kelly FM (2009) The formation and binding of gold nanoparticles onto wool fibres. In: AIP conference proceedings, pp 189–192

Jost K, Dion G, Gogotsi Y (2014) Textile energy storage in perspective. J Mater Chem A, pp 10776–10787. https://doi.org/10.1039/c4ta00203b

Kamali P, Talebian N (2018) Sonochemically sol–gel derived coating of textiles using heterojunction SnO_2/ZnO/chitosan bionanocomposites: in vitro antibacterial evaluation. J Coat Technol Res 15(5):1133–1144

Kangwansupamonkon W et al (2009) Antibacterial effect of apatite-coated titanium dioxide for textiles applications. Nanomed Nanotechnol Biol Med 5(2):240–249

Kaounides L, Yu H, Harper T (2007) Nanotechnology innovation and applications in textiles industry: current markets and future growth trends. Mater Technol 22(4):209–237

Karimi EZ, Ansari M (2018) Comparison of antibacterial activity of ZnO nanoparticles fabricated by two different methods and coated on tetron fabric. Open Biotechnol J 12(1)

Khan MZ et al (2018) Development of UV protective, superhydrophobic and antibacterial textiles using ZnO and TiO_2 nanoparticles. Fibers Polym 19(8):1647–1654

Khose RV et al (2018) Novel approach towards the synthesis of carbon-based transparent highly potent flame retardant. Carbon 139:205–209. https://doi.org/10.1016/j.carbon.2018.06.049

Kim JH et al (2018) Polydopamine-assisted immobilization of hierarchical zinc oxide nanostructures on electrospun nanofibrous membrane for photocatalysis and antimicrobial activity. J Colloid Interface Sci 513:566–574

Konstantinou IK, Albanis TA (2004) TiO_2-assisted photocatalytic degradation of azo dyes in aqueous solution: kinetic and mechanistic investigations: a review. Appl Catal B 49(1):1–14

Kopczyński K, Milczarek G, Lota G (2016) Polysulphides reversible faradaic reactions in supercapacitor application. Electrochem Commun 68:28–31. https://doi.org/10.1016/j.elecom.2016.04.016

Kuo W-S et al (2017) Graphene quantum dots with nitrogen-doped content dependence for highly efficient dual-modality photodynamic antimicrobial therapy and bioimaging. Biomaterials 120:185–194. https://doi.org/10.1016/j.biomaterials.2016.12.022

Lee HJ, Kim J, Park CH (2014) Fabrication of self-cleaning textiles by TiO_2-carbon nanotube treatment. Text Res J 84(3):267–278

Leong KH et al (2018) Physical mixing of N-doped graphene quantum dots functionalized TiO_2 for sustainable degradation of methylene blue. IOP Conf Series Mater Sci Eng 409(1):012009. https://doi.org/10.1088/1757-899x/409/1/012009

Li Y et al (2012) Mechanism of photogenerated reactive oxygen species and correlation with the antibacterial properties of engineered metal-oxide nanoparticles. ACS Nano 6(6):5164–5173

Li R et al (2014) Study on synthesis of ZnO nanorods and its UV-blocking properties on cotton fabrics coated with the ZnO quantum dot. J Nanopart Res 16(9):2581

Liao YF et al (2013) Fabrication of antibacterial and UV protective silk fabrics via in situ generating ZnO nanoparticles by hyperbranched polymer. In: Advanced materials research. Trans Tech Publ, pp. 374–379

Lim PF et al (2018) Solar light harvesting N-graphene quantum dots decorated TiO_2 for enhanced photocatalytic activity. In: Huang YF et al (eds) E3S Web of conferences, vol 65. EDP Sciences, p 05014. https://doi.org/10.1051/e3sconf/20186505014

Liu Y et al (2008) Functionalization of cotton with carbon nanotubes. J Mater Chem 18(29):3454–3460

Liu Y, Chen X, Xin JH (2009) Can superhydrophobic surfaces repel hot water? J Mater Chem 19(31):5602–5611

Liz-Marzán LM (2004) Nanometals: formation and color. Mater Today 7(2):26–31

Long live the NNI (2019) Nature nanotechnology 14(11):995–995. https://doi.org/10.1038/s41565-019-0580-1

Lu X et al (2017) A multi-functional textile that combines self-cleaning, water-proofing and VO2-based temperature-responsive thermoregulating. Sol Energy Mater Sol Cells 159:102–111

Lumbreras-Aguayo A et al (2019) Poly (methacrylic acid)-modified medical cotton gauzes with antimicrobial and drug delivery properties for their use as wound dressings. Carbohyd Polym 205:203–210

Luna M et al (2018) TiO_2-SiO_2 coatings with a low content of AuNPs for producing self-cleaning building materials. Nanomaterials 8(3):177

Mansoori GA (2005) Advances in atomic and molecular nanotechnology. Principles of nanotechnology. World Scientific Publishing Co. Ptc. Ltd, Singapore

Martins NCT et al (2016) N-doped carbon quantum dots/TiO2 composite with improved photocatalytic activity. Appl Catal B Environ 193:67–74. https://doi.org/10.1016/J.APCATB.2016.04.016

McGuffie MJ et al (2016) Zinc oxide nanoparticle suspensions and layer-by-layer coatings inhibit staphylococcal growth. Nanomed Nanotechnol Biol Med 12(1):33–42

Meruvu H et al (2011) Synthesis and characterization of zinc oxide nanoparticles and its antimicrobial activity against *Bacillus subtilis* and *Escherichia coli*. J Rasayan Chem 4(1):217–222

Mills A, Hunte S Le (1997) An overview of semiconductor photocatalysis. J Photochem Photobiol A 108:1–35

Movilla JL et al (2005) Calculation of electronic density of states induced by impurities in TiO_2 quantum dots. Phys Rev B 72(15):153313

Mukhopadhyay SM, Joshi P, Pulikollu RV (2005) Thin films for coating nanomaterials. Tsinghua Sci Technol 10(6):709–717

Nanjappan K, Aarumugam V, Kesavan V (2018) Plasma process for coated fabric materials with Zinc to prepare antibacterial modal fabric. Mater Technol 33(10):635–641

National Science and Technology Council (2018) The national nanotechnology iniative supplement to the president's 2019 budger. Available at: https://www.whitehouse.gov/wp-content/uploads/2018/08/The-National-Nanotechnology-Initiative-Supplement-to-the-President's-2019-Budget.pdf

Neto NFA et al (2017) Photoluminescence and photocatalytic properties of Ag/AgCl synthesized by sonochemistry: statistical experimental design. J Mater Sci Mater Electron 28(16):12273–12281

Neto NFA et al (2018) Increase of antimicrobial and photocatalytic properties of silver-doped PbS obtained by sonochemical method. J Mater Sci Mater Electron 29(22):19052–19062

Neto NFA et al (2019) Effect of temperature on the morphology and optical properties of Ag_2WO_4 obtained by the co-precipitation method: photocatalytic activity. Ceram Int 45(12):15205–15212

Nickoloff BJ (2008) Immunobiology of acute cytotoxic drug reactions. In: Dermatologic immunity. Karger Publishers, pp 53–64

Nourbakhsh S, Montazer M, Khandaghabadi Z (2018) Zinc oxide nano particles coating on polyester fabric functionalized through alkali treatment. J Ind Text 47(6):1006–1023

Onar N et al (2009) Structural, electrical, and electromagnetic properties of cotton fabrics coated with polyaniline and polypyrrole. J Appl Polym Sci 114(4):2003–2010

Ou N-Q et al (2019) Facet-dependent interfacial charge transfer in TiO2/nitrogen-doped graphene quantum dots heterojunctions for visible-light driven photocatalysis. Catalysts 9(4):345

Padmavathy N, Vijayaraghavan R (2008) Enhanced bioactivity of ZnO nanoparticles—an antimicrobial study. Sci Technol Adv Mater 9(3):35004

Pakdel E, Daoud WA, Wang X (2013) Self-cleaning and superhydrophilic wool by TiO_2/SiO_2 nanocomposite. Appl Surf Sci 275:397–402

Pakdel E et al (2017) Enhanced antimicrobial coating on cotton and its impact on UV protection and physical characteristics. Cellulose 24(9):4003–4015

Paladini V et al (2007) Super-capacitors fuel-cell hybrid electric vehicle optimization and control strategy development. Energy Convers Manag 48(11):3001–3008

Pandiyarasan V et al (2017) Hydrothermal growth of reduced graphene oxide on cotton fabric for enhanced ultraviolet protection applications. Mater Lett 188:123–126. https://doi.org/10.1016/J.MATLET.2016.11.047

Pang Y et al (2018) Facile preparation of N-doped graphene quantum dots as quick-dry fluorescent ink for anti-counterfeiting. New J Chem 42(20):17091–17095

Pavlidou S, Paul R (2018) Soil repellency and stain resistance through hydrophobic and oleophobic treatments. In: Williams J (ed) Waterproof and water repellent textiles and clothing. Woodhead Publishing, pp 73–88. https://doi.org/10.1016/b978-0-08-101212-3.00003-4

Pelton R, Geng X, Brook M (2006) Photocatalytic paper from colloidal TiO2—fact or fantasy. Adv Colloid Interface Sci 127(1):43–53
Perelshtein I et al (2008) Sonochemical coating of silver nanoparticles on textile fabrics (nylon, polyester and cotton) and their antibacterial activity. Nanotechnology 19(24):245705
Perelshtein I, Applerot G et al (2009a) Antibacterial properties of an in situ generated and simultaneously deposited nanocrystalline ZnO on fabrics. ACS Appl Mater Interfaces 1(2):361–366
Perelshtein I, Applerot Guy et al (2009b) CuO–cotton nanocomposite: formation, morphology, and antibacterial activity. Surf Coat Technol 204(1–2):54–57
Pinatti IM et al (2019) Rare earth doped silver tungstate for photoluminescent applications. J Alloys Compd 771:433–447
Popescu MC et al (2019) Antibacterial efficiency of cellulose-based fibers covered with ZnO and Al_2O_3 by atomic layer deposition. Appl Surf Sci 481:1287–1298
Porter AL, Youtie J (2009) How interdisciplinary is nanotechnology? J Nanopart Res 11(5):1023–1041
Pozzo RL, Baltanas MA, Cassano AE (1997) Supported titanium oxide as photocatalyst in water decontamination: state of the art. Catal Today 39(3):219–231
Premanathan M et al (2011) Selective toxicity of ZnO nanoparticles toward Gram-positive bacteria and cancer cells by apoptosis through lipid peroxidation. Nanomed Nanotechnol Biol Med 7(2):184–192
Raghupathi KR, Koodali RT, Manna AC (2011) Size-dependent bacterial growth inhibition and mechanism of antibacterial activity of zinc oxide nanoparticles. Langmuir 27(7):4020–4028
Rai M, Yadav A, Gade A (2009) Silver nanoparticles as a new generation of antimicrobials. Biotechnol Adv 27(1):76–83
Ramakrishna S (2005) An introduction to electrospinning and nanofibers. World Scientific
Ramamurthy P et al (2017) Antimicrobial characteristics of pulsed laser deposited metal oxides on polypropylene hydroentangled nonwovens for medical textiles. Fibres & Textiles in Eastern Europe
Ran J et al (2018) Growing ZnO nanoparticles on polydopamine-templated cotton fabrics for durable antimicrobial activity and UV protection. Polymers 10(5):495
Reddy LS et al (2014) Antimicrobial activity of zinc oxide (ZnO) nanoparticle against *Klebsiella pneumoniae*. Pharm Biol 52(11):1388–1397
Reijnders L (2008) Hazard reduction for the application of titania nanoparticles in environmental technology. J Hazard Mater 152(1):440–445
Richardson MJ, Johnston JH (2007) Sorption and binding of nanocrystalline gold by Merino wool fibres—an XPS study. J Colloid Interface Sci 310(2):425–430
Rilda Y et al (2018) The function of cross linker carboxylic acid for TiO_2/Chitosan/SiO_2 coated as self cleaning fabrics. Orient J Chem 34(6):2942
Rivero PJ et al (2015) Nanomaterials for functional textiles and fibers. Nanoscale Res Lett 10(1):501
Rusnano (no date) RUSNANO Annual Report—2017, 2018
Saad SR et al (2016) Self-cleaning technology in fabric: a review. IOP Conf Series Mater Sci Eng 133(1):012028. https://doi.org/10.1088/1757-899X/133/1/012028
Safardoust-Hojaghan H, Salavati-Niasari M (2017) Degradation of methylene blue as a pollutant with N-doped graphene quantum dot/titanium dioxide nanocomposite. J Cleaner Prod 148:31–36. https://doi.org/10.1016/J.JCLEPRO.2017.01.169
Salama M, El-Sayed AA (2014) Imparting permanent antibacterial properties to viscose fabric using zinc oxide nanoparticles and polymeric binders. World Appl Sci J 32(3):392–398
Salat M et al (2018) Durable antimicrobial cotton textiles coated sonochemically with ZnO nanoparticles embedded in an in-situ enzymatically generated bioadhesive. Carbohydr Polym 189:198–203
Samanta AK et al (2017) Fire retardant finish of jute fabric with nano zinc oxide. Cellulose 24(2):1143–1157
Sasaki T, Tour JM (2008) Synthesis of a new photoactive nanovehicle: a nanoworm. Org Lett 10(5):897–900

Schawabl F (2008) Advanced quantum mechanics. Springer, Berlin, Heidelberg. https://doi.org/10.1007/978-3-540-85062-5_1

Seil JT, Webster TJ (2012) Antimicrobial applications of nanotechnology: methods and literature. Int J Nanomed 7:2767

Sen AK (2007) Coated textiles: principles and applications. CRC Press

Shahidi S, Moazzenchi B (2018) Carbon nanotube and its applications in textile industry—a review. J Text Inst 109(12):1653–1666

Shalaev VM, Kawata S (2006) Nanophotonics with surface plasmons. Elsevier

Sharma M, Saravolatz LD (2009) 17 severe skin and soft tissue infections in critical care. Infectious disease and therapy, p 295

Shen J et al (2011) Facile preparation and upconversion luminescence of graphene quantum dots. Chem Commun 47(9):2580–2582. https://doi.org/10.1039/C0CC04812G

Sheshama M et al (2017) Bulk vs. nano ZnO: influence of fire retardant behavior on sisal fibre yarn. Carbohydr Polym 175:257–264

Shim BS et al (2008) Smart electronic yarns and wearable fabrics for human biomonitoring made by carbon nanotube coating with polyelectrolytes. Nano Lett 8(12):4151–4157

Silva IO et al (2019) Multifunctional chitosan/gold nanoparticles coatings for biomedical textiles. Nanomaterials 9(8):1064

Singh NA (2017) Nanotechnology innovations, industrial applications and patents. Environ Chem Lett 15(2):185–191

Soane D, Offord D, Ware W (2005) Nanotechnology applications in textiles. Wiley Online Library, Jurgen Schulte, p 149

Sójka-Ledakowicz J et al (2009) Functionalization of textile materials by alkoxysilane-grafted titanium dioxide. J Mater Sci 44(14):3852–3860

StartNano (no date) Indicators by Countries. Nano-related Indicators. Innovation 2019'. Available at: https://statnano.com/

Sun L-W et al (2012) Lanthanum-doped ZnO quantum dots with greatly enhanced fluorescent quantum yield. J Mater Chem 22(17):8221–8227

Sun H et al (2014) Graphene quantum dots-band-aids used for wound disinfection. ACS Nano 8(6):6202–6210. https://doi.org/10.1021/nn501640q

Sun X et al (2019) Visible-light driven TiO_2 photocatalyst coated with graphene quantum dots of tunable nitrogen doping. Molecules 24(2):344. https://doi.org/10.3390/molecules24020344

Tan LY et al (2019) Funtionalization and mechanical propeties of cotton fabric with ZnO nanoparticles for antibacterial textile application. In: Solid state phenomena. Trans Tech Publ, pp 292–297

Tchapla A et al (2004) Characterisation of embalming materials of a mummy of the Ptolemaic era. Comparison with balms from mummies of different eras. J Sep Sci 27(3):217–234

The New York Times (no date) Responsible party/Dr. David Soane; Armor against Stains. Available at: https://www.nytimes.com/2002/10/13/business/responsible-party-dr-David-soane-armor-against-stains.html

Tour JM (2007) Nanotechnology: the passive, active and hybrid sides-gauging the investment landscape from the technology perspective. Nanotech 4:361

Tuerhong M, XU Y, Yin X-B (2017) Review on carbon dots and their applications. Chin J Anal Chem 45(1):139–150. https://doi.org/10.1016/S1872-2040(16)60990-8

van Lente H, van Til JI (2008) Articulation of sustainability in the emerging field of nanocoatings. J Clean Prod 16(8–9):967–976

Vaseem M et al (2012) Synthesis of ZnO nanoparticles and their ink-jetting behavior. J Nanosci Nanotechnol 12(3):2380–2386

Verbič A, Gorjanc M, Simončič B (2019) Zinc oxide for functional textile coatings: recent advances. Coatings 9(9):550

Vigneshwaran N et al (2006) Functional finishing of cotton fabrics using zinc oxide–soluble starch nanocomposites. Nanotechnology 17(20):5087

Wang Y-W et al (2014) Superior antibacterial activity of zinc oxide/graphene oxide composites originating from high zinc concentration localized around bacteria. ACS Appl Mater Interfaces 6(4):2791–2798

Wijesena RN et al (2015) Slightly carbomethylated cotton supported TiO_2 nanoparticles as self-cleaning fabrics. J Mol Catal A Chem 398:107–114. https://doi.org/10.1016/J.MOLCATA.2014.11.012

Winter M, Brodd RJ (2004) What are batteries, fuel cells, and supercapacitors? ACS Publications

WMWTA—World Markets for Woven Textiles and Apparel (2009) Textile outlook international, 141

Xu B et al (2015) Self-cleaning cotton fabrics via combination of photocatalytic TiO2 and super-hydrophobic SiO_2. Surf Coat Technol 262:70–76. https://doi.org/10.1016/J.SURFCOAT.2014.12.017

Yang M et al (2019) Facile construction of robust superhydrophobic cotton textiles for effective UV protection, self-cleaning and oil-water separation. Colloids Surf A Physicochem Eng Aspects 570:172–181. https://doi.org/10.1016/J.COLSURFA.2019.03.024

Yetisen AK et al (2016) Nanotechnology in textiles. ACS Nano 10(3):3042–3068

Yu Y et al (2017) Highly fluorescent cotton fiber based on luminescent carbon nanoparticles via a two-step hydrothermal synthesis method. Cellulose 24(4):1669–1677. https://doi.org/10.1007/s10570-017-1230-0

Zahid M et al (2018) Fabrication of visible light-induced antibacterial and self-cleaning cotton fabrics using manganese doped TiO_2 nanoparticles. ACS Appl Bio Mater 1(4):1154–1164. https://doi.org/10.1021/acsabm.8b00357

Zeller A, Johnston JH (no date) New fluorescent hybrid materials comprising quantum dots, organic fluorophores and natural fibre substrates

Zhai Y et al (2016) Influence of doping alkali metal ions on the structure and luminescent properties of microwave synthesized $CaMoO_4$: Dy3 + phosphors. J Alloys Compd 688:241–247

Zhang G et al (2013) Application of ZnO nanoparticles to enhance the antimicrobial activity and ultraviolet protective property of bamboo pulp fabric. Cellulose 20(4):1877–1884

Zhao S-W et al (2018) The preparation and antibacterial activity of cellulose/ZnO composite: a review. Open Chem 16(1):9–20

Zhu C et al (2017) Design and characterization of self-cleaning cotton fabrics exploiting zinc oxide nanoparticle-triggered photocatalytic degradation. Cellulose 24(6):2657–2667

Zuo D et al (2019) UV protection from cotton fabrics finished with boron and nitrogen co-doped carbon dots. Cellulose, 1–8. https://doi.org/10.1007/s10570-019-02365-5

Chapter 9
Nanoparticles for Anticancer Therapy

Marcelo Fernandes Cipreste, Gracielle Ferreira Andrade, Wellington Marcos da Silva, and Edesia Martins Barros de Sousa

1 Introduction

In the last decades, scientists around the world have been aiming their efforts to elucidate the unique properties of biocompatible nanoparticles and how to use these materials to develop new approaches for cancer treatment. The nanoparticles can spontaneously leave the bloodstream in neoplastic sites and passively accumulate in the tumor *interstitium*, avoiding their accumulation in normal tissues. This preferential accumulation in neoplasms is caused by the unique vascular architecture of the malignant tumors with nanoscale-fenestrated vessels and excessive production of vascular mediators that make these vessels more permeable (enhanced permeability and retention—EPR) for particles measuring from 10 to 500 nm (Torchilin 2011). Since the discovery of the EPR effect, scientists have been studying many kinds of nanoparticles as drug delivery systems designed to specifically release therapeutic molecules in tumor tissues to increase load of drugs in the tumors and to reduce the side effects caused by drug uptake by normal cells.

One of the major concerns associated with the therapy of many diseases, including cancer, is the low specificity of the chemotherapeutic drugs (Kawasaki and Freire

M. F. Cipreste · G. F. Andrade · W. M. da Silva · E. M. B. de Sousa (✉)
Centro de Desenvolvimento da Tecnologia Nuclear—CDTN, Serviço de Nanotecnologia—SENAN, Av. Antonio Carlos 6.627, Campus da UFMG, Pampulha, Belo Horizonte, MG 31270-901, Brazil
e-mail: sousaem@cdtn.br

M. F. Cipreste
e-mail: mcipreste@gmail.com

G. F. Andrade
e-mail: graciellefandrade@yahoo.com.br

W. M. da Silva
e-mail: wellingtonmarcos@yahoo.com.br

R. F. do. Nascimento et al. (eds.), *Nanomaterials and Nanotechnology*, Materials Horizons: From Nature to Nanomaterials,
https://doi.org/10.1007/978-981-33-6056-3_9

2011). Furthermore, high doses of drugs are required to have desired effect. In this context, nanoparticles have been making considerable contributions to deliver anticancer therapeutic molecules to tumor sites (Chaturvedi et al. 2018). The use of nanotechnology in cancer therapy and diagnostics raises high beliefs for many patients for better, more efficient, secure and low-cost health care (Rosenholm and Mamaeva 2017).

Nanocarriers can lead to augment drug half-life in the systemic circulation, increase tumor-targeting efficiency, decreased side effects and thus enhance the efficacy of cancer therapy (Wicki et al. 2015). In recent years, various nanocarriers with potential for application in cancer therapy are under investigation, including mesoporous silica nanoparticles, calcium phosphates such as hydroxyapatite, nanotubes such as boron nitride nanotubes, magnetic nanostructures such as magnetite and metallic materials such as gold nanoparticles. Studies of therapeutic use of mesoporous silica nanoparticles have shown them as a promising carrier for drug and gene delivery and biomedical imaging due to their large surface area, uniform pore size distribution and high pore volume allowing a high drug load, as well as good biocompatibility (de Freitas et al. 2017). Thus, these nanoparticles can act in pharmacokinetic release profiles, leading to increased bioavailability, target delivery and thereby enhanced therapeutic efficacy (Mamaeva et al. 2013a). Hydroxyapatite nanoparticles have become great candidates to act in the treatment of bone pathologies such as osteosarcomas and Ewing tumors due to their similarity with the mineral phase of bone tissues, showing high biocompatibility, osteoconductivity, and affinity for skeletal structures (Cai et al. 2011; Palazzo et al. 2007). Besides these characteristics, as an open crystalline structure, hydroxyapatite matrix can be doped with radioisotopes and magnetic metals to provide therapeutic and diagnostic properties to this material (Cipreste et al. 2016b; Mokoena et al. 2015). Recently, boron nitride nanotubes (BNNTs) have attracted significant interest for scientific and technological applications due to their unique physical and chemical properties. Besides these properties, the possibility of incorporating various types of molecules and rare-earth (RE) elements on its surface gives it additional properties (Ferreira et al. 2013, 2015a, b, c; da Silva et al. 2018a, b). For example, when bound to target molecules, the BNNTs may, in fact, be used as specific therapeutic agents capable of being internalized by cancer cells and promote tumor death in a boron neutron capture therapy (Barth et al. 2018). Another important feature of nanoparticles is that their properties can be chemically tuned to achieve appropriate biological responses for each kind of tumor.

The nanostructure porosity, size and morphology can be controlled by variations in the synthesis routes, and the surfaces of the nanoparticles can be modified by functionalization process that means the introduction of functional chemical groups in nanoparticle surfaces by covalent or other types of bond (Wang et al. 2018). To act as drug delivery systems, nanoparticles must show long-term circulation in bloodstream and avoid being recognized and captured by the macrophages; this property can be achieved by the nanoparticle surfaces modification with some polymeric molecules such as polyethylene glycol (Torchilin and Trubetskoy 1995). By functionalizing the nanoparticle surfaces with molecules that act as specific ligands to overexpressed receptors in tumor cells membranes, the nanoparticles can be tuned to provide active targeting to specific kinds of tumors cells, promoting the receptor–ligand-mediated endocytosis and allowing a more specific delivery of therapeutic agents directly inside the cells (Vácha et al. 2011). The radiolabeling process of the nanostructures can provide theranostic properties to these materials, allowing to make the diagnosis simultaneously to the treatment (Koziorowski et al. 2017). Finally, it is even possible to conjugate all these properties in on nanostructured platform consisting a multifunctional nanosystem, granting the application of multiple concomitant diagnostic and therapeutic techniques (Bao et al. 2013; Cole and Holland 2015). As a consequence, silica nanoparticles, hydroxyapatite and BNNTs could become a key material for the development of nanovectors, drugs or genes in cancer therapy, and indeed, these nanoparticles can provide significant advances in the area of medical imaging, molecular biology and biomedical technology.

2 Nanoparticles Characteristics

Hydroxyapatite nanoparticles (Fig. 1), $Ca_{10}(PO_4)_6(OH)_2$, is the most widely accepted bioactive material for the repair and reconstruction of osseous defects and treatment of bone pathologies such as osteosarcomas because of its structural similarity with the inorganic part of bone and its inherent ability to enable biological bonding to its surface showing high affinity for skeletal structures (Cai et al. 2011; Palazzo et al. 2007). These features allow the specific delivery of therapeutic agents to bone tumor sites and favor the bone matrix reconstruction in the treated local. Besides these characteristics, as an open crystalline structure, the hydroxyapatite matrix can be doped with radioisotopes and magnetic metals to provide therapeutic and diagnostic properties to this material (Cipreste et al. 2016b; Mokoena et al. 2015).

Among diverse nanoplatforms, silica-based nanomaterials have been exhibited as attractive platforms for the delivery of molecules in cancer diseases due to promising

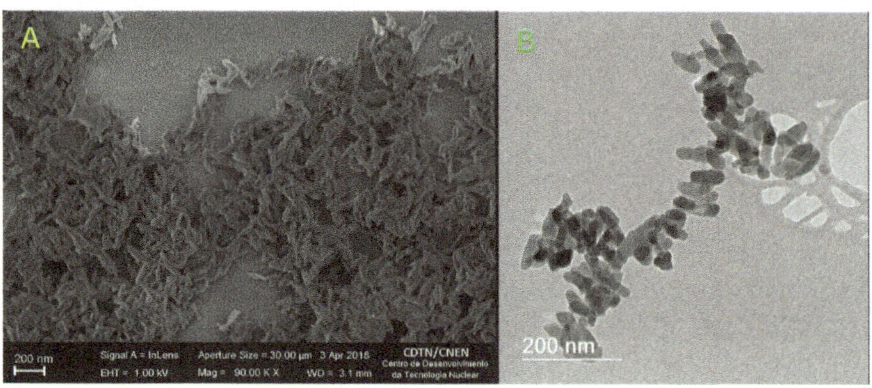

Fig. 1 SEM image (**a**) and **b** TEM image of hydroxyapatite nanoparticles

in vivo results in small-animal disease models (Mamaeva et al. 2013b). The most common synthesis of silica particles through the hydrolysis and condensation process of silicates under basic conditions was introduced by Stöber in 1968. The alkyl silanes act as a source of oxidized silicon, and the basic or acid reagents act as a capping agent for polymerizing –O–Si–O–silanes (Arriagada and Osseo-Asare 1992). The past several decades have witnessed a growth in the synthesis of silica materials especially from the discovery of porous silica particles in 1992 (Hao et al. 2019).

The presence of silica in the body is meanly in bone, cartilage and other supporting tissue. In 1983, Unger and collaborators proposed amorphous silica as a drug delivery carrier (Unger et al. 1983). From then on, many different types of silica materials have been proposed as drug delivery matrixes (Simovic et al. 2011). One common of silica materials proposed for medicine application is the ordered mesoporous silica nanoparticles (Fig. 2).

Ordered mesoporous silica nanoparticles consist of silicate-based compounds that have highly organized pore structures with tuneable pore size about 2–30 nm with narrow distribution (Zhao et al. 1998) and large pore volume that allow for extensive drug loading capacity and easily modified surface.

Many silica nanoparticle synthesis routes provide particles with size sufficiently small (10–100 nm) to penetrate the capillaries and can be taken up in distinct tissues. The use of these cancer-applied nanoparticles is justified by the fact that vicinal endothelial tissues to blood vessels of certain pathological tissues, such as solid tumors, have increased permeability relative to normal tissues (Jain 1999). This physiological characteristic represents, for example, an opportunity for biomolecule-containing nanostructures to be accumulated and retained in tumor regions, allowing diagnoses and/or treatments to be more efficient and selective to pathological regions over non-target healthy tissues.

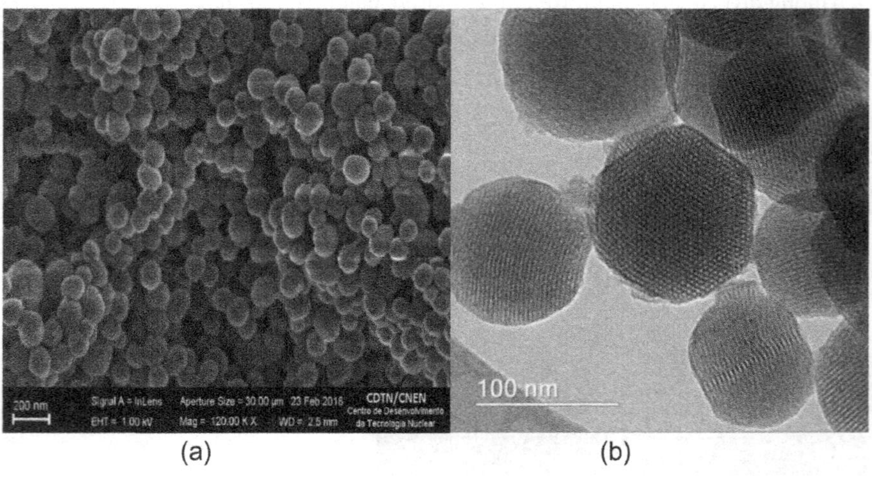

(a) (b)

Fig. 2 Mesoporous silica nanoparticles images: **a** scanning electron microscopy and **b** transmission electron microscopy

Silica nanoparticles are inorganic matrixes that exhibit unique properties such as inertness, stability and biocompatibility (Cherukula et al. 2016). All these aspects make mesoporous silica nanoparticles promising candidates for applications in cancer therapy. In addition, the possibility of selectively modifying these nanomedicine platforms to be used for specific applications provides an extensive area to be explored, taking into account their intrinsic characteristics.

A historical landmark in research into new materials began in 1985 with the discovery of fullerenes by Kroto and Smalley (Kroto et al. 1985). Since then, other carbon nanostructures such as carbon nanotubes (CNT) and graphene have been noted for their different properties an exponential number of scientific publications (Geim and Novoselov 2007; Sumio lljima and Toshinari lchihashi 1993). In this sense, the technological interest for nanomaterials is directly related to the peculiar characteristics that they exhibit. However, other nanostructures, such as boron nitride (BN), have gained prominence in the scientific field. Among them, we can highlight the boron nitride nanotubes (BNNTs) (Fig. 3) that are cylindrical structures of hexagonal boron nitride (h-BN) having diameter in the range of 1–100 nm and length up to several micrometers (Golberg et al. 2010; Terrones et al. 2007) which were theoretically predicted in 1994 (Rubio et al. 1994) and experimentally

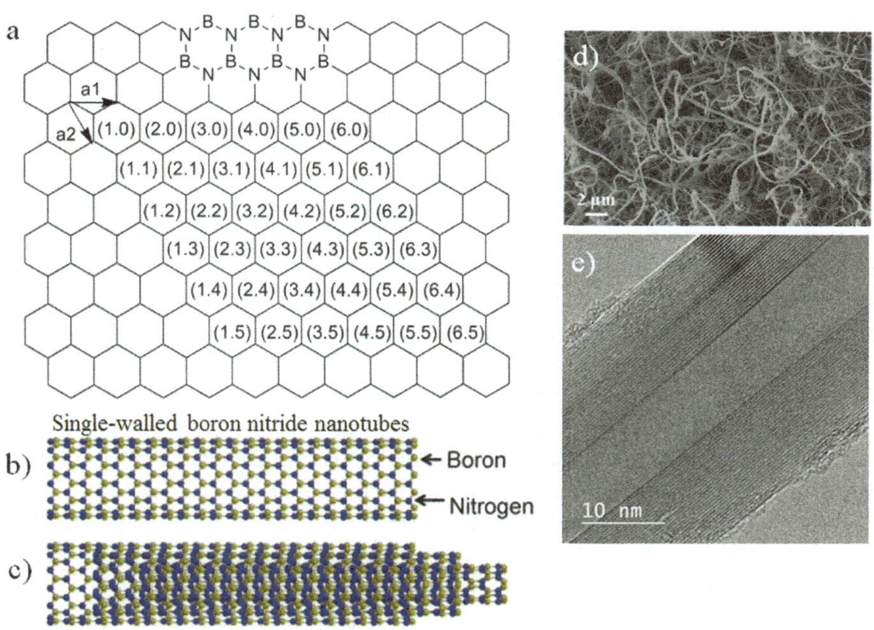

Fig. 3 **a** Vector (*n*, *m*) of single-walled BNNTs on a hexagonal boron nitride sheet (h-BN). Molecular model of **b** single-walled boron nitride nanotubes (SWBNNTs) and **c** multi-walled boron nitride nanotubes (MWBNNTs). *Source* Gao et al. (2014). SEM (**d**) and TEM images (**e**) of MWBNNTs

discovered in 1995 (Chopra et al. 1995). The discovery of BNNTs has opened new possibilities in various areas of knowledge.

BNNT is a highly promising material for applications such as shields/capsules, filler for composites, self-cleaning materials and in biology and medicine (Ciofani et al. 2013b; Pakdel et al. 2011). Unlike other nanoparticles, for example, NTCs and graphene, multi-walled boron nitride nanotubes (MWBNNTs) between 1.5 and 10 microns in length demonstrated no toxic effects for human embryonic kidney cells (Ciofani et al. 2014). This study opened various opportunities for that the BNNTs to become suitable for the development of nanovectors for the delivery of proteins, drugs, or genes beyond hyperthermia treatment for cancer therapy, and indeed, these nanoparticles can provide significant advances in the area of medical imaging, molecular biology, and biomedical technology, contributing, theoretically, to the diagnosis and treatment of several diseases, including cancer (Ciofani and Mattoli 2016), once them can be internalized by cellular targeting (Soares et al. 2012).

3 Surface Functionalization: A Targeting Strategy for Cancer Treatment and Diagnosis

The specific targeting of nanoparticles to tumor sites is one of the most important features of nanotechnology for medical purposes because the lack of specificity of chemotherapeutic and other treatment agents limits their application and causes many side effects in patients. Matsumura and Maeda (Matsumura and Maeda 1986) were the first authors to describe the preferential accumulation of macromolecules in tumor sites. They found that a polymer conjugated to an anticancer protein accumulated more in tumor tissues than did the protein alone. They speculated that the tumoritropic accumulation of these macromolecules resulted because of the hypervasculature and little recovery through either blood vessels or lymphatic vessels, causing the so-called EPR effect to even macromolecules. This was the first mention of the EPR effect. Today, it is known that the irregular angiogenesis in tumor tissues leads to a nanoscale spacing of blood vessel cells, making easy to nanoparticles passively extravasate into the tumor's *interstitium* through its leaky vasculature (Fig. 4) and accumulate more in tumor tissues than in other organs (Ngoune et al. 2016), allowing the reduction of side effects and improving the efficacy of these agents.

Besides the passive targeting strategy of spontaneous accumulation by the EPR effect, the surface modification of nanoparticles can provide active targeting to tumor cells based on the receptor-mediated endocytosis (Vácha et al. 2011).

An important feature of nanoparticles is that their properties can be chemically tuned to achieve appropriate biological responses for each kind of tumor. The nanostructure porosity, size, and morphology can be controlled by variations in the synthesis routes, and the surfaces of the nanoparticles can be modified by functionalization process by means of the introduction of functional chemical groups in the nanoparticle surfaces by covalent or other types of bond.

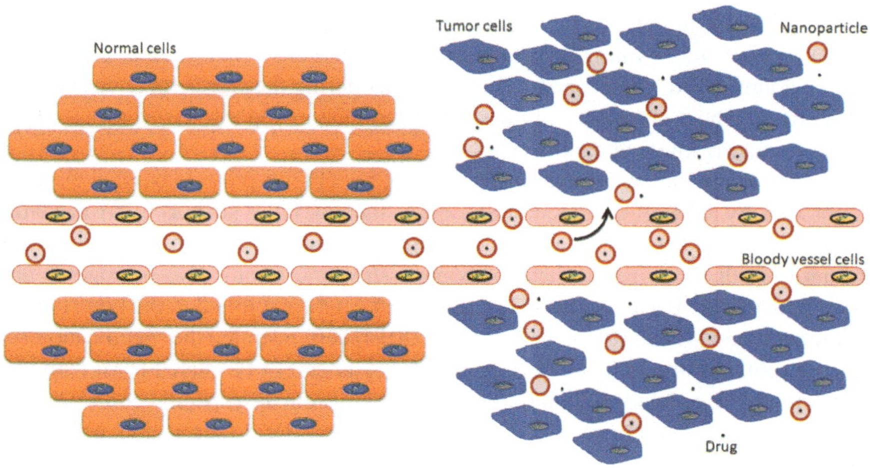

Fig. 4 Enhanced permeability and retention (EPR) effect illustration. Due to the sparse distances between endothelial cells of blood vessels in tumor regions, nanoparticles may passively leave the bloodstream and accumulate preferentially in the cellular *interstitium* of tumor tissues because lymphatic drainage in these tissues is depleted, and in the normal tissues, there is no increased permeability

Stahl and Schwartz (1986) described the receptor-mediated endocytosis as a process, whereby extracellular macromolecules and particles get into the intracellular environment as specific ligands that bind to specific cell surface receptors of the plasma membrane. Thus, the uptake of the foreign macromolecule depends on its interaction with an appropriate receptor on the surface of the tumor cell. The binding of the macromolecule to a receptor leads to the polymerization of a specific protein present in the plasma membrane that triggers the invagination process, resulting in the internalization of the complex receptor–ligand into a coated vesicle (Mak and Saunders 2005). This approach can be exploited to promote the drug or radioisotope-loaded nanoparticles uptake into the tumor cells and allow more specific delivery of chemotherapeutic and radioisotopes to tumor cells. Furthermore, it is also possible to tune the surfaces of the nanoparticles to promote their binding in specific tumor cells by knowing the overexpressed receptors present in different kinds of cancer cells. Large and co-workers (2019) reviewed the most common receptors overexpressed in tumor cells with their ligands and the drug delivery vehicle currently found in the literature. One of these interesting works was conducted by Jang and colleagues (2015) where the authors described the targeting effect of nanographene functionalized with folate and RGD to a human cervix carcinoma cell lineage. Likewise, Ringhieri and colleagues (2017) demonstrated that liposomes externally decorated with P6.1 peptide sequence can act as target-selective delivery systems for cancer cells overexpressing HER-2 receptor-like as breast cancer cells. There are other receptors discussed in that review, like estrogen, transferrin, $\alpha\nu\beta3$ integrin, PSMA, CXCR4, ICAM1, androgen, CD, EGFR, IL, TNF, glycyrrhetinic and VEGF; the authors also discuss about the types of cancer where these receptors are common overexpressed such as lung, prostate, brain, lymphoma, bone, glioma, pancreatic, oral, melanoma and colon cancers (Fig. 5). Considering the works reviewed by Jang and colleges, there are plenty of possibilities to tune nanoparticle surfaces in order to reach specific kinds of cancer cells.

Among many classes of nanomaterials that are able to provide passive target to tumor sites by EPR effect, calcium phosphate nanoparticles such as hydroxyapatite nanoparticles (HAn) stand out as a biocompatible and bioactive material with high similarity with the mineral phase of bone tissues, presenting also osteogenic and osteoinductive properties (Dorozhkin 2015) that can be exploited in the bone tumors treatment. Peng Mi and colleagues (Mi et al. 2014) demonstrated the specific target of HAn in solid tumors. In this work, the authors synthesized PEGlated HAn with Gd-DTPA incorporated and studied the effect of this system as a specific contrast agent for tumors in magnetic resonance imaging systems (MRI). The results of Mi and

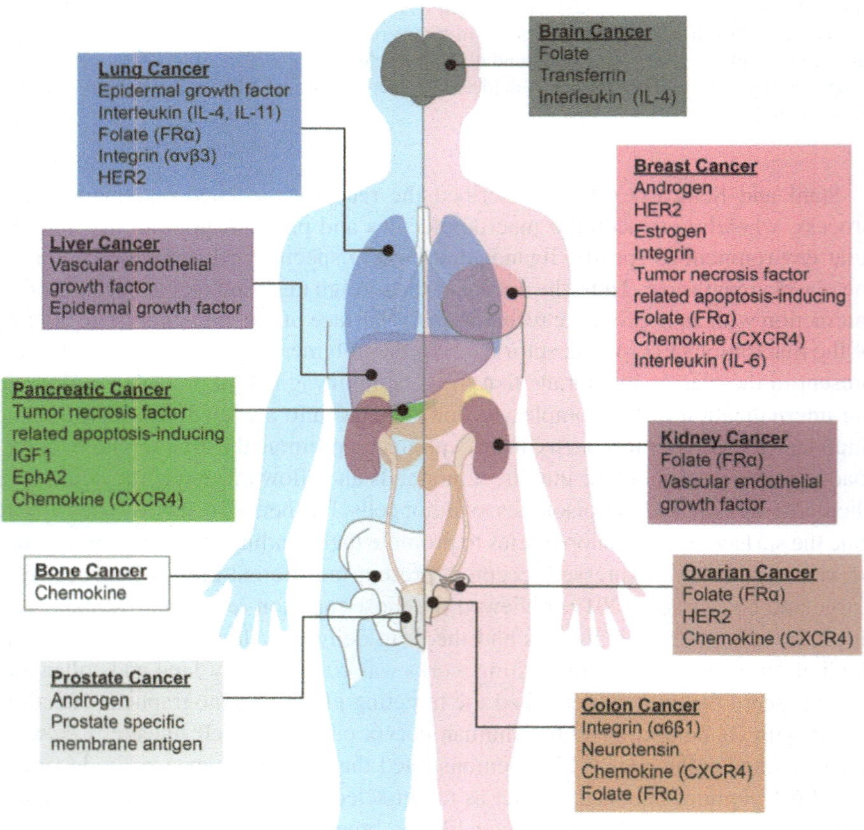

Fig. 5 Common overexpressed receptors in cancer cells and association with types of cancer. *Source* Large et al. (2019)

colleagues suggest that PEGlated HAn can accumulate in a solid tumor, enhancing the tumor signal in imaging systems. Furthermore, the HAn surfaces can be modified with ligand molecules to provide active targeting to cancer cells. Cipreste et al. (2016b) suggest that aminated folic acid can be attached to HAn surfaces with high stability to provide active targeting to osteosarcoma cells. Moreover, Kong and co-workers (2016) described the functionalization of HAn with polyethyleneimine and hyaluronic acid to stabilize the nanoparticles and to provide active targeting to CD44 receptors which are overexpressed on some cancer cells. The results in this work indicate an enhanced cellular uptake of the synthesized material in adenocarcinomic human cells. These works show that hydroxyapatite can be an important nanomaterial in the targeting strategy for cancer treatment and diagnosis.

The surface properties of silica nanoparticles are known to play an important function in determining the interactions between nanoparticles and a biological system (Tang and Cheng 2013). Through the functionalization process, silica nanoparticles can acquire targeted functionality, and the pore structure allows the protection of diverse molecules against enzymatic action during absorption in the human body. Thus, mesoporous silica nanoparticles may serve as obstacles to undesirable degradation or metabolization of the drugs, thus keeping the dose within the range of therapeutic efficacy and leading to a decrease in unwanted effects (Rámila et al. 2003).

Modified silica nanoparticles can promote enhanced stability, improved targeting specificity and increased delivery efficiency of multicomponent, as well as can reduce dosage (Ashley et al. 2011). Due to the ample possibilities for further surface modification and the flexibility of the matrixes, they may act on targeted delivery and controlled release. Both the structure and the chemical composition promote final characteristics and interesting properties for biomedicine applications.

The surface modification of inorganic nanoparticles with specific molecules, such as synthetic compounds, polymers and biomolecules has significantly improved the delivery process to the pathological cell, affecting the biodistribution profile of molecules, as well as reducing the aggregation and enhancing therapeutic efficacy (Gomes et al. 2016).

To get the best control in the release of the drugs and to obtain greater adsorption of the drugs, modifications of the chemical surface on the host matrices are necessary and crucial. The main alternative for it is the functionalization by either co-condensation of organic species during synthesis or subsequent surface modification or grafting. The latter provides good preservation of the mesostructure after treatment and advantageous surface properties, such as a higher selectivity of a specific adsorbent as proved by (Muñoz et al. 2003). Furthermore, it is possible to join the texture and structural properties of mesoporous silica nanoparticles with organic groups bonded to the silica surface. Indeed, an important advantage of mesoporous silica is to adsorb specific compounds. Recent studies have reported that numerous functional groups, including amines, chlorides, thiols, carboxylic acids and phenyl, may be attached successfully to the surface of mesoporous silica via tethering alkyl chains (Stein et al. 2000).

A widely exploited polymer for the functionalization of silica nanoparticles is the polyethylene glycol (PEG). This polymer can be conjugated to the silica nanoparticle surface via a "grafting-to" method prolonging their blood circulation. The PEGylation of nanoparticles has been used as a general and effective way to reduce the nonspecific binding of nanoparticles to blood proteins and cells of the reticuloendothelial system. In this way, the enhanced permeability and retention (EPR) effect can be greatly augmented, and their blood circulation half-lives can be prolonged (He et al. 2010).

According to Freitas and co-workers (2017), the targeting of a nanoparticle is an important strategy to decrease drug delivery to undesirable tissues, in order to reduce the side effects of very toxic drugs. This can be accomplished by the functionalization of nanoparticles with a directing moiety such as folic acid, sugar, antibodies or some peptides such as RGD and CREKA. CREKA (Cys–Arg–Glu–Lys–Ala) is a pentapeptide that presents high affinity to fibrin. It binds to clotted plasma proteins and homes to tumors, since interstitial tissue of tumors (Dvorak et al. 1985) and the vessel walls (Simberg et al. 2007) contain clotted plasma proteins, whereas the vessels in normal tissues do not. The vast majority of studies involving CREKA aims to use this peptide as driver of nanoparticles to target the tumor, micrometastasis, thrombosis and microthrombosis regions (Zhao et al. 2015). In addition, it was shown that nanoparticles functionalized with the CREKA peptide can bind to clotted plasma proteins in tumor vessels and induce additional clotting, thus amplifying their own homing (Simberg et al. 2007).

BNNT is thermal and chemically layered structure more stable than graphitic carbon structures (Silva et al. 2012). Due to its excellent chemical stability, better biological performance is expected (Ferreira et al. 2015a). In this sense, the functionalization or chemical modification of BNNTs (Fig. 6), that is, the covalent or

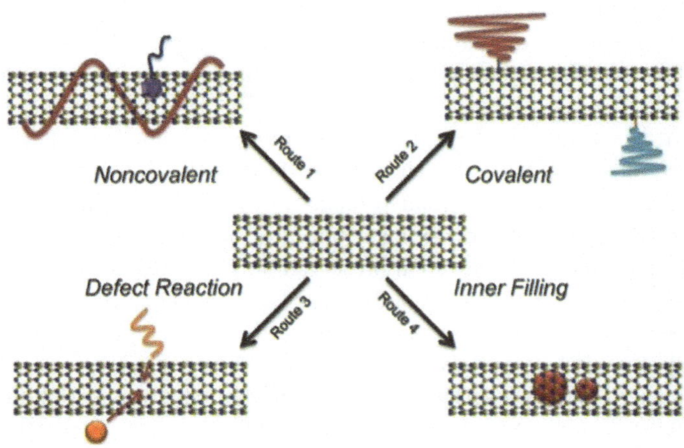

Fig. 6 Functionalization approaches for BNNTs. *Source* Gao et al. (2014)

noncovalent attachment of atoms or molecules to the surface, has been used to introduce chemical specificity and processability in different environments (Ciofani et al. 2012).

The adequate modification process provides specific surfaces that is an important role regarding selectivity, for instance, in biological systems (Ciofani et al. 2013b). BNNTs can be coated with biological molecules, such as folic acid, polymers, oligonucleotides, carbohydrate and other molecules (Del Turco et al. 2013; Emanet et al. 2015; Ferreira et al. 2015a, b, c; Li et al. 2019; Özlem et al. 2017; Rocca et al. 2016; da Silva et al. 2018b) to facilitate their interaction with a living system, binding to a specific target, like tumor cells. Regarding the tumor-targeting strategies, the EPR effect of nanomaterials is a key mechanism for solid tumor targeting.

Several interesting studies have been performed to evaluate the behavior of functionalized BNNTs in some cell lines (Genchi et al. 2018; Horváth et al. 2011). Ferreira and co-works (2015a, b, c) studied the influence of BNNTs functionalized with gum arabic on the differentiation of mesenchymal stem cells. The interactions of the cells with the nanoparticles were further investigated by analyzing the conformation of the cytoskeleton. The differentiation of the mesenchymal stem cells into adipocytes and osteocytes after treatments with safe doses of BNNTs was assessed at both the gene and phenotype levels. The study showed that BNNTs have satisfactory biocompatibility toward stem cells. They also show the ability to affect cellular phenotype by enhancing adipogenic differentiation. Another study demonstrated that pectin-coated BNNTs (P-BNNTs) are nontoxic for macrophages up to 50 μg/mL after 24 h of incubation. According to this study, cytokine expression was not affected by the administration of the P-BNNT, and its uptake by macrophages did not cause any cell membrane impairment, adverse effects or inflammation processes in the cell (Rocca et al. 2016). BNNTs coated with glycol-chitosan (GC) assay were performed to evaluate the BNNT interactions with biological systems and determine their biosafety, using human vein endothelial cells (HUVECs). Various parameters were observed such as cell toxicity, proliferation, cytoskeleton integrity, cell activation and DNA damage. At the highest concentration, only a small reduction in cell viability and the increase of a vascular adhesion molecule (a marker of cell activation) expression were identified. According to that study, these findings show that GC-BNNTs do not affect endothelial cell biology and can be further investigated for vascular targeting, imaging and drug delivery (Del Turco et al. 2013). Genchi and co-works (2018) manufactured piezoelectric films of poly(vinylidenedifluoride-trifluoroethylene) [P(VDF-TrFE)] and of P(VDF-TrFE)/boron nitride nanotubes (BNNTs) by cast-annealing and used for SaOS-2 osteoblast-like cell culture. Films were characterized in terms of surface and bulk features. Preliminary evidence of enhanced differentiation of SaOS-2 osteoblast-like cells was found on composite piezoelectric films prepared with P(VDF-TrFE) and BNNTs upon ultrasound stimulation.

4 Drug Delivery Systems

Parveen et al. (2012) defined drug delivery as the process of releasing a bioactive agent at a specific rate and at a specific site. They also affirm that the application of many types of drugs is limited mainly by their poor solubility, high toxicity, lack of specificity, in vivo degradation and short circulating period. The drug delivery systems have been investigated since 1952 as an alternative approach to overcome these obstacles when Smith Kline & French developed an oral system for 12-h delivery of dextroamphetamine (Park 2014). However, the use of large-sized particles in drug delivery presents major challenges, such as in vivo instability, poor bioavailability and no target specificity (Patra et al. 2018). These problems can be overcome by reducing the size of the particulate carriers to the nanoscale.

Nanomedicines have been extensively exploited in scientific research in the last decades considering that nanostructures could be used for encapsulating or attaching therapeutic drugs and precisely deliver them to specific target tissues with a controllable releasing rate (Jahangirian et al. 2017). The drug-loaded nanoparticles can accumulate in various pathological areas with affected vasculature via the EPR effect; due to this feature, they have been exploited in scientific research for drug delivery into tumors via passive accumulation (Torchilin 2011), enhancing the specificity of carriers to tumor tissues and overwhelming the non-specificity trouble. By tuning the nanoparticle surfaces and porosity, it is also possible to control the specific rate of drug releasing, making an important advance in the pharmaceutical applications (Li 2017). To overcome the trouble of poor bioavailability, nanostructures can stay in the blood circulatory system for a prolonged period, and this feature, together with the controllable releasing rate, results in less plasma fluctuations of the loaded drugs (Patra et al. 2018). The in vivo stability is another problem in pharmaceutical applications that can be improved with the nanocarriers. By promoting specific modifications in nanoparticle surfaces, it is possible to enhance the stability of nanomaterials in biological fluids (Guerrini et al. 2018). Furthermore, the drug-loaded nanoparticle surfaces can be modified with thermal- or pH-sensitive polymers that allow the specific releasing of drugs with temperature or pH stimuli (Louguet et al. 2012). Some of these features have already been used in some clinical applications with the first generation of nanoparticles approved by FDA as drug delivery systems, the lipid systems such as liposomes and micelles (de Villiers et al. 2009).

Many classes of nanobiomaterials have been exploited in drug delivery applications due to their unique features previously discussed above. Hydroxyapatite nanoparticles show up as versatile materials in the drug carrier field due to the possibility of tuning their morphology, pore structure and surface area through different types of chemical routes (Sadat-Shojai et al. 2013). The main techniques described in the literature are based on the use of ionic or nonionic surfactants as template agents during the synthesis process to act as crystal growth drivers. CTAB, F127 and P124 are the most common templates exploited, but the last ones can provide more appropriate characteristics for drug delivery applications, such as narrow pore size distributions and higher surface areas (Mohammad et al. 2014). Palazzo and coworkers (2007) investigate the adsorption and desorption of distinct anticancer drugs in two different shaped HA nanoparticles (plate-shaped or needle-shaped morphologies). The results indicate that adsorption and desorption kinetics are dependent on the specific properties of the drugs and the morphology of the HA nanoparticles, demonstrating that HA nanoparticles can be tailored for specific therapeutic applications. Similarly, Venkatasubbu and colleagues (2013) studied the attaching of paclitaxel in hydroxyapatite nanoparticles functionalized with folic acid and polyethylene glycol, analyzing the in vitro drug release. This work shows that besides particle morphology and porosity tuning, a hydroxyapatite surface modification can provide a controlled drug release of anticancer drugs. Another important research was conducted by Wu et al. (2017) where the authors report the effect of hydroxyapatite nanoparticles loaded with medronate and JQ1 as an inhibitor system that can be used in the osteosarcoma (OS) treatment. Their results showed that 98% of JQ1 was released after 48 h; the effect was pronounced against the OS cells and nonexistent against the healthy fibroblasts, while OS cell invasion was significantly inhibited. Moreover, the internalization of JQ1-loaded HA nanoparticles happened in the majority of the OS cells as confirmed by flow cytometry tests, suggesting that this material provides active targeting for specifically deliver anticancer drugs into osteosarcoma cells.

Ordered mesoporous silica materials exhibit potential features to be used as controlled drug delivery systems. The control of the inner surface properties induced by such surface modification is required for the design of adsorption/desorption media for specific purposes, as drug delivery controlled devices. Reactions using coupling agents may be employed to improve the functionalization efficiency of silica. An example of a functionalization used is the reaction between the folic acid and (3-aminopropyl)triethoxysilane (APTES) using coupling agent, carbodiimide (EDC.HCl), to convert carboxylic acid to amide. Then, the silica can be reacted with the synthesized product between APTES and folic acid by post-synthesis grafting (Fig. 7).

De Freitas et al. (2015) synthesized ordered mesoporous materials with different morphologies, pore sizes and array of mesopores, and functionalized with folic acid by post-synthesis grafting and loaded with the anticancer agent methotrexate (MTX). The adequate combination of different characterization techniques allowed to show that the functional group N-folate-3-aminopropyl (FAP) was linked on the surface of MNPs samples, although the degree of functionalization differs depending on their structural characteristic. The results showed that all the functionalized particles exhibited a kinetic constant (K) values smaller than the corresponding bare materials, suggesting a stronger interaction between MTX and the FAP group when compared to the silanol groups, probably through electrostatic interactions. On the other hand, there were no distinguishable changes in the drug loading between the three different kinds of mesoporous silica, although they present distinct textural and structural properties.

Diverse studies with functionalized silica nanoparticles also have demonstrated that the release profile can be controlled by pH variations (dos Apostolos et al. 2019), temperature change (Monteiro et al. 2019), redox reactions (Guo et al. 2018) and

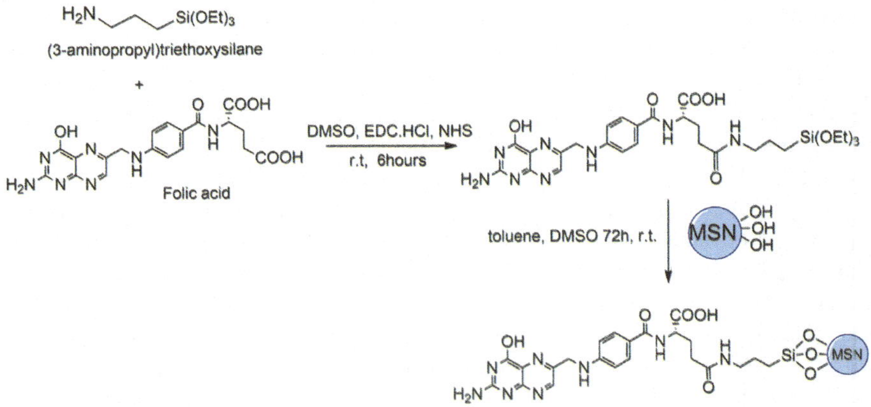

Fig. 7 Functionalization process of mesosporous silica nanoparticles with folic acid targeting ligand

magnetic field (Azevedo et al. 2014). Stimuli-responsive nanoparticles are called smart drug delivery system that can be used to carry chemotherapies that is widely used to treat cancer. Smart nanocarriers can deliver drugs to the target sites with reduced dosage frequency and in a spatially controlled manner to mitigate the side effects (Hossen et al. 2019).

Yuan et al. (2018) synthesized as an intelligent agent carrier, a redox-responsive mesoporous silica nanoparticle, based platform modified with protoporphyrin IX (PpIX), an excellent photosensitizer, and peptides. On the other hand, the weaknesses and constraints of PpIX, such as its unselective for tumors and hydrophilic characteristics, make it more difficult for use in clinical applications. Thus, mesoporous silica nanoparticles were used as platforms for photosensitizer-mediated photodynamic therapy. Thereby, redox-responsive nanosystem could release the agent under a glutathione stimulus. Glutathione is an essential intracellular antioxidant, and cancer cells contain higher levels of glutathione compared with normal cells. Moreover, the multifunctional system produces fluorescence recovery for tumor-specific fluorescence imaging and provides tumor-enhanced photodynamic therapy (Yuan et al. 2018).

Monteiro et al. (2019) synthesized Poly[(N-isopropylacrylamide)-co-(mathacrylic acid)])(P[(N-iPAAm)-co-(MAA)]) hydrogel by free radical polymerization varying the fractions between the monomers on the presence of mesoporous silica forming different thermo- and pH-sensitive hybrid systems aiming bioapplications as highly specific drug delivery systems (DDS). The drug methotrexate (MTX) was loaded into the studied hybrid and MPS samples, and its release behavior was studied using UV–Vis spectroscopy in different media at different temperatures to ensure the sensibility of the DDS to external stimuli. The MTX absorption studies and the latter released in vitro behavior analysis showed that SBA-15/P[(N-iPAAm)-co-(MAA)] is sensitive to both pH and temperature and presents potential to be used as a highly specific DDS targeting cancer treatment.

BNNTs-based drug carries have become a hot spot of research at the interface of nanotechnology and biomedicine because they allow the protection of the drug in the therapeutic system against possible instability in the body, promoting maintenance of plasma levels in constant concentration; increased therapeutic efficacy; the progressive and controlled release of the drug by conditioning to stimuli of the environment in which they are sensitive to pH or temperature variation; the significant decrease in toxicity by reducing plasma peaks of maximum concentration; decreased instability and decomposition of sensitive drugs; the possibility of targeting specific targets (site-specificity); the possibility of incorporating both hydrophilic and lipophilic substances into the devices; decreased therapeutic dose and number of administrations (Ciofani and Mattoli 2016).

5 Radiolabeling Nanoparticles for Theranostic Applications

Nanomaterials have come to contribute to traditional medicine with the introduction of new concepts and methods. In the case of cancer diagnostics, for example, some imaging techniques such as magnetic resonance imaging (MRI), X-ray computed tomography (CT), single-photon emission computer tomography (SPECT), positron emission tomography (PET) and optical imaging are powerful tools. As for therapy methods, photodynamic therapy (PDT), photothermal therapy (PTT), delivery of chemotherapeutical drugs and techniques based on magnetism have been applied (Beckert et al. 2016; Gai et al. 2014).

There are plenty of radionuclides that can be combined or incorporated into nanoparticles to make them specific tumor tracers in medical radiodiagnostic imaging systems like SPECT and PET. Since the beginning of century XX, right after the discovery of nuclear radiation by Marie Currie and Henri Becquerel, scientists start to question themselves about the possibility to detect radionuclides path in the human body. The first diagnostic trial performed using radionuclides on humans dates from 1925, when Hermann Blumgart used a natural mixture of ^{214}Pb and ^{214}Bi, originated from radium decay, to measure the blood flow through a modified Shimizu's dynamic Wilson cloud chamber (Patton 2003). Following these experiments, radiopharmacy started to be investigated at the end of 1940s decade with the experimental use of ^{131}I and ^{32}P to build traced maps of thyroid gland and to treat polycythaemia, respectively (Mccready 2016). The growth of nuclear medicine was boosted in the 1950s with the introduction of ^{99m}Tc generator and the design of the scintillation camera by Hal Oscar Anger (Patidar et al. 2010) that made it possible to develop many kinds of ^{99m}Tc based radiopharmaceuticals for specific organic systems imaging. Nowadays, there is good deal of radiopharmaceuticals based in a variety of short half-life radionuclides with distinct radiation energy and type of emission to make possible the diagnosis and treatment of different types of disease (Bailey et al. 2014). However, there is an emergent need to deliver a maximum radiation dose to tumor areas in a selective and localized manner (Ferro-Flores et al. 2014). The specific targeting of radiopharmaceuticals to tumor tissues can be improved with the advent of nanotechnology.

The radiolabeling technique can be an attractive approach to yield theranostic nanoparticulate materials. The term theranostics is assigned to the combination of therapy and diagnosis; therefore, the theranostic nanoparticles are nanostructures that are able to simultaneously treat and diagnose specific illnesses (Chen et al. 2014). Once inside tumor cells as a result of specific targeting of nanoparticulate carries, the gamma or positron radiations emitted by radionuclides incorporated in nanoparticles can be detected by external types of equipment that are able to localize the spatial region of these emissions and construct a diagnostic image. It is also possible to incorporate anticancer drugs or thermotherapy agents in combination with radionuclides to allow the treatment of cancer cells simultaneously to their imaging (Ferro-Flores et al. 2014). Some radionuclides show theranostic properties themselves, generally those that present both beta and gamma radiation or beta and positron emissions such as ^{131}I and ^{64}Cu, respectively (Choudhury and Gupta 2018; Zhao et al. 2014). The beta particles are considered as high linear energy transfer (LET) radiations that are able to cause several damages to tumor tissues once inside tumor cells, while gamma rays are low LET radiations with high penetrating power that make them reliable for outside detection. Nonetheless, there are other important factors that must be considered apart from the type of emission in order to choose the ideal radioisotope when radiolabeling nanoparticles for medical applications, such as the physical half-life, the chemical properties, the daughter nuclides, the availability and the cost of production (Koziorowski et al. 2017). The radionuclide must be a short half-life emitter, presenting no chemical toxicity with stable daughters, but some radionuclides, even presenting the above characteristics, show costly or impracticable production.

In recent years, researchers have been a great deal focused on the development of nanostructured systems such as pure or functionalized nanoparticles capable of encapsulating radioisotopes for the diagnosis and treatment of various types of cancer (Yaari et al. 2016). These systems have as main objective the delivery and retention of radioisotopes with high specific activity to the tumor cells, depositing in them high doses of radiation and leading them to death. At the same time, these systems aim to minimize the adverse effects of radiation on adjacent healthy tissues.

After the choice of the radionuclide for the desired application, there are some radiolabeling strategies reported in the literature. According to Koziorowski and colleagues, the radioisotopes can be incorporated in nanomaterials during the nanoparticle synthesis, and the nanoparticle constituents can be activated in situ by neutron irradiation of nanomaterials, or the radioisotopes can be incorporated by post-synthesis modifications of the nanoparticles.

Zhao and co-workers (2014) reported the possibility of alloying ^{64}Cu directly into the lattice of gold nanostructures with tunable specific activity, reaching 5.5 GBq/nmol and enabling highly sensitive detection. This material presented high radiolabeled stability up to 48 h which was significantly better than the stability of ^{64}Cu-DOTA (dodecane tetraacetic acid) functionalized gold nanoparticles as reported in the literature (Wang et al. 2012). Moreover, the authors studied the pharmacokinetics of ^{64}Cu-doped gold nanoparticles in BALB/c mice. Until 1-h postinjection, most of the nanoparticles stayed in the systemic circulation with low mononuclear phagocytic system uptake. Furthermore, the passive targeting capability of these radiolabeled nanoparticles was tested in an EMT-6 mouse breast cancer model using a small-animal PET/CT system. The results were consistent with the biodistribution profile, the PET/CT image showed high blood retention and the intra-tumoral uptake.

Cipreste et al. (Cipreste 2017) showed that hydroxyapatite nanoparticles can be activated by neutron irradiation to produce $^{64}Cu/^{32}P$ and $^{159}Gd/^{32}P$ radiolabeled nanoparticles to act in the bone tumors treatment and diagnosis, considering that hydroxyapatite shows high affinity for bone tissues. In this study, Gd-doped hydroxyapatite nanorods and tenorite-hydroxyapatite nanocomposites were functionalized with folic acid and then irradiated with neutron flux in a nuclear reactor. Cu, Gd and structural P of hydroxyapatite were activated to produce dual radioisotope combination. Phosphorous-32 is a pure beta emitter, able to provide therapeutic applications to these materials. The FA/HA-[^{32}P]-[^{159}Gd] nanoparticles as paramagnetic and gamma/beta emitters show ideal properties to be used in the treatment of multimodal image systems, such as nuclear medicine and magnetic resonance imaging, while the FA/HA-$^{64}CuO^{32}P$ nanoparticles present adequate properties to act as a theranostic system, allowing diagnostic imagens through PET modality. The results indicate that the post-synthesis irradiation can be an alternative way to radiolabel nanoparticles.

The theranostics materials have been interesting for several applications, especially in the early diagnosis and treatment of diseases such as cancer (Ma et al. 2019). Within this context, along with the difficulty of treating any type of injury that may affect body tissues, efforts have been directed toward the development of vehicles that can perform targeted delivery of drugs and radioisotopes.

The joint efforts of the researchers in multidisciplinary fields (chemistry, materials science, biology, immunology, and medicine) will strengthen the future of mesoporous silica nanoparticles-based diagnostics for targeted and personalized treatment of tumors. In addition to the development of new nanoparticle platforms that can be used for diagnostic purposes, exciting advances in theranostic nanoparticle constructions are taking place. Silica nanoparticles can be used for both diagnostic (imaging) and therapy (drug delivery) applications, thereby potentially acting as a single device for patients. The multifunctional platforms can contain in the same system a nanoparticles chelator agent, a target peptide, e.g., CREKA, and a loaded drug. Chelate is a metal complex that can be linked to a radioisotope used for therapy and diagnosis (Fig. 8).

de Freitas et al. (2017) investigated the synthesis and characterization of mesoporous silica nanoparticles for multifunctional targeting and controlled drug release. Nanotheranostic system based on silica nanoparticles was prepared through post-synthesis functionalization of mesoporous silica with the tumor-homing peptide CREKA and diethylenetriaminepentaacetic acid (DTPA), a ligand for copper ions. Copper (Cu) can be activated in a nuclear reactor to produce the radioisotope ^{64}Cu,

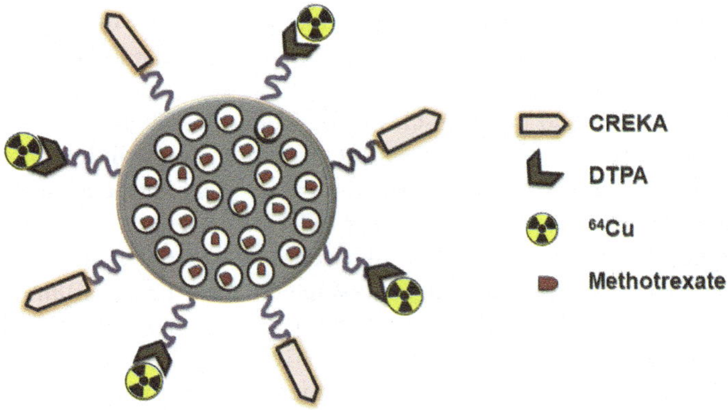

Fig. 8 Schematic illustration of the multifunctional mesoporous silica nanoparticle

a positron (β^+) and beta (β^-) radiations emitter. The radioisotope ^{64}Cu has decay characteristics that make it a promising radionuclide not only for positron emission tomography (PET) imaging but also for cancer radiotherapy. Additionally, the nanoparticles were loaded with the clinically approved anticancer drug methotrexate, and the drug release kinetics profile was investigated. The results demonstrated that the multifunctional silica-based theranostic system could serve as a platform for cancer-targeted, controlled drug delivery and PET imaging to be applied in cancer therapy.

There are several efforts to enhance the quality of images especially those taken from MRI. Notwithstanding, technical advances in nanotechnology are creating novel classes of MRI contrast-enhancing agents offering much higher relativities than most current clinical contrast agents, which translates into greater MRI contrast enhancement. These nanoscale agents also have the potential to revolutionize in vivo applications of contrast-enhanced MRI since they offer multiple advantages of low toxicities, extremely high relaxivities and cell internalization capabilities (Matson and Wilson 2010). One of these nanomaterials presenting good possibilities to enhance MRI contrast is BNNTs containing Fe paramagnetic impurities. One drawback of pristine BNNTs is its high hydrophobicity, and to overcome the low solubility of pristine BNNTs in aqueous solution, they can be wrapped with poly-L-lysine creating a PLL-BNNTs. These soluble BNNTs containing Fe have demonstrated values of transverse relaxivities comparable to commercial superparamagnetic iron oxide nanoparticles suggesting Fe-BNNT as a potential magnetic-enhanced contrast agent for MRI images at a field of 3 T (Menichetti et al. 2011). Gadolinium also is a widely used MRI contrast agent owing to its huge magnetic moment ($\mu^2 = 63\mu_B^2$) and its symmetric electronic ground state, ^{8}S7/2 (Ciofani et al. 2013a).

Radiotherapy is a widely used method for treating patients with various types of cancer. In this sense, when bound to target molecules, the BNNTs may, in fact, be used as specific therapeutic agents capable of being internalized by cancer cells and promote tumor death in a boron neutron capture therapy. This medical approach

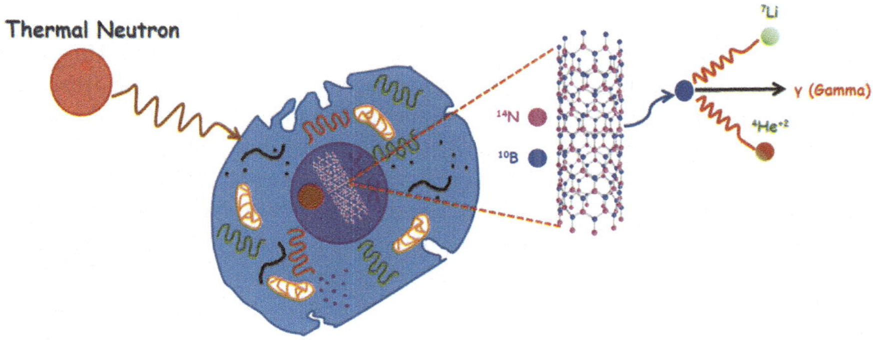

Fig. 9 Schematic representation of the activation of ^{10}B through thermal neutron irradiation

is used in the treatment of brain cancer, and it is based on $^{10}B(n, \alpha)^{7}Li$ neutron (n) capture, where one ^{10}B nucleus captures a low-energy thermal neutron and then decays producing ^{4}He (alpha particles-α) and ^{7}Li (Fig. 9). This event results in dense ionizing radiation capable of destroying cells where the reaction occurs (Ciofani et al. 2013b).

Radioisotopes can be used as an image resource in nuclear medicine and in some cases as a therapy agent too. BNNTs doped with samarium 152, a radioactive isotope from a rare-earth metal of the lanthanide group, have shown low toxicity in MRC-5 fibroblasts suggesting this nanomaterial as a potential nanosized β-emission source for nuclear medicine therapy especially for bone metastasis treatment (da Silva et al. 2018a).

Other applications of BNNTs in the field of cancer therapy due to their physical properties have been proposed. Raffa e co-works (Raffa et al. 2012) investigated the responses of tumor human HeLa cells when subjected to irreversible lethal electroporation (IRE) in the presence of BNNTs. IRE is a new non-thermal ablation modality that uses short pulses of high-amplitude static electric fields (up to 1000 V/cm) to create irreversible pores in the cell membrane, thus causing cell death. Recently, IRE has emerged as a promising clinical modality for cancer disease treatment. This study shows that BNNTs allows a strong reduction of the electric field required for irreversible electroporation of human HeLa cells. Preliminary experimental data confirmed that the phenomenon of IRE could be further enhanced by increasing BNNT concentration in the medium.

6 Magnetohyperthermia

Magnetic nanoparticles, such as magnetite (Fe_3O_4), are very interesting materials for biomedical applications because of their unique magnetic properties. The superparamagnetic regime of these particles imposes a very slow magnetic dynamics, which allows the recovery of a material remnant in a well-controlled manner when an alternating high-frequency magnetic field is applied. In this situation, magnetic nanoparticles release energy, promoting a localized temperature increase. This effect is promising for cancer treatment and is known as magnetic hyperthermia. Treatment is based on localized warming of tumor tissue in the range of 41–45 °C, stimulated by an external alternating magnetic field. In this temperature range, the damage is reversible in healthy cells, but not in tumor cells (Brollo et al. 2016). However, the aggregation of these nanoparticles due to van der Walls forces has become a limiting factor for their application. In this sense, nanoparticles such as mesoporous silica and BNNT can be investigated as potential vectors suitable for hyperthermia.

The introduction of magnetic nanoparticles into the BNNT structure allows the use of this system as a hyperthermia agent (Fig. 10). The superparamagnetic profile of the magnetite nanoparticles enables the system to reach required temperatures when submitted to the AC magnetic field, under specific conditions. Thus, using the magnetohyperthermia technique, the tumor can be treated by increasing the temperature of cancer cells..

In a recent study, Wang and co-works manufactured a hybrid nanocomposite consisting of uniform magnetite (Fe_3O_4) nanoparticles and boron nitride nanospheres (BNNS) with an average diameter of 150 nm. In this system, the BNNS were uniformly coated with dense ultra-small Fe_3O_4. Magnetic measurement by vibrating

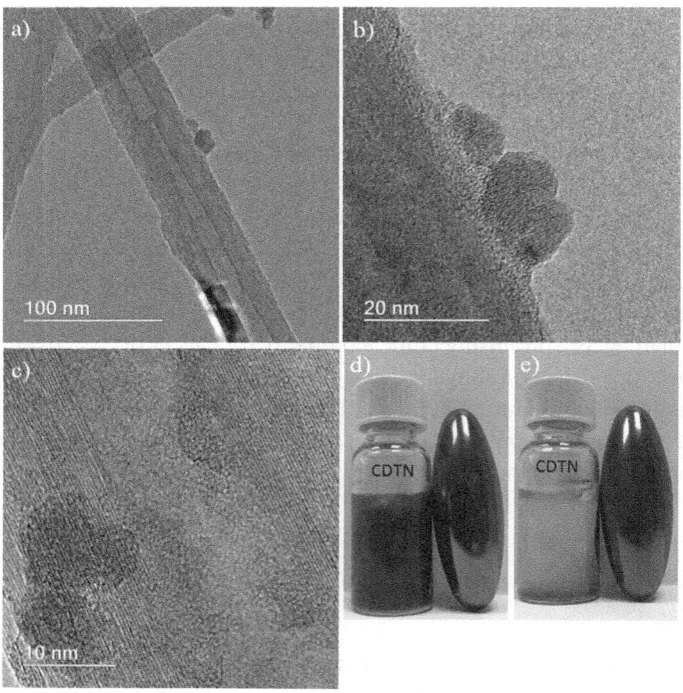

Fig. 10 TEM images of (**a**, **b** and **c**) BNNTs-Fe_3O_4 and under low magnetic field at **d** 0 min and **e** after 30 s

sample magnetometer (VSM) shows the superparamagnetic behavior, and the nanocomposite can be physically manipulated at a low magnetic field. The nanocomposite showed cytocompatibility at low concentration and has little effect on cell viability on MCF-7, MCF-10 and HeLa cell lines (Wang et al. 2017). In other work, Ferreira and co-works manufactured nanocomposites of BNNT-Fe_3O_4 by chemical co-precipitation method. The results obtained show that magnetite nanoparticles are linked to the BNNTs. Magnetic measurements show that coercivity and magnetization were not disturbed after incorporation to the BNNTs. The biological assays of the system demonstrate its good cell viability and the great potential of this nanomaterial as a magnetohyperthermia agent for cancer treatment (Ferreira et al. 2018).

Due to the flexibility of chemical modification in the surface characteristic of mesoporous silica nanoparticles, surface engineering has been adopted to perform robust functionalization, associating different molecules for other purposes in order to direct these particles to the target site (Vallet-Regí et al. 2007). In addition, this material has been studied as a molecule carrier allowing the development of silica nanoparticles for different applications as support for magnetic hyperthermia (Azevedo et al. 2014), gene targeting (Steinbacher and Landry 2014) and contrast agents for biomedical imaging (Benezra et al. 2011). As magnetic nanoparticles present high specific surface areas, unprotected nanoparticles can easily form aggregates and react with oxygen in the air. They can also rapidly biodegrade when directly exposed to biological systems. In this context, a special and facile synthesis route to obtain magnetite nanocomposite highly dispersed in mesoporous silica was reported by Souza et al. (2010). Hyperthermia tests indicated that the nanocomposite can be used as a matrix for localized hyperthermia treatment of cancers. Results indicated that heat generation depends on the intensity of AC magnetic field and the nanoparticle medium.

Azevedo et al. (2014) have shown the combining mesoporous silica–magnetite and thermally sensitive polymers, Poly(N-isopropylacrylamide)—P(*N*-iPAAm), for applications in hyperthermia. Temperature sensitivity is one of the most interesting characteristics in stimulus-responsive polymeric nanocarriers and has been extensively investigated to exploit the hyperthermia condition for drug and gene delivery. The heat generated in the system can lead to the phase transition of the thermosensitive polymer, and the resultant contraction of the gel phase can trigger the release of drugs embedded in the network, favoring treatments associated with hyperthermia. In this case, it was proposed a synergistic effect of hyperthermia and controlled drug delivery for a hybrid system composed of the combination of mesoporous silica nanoparticles, magnetite and P(*N*-iPAAm).

7 Gene Delivery

Silica nanoparticles have been extensively studied for both their toxicity and in vivo biodistribution and have shown good tolerance and favorable clearance profiles(de Barros et al. 2015). The first silica nanoparticles used in the diagnosis of human melanoma have been approved for initial clinical trials in humans by Food and Drug Administration (FDA) (Benezra et al. 2011). Collectively, the biocompatibility and distinctive structure offer a diversity of interaction with the biological environment, which can improve drug delivery, cellular targetability and protection for therapeutic agents (Zhou et al. 2018).

Gene delivery is another application of silica nanoparticle likewise the delivery of small molecules and proteins. The use of silica nanoparticles as non-viral vector for gene transfection has been explored due to their flexible surface modified with cationic molecules, which allow for stable condensation with nucleotides that are highly negatively charged and the protection of them from nuclease in physiological state. Nhavene et al. (2018) used biodegradable polymers, polycaprolactone (PCL) and chitosan (CS), covalently attached by crosslink molecules to the surface of silica mesoporous nanoparticles to act as an anchor for siRNA delivery for cervical cancer cells (HeLa). The biocompatibility in vitro assay and cytotoxicity test suggests the ability of silica functionalized with PCL to support passive cellular uptake and consequently indicated its application as potential non-viral vector in gene delivery systems (Fig. 11).

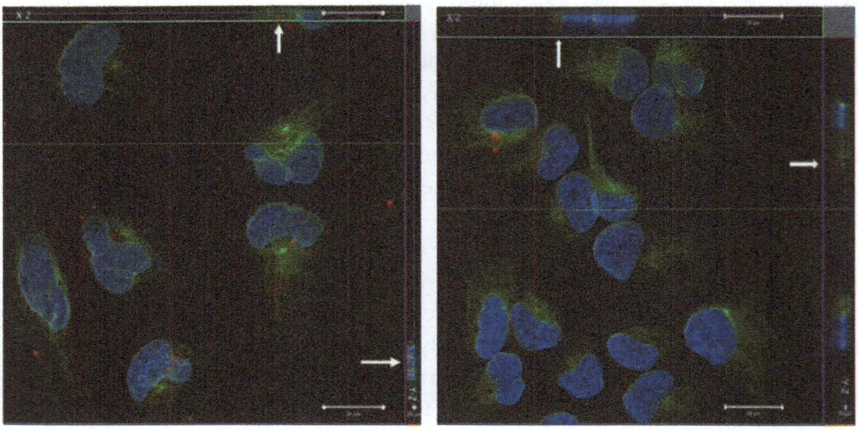

(a) **(b)**

Fig. 11 Cellular internalization of siRNA-MCM-41 + APTES + PCL conjugates. Rhodamine-siRNA- MCM-41 + APTES + PCL is in red, α-tubulin is in green, Lamin B1 is in cyan, and the nucleus is stained with Hoechst in blue. Representative image of serial optical sections collected for three-dimensional reconstruction; x-z sections are shown at the top, and y-z sections are shown at the right of image. siRNA-MCM-41 + APTES + PCL co-localizes with α-tubulin, as indicated by yellow overlay (White arrows). Scale bar = 20 μm. **a** Cellular uptake of fluorescent rhodamine-siRNA-MCM-41 + APTES + PCL in HeLa cells; **b** Cellular uptake of fluorescent rhodamine-siRNA-MCM-41 + APTES + PCL in HeLa cells in another field. *Source* Nhavene et al. (2018)

8 Conclusion

This chapter describes recent studies of nanoparticles for anticancer therapy. Despite considerable progress in the treatment of different forms of cancer, cell resistance to chemotherapeutic agents and damage caused to healthy tissues by radiotherapy are still a challenge that the medical community has faced for many years. In this sense, nanomaterials have come to contribute to traditional medicine with the introduction of new concepts and methods. Nanomaterials such as silica nanoparticles, hydroxyapatite and BNNTs could become a key material for the development of nanovectors, drugs, or genes in cancer therapy, and indeed, these nanoparticles can provide significant advances in the area of medical imaging, molecular biology and biomedical technology.

References

Arriagada FJ, Osseo-Asare K (1992) Phase and dispersion stability effects in the synthesis of silica nanoparticles in a non-ionic reverse microemulsion. Colloids Surf 69:105–115

Ashley CE, Carnes EC, Phillips GK, Padilla D, Durfee PN, Brown PA, Hanna TN, Liu J, Phillips B, Carter MB, Carroll NJ, Jiang X, Dunphy DR, Willman CL, Petsev DN, Evans DG, Parikh AN, Chackerian B, Wharton W, Peabody DS, Brinker CJ (2011) The targeted delivery of multicomponent cargos to cancer cells by nanoporous particle-supported lipid bilayers. Nat Mater 10:389–397

Azevedo RCS, Sousa RG, Macedo WAA, Sousa EMB (2014) Combining mesoporous silica–magnetite and thermally-sensitive polymers for applications in hyperthermia. J Sol Gel Sci Technol 72:208–218

Bailey DL, Humm JL, Todd-Pokropek A, van Aswegen A (eds) (2014) Nuclear medicine physics: a handbook for teachers and students. International Atomic Energy Agency, Vienna

Bao G, Mitragotri S, Tong S (2013) Multifunctional nanoparticles for drug delivery and molecular imaging. Annu Rev Biomed Eng 15:253–282

Barth RF, Mi P, Yang W (2018) Boron delivery agents for neutron capture therapy of cancer. Cancer Commun 38:1–15

Beckert MB, Gallego S, Ding Y, Elder E, Nadler JH (2016) Medical imaging scintillators from glass-ceramics using mixed rare-earth halides. Opt Mater (Amst) 60:513–520

Benezra M, Penate-medina O, Zanzonico PB, Schaer D, Ow H, Burns A, DeStanchina E, Longo V, Herz E, Iyer S, Wolchok J, Larson SM, Wiesner U, Bradbury MS (2011) Multimodal silica nanoparticles are effective cancer-targeted probes in a model of human melanoma find the latest version: technical advance multimodal silica nanoparticles are effective cancer-targeted probes in a model of human melanoma. J Clin Invest 121:2768–2780. https://doi.org/10.1172/JCI45600

Blase X, Rubio A, Louie SG, Cohen ML (1994) Stability and band gap constancy of boron nitride nanotubes. Europhys Lett 28:335–340

Brollo MEF, Orozco-Henao JM, López-Ruiz R, Muraca D, Dias CSB, Pirota KR, Knobel M (2016) Magnetic hyperthermia in brick-like Ag@Fe_3O_4 core–shell nanoparticles. J Magn Magn Mater 397:20–27

Cai L, Guinn AS, Wang S (2011) Exposed hydroxyapatite particles on the surface of photo-crosslinked nanocomposites for promoting MC3T3 cell proliferation and differentiation. Acta Biomater 7:2185–2199

Chaturvedi VK, Singh A, Singh VK, Singh MP (2018) Cancer nanotechnology: a new revolution for cancer diagnosis and therapy. Curr Drug Metab 20:416–429

Chen F, Ehlerding EB, Cai W (2014) Theranostic nanoparticles. J Nucl Med, 1–5
Cherukula K, Manickavasagam Lekshmi K, Uthaman S, Cho K, Cho C-S, Park I-K (2016) Multifunctional inorganic nanoparticles: recent progress in thermal therapy and imaging. Nanomaterials 6:76
Chopra NG, Luyken RJ, Cherrey K, Crespi VH, Cohen ML, Louie SG, Zettl A (1995) Boron nitride nanotubes. Science, 80–
Choudhury PS, Gupta M (2018) Differentiated thyroid cancer theranostics: radioiodine and beyond. Br J Radiol 91:20180136
Ciofani G, Mattoli V (2016) Boron nitride nanotubes in nanomedicine, 1st edn. William Andrew, Oxford
Ciofani G, Genchi GG, Liakos I, Athanassiou A, Dinucci D, Chiellini F, Mattoli V (2012) A simple approach to covalent functionalization of boron nitride nanotubes. J Colloid Interface Sci 374:308–314
Ciofani G, Boni A, Calucci L, Forte C, Gozzi A, Mazzolai B, Mattoli V, (2013a) Gd-doped BNNTs as T2-weighted MRI contrast agents. Nanotechnology, 24
Ciofani G, Danti S, Genchi GG, Mazzolai B, Mattoli V (2013b) Boron nitride nanotubes: biocompatibility and potential spill-over in nanomedicine. Small 9:1672–1685
Ciofani G, Del Turco S, Rocca A, De Vito G, Cappello V, Yamaguchi M, Li X, Mazzolai B, Basta G, Gemmi M, Piazza V, Golberg D, Mattoli V (2014) Cytocompatibility evaluation of gum Arabic-coated ultra-pure boron nitride nanotubes on human cells. Nanomedicine 9:773–788
Cipreste MF (2017) NANOBASTÕES DE HIDROXIAPATITA RADIOMARCADOS COMO AGENTES TERANÓSTICOS PARA OSTEOSSARCOMAS E METASTASES ÓSSEAS. Centro de Desenvolvimento da Tecnologia Nuclear
Cipreste Marcelo Fernandes, Gonzalez I, da Mata Maria, Martins T, Goes AM, de Almeida Augusto, Macedo W, Barros de Sousa EM (2016a) Attaching folic acid on hydroxyapatite nanorod surfaces: an investigation of the HA–FA interaction. RSC Adv 6:76390–76400
Cipreste Marcelo F, Peres AM, Cotta AAC, Aragón FH, Antunes ADM, Leal AS, Macedo WAA, de Sousa EMB (2016b) Synthesis and characterization of 159 Gd-doped hydroxyapatite nanorods for bioapplications as theranostic systems. Mater Chem Phys 181:301–311
Cole JT, Holland NB (2015) Multifunctional nanoparticles for use in theranostic applications. Drug Deliv Transl Res 5:295–309
da Silva WM, Hilário Ferreira T, de Morais CA, Soares Leal A, Barros Sousa EM (2018a) Samarium doped boron nitride nanotubes. Appl Radiat Isot 131:30–35
da Silva WM, Monteiro GAA, Gastelois PL, de Sousa RG, de Macedo WAA, Sousa EMB (2018b) Efficient sensitive polymer-grafted boron nitride nanotubes by microwave-assisted process. Nano Struct Nano-Objects 15:186–196
de Barros ALB, de Oliveira Ferraz KS, Dantas TCS, Andrade GF, Cardoso VN, Sousa EMB De (2015) Synthesis, characterization, and biodistribution studies of ^{99m}Tc-labeled SBA-16 mesoporous silica nanoparticles. Mater Sci Eng C 56:181–188
de Freitas LBO, Bravo IJG, de Macedo WAA, de Sousa EMB (2015) Mesoporous silica materials functionalized with folic acid: preparation, characterization and release profile study with methotrexate. J Sol Gel Sci Technol, 186–204
de Freitas LBO, de Corgosinho LM, Faria JAQA, dos Santos VM, Resende JM, Leal AS, Gomes DA, de Sousa EMB (2017) Multifunctional mesoporous silica nanoparticles for cancer-targeted, controlled drug delivery and imaging. Microporous Mesoporous Mater 242:271–283
de Villiers MM, Aramwit P, Kwon GS (eds) (2009) Nanotechnology in drug delivery. Springer, New York
Del Turco S, Ciofani G, Cappello V, Gemmi M, Cervelli T, Saponaro C, Nitti S, Mazzolai B, Basta G, Mattoli V (2013) Cytocompatibility evaluation of glycol-chitosan coated boron nitride nanotubes in human endothelial cells. Colloids Surf B Biointerfaces
Dorozhkin SV (2015) Calcium orthophosphate bioceramics. Ceram Int 41:13913–13966

dos Apostolos RCR, Andrade GF, da Silva WM, de Assis Gomes D, de Miranda MC, de Sousa EMB (2019) Hybrid polymeric systems of mesoporous silica/hydroxyapatite nanoparticles applied as antitumor drug delivery platform. Int J Appl Ceram Technol, 1836–1849

Dvorak H, Senger D, Dvorak A, Harvey V, McDonagh J (1985) Regulation of extravascular coagulation by microvascular permeability. Science(80–):227:1059–1061

Emanet M, Şen Ö, Çobandede Z, Çulha M (2015) Interaction of carbohydrate modified boron nitride nanotubes with living cells. Colloids Surf B Biointerfaces 134:440–446

Ferreira TH, Ferreira Soares DC, Costa Moreira LM, Da Silva PRO, Dos Santos RG, De Sousa EMB (2013) Boron nitride nanotubes coated with organic hydrophilic agents: stability and cytocompatibility studies. Mater Sci Eng C 33

Ferreira Tiago H, Rocca A, Marino A, Mattoli V, de Sousa EMB, Ciofani G (2015a) Evaluation of the effects of boron nitride nanotubes functionalized with gum arabic on the differentiation of rat mesenchymal stem cells. RSC Adv 5:45431–45438

Ferreira TH, Rocca A, Marino A, Mattoli V, De Sousa EMB, Ciofani G (2015) Evaluation of the effects of boron nitride nanotubes functionalized with gum arabic on the differentiation of rat mesenchymal stem cells. RSC Adv, 5

Ferreira THTH, Marino A, Rocca A, Liakos I, Nitti S, Athanassiou A, Mattoli V, Mazzolai B, de Sousa EMBEMB, Ciofani G (2015c) Folate-grafted boron nitride nanotubes: possible exploitation in cancer therapy. Int J Pharm 481:56–63

Ferreira TH, Faria JAQA, Gonzalez IJ, Outon LEF, Macedo WAA, Gomes DA, Sousa EMB (2018) BNNT/Fe_3O_4 system as an efficient tool for magnetohyperthermia therapy. J Nanosci Nanotechnol 18:6746–6755

Ferro-Flores G, Ocampo-García BE, Santos-Cuevas CL, Morales-Avila E, Azorín-Vega E (2014) Multifunctional radiolabeled nanoparticles for targeted therapy. Curr Med Chem 21:124–138

Gai S, Li C, Yang P, Lin J (2014) Recent progress in rare earth micro/nanocrystals: soft chemical synthesis, luminescent properties, and biomedical applications. Chem Rev 114:2343–2389. https://doi.org/10.1021/cr4001594

Gao Z, Zhi C, Bando Y, Golberg D, Serizawa T (2014) Noncovalent functionalization of boron nitride nanotubes in aqueous media opens application roads in nanobiomedicine. Nanobiomedicine 1:7

Geim AK, Novoselov KS (2007) The rise of graphene. Nat Mater 6:183–191

Genchi GG, Sinibaldi E, Ceseracciu L, Labardi M, Marino A, Marras S, De Simoni G, Mattoli V, Ciofani G (2018) Ultrasound-activated piezoelectric P(VDF-TrFE)/boron nitride nanotube composite films promote differentiation of human SaOS-2 osteoblast-like cells. Nanomed Nanotechnol Biol Med 14:2421–2432

Golberg D, Bando Y, Huang Y, Terao T, Mitome M, Tang C, Zhi C (2010) Boron nitride nanotubes and nanosheets. ACS Nano 4:2979–2993

Gomes MC, Cunha Â, Trindade T, Tomé JPC (2016) The role of surface functionalization of silica nanoparticles for bioimaging. J Innov Opt Health Sci 09:1630005

Guerrini L, Alvarez-Puebla R, Pazos-Perez N (2018) Surface modifications of nanoparticles for stability in biological fluids. Materials (Basel). 11:1154

Guo X, Cheng Y, Zhao X, Luo Y, Chen J, Yuan WE (2018) Advances in redox-responsive drug delivery systems of tumor microenvironment. J Nanobiotechnol 16:1–10

Hao N, Nie Y, Zhang JXJ (2019) Microfluidics for silica biomaterials synthesis: opportunities and challenges. Biomater Sci 7:2218–2240

He Q, Zhang J, Shi J, Zhu Z, Zhang L, Bu W, Guo L, Chen Y (2010) The effect of PEGylation of mesoporous silica nanoparticles on nonspecific binding of serum proteins and cellular responses. Biomaterials 31:1085–1092

Horváth L, Magrez A, Golberg D, Zhi C, Bando Y, Smajda R, Horváth E, Forró L, Schwaller B (2011) In vitro investigation of the cellular toxicity of boron nitride nanotubes. ACS Nano 5:3800–3810

Hossen S, Hossain MK, Basher MK, Mia MNH, Rahman MT, Uddin MJ (2019) Smart nanocarrier-based drug delivery systems for cancer therapy and toxicity studies: a review. J Adv Res 15:1–18

Jahangirian H, Ghasemian lemraski E, Webster TJ, Rafiee-Moghaddam R, Abdollahi Y (2017) A review of drug delivery systems based on nanotechnology and green chemistry: green nanomedicine. Int J Nanomedicine 12:2957–2978

Jain RK (1999) Transport of molecules, particles, and cells in solid tumors. Annu Rev Biomed Eng 1:241–263

Jang C, Lee JH, Sahu A, Tae G (2015) The synergistic effect of folate and RGD dual ligand of nanographene oxide on tumor targeting and photothermal therapy in vivo. Nanoscale 7:18584–18594

Kawasaki Y, Freire E (2011) Finding a better path to drug selectivity. Drug Discov Today 16:985–990

Kong L, Mu Z, Yu Y, Zhang L, Hu J (2016) Polyethyleneimine-stabilized hydroxyapatite nanoparticles modified with hyaluronic acid for targeted drug delivery. RSC Adv 6:101790–101799

Koziorowski J, Stanciu A, Gomez-Vallejo V, Llop J (2017) Radiolabeled nanoparticles for cancer diagnosis and therapy. Anticancer Agents Med Chem 17:333–354

Kroto HW, Heath JR, O'Brien SC, Curl RF, Smalley RE (1985) C60: Buckminsterfullerene. Nature 318:162–163

Large DE, Soucy JR, Hebert J, Auguste DT (2019) Advances in receptor-mediated, tumor-targeted drug delivery. Adv Ther 2:1800091

Li Z (2017) How can we fine-tune nanoparticles to improve drug delivery? Ther Deliv 8:597–600

Li W, Xie X, Wu T, Lin H, Luo L, Yang H, Li J, Xin Y, Lin X, Chen Y (2019) Loading Auristatin PE onto boron nitride nanotubes and their effects on the apoptosis of Hep G2 cells. Colloids Surf B Biointerfaces 181:305–314

Louguet S, Rousseau B, Epherre R, Guidolin N, Goglio G, Mornet S, Duguet E, Lecommandoux S, Schatz C (2012) Thermoresponsive polymer brush-functionalized magnetic manganite nanoparticles for remotely triggered drug release. Polym Chem 3:1408

Ma Z, Wan H, Wang W, Zhang X, Uno T, Yang Q, Yue J, Gao H, Zhong Y, Tian Y, Sun Q, Liang Y, Dai H (2019) A theranostic agent for cancer therapy and imaging in the second near-infrared window. Nano Res 12:273–279

Mak TW, Saunders ME (2005) The immune response: basic and clinical principles, 1st edn. Academic Press

Mamaeva V, Sahlgren C, Lindén M (2013) Mesoporous silica nanoparticles in medicine-Recent advances. Adv Drug Deliv Rev 65:689–702

Matson ML, Wilson LJ (2010) Nanotechnology and MRI contrast enhancement. Future Med, Chem, p 2

Matsumura Y, Maeda H (1986) A new concept for macromolecular therapeutics in cancer chemothrapy: mechanism of tumoritropic accumulation of proteins and the antitumor agent smancs. Cancer Res 46:6387–6392

Mccready R (2016) A history of radionuclide studies in the UK. Springer International Publishing, Cham

Menichetti L, Marchi D De, Calucci L, Ciofani G, Menciassi A, Forte C (2011) Boron nitride nanotubes for boron neutron capture therapy as contrast agents in magnetic resonance imaging at 3 T. Appl Radiat Isot 69:1725–1727

Mi P, Kokuryo D, Cabral H, Kumagai M, Nomoto T, Aoki I, Terada Y, Kishimura A, Nishiyama N, Kataoka K (2014) Hydrothermally synthesized PEGylated calcium phosphate nanoparticles incorporating Gd-DTPA for contrast enhanced MRI diagnosis of solid tumors. J Control Release 174:63–71

Mohammad NF, Othman R, Yee-Yeoh F (2014) Nanoporous hydroxyapatite preparation methods for drug delivery applications. Rev Adv Mater Sci 38:138–147

Mokoena PP, Chithambo ML, Kumar V, Swart HC, Ntwaeaborwa OM (2015) Thermoluminescence of calcium phosphate co-doped with gadolinium and praseodymium. Radiat Meas 77:26–33

Monteiro GAA, da Silva WM, de Sousa RG, de Sousa EMB (2019) SBA-15/P[(N-ipaam)-co-(MAA)] thermo and pH-sensitive hybrid systems and their methotrexate (MTX) incorporation and release studies. J Drug Deliv Sci Technol 52:895–904

Muñoz B, Rámila A, Pérez-Pariente J, Díaz I, Vallet-Regí M (2003) MCM-41 organic modification as drug delivery rate regulator. Chem Mater 15:500–503
Ngoune R, Peters A, von Elverfeldt D, Winkler K, Pütz G (2016) Accumulating nanoparticles by EPR: A route of no return. J Control Release 238:58–70
Nhavene E, Andrade G, Faria J, Gomes D, Sousa E (2018) Biodegradable polymers grafted onto multifunctional mesoporous silica nanoparticles for gene delivery. Chem Engineering 2:24
Özlem Ş, Çobandede Z, Emanet M, Faruk Ö, Çulha M (2017) BBA—General Subjects Boron nitride nanotubes for gene silencing
Pakdel A, Zhi C, Bando Y, Nakayama T, Golberg D (2011) Boron nitride nanosheet coatings with controllable water repellency. ACS Nano 5:6507–6515
Palazzo B, Iafisco M, Laforgia M, Margiotta N, Natile G, Bianchi CL, Walsh D, Mann S, Roveri N (2007) Biomimetic hydroxyapatite-drug nanocrystals as potential bone substitutes with antitumor drug delivery properties. Adv Funct Mater 17:2180–2188
Park K (2014) Controlled drug delivery systems: past forward and future back. J Control Release 190:3–8
Parveen S, Misra R, Sahoo SK (2012) Nanoparticles: a boon to drug delivery, therapeutics, diagnostics and imaging. Nanomed Nanotechnol Biol. Med. 8:147–166
Patidar AK, Patidar P, Tandel TS, Mobiya AK, Selvam G, Jeyakandan M (2010) Current trends in nuclear pharmacy practice. Int J Pharm Sci Rev Res 5:145–150
Patra JK, Das G, Fraceto LF, Campos EVR, Rodriguez-Torres MDP, Acosta-Torres LS, Diaz-Torres LA, Grillo R, Swamy MK, Sharma S, Habtemariam S, Shin H-S (2018) Nano based drug delivery systems: recent developments and future prospects. J Nanobiotechnol 16:71
Patton DD (2003) The birth of nuclear medicine instrumentation: Blumgart and Yens, 1925. J Nucl Med 44:1362–1365
Raffa V, Riggio C, Smith MW, Jordan KC, Cao W, Cuschieri A (2012) BNNT-mediated irreversible electroporation: its potential on cancer cells. Technol Cancer Res Treat 11:459–465
Rámila A, Muñoz B, Pérez-Pariente J, Vallet-Regí M (2003) Mesoporous MCM-41 as drug host system. J Sol Gel Sci Technol 26:1199–1202
Ringhieri P, Mannucci S, Conti G, Nicolato E, Fracasso G, Marzola P, Morelli G, Accardo A (2017) Liposomes derivatized with multimeric copies of KCCYS lpeptide as targeting agents for HER-2-overexpressing tumor cells. Int J Nanomed 12:501–514
Rocca A, Marino A, Del Turco S, Cappello V, Parlanti P, Pellegrino M, Golberg D, Mattoli V, Ciofani G (2016) Pectin-coated boron nitride nanotubes: in vitro cyto-/immune-compatibility on RAW 264.7 macrophages. Biochim Biophys Acta Gen Subj 1860:775–784
Rosenholm J, Mamaeva V (2017) Nanoparticles in targeted cancer therapy : Mesoporous silica nanoparticles entering preclinical development. Review, 111–120
Sadat-Shojai M, Khorasani M-T, Dinpanah-Khoshdargi E, Jamshidi A (2013) Synthesis methods for nanosized hydroxyapatite with diverse structures. Acta Biomater 9:7591–7621
Silva WM, Ribeiro H, Seara LM, Calado HD, Ferlauto AS, Paniago RM, Leite CF, Silva GG (2012) Surface properties of oxidized and aminated multi-walled carbonnanotubes. J Braz Chem Soc 23(6):1078e1086.
Simberg D, Duza T, Park JH, Essler M, Pilch J, Zhang L, Derfus AM, Yang M, Hoffman RM, Bhatia S, Sailor MJ, Ruoslahti E (2007) Biomimetic amplification of nanoparticle homing to tumors. Proc Natl Acad Sci 104:932–936
Simovic S, Ghouchi-Eskandar N, Moom Sinn A, Losic DA, Prestidge C (2011) Silica materials in drug delivery applications. Curr Drug Discov Technol 8:250–268
Soares DCF, Ferreira TH, Ferreira CDA, Cardoso VN, De Sousa EMB (2012) Boron nitride nanotubes radiolabeled with ^{99m}Tc: preparation, physicochemical characterization, biodistribution study, and scintigraphic imaging in Swiss mice. Int J Pharm 423:489–495
Souza KC, Mohallem NDS, Sousa EMB (2010) Mesoporous silica-magnetite nanocomposite: facile synthesis route for application in hyperthermia. J Sol Gel Sci Technol 53:418–427
Stahl P, Schwartz AL (1986) Receptor-mediated Endocytosis. TJ Clin Invest 77:657–662

Stein A, Melde BJ, Schroden RC (2000) Hybrid inorganic-organic mesoporous silicates—nanoscopic reactors coming of age. Adv Mater 12:1403–1419
Steinbacher JL, Landry CC (2014) Adsorption and release of siRNA from porous silica. Langmuir 30:4396–4405
Stober W (1968) Stober Method.Pdf 69:62–69
Sumio I, Toshinari I (1993) Single-shell carbon nanotubes of 1-nm diameter. Nature 363:603–604
Tang Li, Cheng Jianjun (2013) Nonporous silica nanoparticles for nanomedicine application. Nano Today 8:290–312
Terrones M, Romo-Herrera JM, Cruz-Silva E, López-Urías F, Muñoz-Sandoval E, Velázquez-Salazar JJ, Terrones H, Bando Y, Golberg D (2007) Pure and doped boron nitride nanotubes. Mater Today 10:30–38
Torchilin V (2011) Tumor delivery of macromolecular drugs based on the EPR effect. Adv Drug Deliv Rev 63:131–5
Torchilin VP, Trubetskoy VS (1995) Which polymers can make nanoparticulate drug carriers long-circulating? Adv Drug Deliv Rev 16:141–155
Unger K, Rupprecht H, Valentin BKW (1983) The use of porous and surface modified silicas as drug delivery and stabilizing agents 9:69–91
Vácha R, Martinez-Veracoechea FJ, Frenkel D (2011) Receptor-mediated endocytosis of nanoparticles of various shapes. Nano Lett 11:5391–5395
Vallet-Regí M, Balas F, Arcos D (2007) Mesoporous materials for drug delivery. Angew Chem Int Ed Engl 46:7548–7558
Venkatasubbu GD, Ramasamy S, Avadhani GS, Ramakrishnan V, Kumar J (2013) Surface modification and paclitaxel drug delivery of folic acid modified polyethylene glycol functionalized hydroxyapatite nanoparticles. Powder Technol 235:437–442
Wang Y, Liu Y, Luehmann H, Xia X, Brown P, Jarreau C, Welch M, Xia Y (2012) Evaluating the pharmacokinetics and in vivo cancer targeting capability of au nanocages by positron emission tomography imaging. ACS Nano 6:5880–5888
Wang W, Lin J, Xing C, Chai R, Abbas S, Song T, Tang C, Huang Y (2017) Fe 3 O 4 nanoparticle-coated boron nitride nanospheres: synthesis, magnetic property and biocompatibility study. Ceram Int 43:6371–6376
Wang X, Ramalingam M, Kong X, Zhao L (eds) (2018) Nanobiomaterials: classification, fabrication and biomedical applications, 1st edn. Wiley, Boschstr
Wicki A, Witzigmann D, Balasubramanian V, Huwyler J (2015) Nanomedicine in cancer therapy: challenges, opportunities, and clinical applications. J Control Release 200:138–157
Wu VM, Mickens J, Uskoković V (2017) Bisphosphonate-functionalized hydroxyapatite nanoparticles for the delivery of the bromodomain inhibitor JQ1 in the treatment of osteosarcoma. ACS Appl Mater Interfaces 9:25887–25904
Yaari Z, Da Silva D, Zinger A, Goldman E, Kajal A, Tshuva R, Barak E, Dahan N, Hershkovitz D, Goldfeder M, Roitman JS, Schroeder A (2016) Theranostic barcoded nanoparticles for personalized cancer medicine. Nat Commun 7:1–10
Yuan F, Li JL, Cheng H, Zeng X, Zhang XZ (2018) A redox-responsive mesoporous silica based nanoplatform for: In vitro tumor-specific fluorescence imaging and enhanced photodynamic therapy. Biomater Sci 6:96–100
Zhao D, Huo Q, Feng J, Chmelka BF, Stucky GD (1998) Nonionic triblock and star diblock copolymer and oligomeric surfactant syntheses of highly ordered, hydrothermally stable, mesoporous silica structures. J Am Chem Soc 7863:6024–6036
Zhao Y, Sultan D, Detering L, Cho S, Sun G, Pierce R, Wooley KL, Liu Y (2014) Copper-64-alloyed gold nanoparticles for cancer imaging: improved radiolabel stability and diagnostic accuracy. Angew Chemie Int Ed 53:156–159
Zhao J, Zhang B, Shen S, Chen J, Zhang Q, Jiang X, Pang Z (2015) CREKA peptide-conjugated dendrimer nanoparticles for glioblastoma multiforme delivery. J Colloid Interface Sci 450:396–403

Zhou Y, Quan G, Wu Q, Zhang X, Niu B, Wu B, Huang Y, Pan X, Wu C (2018) Mesoporous silica nanoparticles for drug and gene delivery. Acta Pharm Sin B 8:165–177

Chapter 10
Nanoparticles by Ultrasound Irradiation: Organic and Inorganic Materials

Lillian Maria Uchoa Dutra Fechine, Fernando Lima Menezes, Letícia Nogueira Xavier, Aldenor Souza de Oliveira, and Pierre Basílio Almeida Fechine

1 Introduction

Organic/inorganic-based nanomaterials have shown great versatility mainly due to their unique properties from their bulk counterparts. For instance, the reduced size of particles rises up the reaction rate once the surface area is increased, which constitutes the main mean where the reagents can interact with each other in chemical reactions (Islam et al. 2019). Hence, the synthesis of this class of materials requires more efficient methods which are capable of controlling some parameters, such as size, morphology and composition (Bang and Suslick 2010).

Regarding applications, these materials are well applied in electronic gadgets, e.g. thermoplastic polyurethane (TPU) and thermally reduced and annealed graphene sheets (TRG) producing a TPU/TRG nanocomposite that presents the basic requirements of the commercial electrical applications against electromagnetic pollution, mainly because of its properties, such as high electrical conductivity, dielectric constant and tangent losses (Joshi et al. 2018); drug delivery systems, e.g. gold nanoparticles that are considered to present microbial and anti-carcinogenic activities besides inertness and low levels of cytotoxicity (Anniebell and Gopinath 2018); catalysis, e.g. graphene-based nanomaterials applied especially for energy conversion and environmental protection (Hu et al. 2017); diagnosis, e.g. the detection of tuberculosis antigens and DNA by nanomaterials-based electrochemical biosensors (Phan et al. 2019); imaging agents, e.g. iron and manganese oxides-based nanomaterials allow controllable design and surface modification of the contrast agent, which affect their magnetic properties and turn them suitable to help in MRI (Anderson et al. 2019).

L. M. U. D. Fechine · F. L. Menezes · L. N. Xavier · A. S. de Oliveira · P. B. A. Fechine (✉)
Group of Chemistry of Advanced Materials (GQMat), Department of Analytical Chemistry and Physical-Chemistry, Federal University of Ceará – UFC, Campus do Pici, CP 12100, CEP, 60451-970 Fortaleza, CE, Brazil
e-mail: fechine@ufc.br

R. F. do. Nascimento et al. (eds.), *Nanomaterials and Nanotechnology*,
Materials Horizons: From Nature to Nanomaterials,
https://doi.org/10.1007/978-981-33-6056-3_10

The relevance of choosing an appropriate synthetic route for designing nanostructure materials is due to strongly method dependence of their properties and, consequently, applications (Bang and Suslick 2010). There are quite of methods that are already implemented in synthetic pathways of organic, inorganic and hybrid materials at "nanorange", including the utilization of sol-gel technique (Vennela et al. 2019), colloidal methods (Murray et al. 2001), self-propagating high-temperature combustion method (Levashov et al. 2016), microwave-assisted technique (Elazab 2018), hydrothermal (HT) synthesis, spray pyrolysis (Tovar-martinez et al. 2018) and microemulsion technique (Li et al. 2017).

Despite the common applications, the ultrasound-assisted chemistry synthesis, well called as sonochemistry, involves the generation of extreme transient conditions by the acoustic cavitation phenomenon, which can induce physical, thermal and chemical effects in solutions (Chatel 2018). The ultrasonic irradiation appears as an approach to synthesize organic and inorganic materials under ambient and mild conditions, and also produce reactions at the liquid–liquid interface between reagents that are immiscible in each other, i.e. this method allows reactions that cannot be performed in homogeneous medium (Piradashvili et al. 2016). Moreover, compared to traditional methods, the ultrasonic irradiation is advantageous due to the unconventional provided reaction conditions that decreases the duration required to perform the reaction and the consumption of surfactants. Besides, the precise control over chemical reactions by the manipulation of some parameters including time and input energy into the system and shows a considerable improvement in the yield of the desired product (Bang and Suslick 2010).

In brief, sonochemistry consists in one of the most successful ultrasound-assisted synthetic methods that consist in a versatile tool for the preparation of nanostructured materials, i.e. it is the research area that studies enabled chemical reactions by the application of ultrasound irradiation (Mason and Cintas 2002; Gedanken 2004).

2 Ultrasound Irradiation: Overview

The "ultrasound" (US) is described as a longitudinal compression sound wave type beyond human hearing, which is generated in a frequency higher than 20 kHz (i.e. >20,000 Hz). Normally, ultrasound waves are generated by specific ceramic materials with piezoelectric property, which can in-depth describe about the theoretical concept of electricity conversion to ultrasound wave generation. Briefly, the ceramic material, transducer, expands or contracts when an alternating voltage is through applied. This behaviour provides a continuous ultrasound wave irradiation at the same frequency.

Indeed, the frequency itself plays an essential rule since it is responsible to tailor the type of sound wave. Basically, ultrasound irradiation in a medium (gas or liquid) can be tailored by (1) frequency (f, cycles/s, Hz), which is defined as the number of cycles completed by the wave per time unit (t, s), as shown in the following equation;

$$f = \frac{1}{t} \tag{1}$$

and (2) intensity (I), which is defined as the average energy transmitted through a unit area that is perpendicular to the direction of the wave propagation, as shown in Eq. 2.

$$I = \frac{p^2}{2\rho C} \tag{2}$$

where I is the acoustic intensity (W/m^2), p the maximum pressure of the wave, ρ the density of the through medium (kg/m^3) and C is the wave velocity in the medium (m/s). Herein, acoustic power can be defined by the total energy irradiated by a source, in this case, piezoelectric ceramic, per unit of area (Eq. 3).

$$W = I \times S \tag{3}$$

where S represents the surface irradiant area (m^2), I is the acoustic intensity (W/m^2) and W the acoustic power itself (W) (Cabrera-Trujillo et al. 2016).

Sound waves propagation, i.e. speed, depends on the nature of the medium, for instance, the speed of sound (20–20,000 Hz) is different comparing air and water, 342 m/s and 1440 m/s, respectively. In the case of US irradiation, the nature of medium is also important, being liquid or gas the ultrasound waves can pass through with pressure cycles of compression (high pressure) and rarefaction (low pressure) providing the effect of acoustic cavitation phenomenon. Indeed, cavitation has been identified as transitory at low ultrasonic frequencies (20 kHz $< f <$ 100 kHz), where cavitation causes the rapid growth of bubbles, leading them to collapse. This cavitation generates high temperatures and pressures in the medium, well known as "hot spot" regions. It has been proved that US propagation, and consequently, the acoustic cavitation phenomenon is the main responsible for all versatilities and innumerous applications of ultrasound irradiation, including nanomaterials synthesis (Freire et al. 2016).

The literally concept of "cavitation" phenomenon is the creation of cavities in a liquid, i.e. bubble growth, and in mostly liquids used in laboratory experiments contain dissolved gas molecules, which under US irradiation, the gas bubbles grow and further collapse. This behaviour is well known as "rectified diffusion" process, which is deeper described by Ashokkumar (2018). As bubbles surface area expands and compresses, the amount of gas/vapour *in* and *out* of the bubbles also changes, where the diffusion into the bubble overcome the diffusion *out*, resulting in a specific ratio between ultrasonic frequency (Hz) and bubble radius (m), as seen in Eq. (4).

$$f \times r \sim 3 \tag{4}$$

Herein, as the ultrasound frequency decreases, the radius of the cavitation bubbles grows, resulting in the inertial collapse, and then generating extreme high temperatures, from 1000 to 5000 K, and pressures around 100 atm.

These extreme environments and physical phenomena are responsible for inorganic and organic synthesis under US irradiation such as surface functionalization and inorganic/organic nanoparticles which will be deeper described in next sections.

3 Inorganic Nanoparticles: The Concepts of Sonochemistry and Applications

Sonochemistry can be briefly defined as a chemical reaction driven by ultrasonic irradiation, where the energy involved into materials synthesis is firstly given from US acoustic power itself, which is generated into "hot spot" regions thought a liquid due to acoustic cavitation phenomenon. Despite green chemistry overview, the sonochemical approach has become a great alternative to those materials through non-traditional method. Herein, new synthetic routes and mechanism reactions have been proposed since materials synthesis under US irradiation allows controlling of size, morphology and some tuneable properties, as can be further seen in proposed methods and applications.

US approach has been used to obtain many inorganic nanomaterials, such as metallic (Kamali et al. 2019; Hu et al. 2019), alloys (Okoli et al. 2018), phosphates (Xing et al. 2019), carbides (Argüelles-Pesqueira et al. 2018), sulphides (Kristl et al. 2017) and oxides (Bhosale and Bhanage 2016; Andrade Neto et al. 2017) nanoparticles. Therefore, it has been shown as a versatile technique to synthesize advanced nanomaterials and applications. In Table 1, it is presented different NP types obtained through US in last 10 years.

Zheng and co-workers used ultrasonic irradiation to reduce Pd(II) and H_2PtCl_6 to obtain core-shell metallic Pd@Pt nanoparticles with excellent methanol tolerance and selectivity to oxygen reaction, in order to be applied in catalysis of methanol fuel cells (Zheng et al. 2019). The schematic illustration of the synthesis is shown in Fig. 1.

One-step synthesis was performed under US irradiation, where a radicular mechanism was proposed since the dissociation of H_2O molecules into radical species of hydrogen ($H^{\cdot}$) and hydroxyl ($OH^{\cdot}$) takes place due to high temperature and pressure of 5000 K (hot spots) and 2000 atm from bubble collapses (Eq. 5). Further, a sequential generation of the other radical molecules can be assigned as the following equations:

$$H_2O \rightarrow H^{\cdot} + OH^{\cdot} \tag{5}$$

$$RH + OH^{\cdot}(H^{\cdot}) \rightarrow R^{\cdot} + H_2O(H_2) \tag{6}$$

Table 1 Examples of different types of NPs, US regime, particle size and properties in last 10 years

NPs by US	US regime	Size/properties	References
Nanocrystals of spinel Co_3O_4 and Mn_3O_4	Ultrasonic horn, 1.23 cm^2 Ti horn at 20 kHz and power output of 600 W	NPs around 19–21.5 nm, where US reduced NPs sizes and increased their crystallinity	Askarinejad and Morsali (2009)
Polymethylacrylic acid (PMMA)-coated Ag nanoclusters	Ultrasonic horn, 1 cm^2 Ti horn at 20 kHz and 25 W/cm^2 Time of reaction: 10–180 min under Ar flow	Clusters of 2 nm, with optical and fluorescence properties controlled by sonication time, polymer and Ag/COO^- ratio	Xu and Suslick (2010)
Er_2O_3 NPs	Ultrasonic horn, 0.7 in. Ti horn at 20 kHz and 29 $W \bullet cm^{-2}$. The temperature of reaction was kept at 40 °C in a reactor cell	3 nm with hexagonal and spherical geometries. Er carboxioxide NPs on carbon nanotubes showed electromagnetic emission in the visible region and the photoluminescence of hexagonal erbium oxide NPs was long lived	Radziuk et al. (2011)
Carbonated hydroxyapatite nanopowders	Ultrasonic horn, 10 mm diameter titanium at 25 kHz and 900 W working in a pulse regime cycle of 2s *on* and 1s *off*	Rod-like crystals with a diameter of 8 nm and length of 30 nm were obtained at high yield, where had higher specific surface area as well as better dispersibility	Zou et al. (2012)
CdTe/PbS core/shell nanocrystals	Ultrasonic horn, 0.196 cm^2 Ti horn at 20 kHz and 220 W, at 25 °C, 1.4 atm under Ar flow	8–13 nm CdTe NPs were covered with 1–2 nm depth PbS to obtain a lower band-gap luminescent material	Park et al. (2013)
PdM NPs (M = Ni, Co, Fe, Mn)	Ultrasonic horn, 13 mm solid probe at 20 kHz and 500 W with 30% of amplitude for 3 h under an Ar environment at room temperature.=	NPs with particle size from 6.8 nm (PdFe/C) to 7.8 nm (Pd/C) were obtained. PdM alloy NPs showed enhanced activity and durability in FAO electrocatalyst when compared to Pd	Matin et al. (2014)

(continued)

Table 1 (continued)

NPs by US	US regime	Size/properties	References
Mn–Zn ferrite NPs	Ultrasonic horn, 5 cm^2 Ti probe at 20 kHz with a maximum power of 1500 W for 75 min	NPs with like-cubic shapes and size from 19.5 ($Mn_{0.8}Zn_{0.2}Fe_2O_4$) to 22.8 nm ($Mn_{0.5}Zn_{0.5}Fe_2O_4$), showing high crystallinity and monodispersity without any post-calcination reaction	Abbas et al. (2015)
Chitosan-functionalized Fe_3O_4 NPs	Ultrasonic horn, Ti horn, 50% amplitude, in a pulse regime of 20s *on*—10s *off* for 2 min	NPs with average crystallite sizes from 8.37 nm to 12.67 nm with high crystallinity and saturation magnetization, which showed a great potential as electrochemical sensor, with signal exceptionally stable and increased electroactive area	Freire et al. (2016)
Nickel and cobalt sulphide NPs	Ultrasonic horn, 1 cm^2 Ti horn at 20 kHz and 600 W, working at ambient atmosphere, different amplitudes of 30, 50, 70 and 90%; Time of reaction: from 30 to 120 min	NiS NPs were obtained with a crystallite size from 13 to 24 nm, showing a reduced band-gap from 3.8 to 3.3 eV. Crystallite size of Co ranged from 22 to 30 nm	Kristl et al. (2017)
$Ni(OH)_2$ NPs	Ultrasonic horn, Ti horn at 20 kHz and 130 W, working at room temperature with amplitude of 20% during 5 min	Particle size ranged from 5 to 10 nm. US irradiation increased ζ-potential of $Ni(OH)_2$ NPs. The materials showed remarkable electrochemical stability	Lorenzen et al. (2018)
Pd@Pt NPs	Ultrasonic horn, Ti horn at 20 kHz and 500 W working at 25 °C. Time of reaction: from 20 to 40 min	Pd_3Pt and Pd_4Pt NPs were obtained with size around 3–4 nm and 4–5 nm, respectively The materials exhibited excellent methanol tolerance and selectivity to oxygen reaction on direct methanol fuel cells	Zheng et al. (2019)

(continued)

Table 1 (continued)

NPs by US	US regime	Size/properties	References
Hg/Pd Alloy	Ultrasonic horn, Ti horn at 20 kHz and 750 W working at 65 °C for 60 min. Hg/Pd ratio was varied (1:1, 2.5:1 and 4:1)	Particle size ranged from 2 to 3 nm (1:1 alloy) to 8 nm (4:1 Hg/Pd alloy). The NPs exhibited excellent catalytic activity and good stability with reusability	Harika et al. (2020)

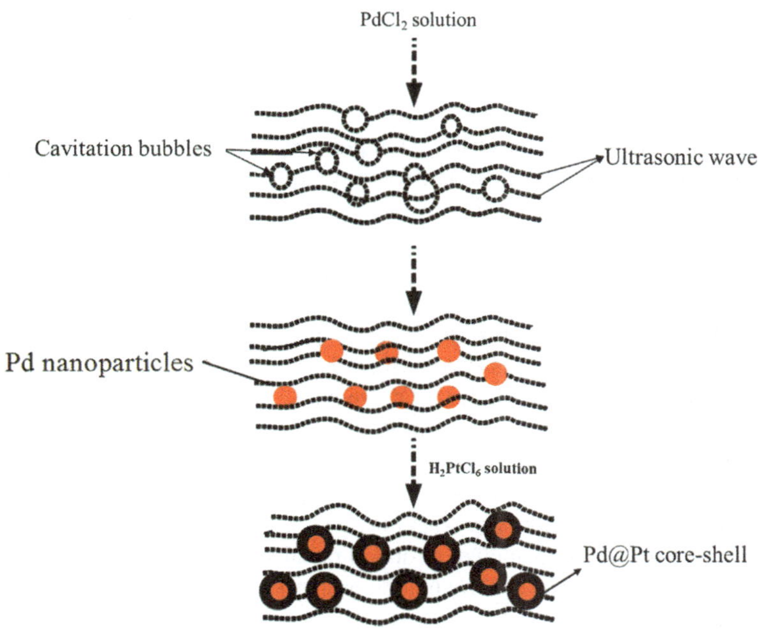

Fig. 1 Schematic illustration for the synthesis of Pd@Pt core-shell using ultrasonication (Zheng et al. 2019)

$$xM^{n+} + xn\mathrm{H}^{\cdot} \rightarrow xM^0 + xn\mathrm{H}^+ \tag{7}$$

$$xM^{n+} + xnR^{\cdot} \rightarrow xM^0 + xnR' + \mathrm{H}^+ \tag{8}$$

where RH denotes surfactant or alcohol, and *M* and *R* represents the metal and the secondary radical formed from RH molecule, respectively. Those species were generated during the high-energy bubble collapse due to cavitation phenomenon, since the authors performed a low frequency US irradiation of 20 kHz around 20–40 min.

A similar mechanism was used by Okoliand co-workers to obtain Pd_3Mo alloy material at nanoscale range to be applied as electrode and electrocatalyst (Okoli et al. 2018). The authors evaluated the effect of the solvent, organics and water, in the physical-chemical properties of the final nanomaterial.

All alloys were obtained under US irradiation using a Misonix 3000 ultrasonic liquid processor at average intensity of 84 (W/cm^2) at 40 °C during 30 min. The well description of the sonolysis methodology can be found in the Okoli et al. (2018). Briefly, the proposal mechanism started from the metallic precursors reduction, where molybdenum hexacarbonyl $Mo(CO)_6$ and palladium (II) acetylacetonate ($Pd(acac)_2$) cleavage in radicular species, reducing in a mixture of ionic liquid, ethanol and water.

The proposed method greatly produced smaller particles sizes (3.00 $\pm$ 0.05 nm) and more uniform size distribution particles for all tested solvents, despite PEG 400 solution (4.00 $\pm$ 1 nm) and hexadecane with polyvinylpyrrolidone (PVP) (14.00 $\pm$ 1 nm).

The study also discusses the effect of temperature, sonication, time and precursors on the obtention of the nanomaterial, where the parameter using low temperature (40 °C) and 30–40 min of sonication was chosen to be a helpful controlling particles size growth, and higher catalytic activity (Okoli et al. 2018).

The ultrasound effect on the size of materials was also studied by Xing (Xing et al. 2019), in the obtention of heparin-coated Eu^{3+}-doped hydroxyapatite (HAP) nanoparticles by co-precipitation reaction. In just 6 min of high intensity of US irradiation, the authors were able to reduce the average particle size from 170 to 110 nm, and the polydispersity index (PDI) from 0.23 to 0.17. The ultrasound-assisted synthesis also promoted the crystallization of HAP, as well as its photoluminescence and stabilization with heparin for use as a bioimaging agent. The US irradiation could act as particle size controller and also promoted a stable nanoparticle dispersion in water with zeta potential around −47.5 mV.

Sonochemistry was also applied to obtain metal sulphide-based nanoparticles at low cost, since those materials have unique properties to be used as semiconductors and catalysts. Kristl et al. (2017) proposed to synthesize nickel and cobalt sulphide nanoparticles under US irradiation, avoiding traditional solid-state methods, or even ions solution reaction. Besides, the authors also evaluated the effect of different low frequency US (20 kHz) conditions on the size, morphology and optical band-gap energy of the nanoparticles. Basically, the methodology was performed adding $Ni(CH_3COO)_2{\cdot}4H_2O$ or $Co(CH_3COO)_2{\cdot}2H_2O$ and C_2H_5NS into aqueous solution, and further, the mixture was sonicated during 1 h at different amplitudes. As expected, the increase of the acoustic power greatly decreased the band-gap of Ni sulphide nanoparticles from 3.8 to 3.3 eV, as well as positively contributed to improve the physical-chemical properties of both synthesized materials, Co and Ni sulphide nanoparticles (Kristl et al. 2017).

Andrade Neto et al. (2017) used ultrasonic irradiation to synthesize and functionalize Fe_3O_4 nanoparticles in a short-time reaction with superparamagnetic properties to be applied as image contrasts in magnetic resonance image (MRI). The 12 min' methodology allowed to obtain particles with higher crystallinity and saturation

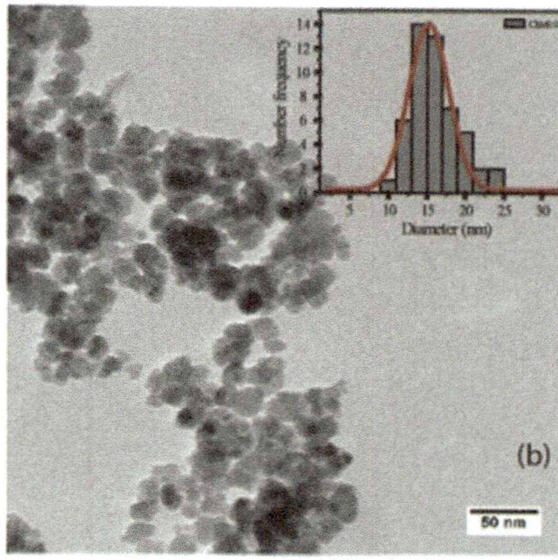

Fig. 2 Transmission electronic microscopy of chitosan@magnetite and particle diameter histogram (inset) (Freire et al. 2016)

magnetization (M_s) compared to traditional co-precipitation method. The sonochemical approach proved to be a versatile technique in the obtention of Fe_3O_4 functionalized with different molecules, polyethyleneimine, sodium citrate, sodium polyacrylate and sodium oleate. Actually, the US-assisted syntheses provided monodisperse particles with high colloidal stability and also showing the efficiency in the surface modification of Fe_3O_4 (Andrade Neto et al. 2017).

Previously, the same research group also obtained chitosan@Fe_3O_4 NPs in just 2 min by an ultra-fast in situ reaction using sonochemistry (Freire et al. 2016). Chitosan-covered iron oxide NPs with high crystallinity were obtained, where showed a crystallite size around 8.37–12.67 nm. The authors also evaluated the polymer concentration influence into magnetite physical-chemical structure and properties, as well as in the synthesis itself of the nanocomposite. As expected, despite chitosan@Fe_3O_4 NPs have been synthesized by conventional co-precipitation reaction, both polymer and fast-assisted US approach could act in the composite size controlling, and consequently increase the inorganic dispersion into polymer matrix (Fig. 2).

Besides, those particles also showed a superparamagnetic behaviour with saturation magnetization around 32–57 emu g^{-1} and great potential to be applied as a modified electrode sensor (Freire et al. 2016). Figure 3 shows the schematic representation of nanocomposite preparation and final structural obtention.

It is more than proved that sonochemistry has been useful for synthesizing different inorganic nanoparticles with outstanding efficiency regarding energy-consuming and materials properties. However, sonochemistry can be considered beyond chemical reactions. This technique was also used for improvement of already synthesized materials. For instance, Barbosa et al. (2018) used US irradiation to cover TiO_2 nanoparticles with SiO_2, Al_2O_3, ZrO_2 and sodium polyacrylate (PAANa) to be applied in sunscreen formulation. The sonication process provided a significant improvement of colloidal stability (lower hydrodynamic size and higher zeta potential) of commercial TiO_2 particles. Interesting, a better aesthetic appearance of the sunscreen formulation was observed, and the incorporation of the sonochemical-coated TiO_2 particles

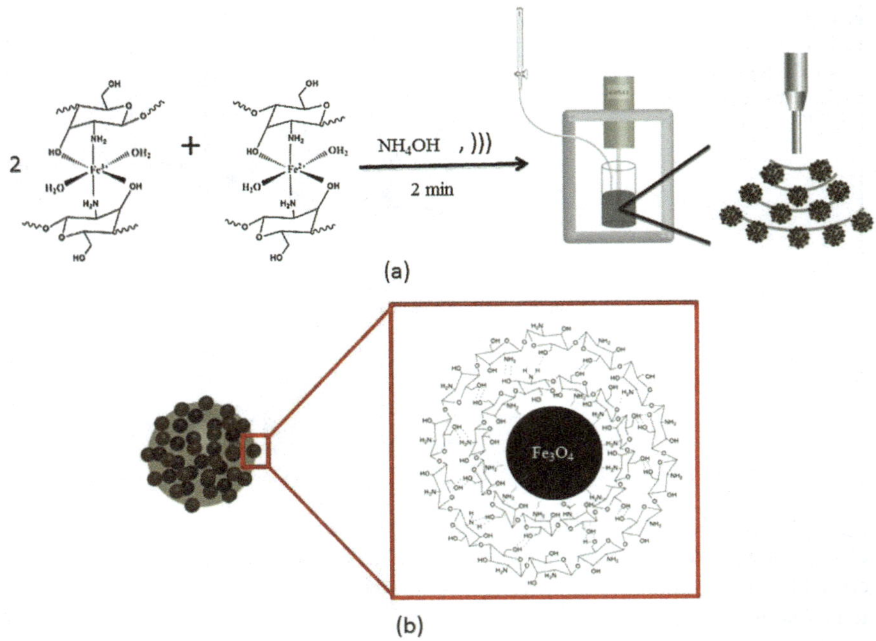

Fig. 3 **a** Schematic representation for the preparation of Fe_3O_4-chitosan nanocomposite by sonochemistry and **b** proposed final structural arrangement (Freire et al. 2016)

into formulation also promoted a greater health security due to its lower free radicals production (Barbosa et al. 2018), as seen in Fig. 4.

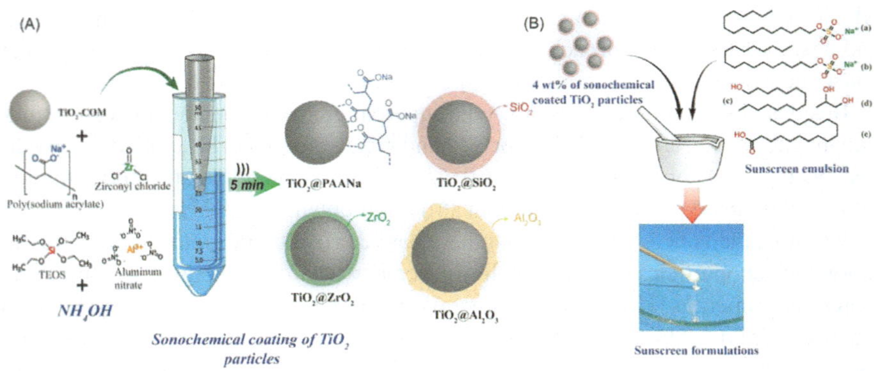

Fig. 4 **a** Scheme of commercial TiO_2 coating under US and **b** TiO_2 incorporation into sunscreen formulation (Barbosa et al. 2018)

4 Organic Nanoparticles: Miniemulsion Technique and Applications

As already discussed, the versatility of US irradiation comes from the acoustic cavitation phenomenon. For example, in the case of two immiscible liquids, if a bubble collapses near the phase boundary of the liquid, the resulting shock wave can provide a very effective mixing of the layers, producing emulsions within nanoscale range droplets and a narrow size distribution (Cárcel et al. 2012). Table 2 shows some examples of organic nanoparticles/capsules obtained through miniemulsion in last 10 years.

Thus, the ultrasonic irradiation applications have been considered to be a powerful and emerging technology, and it is considered a green methodology that offers great potential for a variety of nanomaterials synthesis such as drug delivery systems and polymer-based nanoparticles/capsules (Gallego-Juárez et al. 2010).

4.1 Miniemulsion Concept

Miniemulsion (also referred as nanoemulsion) consists of a heterogeneous system composed of two immiscible liquids that can be classified according to the composition as oil-in-water (direct miniemulsion) or water-in-oil (inverse miniemulsion), using an appropriate surfactant. The dispersion and the size of the stable droplets (narrow size distribution in a range of 30–500 nm) are created by applying a shear field to the system caused by the ultrasound waves through the liquid medium, retaining each single droplet without nucleation or mass transfer processes involved (Landfester 2001a).

Miniemulsions are typically prepared by ultrasound (normally at low frequency) in a two-step mechanism, where a macroemulsion is first prepared, and then it is converted to a nanoemulsion in a second step: (1) the application of an acoustic field produces a combination of interfacial waves which leads to an instability that causes an eruption of the dispersed phase into the continuous phase into droplets form; and (2) the acoustic cavitation phenomenon causes the growth and implosion of short life-time microbubbles by the pressure fluctuations.

Each bubble collapse occurs when the applied shear is greater than the *Laplace* pressure of the droplets and, consequently, produces highly localized turbulence that act in the breaking up of dispersed phase into nanodroplets (Kentish et al. 2008). This extreme implosion at the interface of a liquid-liquid surface promotes a disruption between their interface, enhancing liquid–liquid dispersion, and thus improves interfacial area and mass transfer coefficient between two phases (Wilhelm et al. 2010).

The major differences between nanoemulsion and microemulsion and macroemulsion are in droplet size range and stability characteristics. Macro and nanoemulsions are both thermodynamically unstable, i.e. given sufficient time, phase separation

Table 2 Nanocapsules and nanoparticles by miniemulsion technique: US regime, particle size and properties in the last 10 years

NPs by miniemulsion	US regime	Size/properties	References
Nanocapsules of methacrylate/lecithin via miniemulsion	Ultrasonic horn during 4 min with an amplitude of 60% at room temperature	NCs morphology with a particle size distribution of 200 nm, where the capsules formed a droplet/template for topical application	Romio et al. (2009)
Cyclodextrin nanospheres	Ultrasonic horn during 2 min with an amplitude of 68% at 0 °C	Nanospheres with size around 200 nm, presenting high ability to absorb aromatic organic molecules from water	Baruch-Teblum et al. (2010)
Poly(urethane–urea) nanocapsules	Ultrasonic horn of 25 mm (750 W, 20 kHz) during 1 min under 20% of amplitude	Nanocapsules with size around 150 nm, where can act as "nanoreactors" for interface reactions	Gaudin and Sintes-Zydowicz (2011)
Nanocapsules of N-isopropyl acrylamide (NIPAM), N,N′-methylene bisacrylamide (MBA) and a functional monomer, 4-vinyl pyridine (4-VP)	120s under US irradiation using a Branson 450 W digital sonifier at 90% of amplitude	Nanocapsules with Z-average size of 276 nm, showing pH-thermosensitivity property	Cao et al. (2012)
Polypyrrole – zinc oxide (PPy/ZnO) functional latex	Ultrasonic horn of 13 mm (750 W, 22 kHz) operated at 40% of amplitude	The TEM analysis showed narrow size dispersion with spherically shaped ZnO nanoparticles in the polypyrrole matrix presenting great response for sensor application	Barkade et al. (2013)

(continued)

Table 2 (continued)

NPs by miniemulsion	US regime	Size/properties	References
Poly(urea-urethane)/Fe_3O_4 nanoparticles	Ultrasonic horn (400 W) for 180s at 70% of power intensity	Magnetic NPs were synthesized with a diameter between 5 and 15 nm, showing high magnetic response, and TEM presented NPs with high encapsulation level	Chiaradia et al. (2015)
Lignin-based nanocapsules	Ultrasonic horn for 3 min (70% of 300 W power under pulse regime)	Nanocapsules have a particle size range from 100 to 400 nm, exhibiting excellent thermal stability and could effectively encapsulate hydrophobic charge for potential application in medical industries	Chen et al. (2016)
Nanocapsules of graphene and folic acid	Ultrasonic horn (500 W/cm^2, 20 kHz) for 5 min	Nanocapsules with diameter around 800 nm of graphene oxide, showing attractive reduction response and controlled release ability for hydrophobic drugs	Cui et al. (2017)
Polymeric nanocapsules of the sodium dodecyl sulphate (SDS) and n-hexadecane, (HD)		Nanocapsules showing an average diameter of 220 nm and potential application for encapsulation of different types of materials, such as drugs, dyes or insecticides	Artusio et al. (2018)
Poly(styrene-co-methyl methacrylate) nanospheres	US horn (20 kHz) operating in a pulse mode (3s *on* and 1s *off*) at 55% amplitude to deliver 275 W	US-assisted synthesis showed a notable effect on the NPs size with diameter less than 100 nm	Buruga and Kalathi (2019)

(continued)

Table 2 (continued)

NPs by miniemulsion	US regime	Size/properties	References
Nanocomposite phase change material with double phase change points by encapsulations of n-ODE and n-ESE	US horn with 50% of amplitude	It was observed that the nanocapsules were smooth and intact spherical which were not damaged after the double phase change energy storage materials (DPCESM) preparation	Zhao et al. (2020)

occurs; whilst microemulsions are thermodynamically stable systems with an interfacial tension at the oil/water interface close to zero. However, because of the small size of nanoemulsion, it can be kinetically stable over long time scale (Gupta et al. 2016). Briefly, in comparison with microemulsion, miniemulsion requires an energy input to be produced. Indeed, the given energy is used to overcome the surface free energy required to increase the interfacial area between the two immiscible phases, and finally disperse one phase into another (Delmas et al. 2011).

Miniemulsions are kinetically stable and at any moment, i.e. further the emulsion will naturally separate into different phases. The stability improvement occurs due to the thickness of the emulsifier layer on the droplet interface and is comparable to the droplet size. In other words, the steric stabilization increases the repulsive maximum of the droplet–droplet interaction potential, which in turn stabilizes the miniemulsion (Gupta et al. 2016).

Indeed, several different destabilization phenomena may occur, including the droplet adhesion that is generally reversible (creaming) and other related to particle size evolution that are irreversible, such as coalescence and Ostwald ripening (Delmas et al. 2011), which will not be deeper described in this chapter. However, these kinetically stable nanodroplets can be generated by suppression of those two processes: (1) coalescence, which can be controlled with the addition of an efficient surfactant; and (2) Ostwald ripening, which can be achieved by the addition of a co-surfactant (an osmotic pressure agent) dissolved only in the dispersed phase, usually named as hydrophobe or lipophobe for direct and inverse miniemulsion, respectively (Landfester 2001a).

Therefore, surfactants play an essential role in the improvement of miniemulsion stability at both stages, during the emulsification process and the further storage. A wide range of anionic, cationic and nonionic surfactants can be used, resulting in differently charged dispersions that, in turn, exhibit different stabilities. Anionic and cationic surfactants have been employed for the formation of monodisperse droplets displaying sizes between 30 and 200 nm, whilst nonionic oligomeric or polymeric surfactants are suitable for the formation of droplets between 100 and 800 nm (van Zyl et al. 2004; Zanetti-ramos et al. 2008; Hecht et al. 2011).

Until now, we briefly described the theoretical miniemulsion formation and stability. Indeed, due to its kinetically stability, each miniemulsion droplets' can perform parallel reactions after ultrasound irradiation process, being the main responsible of miniemulsion versatility and applications. Herein, the nanodroplets can be considered as a "nanoreactor" since allows a large number of independent reactions performances, in average 10^{18}–10^{20} "nanocompartments" are created, which are separated from each other by the continuous phase (Landfester 2001a). Therefore, the use of more reduced size and more homogeneously distributed droplets are of high interest regarding nanomaterials syntheses (Musyanovych et al. 2005).

The miniemulsion method could also offers mechanical strength by interfacial reaction on the surface of nanodroplet and also post-modification surface possibility, i.e. the surface of nanoparticle could be tailored according to its specific targeting site. For example, targeting chemotherapeutic drugs to cancer cell, improving its regioselectivity, by the aid of nanocapsules. This technique also enables a whole amount of reactions type, including polymerization and interfacial reactions, e.g. monomers can be dissolved in both phases, dispersed and/or continues, and polymerize producing polymer-based nanoparticles (Piradashvili et al. 2016).

It is important to mention that miniemulsion technique is not restricted for syntheses of polymer nanoparticles, where the polymerization reaction occurs into nanodroplet core. As a "nanoreactor", it could also be used for promoting interfacial reactions, i.e. onto nanodroplets interface, by adding molecules for a sequential and specific desired reaction, such as interfacial polymerization or crosslinking reactions, for obtention of polymer-based nanocapsules. Therefore, in the next section, the mostly used syntheses of polymeric-based nanocapsules and particles, by aid of US, are well-described, principally highlighting their chemical reactions, advantages and versatility applications.

4.2 *Nanocapsules X Nanoparticles: What Is the Difference?*

Firstly, what does make polymers so special? Well, it is known that polymers are giant molecules which their molecular weight normally ranges from thousands to millions of g/mol, showing a huge difference from low molecular weight molecules. Another important characteristic is the composition of the molecule, where small moieties, so-called monomers, are combined, repeating along the chain, to form the final molecule. Herein, both features together make polymers unique into chemistry science, furthermore, polymeric-based nanomaterials also provided the surface property from nanoscale range (Causin 2015).

Numerous biocompatible polymeric-based nanocarriers, such as drug, cell and gene delivery systems, present polymer dependence, where mostly are structurally formed by nanoparticle and nanocapsule morphologies. Well, what about capsules and particles difference? The difference can be summarized in either synthetic route or further inner cavity of the particles. In general, polymeric nanocapsules (PNCs) present a hollow core-shell structure made by a covalently binding shell, which can be tailored to enhance their function and applicability. On the other hand, nanoparticles are whole polymer composed, i.e. also presenting a polymer matrix core. However, the more robust PNC structure is well suited for applications as nanocarriers with variable shapes and dimensions. For instance, PNCs have less mass, i.e. their core-shell structure provides an empty core environment, promoting a larger amount of loading-drug molecules when compared to polymeric nanoparticles of the same composition and outer dimensions (Sun et al. 2016).

Normally, for PNC synthesis, it is used a templating accuracy, essentially determined by stability during the reaction process (Sun et al. 2016). Classified terms of the types of templates are given as follow: (1) cavitation of shell-crosslinked nanostructures; (2) vesicle-based crosslinking; and (3) emulsion interfacial crosslinking. In general, nanocapsules contain a lipophilic core composed of volatile organic solvents that are extracted by evaporation. Additionally, vegetable oils have emerged as an alternative to compose the core of the capsules, as a strategy to improve pharmacological potentialities and antioxidant properties of formulations (Ramos et al. 2019). However, regarding US irradiation and kinetically stability of miniemulsion, only the emulsion interfacial crosslinking PNCs template will be deeper described in this chapter.

4.3 Nanocapsules via Interfacial Crosslinking Template

Besides, all template techniques for synthesizing PNCs, the miniemulsion technique embraces the concept of interfacial reactions to polymer science, which is one of the most well-established methods in preparation of nanocapsules. This method allows the utilization of all types of monomers and soluble in distinct phases and is also capable of encapsulating organic and inorganic materials into a polymeric shell. Indeed, it is interesting the generation of polymeric shells since can protect desired inner materials from aggressive environments, such as drugs from physiologic medium (Landfester 2009; Odrobinska et al. 2019).

The general idea of interfacial polymerization by miniemulsion is to initiate the reaction in each of the stabilized nanodroplets without changing the particle identity (Landfester 2001a, b). For instance, hybrid nanocapsules can be obtained by miniemulsion when a required material must compose the core of the capsule is added in the dispersed phase, where the monomer itself comprehends the organic phase and then undergoes in a miniemulsion process (Landfester 2008). After US irradiation, a stable miniemulsion is obtained, and then a second reagent, a different monomer, is added to the system (Fig. 5). The reaction is performed in the interface liquid–liquid of the immiscible phases, until final obtention of a PNC core-shell structure.

However, as this polymerization proceeds at the oil-water interface, some issues can be driven by low solubilities of lipophilic monomers and reluctance of each monomer to homopolymerize. Therefore, the reaction is limited by the ability of the monomers to diffuse and react, and, as a consequence, full conversion into polymeric

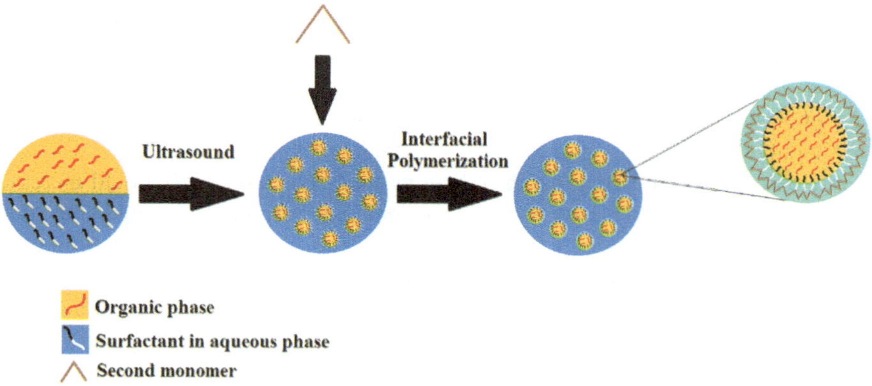

Fig. 5 Generically schematic representation of miniemulsion and interfacial reactions

nanocapsule is only possible at low monomer concentrations showing an impact on the capsule shell thickness and permeability (Weiss and Landfester 2010; Piradashvili et al. 2016). Besides, several PNCs have been obtained by interfacial reaction via radical polymerization, since miniemulsion technique eliminates the drawback of the bulk syntheses, which present high polydispersity and a number of polymer chains in the high conversion stages of polymerization (Landfester 2001b).

Actually, miniemulsion technique presents an outstanding particularity since it is not limited to radical polymerization as microemulsion polymerization. Indeed, other categories of polymerization can also be accomplished, such as polycondensation, polyaddition and ionic polymerization.

4.3.1 Polycondensation and Polyaddition

Polycondensation and polyaddition are associated with step-growth polymerizations and comprehend two most favourable types of interfacial polymerization via miniemulsion technique. Both reactions synthesize polymers with high functionality since the reactants present additional reactive groups, i.e. groups that do not are involved in the polymerization step (Crespy and Landfester 2010).

The polymeric nanocapsule generation by interfacial polycondensation can be divided into three stages: (1) the opening polycondensation stage, (2) initial membrane formation and (3) consecutive shell thickening (Crespy and Landfester 2010). However, the presence of water in polycondensation reaction seems to be contradictory once it is known that the process requires high temperature and water removal. Indeed, in biphasic systems is generated, a localized high hydrophobic region into nanodroplets, providing a free-water environment (Landfester 2009). This approach is interesting for synthesizing polymers that demands high polymerization temperatures (Crespy and Landfester 2010).

In the case of polyaddition reaction, both direct and inverse miniemulsions are performed to encapsulate oil or water liquids as dispersed phase, such as physiological solutions for biomedical applications (Weiss and Landfester 2010). In the case of inverse miniemulsion, after stable aqueous nanodroplets have been generated, a polymeric shell is synthesized by interfacial polyaddition reaction, creating a physic barrier and protecting the drug-loading molecule, which also could increase the blood time circulation at in vivo performance. Furthermore, as a sequential step, it is also possible to obtain an aqueous core nanocapsule into aqueous medium by aid of hydrophilic surfactants (Weiss and Landfester 2010).

In direct miniemulsion, polycondensations are even more demanding compared to polyaddition reactions, since the reaction site must be water free (Crespy and Landfester 2010). Additionally, for both type reactions, the controlling over size and thickness of polymeric shell capsules could be obtained by the speed of second monomer addition. For instance, faster addition contributes to design smaller nanocapsules, whilst slower additions lead to increase droplets size, where a time-consuming monomer addition promotes a rearrangement and diffusion process on the droplet surface (Weiss and Landfester 2010; Piradashvili et al. 2016).

Several groups have reported that hydrophobic compounds could be efficiently encapsulated into thin shells made by free-radical polymerization, where an active surface initiator can be used to initiate an alternate co-polymerization between a monomer present in the dispersed oil phase and a monomer in the continuous aqueous phase.

For instance, Crespy and colleagues synthesized hollow nanocapsules with an aqueous core-shell morphology made by polyurea (amino-monomer) and polyurethane (hydroxyl-monomer) linkages obtained via inverse miniemulsion process through polycondensation reaction with the second monomer (diisocyanate). The capsules could actually act as nanoreactors to reduce silver nitrate in silver nanoparticles at aqueous medium, showing the accuracy of the interfacial reaction template in PNCs synthesis. See micrographs of polyurea-silver nanocapsules are shown in Fig. 6 (Crespy et al. 2007).

In general, the capsules shape was considered spherical, where nanoscale size depended of surfactant concentration and time of the second monomer addition, and the wall thickness was controlled by the amount of added monomer (Crespy et al. 2007).

Thongchaivetcharat and colleagues obtained dextran derivated-based nanocapsules by interfacial crosslinking via inverse miniemulsion and subsequently redispersion in aqueous solutions (Thongchaivetcharat et al. 2019). They investigated the pH dependence of the release kinetics of corrosion inhibitors (benzotriazole and nicotinic acid), where it was found a high pH response of the PNCs.

Dextran nanocapsules were synthesized using two dextran derivates, carbonate and carbamate, by interfacial crosslinking reaction with different crosslinkers, diisocyanate and glutaraldehyde, at the interface of nanodroplets. Firstly, dextran carbonate was obtained by nucleophilic substitution of dextran hydroxyl groups by phenyl chloroformate. Then, the carbonate was subjected to a nucleophilic

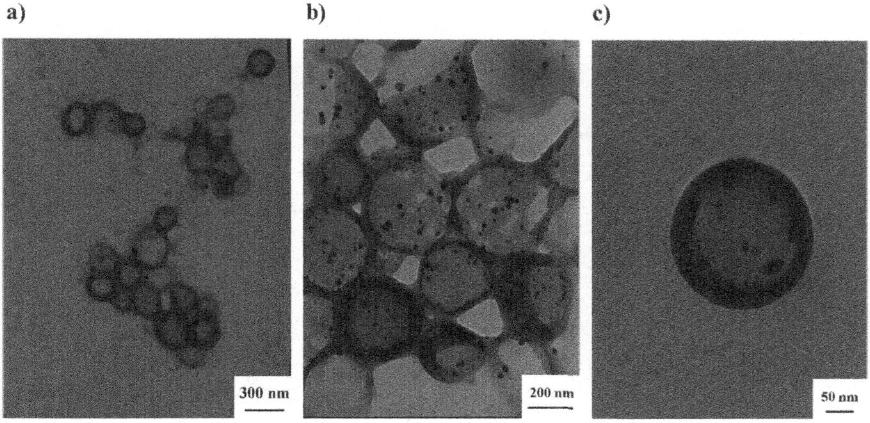

Fig. 6 TEM micrographs of polyurea capsules loaded with different amounts of silver nanoparticles (Crespy et al. 2007)

attack by dimethylaminopropylamine to produce dextran carbamate. For nanocapsule synthesis step, both dextran derivates were solubilized in aqueous phase, and then emulsified in oil containing PGPR by ultrasound (50% US amplitude and 3 min in a pulse regime of *1s on-1s off*) to form nanodroplets. Dextran or dextran carbamate nanocapsules were formed after the addition of the crosslinker to the dispersions, as shown in Figs. 7 and 8.

During dextran nanocapsule synthesis, a solvent emulsion evaporation step is also required, in order to remove all organic solvent, providing a biodegradable nanocapsule. The core-shell structure of nanocapsules was confirmed by scanning electron microscopy (SEM) and transmission electron microscopy (TEM).

The dextran and dextran carbamate nanocapsules exhibited a well-defined core-shell structure. The corrosion inhibitors benzotriazole and nicotinic acid were directly encapsulated during nanocontainers formation by dissolving them in the dispersed aqueous phase. Both synthesized nanocapsules exhibited slow release at low pH values and rapid release at high pH values. Furthermore, the release kinetics of two evaluated corrosion inhibitors were then classified as pH dependence (Fig. 9).

5 Final Remarks

As an overview, the synthesis and application of nanomaterials have made extraordinary achievements principally due to improvement in synthetic routes at nanoscale. The sonochemical approach promoted the production of novel materials and even though already known materials under slightly environments, such as pressure, quick reaction time, temperature and solvent free. Furthermore, this technique also provides nanomaterials with greater surface properties like colloidal stability and

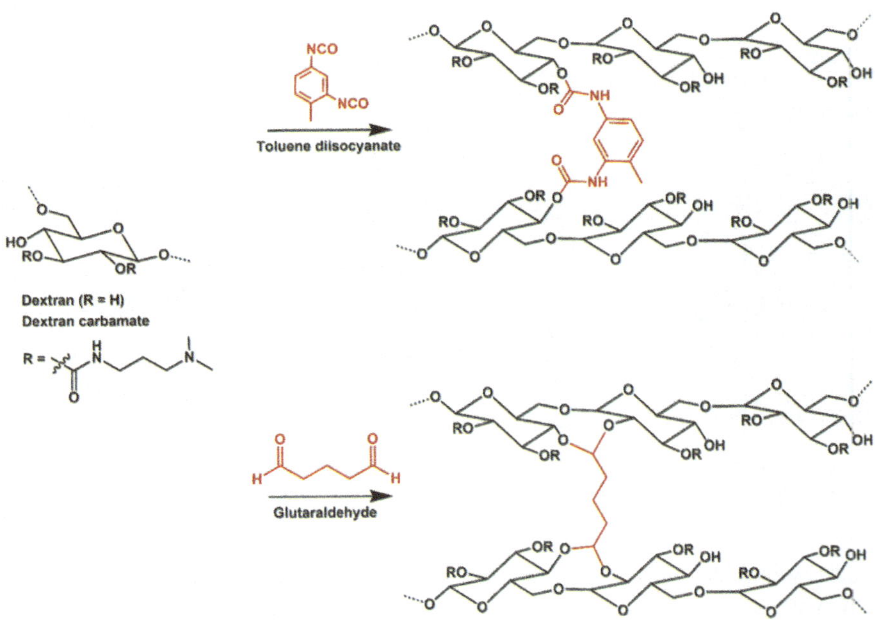

Fig. 7 Crosslinking reactions between dextran and dextran carbamate with toluene diisocyanate and glutaraldehyde (Thongchaivetcharat et al. 2019)

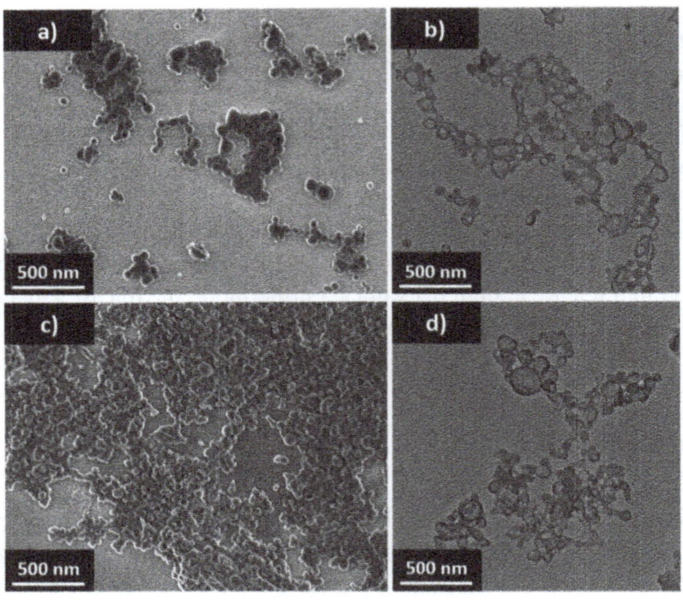

Fig. 8 SEM and TEM micrographs of (**a**, **b**) dextran and (**c**), **d** dextran carbamate nanocapsules (Thongchaivetcharat et al. 2019)

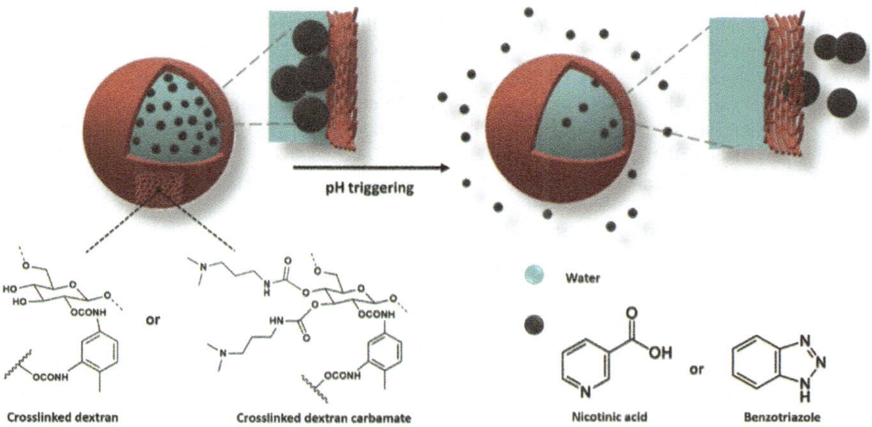

Fig. 9 Illustration of corrosion inhibitors release (Thongchaivetcharat et al. 2019)

crystalline structure, where both chemical and physical phenomena associated with high-intensity ultrasound are responsible for the production or modification of nanomaterials. Herein, the aim of this chapter was to summarize currently examples of US-based methodologies of synthesis and also showed as a versatile tool for both organic and inorganic nanomaterials obtention. Literally, sonochemistry deserved attention of science community, since can control the design of nanostructures with tuneable geometries, size and properties, opening the "door" for different applications regarding nanotechnology, such as medicine, catalysis, sensor, environmental remediation, electronic devices and others.

Acknowledgements We gratefully acknowledge the financial support of the Brazilian agencies for scientific and technological development: CNPq (408790/2016-4), CAPES (Finance Code 001) and Funcap (PNE-0112-00048.01.00/16).

References

Abbas M, Torati SR, Parvatheeswara Rao B et al (2015) Size controlled sonochemical synthesis of highly crystalline superparamagnetic Mn-Zn ferrite nanoparticles in aqueous medium. J Alloys Compd 644:774–782. https://doi.org/10.1016/j.jallcom.2015.05.101

Anderson D, Anderson T, Fahmi F (2019) Advances in applications of metal oxide nanomaterials as imaging contrast agents. Phys Status Solid 216:1–16. https://doi.org/10.1002/pssa.201801008

Andrade Neto DM, Freire RM, Gallo J et al (2017) Rapid sonochemical approach produces functionalized Fe_3O_4 nanoparticles with excellent magnetic, colloidal, and relaxivity properties for MRI application. J Phys Chem C 121:24206–24222. https://doi.org/10.1021/acs.jpcc.7b04941

Anniebell S, Gopinath SCB (2018) Polymer conjugated gold nanoparticles in biomedical applications. Curr Med Chem 25:1433–1445. https://doi.org/10.2174/0929867324666170116123633

Argüelles-Pesqueira AI, Diéguez-Armenta NM, Bobadilla-Valencia AK et al (2018) Low intensity sonosynthesis of iron carbide@iron oxide core-shell nanoparticles. Ultrason Sonochem 49:303–309. https://doi.org/10.1016/J.ULTSONCH.2018.08.017

Artusio F, Bazzano M, Pisano R et al (2018) Polymeric nanocapsules via interfacial cationic photopolymerization in miniemulsion. Polymer (Guildf) 139:155–162. https://doi.org/10.1016/j.polymer.2018.02.019

Ashokkumar M (2018) Introductory text to sonochemistry. Chem Texts 4:1–9. https://doi.org/10.1007/s40828-018-0061-4

Askarinejad A, Morsali A (2009) Direct ultrasonic-assisted synthesis of sphere-like nanocrystals of spinel Co_3O_4 and Mn_3O_4. Ultrason Sonochem 16:124–131. https://doi.org/10.1016/j.ultsonch.2008.05.015

Bang JH, Suslick KS (2010) Applications of ultrasound to the synthesis of nanostructured materials. Adv Mater 22:1039–1059. https://doi.org/10.1002/adma.200904093

Barbosa JS, Neto DMA, Freire RM et al (2018) Ultrafast sonochemistry-based approach to coat TiO_2 commercial particles for sunscreen formulation. Ultrason Sonochem 48:340–348. https://doi.org/10.1016/J.ULTSONCH.2018.06.015

Barkade SS, Pinjari DV, Singh AK et al (2013) Ultrasound assisted miniemulsion polymerization for preparation of polypyrrole-zinc oxide (PPy/ZnO) functional latex for liquefied petroleum gas sensing. Ind Eng Chem Res 52:7704–7712. https://doi.org/10.1021/ie301698g

Baruch-Teblum E, Mastai Y, Landfester K (2010) Miniemulsion polymerization of cyclodextrin nanospheres for water purification from organic pollutants. Eur Polym J 46:1671–1678. https://doi.org/10.1016/j.eurpolymj.2010.05.007

Bhosale MA, Bhanage BM (2016) A simple approach for sonochemical synthesis of Cu_2O nanoparticles with high catalytic properties. Adv Powder Technol 27:238–244. https://doi.org/10.1016/J.APT.2015.12.008

Buruga K, Kalathi JT (2019) Synthesis of poly(styrene-co-methyl methacrylate) nanospheres by ultrasound-mediated pickering nanoemulsion polymerization. J Polym Res 26. https://doi.org/10.1007/s10965-019-1871-9

Cabrera-Trujillo MA, Sotelo-Díaz LI, Quintanilla-Carvajal MX (2016) Efecto de la amplitud y pulsaci6n en ultrasonido de sonda a baja frecuencia sobre emulsiones aceite/agua. DYNA 83:63–68. https://doi.org/10.15446/dyna.v83n199.56192

Cao Z, Landfester K, Ziener U (2012) Preparation of dually, pH- and thermo-responsive nanocapsules in inverse miniemulsion. Langmuir 28:1163–1168. https://doi.org/10.1021/la2041357

Cárcel JA, García-Pérez JV, Benedito J, Mulet A (2012) Food process innovation through new technologies: use of ultrasound. J Food Eng 110:200–207. https://doi.org/10.1016/j.jfoodeng.2011.05.038

Causin V (2015) Polymers on the Crime Scene. In: Polymers on the Crime Scene. Springer, Cham. https://doi.org/10.1007/978-3-319-15494-7_4

Chatel G (2018) The use of ultrasound from the catalyst synthesis to the catalytic reaction. Curr Opin Green Sustain Chem 15:1–6. https://doi.org/10.1016/j.cogsc.2018.07.004

Chen N, Dempere LA, Tong Z (2016) Synthesis of pH-responsive lignin-based nanocapsules for controlled release of hydrophobic molecules. ACS Sustain Chem Eng 4:5204–5211. https://doi.org/10.1021/acssuschemeng.6b01209

Chiaradia V, Valério A, Feuser PE et al (2015) Incorporation of superparamagnetic nanoparticles into poly(urea-urethane) nanoparticles by step growth interfacial polymerization in miniemulsion. Colloids Surfaces A Physicochem Eng Asp 482:596–603. https://doi.org/10.1016/j.colsurfa.2015.06.035

Crespy D, Landfester K (2010) Miniemulsion polymerization as a versatile tool for the synthesis of functionalized polymers. Beilstein J Org Chem 6:1132–1148. https://doi.org/10.3762/bjoc.6.130

Crespy D, Stark M, Hoffmann-Richter C et al (2007) Polymeric nanoreactors for hydrophilic reagents synthesized by interfacial polycondensation on miniemulsion droplets. Macromolecules 40:3122–3135. https://doi.org/10.1021/ma0621932

Cui X, Dong L, Zhong S et al (2017) Sonochemical fabrication of folic acid functionalized multistimuli-responsive magnetic graphene oxide-based nanocapsules for targeted drug delivery. Chem Eng J 326:839–848. https://doi.org/10.1016/j.cej.2017.06.045
Delmas T, Piraux H, Couffin A-C et al (2011) How to prepare and stabilize very small nanoemulsions. Langmuir 27:1683–1692. https://doi.org/10.1021/la104221q
Elazab HA (2018) Microwave-assisted synthesis of palladium nanoparticles supported on copper oxide in aqueous medium as an efficient catalyst for Suzuki cross-coupling reaction. Adsorpt Sci Technol 36:1352–1365. https://doi.org/10.1177/0263617418771777
Freire TM, Dutra LMU, Queiroz DC et al (2016) Fast ultrasound assisted synthesis of chitosan-based magnetite nanocomposites as a modified electrode sensor. Carbohydr Polym 151:760–769. https://doi.org/10.1016/j.carbpol.2016.05.095
Gallego-Juárez JA, Rodriguez G, Acosta V, Riera E (2010) Power ultrasonic transducers with extensive radiators for industrial processing. Ultrason Sonochem 17:953–964. https://doi.org/10.1016/j.ultsonch.2009.11.006
Gaudin F, Sintes-Zydowicz N (2011) Poly(urethane-urea) nanocapsules prepared by interfacial step polymerisation in miniemulsion. The droplet size: a key-factor for the molecular and thermal characteristics of the polymeric membrane of the nanocapsules? Colloids Surf A Physicochem Eng Asp 384:698–712. https://doi.org/10.1016/j.colsurfa.2011.05.050
Gedanken A (2004) Using sonochemistry for the fabrication of nanomaterials. Ultrason Sonochem 11:47–55. https://doi.org/10.1016/j.ultsonch.2004.01.037
Gupta A, Eral HB, Hatton TA, Doyle PS (2016) Nanoemulsions: formation, properties and applications. Soft Matter 12:2826–2841. https://doi.org/10.1039/c5sm02958a
Harika VK, Sadhanala HK, Perelshtein I, Gedanken A (2020) Sonication-assisted synthesis of bimetallic Hg/Pd alloy nanoparticles for catalytic reduction of nitrophenol and its derivatives. Ultrason Sonochem 60. https://doi.org/10.1016/j.ultsonch.2019.104804
Hecht LL, Wagner C, Landfester K, Schuchmann HP (2011) Surfactant concentration regime in miniemulsion polymerization for the formation of MMA nanodroplets by high-pressure homogenization. Langmuir 27:2279–2285
Hu M, Yao Z, Wang X (2017) Graphene-based nanomaterials for catalysis. Ind Eng Chem Res 56:3477–3502. https://doi.org/10.1021/acs.iecr.6b05048
Hu Q, Zhao C, Zhang Z et al (2019) Sonochemical synthesis of silver nanoparticles coated copper wire for low-temperature solid state bonding on silicon substrate. Chinese Chem Lett 30:1455–1459. https://doi.org/10.1016/J.CCLET.2019.04.050
Islam H, Paul MTY, Burheim OS, Pollet BG (2019) Recent developments in the sonoelectrochemical synthesis of nanomaterials. Ultrason Sonochemistry 59:1–8. https://doi.org/10.1016/j.ultsonch.2019.104711
Joshi M, Bansala T, Mukhopadhyay S (2018) Innovations in polymer nanocomposite coatings for EMI Shielding applications. Mater Sci Eng 460:1–8. https://doi.org/10.1088/1757-899X/460/1/012012
Kamali M, Costa MEV, Otero-Irurueta G, Capela I (2019) Ultrasonic irradiation as a green production route for coupling crystallinity and high specific surface area in iron nanomaterials. J Clean Prod 211:185–197. https://doi.org/10.1016/J.JCLEPRO.2018.11.127
Kentish S, Wooster TJ, Ashokkumar M et al (2008) The use of ultrasonics for nanoemulsion preparation. Innov Food Sci Emerg Technol 9:170–175. https://doi.org/10.1016/j.ifset.2007.07.005
Kristl M, Dojer B, Gyergyek S, Kristl J (2017) Synthesis of nickel and cobalt sulfide nanoparticles using a low cost sonochemical method. Heliyon 3:e00273. https://doi.org/10.1016/J.HELIYON.2017.E00273
Landfester BK (2001a) The generation of nanoparticles in miniemulsions. Adv Mater 13:765–768
Landfester K (2009) Miniemulsion polymerization and the structure of polymer and hybrid nanoparticles. Angew Chemie Int Ed 48:4488–4507. https://doi.org/10.1002/anie.200900723
Landfester K (2001b) Polyreactions in miniemulsions. Macromol Rapid Commun 22:896–936

Landfester K (2008) Minimulsion droplets as nanoreactors. Nanoreactor engineering for life sciences and medicine, 1st edn. Artech House, Boston, pp 47–96

Levashov EA, Mukasyan AS, Rogachev AS, Shtansky DV (2016) Self-propagating high-temperature synthesis of advanced materials and coatings. Int Mater Rev 62:203–239. https://doi.org/10.1080/09506608.2016.1243291

Li X, Naguib YW, Cui Z (2017) In vivo distribution of zoledronic acid in a bisphosphonate-metal complex-based nanoparticle formulation synthesized by a reverse microemulsion method. Int J Pharm 526:69–76. https://doi.org/10.1016/j.ijpharm.2017.04.053

Lorenzen AL, Rossi TS, Riegel-Vidotti IC, Vidotti M (2018) Influence of cationic and anionic micelles in the (sono)chemical synthesis of stable Ni(OH) 2 nanoparticles: "In situ" zeta-potential measurements and electrochemical properties. Appl Surf Sci 455:357–366. https://doi.org/10.1016/j.apsusc.2018.05.198

Mason JM, Cintas P (2002) Sonochemistry. In: Clark JH, Macquarrie D (eds) Handbook of green chemistry and technology, 1st edn. Blackwell Science Ltd, Oxford, pp 372–396

Matin MA, Jang JH, Kwon YU (2014) PdM nanoparticles (M = Ni Co, Fe, Mn) with high activity and stability in formic acid oxidation synthesized by sonochemical reactions. J Power Sources 262:356–363. https://doi.org/10.1016/j.jpowsour.2014.03.109

Murray CB, Doyle H, Kagan CR (2001) Colloidal synthesis of nanocrystals and nanocrystal superlattices. IBM J Res Dev 45:47–56. https://doi.org/10.1147/rd.451.0047

Musyanovych A, Mailänder V, Landfester K (2005) Miniemulsion droplets as single molecule nanoreactors for polymerase chain reaction. Biomacromol 6:1824–1828

Odrobinska J, Gumieniczek-Chłopek E, Szuwarzynski M et al (2019) Magnetically navigated core – shell polymer capsules as nanoreactors loadable at the oil/water interface. ACS Appl Mater Interfaces 11:10905–10913. https://doi.org/10.1021/acsami.8b22690

Okoli CU, Kuttiyiel KA, Cole J et al (2018) Solvent effect in sonochemical synthesis of metal-alloy nanoparticles for use as electrocatalysts. Ultrason Sonochem 41:427–434. https://doi.org/10.1016/J.ULTSONCH.2017.09.049

Park J, Jung WM, Song M et al (2013) Preparation of PbS-coated CdTe nanocrystals through sonochemical reaction. Bull Korean Chem Soc 34:680–682. https://doi.org/10.5012/bkcs.2013.34.2.680

Phan LMT, Tufa LT, Kim H et al (2019) Trends in diagnosis for active tuberculosis using nanomaterials. Curr Med Chem 26:1946–1959. https://doi.org/10.2174/0929867325666180912105617

Piradashvili K, Alexandrino EM, Wurm FR, Landfester K (2016) Reactions and polymerizations at the liquid–liquid interface. Chem Rev 116:2141–2169. https://doi.org/10.1021/acs.chemrev.5b00567

Radziuk D, Skirtach A, Geßner A et al (2011) Ultrasonic approach for formation of erbium oxide nanoparticles with variable geometries. Langmuir 27:14472–14480. https://doi.org/10.1021/la203622u

Ramos PT, Pedra NS, Soares MSP et al (2019) Ketoprofen-loaded rose hip oil nanocapsules attenuate chronic inflammatory response in a pre-clinical trial in mice. Mater Sci Eng C 103:109742. https://doi.org/10.1016/j.msec.2019.109742

Romio AP, Bernardy N, Lemos Senna E et al (2009) Polymeric nanocapsules via miniemulsion polymerization using redox initiation. Mater Sci Eng C 29:514–518. https://doi.org/10.1016/j.msec.2008.09.011

Sun H, Chen CK, Cui H, Cheng C (2016) Crosslinked polymer nanocapsules. Polym Int 65:351–361. https://doi.org/10.1002/pi.5077

Thongchaivetcharat K, Jenjob R, Seidi F, Crespy D (2019) Programming pH-responsive release of two payloads from dextran-based nanocapsules. Carbohydr Polym 217:217–223. https://doi.org/10.1016/j.carbpol.2019.04.023

Tovar-martinez E, Moreno-torres JA, Cabrera-salazar JV, Reyes-reyes M (2018) Synthesis of carbon nano-onions doped with nitrogen using spray pyrolisis. Carbon N Y 140:171–181. https://doi.org/10.1016/j.carbon.2018.08.056

van Zyl AJP, de Wet-roos D, Sanderson RD, Klumperman B (2004) The role of surfactant in controlling particle size and stability in the miniemulsion polymerization of polymeric nanocapsules. Eur Polym J 40:2717–2725. https://doi.org/10.1016/j.eurpolymj.2004.07.021

Vennela AB, Mangalaraj D, Muthukumarasamy N et al (2019) Structural and optical properties of Co_3O_4 nanoparticles prepared by sol-gel technique for photocatalytic application. Int J Electrochem Sci 14:3535–3552. https://doi.org/10.20964/2019.04.40

Weiss CK, Landfester K (2010) Miniemulsion polymerization as a means to encapsulare organic and inorganic materials. In: van Herk AM, Landfester K (eds) Hybrid latex particles: preparation with (mini)emulsion polymerization. Springer, Berlin, pp 186–227

Wilhelm A-M, Laugier F, Kidak R et al (2010) Ultrasound to enhance a liquid–liquid reaction. J Chem Eng Japan 43:751–756

Xing Q, Zhang X, Wu D et al (2019) Ultrasound-assisted synthesis and characterization of heparin-coated Eu^{3+} doped hydroxyapatite luminescent nanoparticles. Colloid Interface Sci Commun 29:17–25. https://doi.org/10.1016/J.COLCOM.2019.01.001

Xu H, Suslick KS (2010) Sonochemical synthesis of highly fluorescent Ag nanoclusters. ACS Nano 4:3209–3214. https://doi.org/10.1021/nn100987k

Zanetti-ramos BG, Lemos-Senna E, Cramail H et al (2008) The role of surfactant in the miniemulsion polymerization of biodegradable polyurethane nanoparticles. Mater Sci Eng C 28:526–531. https://doi.org/10.1016/j.msec.2007.04.041

Zhao M, Zhang X, Kong X (2020) Preparation and characterization of a novel composite phase change material with double phase change points based on nanocapsules. Renew Energy 147:374–383. https://doi.org/10.1016/j.renene.2019.08.117

Zheng H, Matseke MS, Munonde TS (2019) The unique Pd@Pt/C core-shell nanoparticles as methanol-tolerant catalysts using sonochemical synthesis. Ultrason Sonochem 57:166–171. https://doi.org/10.1016/J.ULTSONCH.2019.05.023

Zou Z, Lin K, Chen L, Chang J (2012) Ultrafast synthesis and characterization of carbonated hydroxyapatite nanopowders via sonochemistry-assisted microwave process. Ultrason Sonochem 19:1174–1179. https://doi.org/10.1016/j.ultsonch.2012.04.002

Chapter 11
Titanates Nanotubes and Nanoribbons Applied in Dye-Sensitized Solar Cells

Antonio Paulo Santos Souza, Ana Fabíola Leite Almeida, Francisco Nivaldo Aguiar Freire, Vanja Fontenele Nunes, and Francisco Marcone de Lima

1 Introduction

Dye-sensitized solar cells, or, simply DSSCs, had its origin traced back to the end of the nineteenth century, when Moser (Unger 2012) verified the increase of the photocurrent generated on halogenated silver plates when sensitized by dye. However, only in 1991 those cells received the proper attention, when O'Regan and Grätzel published an article displaying DSSCs with high efficiency (1991). Since then, this work generated thousands of citations, creating a new field of investigation, the third generation of solar cells.

These cells are conventionally built with a nanoporous layer of a semiconductor material of large *gap*, usually TiO_2, ZnO, or SnO_2, covered by a layer of photoexcited dye, generally the dyes with Ruthenium (II) polypyridine complex (Chang and Chow 2011; Mende et al. 2005). However, the photovoltaic effect can also be observed in numerous biological molecules, such as carotenes, chlorophylls and other porphyrins,

A. P. S. Souza (✉) · A. F. L. Almeida · F. N. A. Freire
LAFFER – Thin Films Laboratory in Renewable Energy, Mechanics Engineering Department, Ceará Federal University—UFC CEP, Fortaleza, CE 60455-900, Brazil
e-mail: antonio.souza@fisica.ufc.br

A. F. L. Almeida
e-mail: anfaleal@yahoo.com.br

F. N. A. Freire
e-mail: nivaldo@ufc.br

V. F. Nunes · F. M. de Lima
Department of Engineering and Materials Science, Ceará Federal University—UFC CEP, Fortaleza, CE 60455-900, Brazil
e-mail: vanjafnunes@gmail.com

F. M. de Lima
e-mail: marconeufc@gmail.com

R. F. do. Nascimento et al. (eds.), *Nanomaterials and Nanotechnology*, Materials Horizons: From Nature to Nanomaterials,
https://doi.org/10.1007/978-981-33-6056-3_11

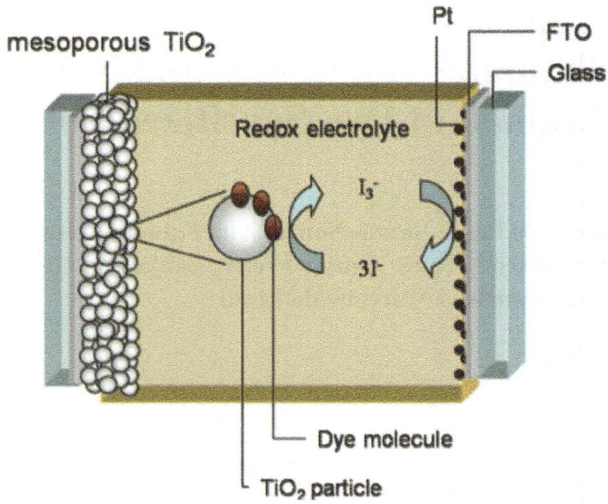

Fig. 1 Illustration of a DSSC. *Source* Dye-sensitized solar cells (2010) (Hagfeldt et al. 2010)

including in some phthalocyanines structures (Bavykin and Walsh 2010; Campbell et al. 2007; Chang and Chow 2011; Cid et al. 2007; Yella et al. 2014). The work electrode is connected to a counter electrode of catalytic material (Platinum, gold, or activated coal) through an electrolyte with a iodide/triiodide redox pair (usually $3I^-/I_3^-$) which holds good stability and reversibility (Lima 2015).

An illustrative design of the operating Grätzel cell is displayed in Fig. 1.

The dye-sensitized solar cells are excitable cells, meaning, the generated photocurrent in these devices happens, mainly, due to the separation of charges in the dye/semiconductor oxide interface (Lima 2015). During this process, the charge dissociation energy is related proportionally to the binding energy of the dye excitons. As such, one of the essential factors which affect the DSSC efficiency is the exciton separation process.

Figure 2 shows the electrons transfer process which happens in the inside of a dye-sensitized solar cell.

For this process, when the incident light reaches the dye, the dye turns into an oxidized state, injecting electrons into the conduction band of the semiconductor oxide. For the dyes based on Ruthenium, the charge transfer happens speedy (10^{-10}–10^{-15} s^{-1}) (Hagfeldt et al. 2010). For an effective charge transfer, the time of repositioning of the dye to its fundamental state must be less than the time for the injection of the electrons into the conduction band of the semiconductor oxide. The time estimated for the Ruthenium dye is (2×10^{-8}–6×10^{-8} s^{-1}) (Lima 2015).

The oxidized dye can be regenerated, through reduction, by the redox pair iodide/triiodide within 10^{-8} s^{-1}. On the same time scale, the oxidized dye captures an iodide electron (I^-) from the regenerator electrolyte, oxidizing to a triiodide (I_3^-). The triiodide recovers its electron through the counter electrode, when the device is connected to an external load.

In the work electrode of semiconductor oxide, or photoanode, the transition process happens through diffusion within an estimated time of (10^{-3}–10^0 s^{-1}). The

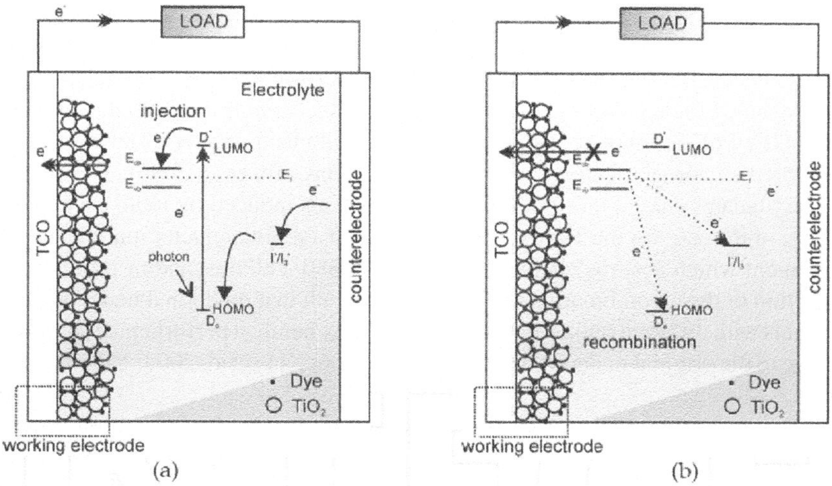

Fig. 2 Main process of electron transfer in the DSSCs; **a** electron immigration through the solar cell from the dye and **b** possible electron recombination process. *Source* William et al. (2011)

recombination process which happens between the photo-injected electrons on the conduction band of the semiconductor and the oxidized form of the electrolyte [triiodide (I_3^-)] happens in different time scales (10^{-6}–10^{-3} s^{-1}), changing according to the concentration of electrons in the conduction band and the incident light intensity (Lima 2015). The open circuit voltage delivered by the device under illumination is calculated by the difference of energy between the level of quasi-Fermi of the electron in the semiconductor oxide and the Nernst potential of the mediator electrolyte.

The TiO_2 nanoparticles deposited randomly on the photoanode of the conventional dye-sensitized solar cells have large superficial areas, assuring sufficient charge. However, the great charge loss caused by recombination on the outline of these deposited nanoparticles can limit the efficiency of light dissipation. To lessen this loss, materials with one-dimensional structures have received great attention for its capacity to create shortcuts on the way of the electrons and contribute to the process of electrical diffusion, which occurs frequently on these cells (Gil et al. 2006; Pang et al. 2011, 2013; Zhao et al. 2009).

Recently, solar cells with electrodes carrying nanotubes or nanoribbons of titanate, NaTiNT and NaTiNT-Ribbons respectively, have attracted interest due to its important properties. The combination of a continuous path for the electrons photogenerated, together with the mesoporous structure of the nanotubes, allied with the great specific superficial area, the efficient ionic exchange properties and the good protonic and electronic conductivity, makes theses nanostructure materials promising for many applications (Hara et al. 2001; Nàdia and João 2012; Tanaka et al. 1997; Tereza et al. 2013).

The advantage of the use of the titanate nanotubes or nanoribbons compared to other materials for electrodes manufacturing in dye-sensitized solar cells abides essentially in its great capacity for the absorption of dye ions (positive) into its surface (negative). It is estimated the possibility of creating a dye layer with about 1000 molecules along the surface of a titanate nanotube. This monolayer of dye improves the efficiency of radiation absorption, which combined with the elongated morphology of the titanate nanotubes (TiNTs), allows a greater efficiency for the transport and storage of electrons in the DSSCs (Bavykin et al. 2006).

The titanates can go through structural alterations induced by ionic exchange or doping, or the case for the TiNTs combination with a semiconductor material or with an element which absorbs visible radiation ($\lambda = 380$–780 nm) and/or promotes the restriction of the recombination rate, which can result in a functional nanocomposite materials with different optical properties and photochemical performances improved relative to the original materials (Bavykin et al. 2006; Xiao et al. 2008; Ylhainen et al. 2008).

The titanate materials, generally speaking, can be synthesized through many methods, such as sol-gel (Baiju 2007), combustion synthesis (Sivalingam 2004), electrochemical deposition (Zhitomirsky 1998), vapor chemical deposition (VCD), physical vapor deposition (PVD), anodizing, hydrothermal treatment (Dresselhaus et al. 2003; Edisson et al. 2009; Ferreira 2006; Kim et al. 2006), by decomposition of Na_2TiO_3 and ultrasonication, among others. However, in 1998, Kasuga and collaborators demonstrated a simple and efficient alternative for the production of nanostructures based on TiO_2. The proposed method is based on the hydrothermal treatment of titanium dioxide in a high alkaline solution (NaOH), submitted to a heating under pressure for a long period. The resulting materials present nanostructures with an average diameter of about 9 nm and internal average diameter of 5 nm, besides length superior to 100 nm (Santos et al. 2013; Viana et al. 2011).

Thus, this chapter will focus on the synthesis by the hydrothermal method of nanotubes and nanoribbons of titanate and its application as a photoanode of dye-sensitized solar cells.

2 Titanium Dioxide TiO_2

The crystalline titanium dioxide, which is a semiconductor with energy *gap* in the ultraviolet region, owns excellent optical and electrical properties and to this day is vastly studied for numerous applications. Due to the structural changes and the nanometer dimensions, the nanotubes based on titanium present properties which differ from the ones found on the titanium dioxide, such as high superficial area, making it more efficient for different applications (Yella et al. 2011).

The titanium dioxide is a semiconductor with energy *gap* in the ultraviolet region, about 3.05 eV for the rutile phase (Marszalek 2011) and 3.4 eV for the anatase phase (Sauvage 2011), which justify its appearance in other spectra. Due to its strong absorbance in this region, the TiO_2 is largely used for cosmetic production for ultraviolet rays' protection.

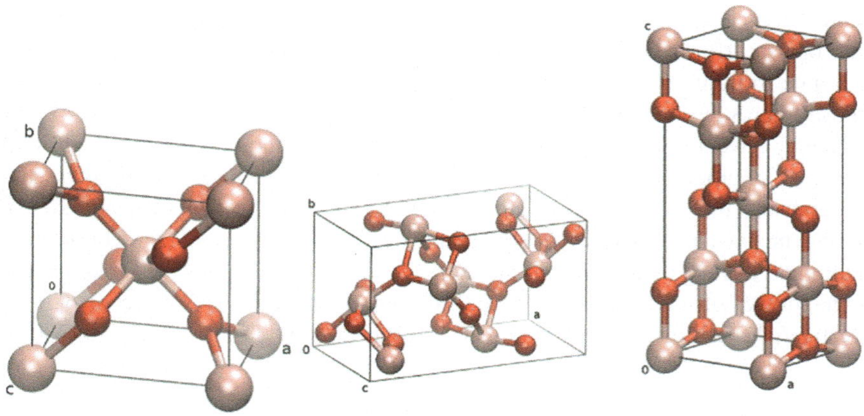

Fig. 3 Stick-and-ball representation of TiO_2 crystal structures: rutile, brookite, and anatase (left to right) (Sauvage 2011). *Source* Sauvage (2011)

In the rutile and anatase phase, the pure and stoichiometric titanium dioxide shows high refraction numbers, about 2.55 for the anatase and 2.73 for the rutile (Liu 2011). This property, together with the particles measuring about half of the size of the wavelength to be scattered (Liu 2011), makes more than 95% of light to be reflected (Marszalek 2011), which makes to be the main white pigment available in the which, which is, today, its leading application. Figure 3 shows the structure phases of the TiO_2.

3 Nanostructures

The nanostructures can be made, basically, from three ways: "top-down," "bottom-up" and the traditional chemical and materials science techniques from molecular components.

Today, with the results from huge efforts, many nanomaterials and artificial nanostructures have been produced with a higher level of control. The improvement and increase access of characterization techniques such as the electronic microscopy have allowed a better understanding of these material properties in the nanometric scale and providing a great help in the materials synthesis process.

The nanostructures can be presented in different morphologies, including structures similarly spherical (nanoparticles), filamentary (nanoribbons, nanowires, etc.), and lamelar (nanosheets) (Ferreira 2006; Morgado 2007a; Bavykin and Frank 2010). Particularly, nanostructures (almost) one-dimensional (1D) like wires, rods, ribbons, and tubes can be seen as ideal systems for the investigation of the effects in the decreasing of the quantic dimensionality and confinement over electrical and thermal transportation properties (Dresselhaus et al. 2003). Besides, due to the tendency to a continuous decrease in the size of the electronic devices, aiming the improvement of its performance and the cut of the costs, 1D nanostructures show up as a candidate to be used in the fabrication of those devices (Law et al. 2004). These nanostructures have high superficial area per gram of material, as an example, 1 g of TiNT may have 434 m^2 of area (Baik 2008). The combination of a high superficial area with an efficient electronic (or ionic) transport toward a direction allows the 1D nanostructures to be used as chemical and biological sensors (Yamada 2006). Among the 1D nanostructures, doubtless, the most known and studied are the carbon nanotubes (Kolmakov and Moskovits 2004; Iijima 1991). Besides these, nanowires and nanotubes of semiconductors oxides like ZnO, SnO_2, and TiO_2 have, also, been investigated (Ferreira 2006; Morgado 2007a; Bavykin and Frank 2010; Baik 2008).

4 Titanate Nanostructures

One of the great contributions of nanotechnology has been the research of titanate nanotubes. Due to its electrical, photocatalytic and sensor properties, these nanotubes present many applications, such as photovoltaic devices, hydrogen sensors, electronic devices, effluent treatment, optical devices, and others (Ferreira 2006; Morgado 2007a).

The nanostructures, generally, can be synthesized by different methods like sol-gel (Baiju 2007), combustion synthesis (Sivalingam 2003), electrochemical deposition (Zhitomirsky 1998), chemical vapor deposition (CVD), physical vapor deposition (PVD), anodizing, hydrothermal treatment (Dresselhaus et al. 2003; Ferreira 2006; Kim et al. 2006; Morgado 2007a), and others. Some of these methods of synthesis generally guides to the formation of spherical nanocrystals and similar. After the discovery of the carbon nanotubes in 1990, the one-dimensional nanostructures became more relevant, for representing a class of materials with unique properties (Iijima 1991). Thus, many research groups have been fabricating not only carbon-based nanotubes, but, also, based on oxide materials, such as titanium. In 1996, a synthesis technique for TiO_2 nanostructures was demonstrated, using alumina nanoporous for the production of frames to be used as a guide for the electrochemical deposition of TiO_2, resulting in nanotubes of 70–100 nm of internal diameter (Hoyer 1996).

However, the synthesis method by frames is too complex and expensive, because these frames need to be destroyed at the end of the process of formation of nanotubes, making it economically unavailable. Then, in 1998, Kasuga et al. demonstrated a simple alternative and efficient for the 1D nanostructure TiO_2 production. The proposed method is based on the hydrothermal treatment of the titanium dioxide on a high alkaline solution (NaOH), submitted to heating under pressure for a long period. The resulting material had very regular morphology, made of tubes with 7–10 nm of internal diameter and a length of about 100 nm. This pioneer work opened the doors for many other works, which has investigated the characterization of the produced nanostructures and the mechanisms involved on the synthesis, as well as its application.

4.1 Formation of Nanotubes Titanate by Hydrothermal Synthesis

The discovery of carbon nanotubes by Iijima (1991) awakened the interest for a new class of materials, whose morphology, crystalline structure and size effects, such as high superficial area, could cause new and excellent physics properties. Since then, nanotubes have been obtained from different oxides, like TiO_2, SiO_2, Al_2O_3, and MoO_3 through different synthesis routes, and some are very simple and low cost, like the alkaline hydrothermal synthesis. This method consists on the hydrothermal treatment of conventional titanium oxide, in aqueous alkaline medium, followed by washing with water and diluted acid. Many authors (Bavykin and Frank 2010; Ferreira 2006; Morgado 2007a; Vanessa 2011) suggest that the product of this synthesis has features of lamellar titanate with general formula $Na_{2-x}H_xTi_3O_7{\cdot}nH_2O$ with ($0 \leq x \leq 2$) and can be presented in the shape of tubes, wires, rods and sheets, and, due to its organized structure can favor the electron transport, and the increase of the efficiency for many devices. Despite the simplicity of the process in one single reactionary step, the mechanism of formation of the titanate nanotubes by this method has not been completely clarified. Because it is a simple method using autoclave, to realize measurements during the hydrothermal process is complicated. For this, the mechanism of formation of titanate nanotubes is still very controversial. Despite the difficulties, many models can explain this formation, where many of those are supported by the measurements realized in intermediary stages of the process, such as submitting the titanium dioxide to the hydrothermal treatment for short time frames until a good yield in the tubes preparation.

The biggest problem for this method is to reach a good yield on the nanotubes production changing a large number of variables, parameters many times neglected, or not informed by the authors of the literature. A large number of those formation models have in common the break of the Ti–O–Ti bonds, produced by the reaction with the alkaline solution of the precursor crystallites and the formation of an intermediary bi-dimensional phase, which it would become the nanotubes (Ferreira 2006; Morgado 2007a; Silva 2012). According to Ferreira (2006), the use of the reactor affects directly on the after treatment product. Besides, the washing process is not a central step on the nanotubes formation, which differs from the talks of Kasuga et al. (1998) On the other hand; Morgado et al. (2007a) state that the formation of bi-dimensional nanostructures titanium dioxide-based happens by the dissociation of the TiO_2 precursor, creating a kind of titanate, which crystallizes in intermediary nanosheets. These nanosheets are subjected to two phenomena, which compete during the reaction, the growth and stacking of multilamellar nanostructures and the curving and/or the nanotubes roll up with multiple walls. Following the author, titanate oxides with great reactivity with the synthesis surroundings dissolve more quickly, increasing the availability of the construction units, favoring the growth and stacking of multilamellar structures to the curving and roll up, creating nanowires and rods, instead of nanotubes. The slower dissolution allows the lamellar nanostructures more time to curve before too much growth, forming nanotubes.

According to Ferreira (2006), from a structural standpoint, obtaining one-dimensional systems from bi-dimensional ones (lamellar), has features of a general mechanism, and, likewise, can be expanded to other ci-dimensional compounds. For this author, nanotubes can be formed from bi-dimensional structures through the curving of the lamellar units, forming tubes of the scroll type, or through the elimination of the random bindings of the lamellar units with few nanometers proportion. Also states that the reactionary medium containing the lamellar units would be responsible to provide for the "gradient force" for the curving of the lamellar units, causing the formation of nanotubes, or to the rupture of preferential directions, taking to forming nanorods. In his work, Ferreira (2006) details the specific manner which titanium dioxide nanotubes (titanate) can be obtained from a lamellar precursor ($Na_2Ti_3O_7$), formed in situ, by the reaction of TiO_2 with NaOH in conditions of hydrothermal treatment, though a mechanism from which the lamellae are rolled up forming nanotubes of the scroll type with wall structures similar to the $Na_2Ti_3O_7$ lamellae bulk with composition for the nanotubes, as prepared, close to $Na_2Ti_3O_7{\cdot}nH_2O$. He supports, though, that depending on the washing conditions (acidic solution, as example), the sodium content on the interwall area of the nanotubes can be lowered due to ionic exchange Na^+ for H^+, being proposed the following formula for the titanate nanotube: $Na_{2\text{-}x}H_xTi_3O_7{\cdot}nH_2O$, $0 \leq x \leq 2$, where the x changes with the washing conditions.

Because it is a simple method using autoclave, measurements during the hydrothermal method can be complicated. For that, the formation mechanism of the titanate nanotubes is still debatable. Despite the complications, many models have been elaborated to explain this formation, where a great part are based on measures realized during the intermediary steps of the process, such as, submitting the titanium dioxide through the hydrothermal treatment for short periods of time until it is achieved a good yield for the tubes production. The biggest problem for this method is the many parameters, which need to be manipulated to achieve a good yield. Many of those parameters are disregarded by the authors or not mentioned by the literature. Many of those models of formation have in common the formation of the Ti–O–Ti bonds, produced by the reaction with the alkaline solution of the precursor's crystallites and the formation a bi-dimensional intermediary phase, which would curve for the nanotubes formation.

Kasuga et al. (1999) produced nanotubes and defined the structures as the titanium dioxide on the anatase phase, from the TiO_2 in the rutile phase. The authors proposed the following mechanism for the formation of the nanotubes. The treatment of the titanium dioxide in sodium hydroxide solution breaks some of the Ti–O–Ti bindings and new bindings Ti–O–Na and Ti–O–H are formed. The washing of the material with water and a solution of chloridric acid creates Ti–O–Ti bonds by the reaction of the acidic solution with the hydroxyl and the terminal sodium atoms from the

Ti–O–Na and Ti–OH structures, forming a type of lamellar anatase metastable with excess of superficial Na^+. In sequence, the washing with distilled water causes a decreasing on the superficial charge excesso and Ti–O–Na bonds are converted to Ti–OH, however, with an excess of superficial charge in areas of the material. The addition of chloridric acid dehydrates the Ti–OH bonds and new bonds of Ti–O–Ti or Ti–O–H–O–Ti are formed, lessen the distance between the bindings of near titanium atoms, resulting in the curving of the sheets in tubes shape.

A controversial aspect about the nanotubes produced by the hydrothermal method is the importance of the washing process in the formation of the material, since some authors regard the acidic wash as been responsible for the rolling of the nanosheets into nanotubes. Aiming to end this discussion, Sun and Li (2003), without previous submitting the samples, produced by the hydrothermal method, to a washing process, used ethyl alcohol and acetone to disperse the samples measured by electron microscopy. The authors observed the yield of formation of nanotubes is higher for the samples in alcohol than the ones in acetone. Besides, they also noticed that, once the nanotubes were formed, the dispersion in acetone did not change the morphology of the samples. They concluded the hydrothermal process results in sheets which, depending on the washing treatment, roll up to nanotubes. Additionally, they concluded that the acidic washing is not a step responsible for the roll up of the sheets in nanotubes.

Wang et al. (2002) suggested that the nanotubes based on titanium dioxide are formed by the cleavage of the anatase in the plane (1 0 1) due to the reaction with the sodium hydroxide solution, forming a bi-dimensional structure with a great number of ions O^- or Ti^{3+} in the surface. This excess of surface charges would be enough to cause the roll up of this bi-dimensional structure in the anatase direction [0 1 0] and form the tubes.

A more detailed mechanism was presented by Yang et al. (2003). According to the authors, only the more lengthy Ti–O bonds in the anatase phase are broken by the reactions with the OH^- ions, causing the exfoliation of the titanium dioxide crystallites. The lossen fragments would connect through the O–Na–O ionic bonds. A scheme of the mechanism proposed by the authors is presented in Fig. 4. The authors proposed that each layer of the tube was formed by a sheet and its limits were connected by a covalent bond of the terminal groups. This way, they proposed that a tube formed by a single layer would be used as the frame for the sequence layers built over it, as it is shown in Fig. 5c, with the Na^+ and OH^- between the walls.

Different from what was proposed by Yang et al. (2003), the majority of the works which investigate the subject, adopt models where the tube is formed, either by the roll up of a package of stacked sheets (Fig. 5b) or the curving of a single sheet (Fig. 5c). In the first case, the transversal section of the tubes has the appearance of an onion, as it is showed on the micrograph in Fig. 6b. In the second, the transversal section of the tubes makes a spiral, as in Fig. 6a. The main evidence of these two models are the electronic microscopy images of the nanotubes transversal sections, in Fig. 6, and the fact that one nanotube presents at the same time different numbers of walls at opposite sides.

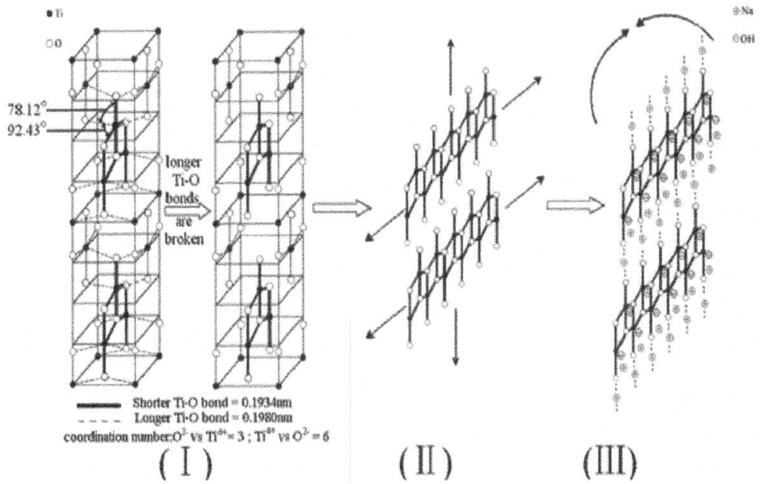

Fig. 4 Representative model of the formation of titanate tubes proposed by Yang et al. *Source* Yang et al. (2003)

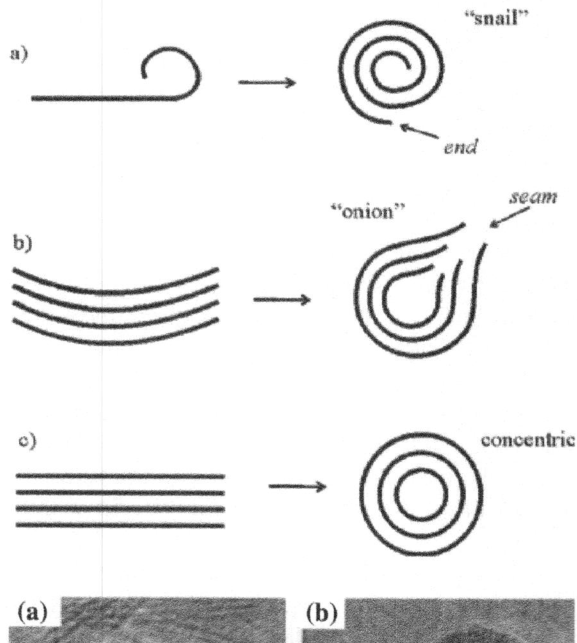

Fig. 5 Scheme showing the possible mechanisms for the formation of nanotubes with multiple walls. **a** Round up of a single sheet, **b** roll up of a package of stacked sheets, and **c** sheets connected by the edge. *Source* Bavykin et al. (2004)

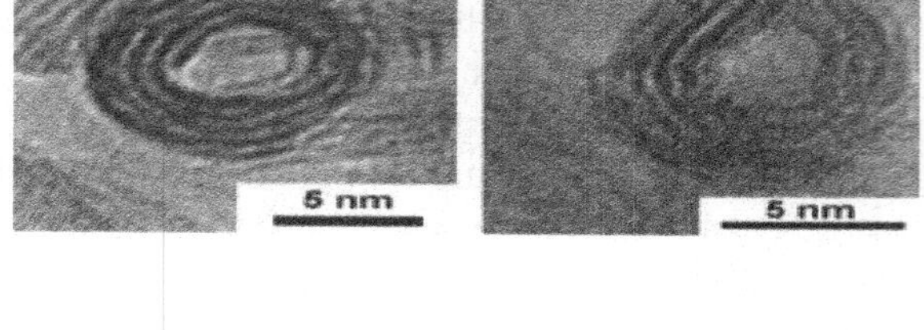

Fig. 6 Electronic transmission microscopy of the transversal section of the titanate nanotubes with **a** spiral form and **b** onion head. *Source* Bavykin et al. (2004)

Kukovecz et al. (2005), through the images of the electronic transmission microscopy of high resolution in hydrothermal synthesized samples during the 1, 2 and 3 h, observed the formation of small curved objects with about 10 nm of length, making a type of spiral on the surface of the anatase crystals. The final sample synthesized during 72 h did not present those nanospirals. In the face of this observation, the authors suggested that the tubes would be formed through a small quantity of materials, which, once removed from the surface of the anatase crystals, would recrystallize in trititanate sheets ($Na_2Ti_3O_7$), which would round making these nanospirals and would be the seed for the oriented growth of the nanotubes.

Ma et al. (2004) produced nanosheets of $Ti_{0.91}O_2$ in colloidal suspension, then, those sheets were stacked through the addition of sodium hydroxide on the solution. Afterward, the material was washed with distilled water many times, causing the formation of nanotubes. The authors suggested that the addition of sodium hydroxide made the colloidal suspension unstable, causing the staking of the sheets, intercalating Na^+ ions between them, linking a sheet with the adjacent sheets around. When water is added to the solution, it penetrates on the region between the sheets, extracting gradually the Na^+ ions, which cause the increase of the distance and decrease of the electrostatic interaction between adjacent sheets, inducing the separation of the package to individual sheets. The decrease of the electrostatic interaction between the adjacent sheets happens mainly on the lateral edges of the sheets, making it to curve gradually over itself, creating then the nanotubes with multiple layers.

Zhang et al. (2003) observed through electronic transmission microscopy the reaction between TiO_2 with NaOH to create a very random phase, similar to the one observed by Kasuga et al. (1999), from which some of the titanate sheets were observed to grow over each other. After 3 days of synthesis, no sheet was noticed, only the nanotubes, which, according to the authors have the formula $H_2Ti_3O_7$, caused by the roll up of a single sheet. Through the calculus, the authors verified the reaction between the H^+ ions in the surface of the sheets with the OH^- in solution, due to the alkaline environment, rather than the reaction of the H^+ between the sheets with the same OH^-. The lack of superficial H^+ would cause a tension which makes the sheet to curve and, under proper circumstances, to lose itself from the inferior layers, forming, then, nanotubes with the formula $H_2Ti_3O_7$.

Chen et al. (2002) suggested that the intermediary lamellar phase, formed by octaedres of TiO_6, was formed with the shape of a trititanate $H_2Ti_3O_7$ (Fig. 7a). The tube would be, then, formed by the enrolling of this structure along the axis [0 1 0] (Fig. 7b). This model was used in a simulation of X-ray diffraction, with good agreement with the experimental results.

The mechanism of nanotubes formation presented by Ma et al. (2004), Zhang et al. (2003) and Chen et al. (2002) presented a great deficiency. It is not capable of explaining the onion shape for the tubes, as in Fig. 6b, and the nanotubes with a number of walls in opposite sides greater than one.

Through calculations, Bavykin et al. (2004) comproved the possible energetic advantage during the process of rolling the multiple stacked sheets of titanate for the nanotubes formation. The authors associated the difference between the dimensions of the sheets in the same stack to a surface energy. The difference between sheets size

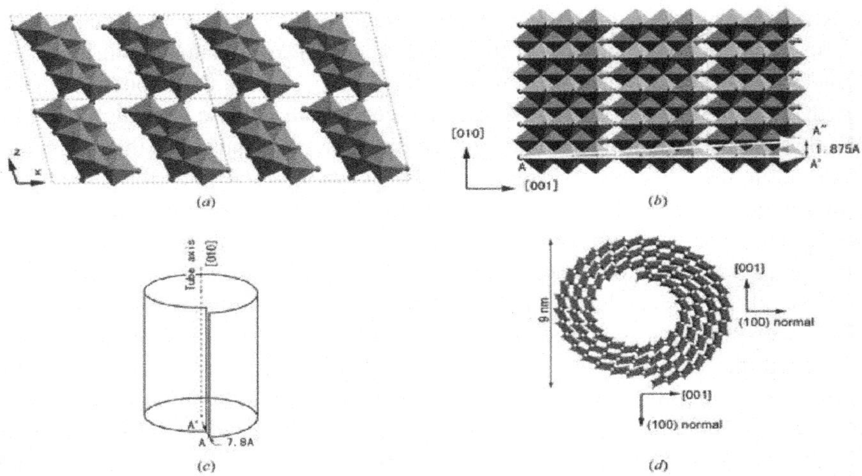

Fig. 7 Scheme of formation of the titanate nanotubes proposed by Chen et al. *Source* (Adapted) from Bavykin, Dmitry & Friedrich, Jens & Walsh, Frank. (2006). Protonated Titanates and TiO_2 Nanostructured Materials: Synthesis, Properties, and Applications. Advanced Materials. 18. 2807–2824. https://doi.org/10.1002/adma.200502696.39 (Bavykin and Frank 2010)

would make it to bend, as it shows in Fig. 8, reducing the surface energy and creating an elastic energy, due to the encurving. The calculation demonstrated that the curving would be five hundred times less energetic than the existence of stacked sheets of

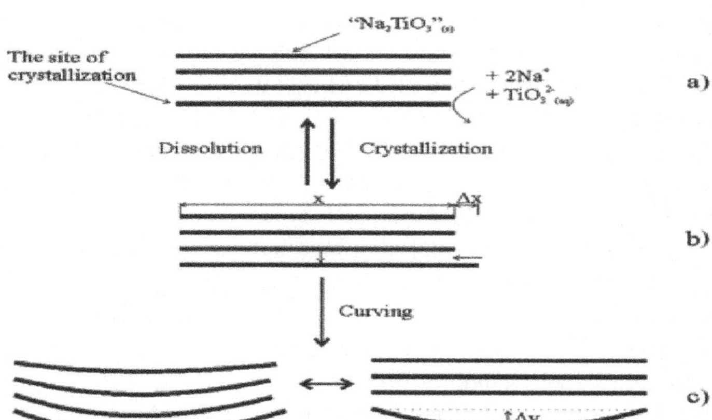

Fig. 8 Graphic representation proposed by Bavykin and Frank (2010) for the formation of titanate nanotubes. **a** Stacking of titanate nanosheets, **b** stacking of one nanosheet greater than the others, **c** nanosheet bended to decrease the system energy. *Source* (Adapted) from Bavykin, Dmitry & Friedrich, Jens & Walsh, Frank. (2006). Protonated Titanates and TiO_2 Nanostructured Materials: Synthesis, Properties, and Applications. Advanced Materials. 18. 2807–2824. https://doi.org/10.1002/adma.200502696.39 (Bavykin and Frank 2010)

different sizes. Then, the stacked sheets would bend, explaining the formation of tubes with onion-shaped transversal section (Fig. 6b).

The authors created the following steps to explain the thermodynamics and kinetic formation of the titanate nanotubes:

Dissolution of the precursor titanium dioxide:

$$TiO_2 \text{ precursor} + 2NaOH \rightarrow Na^+ + TiO_3^{2-} + H_2O$$

Crystallization of the nanosheets:

$$2Na^+ + TiO_3^{2-} \leftrightarrow [Na_2TiO_3] \text{ sheets}$$

Nanosheets bending:

$$[Na_2TiO_3]\text{sheets} + TiO_3^{2-} + 2Na^+ \leftrightarrow [Na_2TiO_3]\text{tube}$$

Ionic exchange ($Na^+ \rightarrow H^+$):

$$[Na_2TiO_3]\text{tube} + 2H_2O \leftrightarrow [H_2TiO_3]\text{tube} + 2NaOH$$

Following Morgado et al. (2007b), the formation of bi-dimension nanostructures based on titanium dioxide happens by the dissociation of precursor TiO_2, creating a type of titanate which recrystallizes on intermediary nanosheets. These nanosheets are subject to two phenomena, which compete during the reaction, the growth and stacking of multilamellar nanostructures and the curving and/or rolling of nanotubes with multiple walls. Still following the author, the titanium dioxide more reactive with the synthesis environment are more easily dissolved, increasing the availability of the construction units, favoring rather the growth and the stacking on multilamellar structures than the curving and roll up, shaping nanoribbons and rods, instead of nanotubes. The slower dissolution allows the lamellar structures to curve before too much growth forming then nanotubes.

Other researchers, such as Wang et al. (2002) also attributed a structure of the anatase type to the nanotubes walls (Fig. 9).

Therefore, in the face of the difficulties found by the electrons on the TiO_2 particles and the unique features of the nanostructures synthesized from these particles, it gives rise to employ the alkaline hydrothermal method or simply the Kasuga method (Kasuga et al. 1999), which is a simple method used to solve the problem of the photocatalytic layer of the solar cell.

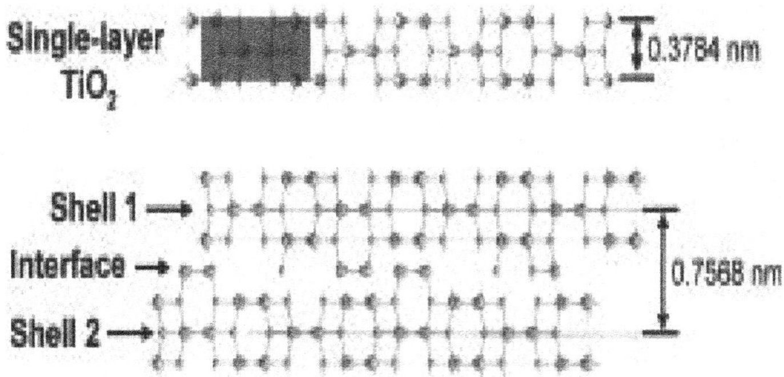

Fig. 9 Graphic representation of the nanotubes walls proposed by Yao et al. *Source* Yao et al. (2003)

4.2 *Nanostructured Photoelectrodes for the Dye-Sensitized Solar Cells*

The nanotechnology has opened many doors to the research and development of new materials, improving the quantity of nanostructures used for dye-sensitized solar cells. As already cited, the TiO_2 nanoparticles with great surface area have been the main material used as the dye adsorption layer for these devices. However, decreasing the recombination rate on these cells layers is the great challenge faced by the scientists. One of the tools to reduce recombination is the use of one-dimension structures with the morphology of tubes, ribbons, wires, and rods that have great surface areas and are capable of improving the adsorption and the electronic transport on those devices

Recently, the number of works, which relate to the use of different morphologies on the DSSCs has increased exponentially. Between these materials, the one-dimensional nanostructures have received attention because the capability to create alternative pathways for the electrons and contribute to the electronic diffusion process on the cells.

It is necessary to grow the thickness of the TiO_2 layer to improve the conversion efficiency. However, due to the lack of a depletion layer at the nanocrystalline surface of the TiO_2, an increase of the TiO_2 layer thickness causes recombination during the charges transportation (Hodes et al. 1992; Ito 2009). Shalan et al. (2012) demonstrated that to improve the properties of electronic transportation in the photoelectrode, an apparatus highly well-disposed of TiO_2 nanorods vertically orientated and well distributed can give better results than the TiO_2 nanoparticles films used in the DSSCs. Figure 10a, b shows the electrons displacement on electrodes with TiO_2 nanoparticles and nanorods.

Law et al. (2005) demonstrated that the nanoribbons of ZnO with a diameter of about 130 nm used in DSSCs photoanodes (Fig. 11) presented almost three times higher electron diffusion coefficient than the TiO_2 nanoparticles. Due to the higher

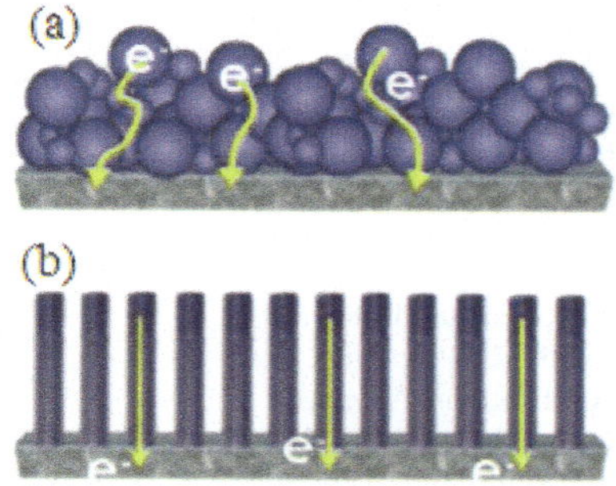

Fig. 10 Scheme of the electron transport on **a** TiO_2 nanoparticles and **b** TiO_2 nanorods. *Source* Shalan et al. (2012)

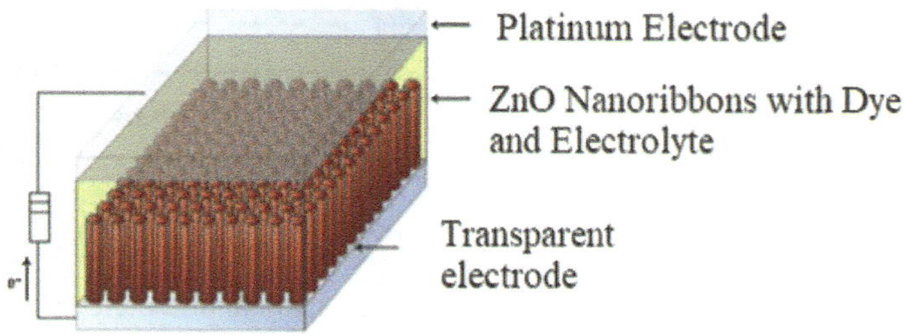

Fig. 11 Scheme for the DSSC with ZnO nanoribbons on the photoelectrode. *Source* Nanowire dye-sensitized solar cells (2005). (Law et al. 2005)

dye adsorption area and more organized structure, the nanoribbons had electron diffusion coefficient of 0.05–0.5 $cm^2\ s^{-1}$, meanwhile, for the TiO_2 film, the coefficient was between $5 \times 10^{-5}\ cm^2\ s^{-1}$ (Law et al. 2005; Zhang and Guozhong 2011).

Besides these nanostructures, the titanate nanotubes have been largely reported in the literature. The elongated morphology and the mesoporous structure of the titanate nanotubes, allied with the high specific surface area, the efficient ionic exchange properties and the relatively good protonic and electronic conductivity, make these nanostructures good materials for many applications (Tanaka et al. 1997; Tereza et al. 2013). The possible applications include many fields, such as catalysis and photocatalysis, fuel cells, electrodes for lithium battery (rechargeable batteries), condensers of high electrical capacity, hydrogen storage, sensors and biosensors, and for dye-sensitized solar cells (DSSC, Grätzell cells) (Agnaldo et al. 2006; Chagas 1984; Lima 2013; Tanaka et al. 1997).

The advantage in the use of titanate nanotubes instead of other materials for the construction of electrodes in the dye-sensitized solar cells resides essentially in its greater capacity for the absorption of the dye ions (positive) on the titanate surface (negative). It is estimated the possibility of making a layer of about 1000 molecules of dye along a titanate nanotube surface. This dye monolayer improves the efficiency of the radiation absorption which, combined with the elongated morphology of the TiNTs, allows better efficiency in the transport and storage of electrons in the DSSC

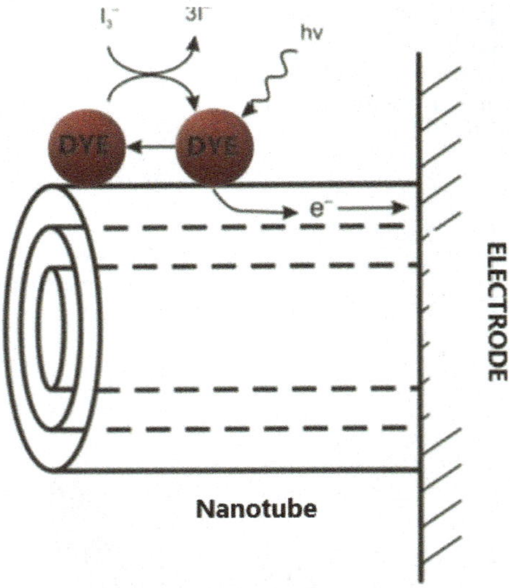

Fig. 12 Working scheme of the titanate nanotube electrode. *Source* (Adapted) from Bavykin, Dmitry & Friedrich, Jens & Walsh, Frank. (2006). Protonated Titanates and TiO_2 Nanostructured Materials: Synthesis, Properties, and Applications. Advanced Materials. 18. 2807–2824. https://doi.org/10.1002/adma.200502696. Titanate and Titania Nanotubes Synthesis, Properties and Applications (2010) (Bavykin and Frank 2010)

(Vanessa 2011). Figure 12 shows the working titanate nanotube electrode with the dye photosensibilization in its surface and the path for the photogenerated electrons.

However, the present limitation of the titanate nanotubes as photoactive materials resides in the weak capacity for the radiation absorption in the visible range. Yet, theses materials seem to present a recombination rate of the electrons/lacunes photogenerated pair lower than for the titanium dioxide (TiO_2) and, for that, are considered to be nanostructures with future impact in the development of photoactive materials with high efficiency. The titanates can go through structure alterations induced by ionic exchange or doping. Also, for the case of the TiNTs combination with a semiconductor material or an element which absorbs visible radiation ($\lambda =$ 400–700 nm) and/or promotes the recombination rate decrease, resulting, possibly, in a nanocomposite functional material with distinct optical properties and photochemical performances, improved relative to the materials which it originated from (Bavykin and Walsh 2009; Xiao et al. 2008; Ylhäinen et al. 2008).

4.3 Properties of the Titanate Nanostructures

The titanate nanostructures obtained by the alkaline hydrothermal method have well-defined diffraction standards. However, there are difficulties to define the crystalline structure of these materials. The great number of alterations from crystals to pure TiO_2 (anatase, rutile, or brookite) and protonated forms, also the small size of the crystals and the enlarging of the diffraction, peaks causes problems for the identification of the structures (Bavykin et al. 2006).

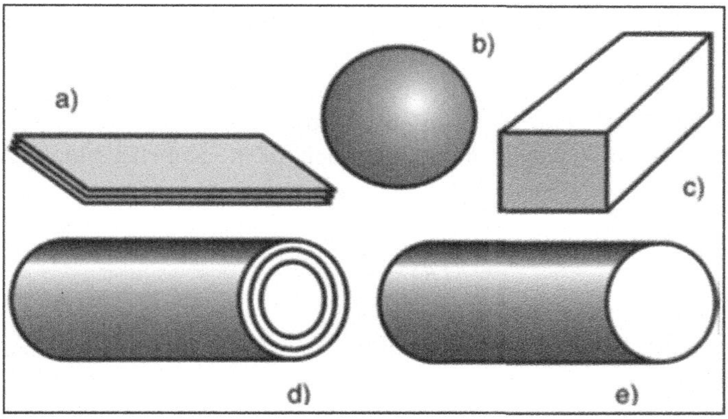

Fig. 13 Illustrative images for the titanate structures: Nanosheets (**a**), nanospheres (**b**), nanoribbons in rectangular section (**c**), nanotubes (**d**) and nanowires or nanorods (**e**). *Source* (Adapted) from Bavykin, Dmitry & Friedrich, Jens & Walsh, Frank. (2006). Protonated Titanates and TiO_2 Nanostructured Materials: Synthesis, Properties, and Applications. Advanced Materials. 18. 2807–2824. https://doi.org/10.1002/adma.200502696 "Protonated Titanates and TiO_2 Nanostructured Materials" (2006) (Bavykin and Frank 2010)

As there are difficulties for the identification of the crystal phases of the titanate nanotubes, there are, also, controversies on the morphology of these materials. Figure 13 shows five different morphologies for the observed titanates, following Bavykin et al. (2010), during the synthesis of the titanate nanotubes obtained by the hydrothermal method.

The nanotubes are, in general, formed by multilayers of nanosheets as in (a) combined by many titanate planes (100). The nanosheets are thin, allowing 10 nm thickness and more than 100 nm of height and width. The nanowires or nanorods in (e) are solid and long cylinders with circular base and without internal layers structures. Nanowires can turn out heating the nanotubes at 450 °C. The long titanates in the parallelepiped form as in (c) can be nominated nanoribbons, in the literature. These structures have good crystallinity with different dimensions in the orientated axis. Nanoribbons can have lengths of dozens of micrometers and width in between 10 and 100 nm. During the hydrothermal treatment, the individual morphological shape of the titanates is driven to conglomerate forming secondary particles.

4.4 Optical and Electrical Properties of the Titanate Nanostructures

The titanate nanotubes are semiconductor materials of large *gap*. Bavykin et al. (2006) estimated the *band gap* energy of the titanate nanotube colloids, processed at room temperature, to be about 3.87 eV. A close value of 3.84 eV was verified for titanate nanosheets. However, both are higher than the 3.2 eV for the bulk TiO_2 (Bavykin et al. 2006). Mozia et al. (2010) estimated for the titanate nanotubes synthesized by different temperatures and deposition times, *band gap* about 3.28 eV. In the same work, 3.10 eV was found for the TiO_2 P25 (pure anatase) used as precursor. This last author estimated the *band gap* using diffuse reflectance (R) and obtaining energy values from $(F(R)h\nu)^{1/2}$ versus $h\nu$ with the help of the Kubelka-Munk (Zeit 1931) method, where:

$$F(R) = \frac{(1-R)^2}{2R} \quad (1)$$

Studies report the effect of the size of the titanate nanostructures to facilitate the energy levels above *band gap*.

Photoluminescence spectrum of titanate nanotubes can detect many energy bands, in short and long wavelengths, which suggests high electron holes, pair density.

4.5 Conductivity Properties

According to the crystalline structure of the titanate nanotubes, the protons occupy the cavities between the layers of TiO_6 octaedre. Bavykin et al. (2006) suggest that the materials conductivity happens through the positive charges. These authors verified, by AC impedance in titanate, the increase of the protonic conductivity with the rise of the temperature of the thermal treatment from 30 to 130 °C. Verified, also, beyond that temperature range, the increase in temperature causes decrease of conductivity, due to the remotion of water from the porous of the nanotubes. This lowered conductivity represents the electron conductivity of the titanate nanotubes. This conductivity relies positively on the temperature, with an apparent activation energy of 0.57 eV, and represents a greater value than the conductivity for the anatase or rutile (Bavykin et al. 2006). As an example, at 225 °C, the conductivity for the titanate nanotubes is about 7.9×10^{-7} Scm^{-1} (Thorne et al. 2005), meanwhile, for the anatase nanoparticles with 6 nm diameter, the value is from the magnitude of 10^{-9} Scm^{-1} (Dittrich et al. 2000).

Recent interest in ferromagnetic semiconductors at room temperature and magnetic materials of nanometer sizes (motivated by semiconductors devices based on spins) stimulated the research on the synthesis and characterization of magnetic materials based on TiO_2 nanotubes (Zeit 1931).

It was demonstrated that titanate nanotubes doped with 4 mol/L % of cobalt prepared by Co^{2+} during thermal treatment held ferromagnetic properties at room temperature. On the other hand, the ionic exchange of the titanate nanotubes with Co^{2+} with rates of (Co/Ti)/(1/3.5) held antiferromagnetic properties at room temperature (Wu et al. 2005).

4.6 Preparation of the Titanate Nanotubes and Nanoribbons

The titanate nanotubes synthesis (TiNTs) was carried out according to Ferreira (Ferreira 2006; Souza 2019). In a regular synthesis, about 2 g of commercial TiO_2 anatase (titanium dioxide IV) is suspended in aqueous solution of sodium hydroxide 10.0 mol/L (Dinâmica). The formed suspension is transferred to a Teflon becker of 60 mL, placed in a reactor of stainless steel and kept at 160 °C for times between 12 and 120 h in conventional stove. For the nanoribbons preparation, the above-mentioned system has a temperature of 190 °C and times between 12 and 120 h. After cooling until reached room temperature, the resulting white solids are separated by centrifugation and washed by distilled water until the pH is between 11 and 12. After drying for 24 h in conventional oven (kiln), at 80 °C, titanates rich in sodium are obtained. For protonated systems ($Na^+ \rightarrow H^+$), the washing of the nanotubes is carried in acid environment.

In this chapter, the following samples were synthesized: NaTiNT (nanotubular) and NaTiNT (nanoribbons) from the anatase TiO_2. Figure 14 shows a fluxogram of the TiNTs synthesis process.

Such nanostructures have an average diameter of about 9 nm (Chen et al. 2002; Ferreira 2006; Morgado 2007a; Silva 2012). The average internal diameter of the tubes is about 5 nm with length superior to 100 nm. According to Ferreira (2006),

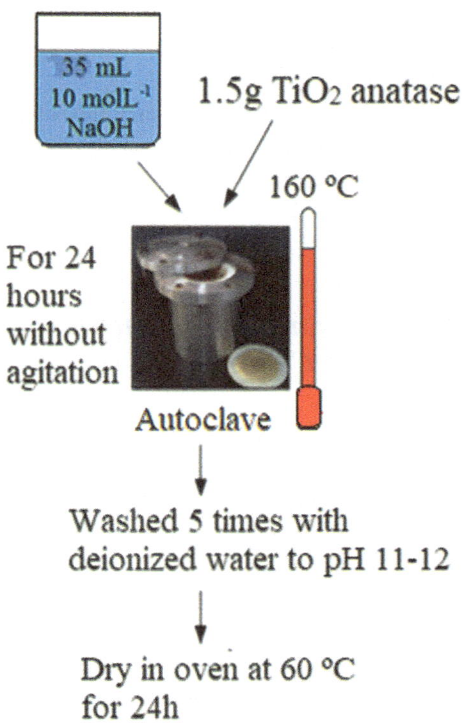

Fig. 14 Scheme of the synthesis process for the titanate. *Source* Own author

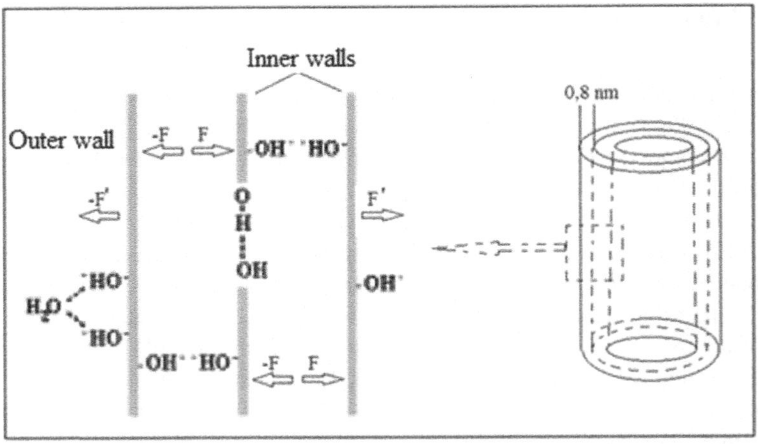

Fig. 15 Adapted model of the TiO_2 nanotube thermally treated in alkaline medium. *Source* Adapted from Zhang et al. 2003)

as the difference between the internal and external diameter is about 4 nm, there are indicators of formed nanotubes with multiple walls.

Zhang et al. (2003) state the OH^- groups present in the nanotubes wall cause repulsive forces (−F, F, −F', F') between itself, making the distance between adjuncts walls to be 0.8 nm. Figure 15 shows an adapted model proposed by this author for the TiO_2 nanotubes thermally treated with NaOH solution and forces acting on the walls.

For Zhang (2003) model, nanotubes with 9.3 nm of external diameter and 6.4 nm of intern diameter had 4 walls with 0.8 nm of distance between the walls. Based on the model proposed by this author, it is suggested a distance between adjunct walls of the TiNT-Na^+ to be around 0.8 nm.

4.7 Applying the Titanate Nanostructures in Dye-Sensitized Solar Cells

The solar cells with effective area of 1.0 cm^2 can be constructed using the previously mentioned films.

The electrodes with the TiO_2, NaTiNT, and NaTiNT-Ribbons nanostructures, with and without thermal treatment, were immersed in an ethanol solution (3×10^{-4} mol L^{-1}) of the dye "*Ruthenizer 535-bisTBA*" (N719) (©Solaronix. Abr 2014) for 24 h to make sure all the sites of anchorage of the semiconductor were filled with the sensitized dye. After that, the electrodes with the adsorbed sensitized dye were washed by ethanol to remove the excess of dye, which was not adsorbed. Each electrode was connected through a redox pair commercial electrolyte (iodide/triiodide)

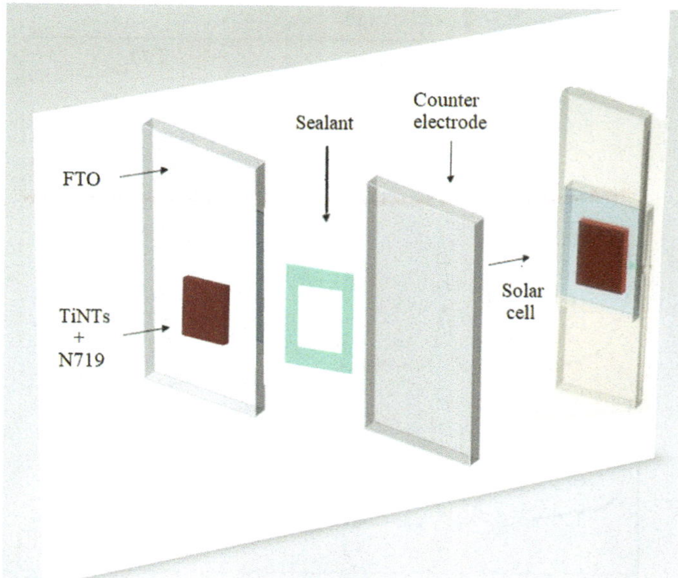

Fig. 16 Illustration of the cell assemble. *Source* Own author

(©Solaronix. Abr 2014), with a counter electrode, also commercial, holding a thin platinum layer deposited on FTO. A commercial sealant was used with the help of pressure clips to avoid the leakage of the electrolyte in the cells. The assembled cells are similar to the ones presented in Fig. 16.

5 Structural Characterization

The photovoltaic devices are characterized by many physical-chemical techniques:

5.1 X-Ray Diffraction

The identification of the crystalline phases of TiO_2 and the titanate can be done in the powder form.

The XRD results of the samples of nanotubes, nanoribbons, and TiO_2 are shown in Fig. 17.

From these diffractograms, it is observed that the differences between the diffraction patterns of the precursor TiO_2 and the titanate nanostructures. The large and asymmetrical peaks seen in the diffraction pattern of the NaTiNT are attributed to the curving of some atomic planes due to the tubular morphology (Cao 2004;

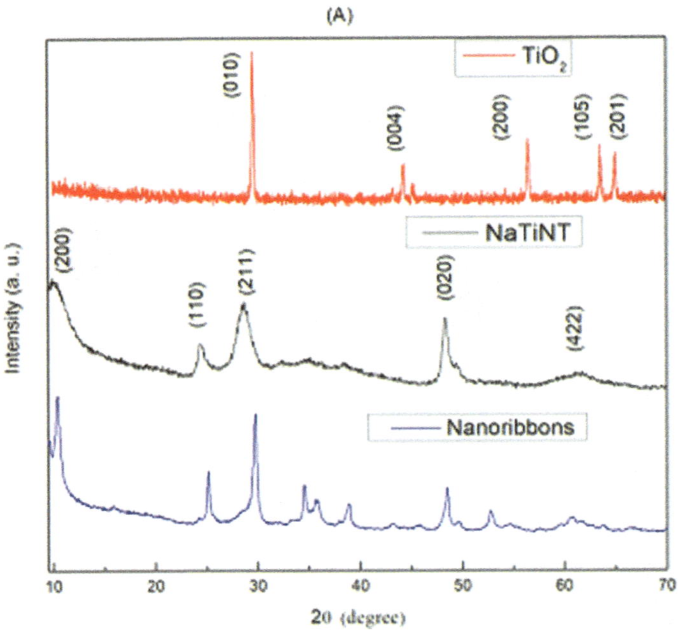

Fig. 17 X-ray diffractograms for the TiO_2, NaTiNT and NaTiNT-ribbons samples. *Source* Souza (2018)

Besra and Liu 2007). Ferreira (2006) asserts that the diffraction pattern is of a lamellar composite, suggesting the nanotubes are of multiple walls constituted by $(Ti_3O_7)^{2-}$ titanates negatively charged. The peaks around $2\theta = 10°, 24°, 28°$ rely on the washing process and are attributed to the interwalls distance, the presence of $H_2Ti_3O_7$ and $Na_2Ti_3O_7$, respectively (Edisson et al. 2009; Ferreira 2006; Souza 2019). The nanoribbons diffraction pattern is very similar to the ones found on the literature (Edisson et al. 2009; Ferreira 2006) with less wide peaks and more intense ones, comparatively to the nanotubes.

6 Raman Spectroscopy

The Raman spectroscopy has this name because is based on the Raman effect which is an inelastic phenomenon of light dispersion, allowing the study of the molecular rotation and vibration (Dresselhaus et al. 2003). Its analyses are based on light dispersed by a material when struck by a monochromatic light beam. This light allows small changes of specific frequency of the material, which was analyzed.

The identification of the vibrational modes of the nanostructures can be made on the dry samples in the powder form. Figure 18 shows the Raman spectrum for

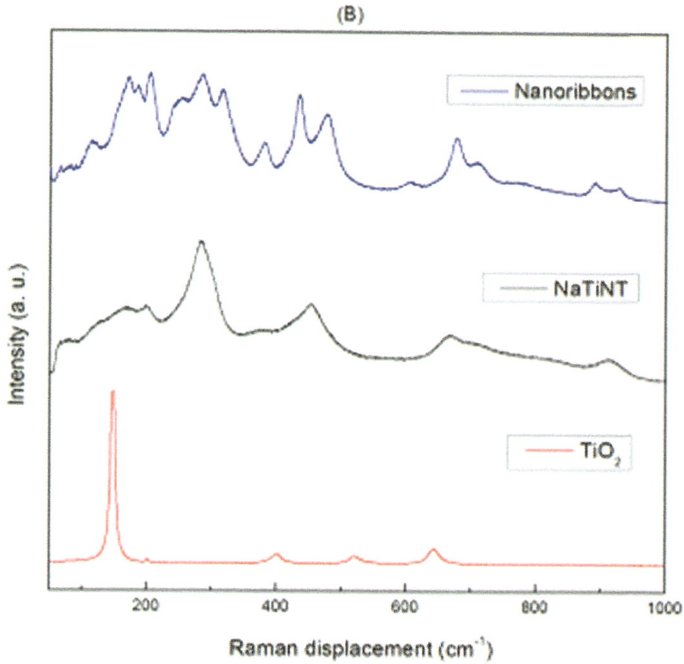

Fig. 18 Raman spectroscopy for the NaTiNT, NaTiNT-ribbons and TiO_2 samples. *Source* Souza (2018)

the NaTiNT, NaTiNT-ribbons, and TiO_2 samples. The difference between the spectrums suggests the structural conversion from TiO_2 nanoparticles to nanotubes and nanoribbons of titanate.

The Raman spectroscopy measures from the anatase phase TiO_2 identify five vibrational modes. It is observed from Fig. 18 the bands in 148, 205, 401, 520 cm^{-1}. Raman spectroscopy measures done by Giarola et al. in monocrystals of TiO_2 in the anatase phase on different polarizations identified six active modes expected in Raman. They identified the bands in 144, 197, 639, 399 e 516 cm^{-1}. These results were also observed by other authors (Thiago 2010; Giarola et al. 2010; Ferreira 2006; Souza 2019).

From a regular Raman spectrum for the titanate nanotubes, there are two intense peaks in 290 e 448 cm^{-1}, one multiple peak centered in 668 cm^{-1} and less intense peaks in 388, 827 e 917 cm^{-1} (Zhang et al. 2003). Bavykin et al. (2006) affirm that an exact reference for these vibrational peaks is up to debate, however, some of those peaks can be seen as Ti–O–Ti in 448 cm^{-1} and 668 cm^{-1} or vibrational Ti–O–Na associated to the peak in 917 cm^{-1}. For the nanoribbons, it is possible to identify many vibrational modes. Many authors (Abed et al. 2017; Ferreira 2006; Giarola et al. 2010; Souza 2019) suggest the peaks 141, 153 cm^{-1} are associated to the vibrational modes in the anatase phase, and the peaks 179, 195 cm^{-1} are attributed to the Na–O–Ti modes, and the 284, 300 cm^{-1} peaks relates to the Ti–O modes.

7 Morphology

Figure 19a–d shows the nanostructure morphology for TiO_2 and NaTiNT analyzed in powder form by SEM and TEM. In the images, it is possible to see the morphological distinction between these structures. In Fig. 19d, it is highlighted the image from one single nanotube. In this image, it is clear that the NaTiNT are actually made with multiple walls and nanometers internal and external diameters.

7.1 Morphology of the Nanotubes and Nanoribbons Films of Titanate and TiO_2 on FTO

Figure 20a–f shows the morphology by SEM of the TiO_2, NaTiNT and NaTiNT-Ribbons films with and without thermal treatment at 450 °C, respectively.

From Fig. 20e and f, it is noticed the preservation of the elongated nanostructures of titanate and the nanoribbons even after thermal treatment. However, these new structures may have nanorods or nanowires morphologies due to, in this case, the collapse of the tube walls.

Figure 21a shows that NTiNT have a tubular morphology and they are multi-walled with an average outer (inner) diameter of 9 nm (5 nm) and a length of several tens of nm. The nanotubes are open ended (see inset to Fig. 21a) with a uniform diameter distribution. In Fig. 21b, we can observe that the nanoribbons samples have a ribbon-like morphology. The NaTiNT-ribbons are long, uniform and multi-walled with a typical average width of approximately 100 nm, and a length of some micrometers. The nanoribbons present a wide width distribution varying from 20 to 200 nm.

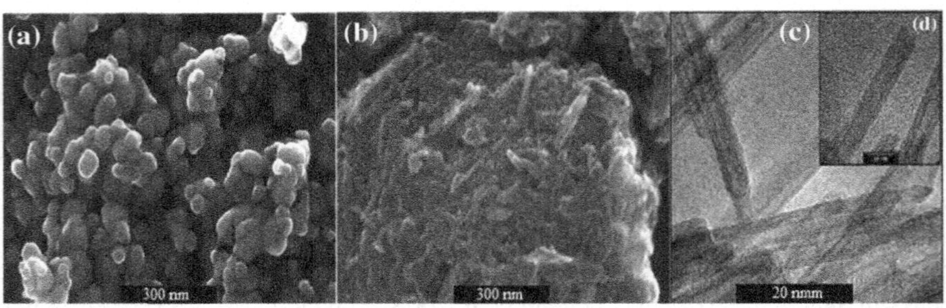

Fig. 19 SEM: **a** Nanoparticles of TiO_2, **b** NaTiNT in powder form, (**c**) and (**d**) TEM images of the nanotubes in powder form. *Source* Own author

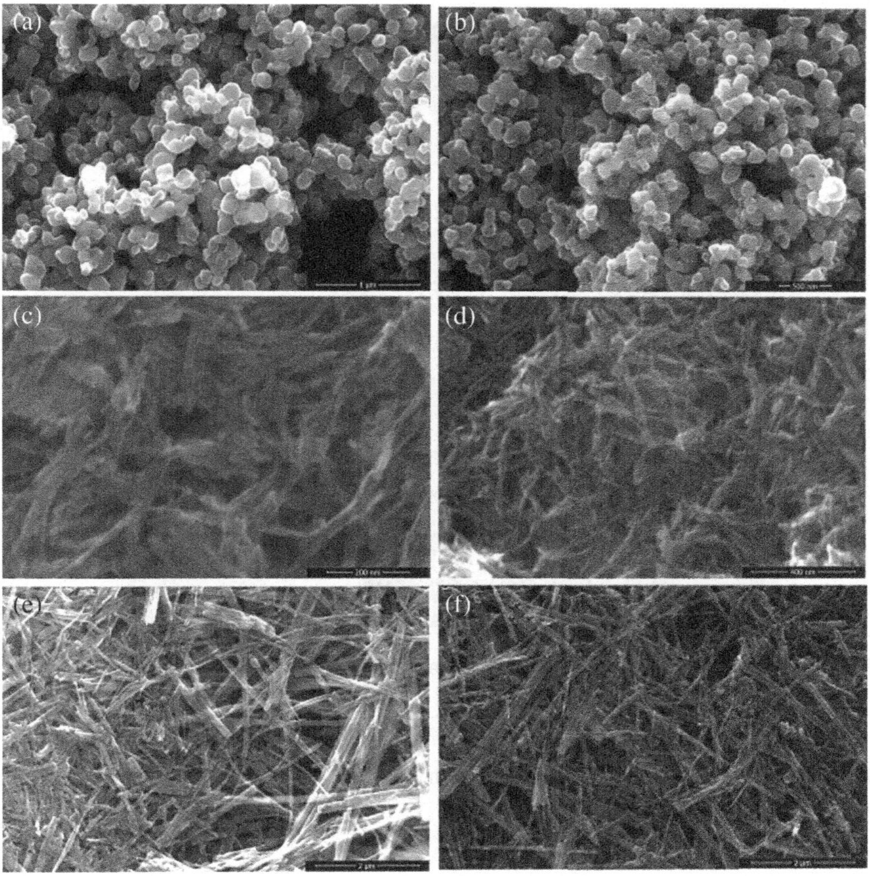

Fig. 20 SEM: **a** Film of TiO_2, **b** film of NaTiNT and **c** film of NaTiNT-Ribbons without thermal treatment and **d** film of TiO_2, **e** film of NaTiNT-Ribbons, and **f** film of NaTiNT with thermal treatment. *Source* Souza (2018)

7.2 *Optical Characterization of the TiO_2 and Titanate Films with and Without Thermal Treatment*

The spectrophotometry measures by UV-Vis were done in the films deposited by electrophoresis (Souza 2018). Figure 22a–c shows the absorbance of the investigated films before and after the thermal treatment. The shift in the absorption spectrum in the visible area is associated with a better absorption in this region, which is explained by the decrease of the band gap of these materials. Table 1 shows the energy values for the characteristic band gap of these materials.

The new structures formed after the thermal treatment have greater capacity for light absorption in the regions near the visible. Studies show a loss of water in the

Fig. 21 TEM images of as-prepared **a** titanate nanotubes (NaTiNT) and **b** titanate nanoribbons (NaTiNT-ribbons). *Source* Adapted of Souza (2018)

thermal treatment of the nanotubes and nanoribbons. In Fig. 22d, there is the overlap of the absorption spectrum of the films.

7.3 Photoelectrochemistry Characterization of the Cells Assembled with the Films of TiO_2, NaTiNT, and NaTiNT-Ribbons with and Without Thermal Treatment

The films with TiO_2 and with nanotubes and nanoribbons deposited by electrophoresis and sensitized by the N719 dye acted, in this chapter, as the working electrodes in the dye-sensitized solar cells. Figure 23A, A1, B, B1, C, C1 present the plots of current density-tension (J–V) of the photovoltaic devices. Table 2 presents the main photoelectrochemical parameters from these cells.

The data in Table 2 shows that, except for the cell with TiO_2, there was an increase of the efficiency of the cells after the thermal treatment. It is noticed important increase on the parameters of current and tension for the cells with nanotubes and nanoribbons after thermal treatment. The gain in the series resistance Rs causes a decrease in the current obtained from the cells. It can be interpreted that the loss in

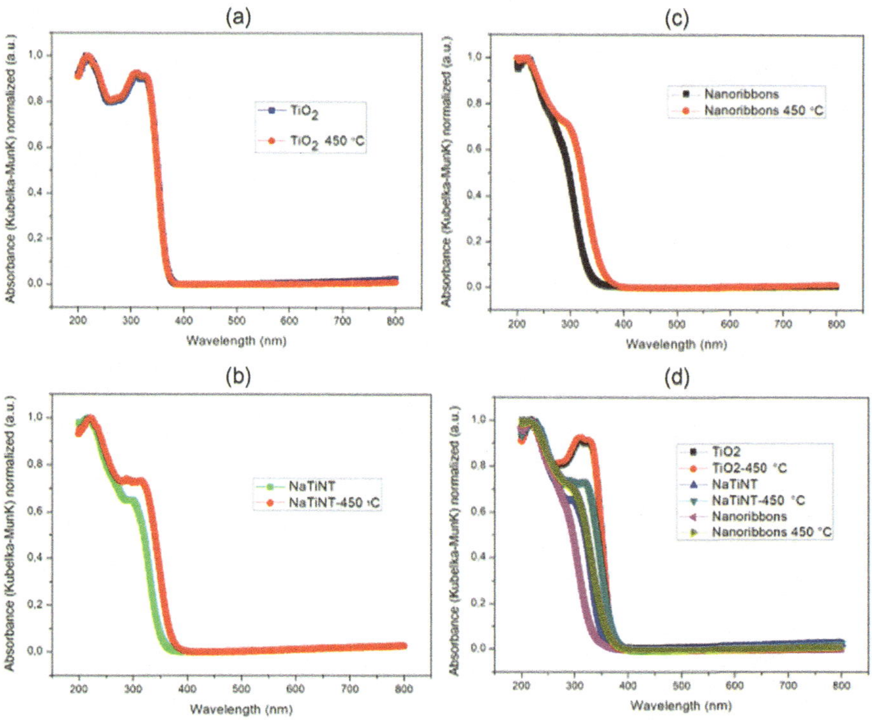

Fig. 22 Absorbance of the films: **a** TiO_2, **b** NaTiNT, **c** NaTiNT-Ribbons, and **d** overlap of the films before and after thermal treatment. *Source* Souza (2018)

Table 1 Estimation of the *band gap* energies by extrapolation

TiO_2 (eV)	TiO_2-450 °C (eV)	NaTiNT (eV)	NaTiNT-450 °C (eV)	NaTiNT-ribbons (eV)	NaTiNT-ribbons-450 °C (eV)
3.32	3.31	3.4	3.33	3.75	3.43

Source Adapted of Souza (2018)

the fill factor (FF) is associated to the increment of the *R*s and decrease in the shunt resistance *Rp* and, therefore, the loss in the cell efficiency. These parameters were favored after the heating of the films.

The efficiency of the TiO_2 had a slight loss after thermal treatment. This happens, possibly, to a greater compaction of the films, which causes a reduction of pores and lower dye adsorption. The pores reduction and also the lower dye adsorption cause a lower photogeneration of cell current. This did not happen for the TiNTs because these have greater surface areas than the TiO_2 nanoparticles, even after thermal treatment.

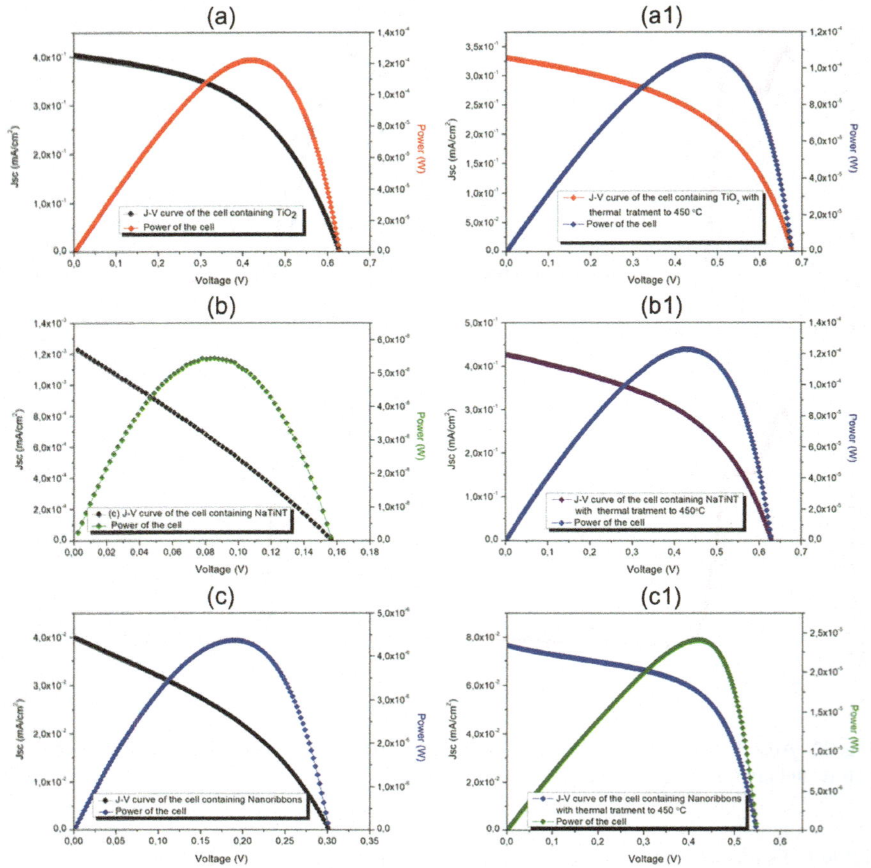

Fig. 23 *J–V* Plots: **A** TiO_2, **A1** TiO_2-450 °C, **B** NaTiNT, **B1** NaTiNT-450 °C, **C** NaTiNT-ribbons and **C1** NaTiNT-ribbons-450 °C. *Source* Souza (2018)

Table 2 Photoelectrochemical parameters of the cells: (A) TiO_2, (A1) TiO_2-450 °C, (B) NaTiNT, (B1) NaTiNT-450 °C, (C) NaTiNT-ribbons and (C1) NaTiNT-ribbons-450 °C

Electrical parameters/cells	Jsc (mA/cm^2)	Voc (V)	FF (%)	*Rs*	*Rp* (kΩ)	Irradiance (mW/cm^2)	η (%)
(A)	0.4028	0.6274	48.5	364.7 Ω	7.7	100	0.122
(A1)	0.3310	0.6762	48.0	388.5 Ω	7.8	100	0.107
(B)	0.0012	0.1562	28.5	88.1 kΩ	140.6	100	0.0000547
(B1)	0.4269	0.6298	45.9	331.5 Ω	4.5	100	0.123
(C)	0.0398	0.3027	36.4	2.6 kΩ	12.4	100	0.0043
(C1)	0.0765	0.5493	57.4	938.5 Ω	25.7	100	0.0241

Source Souza (2018)

However, the closer efficiencies between the cells with TiO_2 films, with and without thermal treatment, may happen due to the few changes in the nanostructure after heating at 450 °C.

The lower and greater efficiency was observed for the NaTiNT film before and after thermal treatment, respectively. The important increase may be related to a reduction in electrical resistance in the film, greater diffusion of the electrons and lower charge recombination. The water evaporation in the nanotubes interstices also favored the electron transport, because the OH^- groups responsible for the H^+ ions diffusion in the nanotubes evaporated with heating. It is suggested that the current generated by these ions goes against the current photogenerated by the cell.

The analysis of the electrochemical impedance spectroscopy (EIS) also was used to study the resistance to the transfer and transport of charges related to the electron recombination in the interface semiconductor/dye/electrolyte. The cells with the films before and after thermal treatment were analyzed. Figure 24A, A1, Fig. 25B, B1, and Fig. 26C, C1 shows the Nyquist diagrams of the cells obtained experimentally

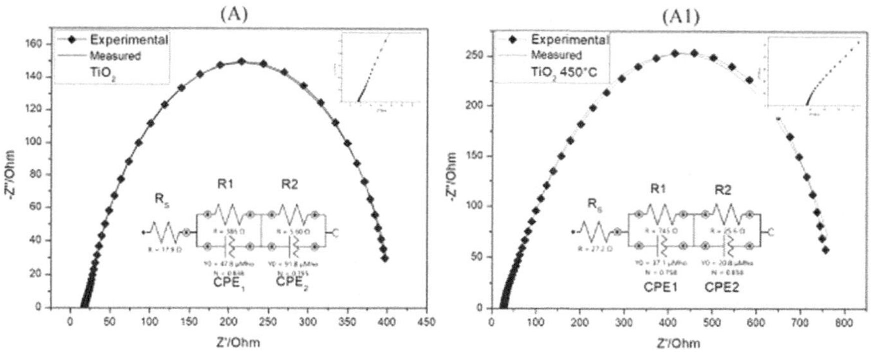

Fig. 24 Nyquist graphs of the cells with: **A** TiO_2 and **A1** TiO_2-450 °C. *Source* Souza (2018)

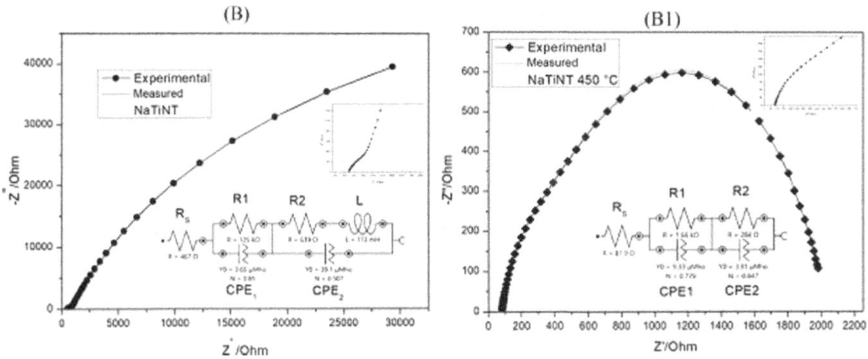

Fig. 25 Nyquist graphs of the cells with: **B** NaTiNT and **B1** NaTiNT-450 °C. *Source* Souza (2018)

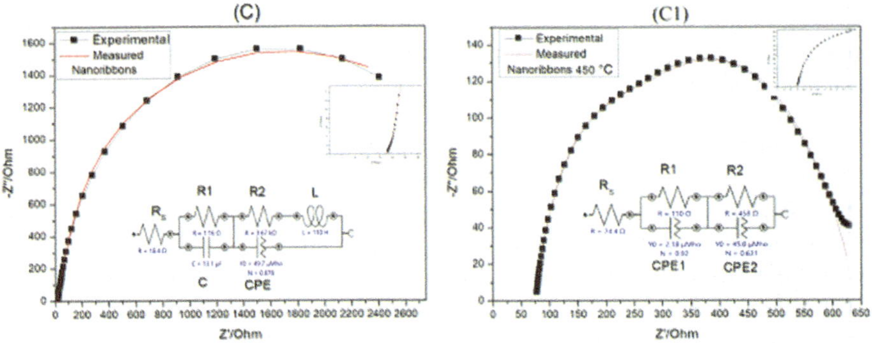

Fig. 26 Nyquist graphs of the cells with: **C** NaTiNT and **C1** NaTiNT-Ribbons-450 °C. *Source* Souza (2018)

and adjusted to the calculation done through the equivalent circuits. The equivalent circuits are represented in Figs. 24, 25, and 26.

In the impedance images, the intersection in the real axis in high frequency is the ohmic series resistance (R_S) which are represented in the insets of the figures. CPE1 and CPE2 are phases elements of the constant phases related to the first and second arcs, respectively. The constant phase element (CPE) is frequently used as a substitute of the capacitor element to adjust the impedance behavior related to the double electrical layer, when it does not behave as an ideal capacitor. The CPE is defined mathematically as:

$$Z_{CPE} = \frac{1}{Y0}(j\omega)^N \quad (2)$$

where, Y_0 is the admittance of the CPE and N can assume values between 0 and 1 (CPE is a resistor for $N = 0$ or ideal capacitor for $N = 1$) (El-Mahdy et al. 2013).

The Nyquist diagrams show two semicircles, one with smaller circumference situated in regions of higher frequency, which is related to charge transfer in the interface counter electrode/electrolyte ($R1$) and the other with bigger circumference situated in regions of lower frequency, related to charge transfer in the interface semiconductor/dye/electrolyte ($R2$) (Wang et al. 2005).

Observing the data from Table 3, it is noticed the decrease in the resistance Rs, R_1 and R_2 for the cells with the nanostructures of titanate NaTiNT and NaTiNT-Ribbons, after the thermal treatment. This reduction on the resistances of transport and charge transfer favors the increase in the photocurrent generated by the cell. This increase happens, mainly, by the reduction in the electron recombination in the interface semiconductor/electrolyte. It also may have been favored by the elimination of the effects produced by the OH^- groups in the titanate films after thermal treatment. Besides, the reduction on the band gaps verified previously helps to improve the electrical conduction.

Table 3 Parameters for the elements of the circuit for the TiO_2 cell (A), TiO_2-450 °C (A1), NaTiNT (B), NaTiNT-450 °C (B1), NaTiNT-Ribbons (C) and NaTiNT-Ribbons-450 °C (C1)

Cell	R_S (Ω)	*R*1	CPE1 Y_O (μMho)	N	*R*2	CPE2 Y_O (μMho)	N
A	17.9	386.0 Ω	47.8	0.838	5.60 Ω	91.8	0.795
A1	27.2	745.0 Ω	37.1	0.758	25.6 Ω	20.8	0.858
B	467	125.0 kΩ	3.65	0.85	639.0 Ω	39.1	0.507
B1	81.9	1.66 kΩ	9.33	0.779	284.0 Ω	3.31	0.847
C	18.4	1.16 Ω	13.1 μF	–	3.67 kΩ	49.7	0.878
C1	74.4	110.0 Ω	2.18	0.92	458.0 Ω	45.0	0.631

Source Souza (2018)

7.4 Final Considerations

The NaTiNT and NaTiNT-Ribbons were obtained through the alkaline hydrothermal method. The studied TiO_2, NaTiNT, and NaTiNT-Ribbons nanostructures of this chapter were dispersed and deposited on FTO successfully through the electrophoresis technique. These structures were characterized in the powder and films form and demonstrated structures and morphologies characteristic of nanospheres, nanotubes, and nanoribbons.

The shift in the absorption spectrum of the films for the region next to the visible, after thermal treatment, manifested clearly for the NaTiNT and NaTiNT-Ribbons nanostructures, probably due to the structural changes after the collapse of the nanostructures. The thermal treatment causes a reduction in the band gaps of these nanostructures, making it to absorb next to the visible.

The photoelectrochemical characterization provides the main parameters to obtain the efficiency for photovoltaic devices. The results of the *J–V* plots of the cells show the cell with the NaTiNT film after thermal treatment to be the one with better improvement for the efficiency. This improvement is due to the reduction in the parasitic resistances. The analysis of the impedances for the NaTiNT nanostructures show that this film has its resistance to charge transfer and transport reduced. These reductions are associated, mainly, to the evaporation of the OH^- groups and oxygen vacancies responsible for the diffusion of H^+ ions and for the production of currents opposed to the photocurrent generated by the cell.

Finally, preliminary studies of this chapter show that these materials are capable to contribute in the future for the development of the photovoltaic devices of high efficiency.

References

Abed GM, Alsammarraie AMA, Al-Abdaly BI (2017) Cr Gd co-doped TiO_2 nanoribbons as photoanode in making dye sensitized solar cell. Nanosci Nanometrol 3:27–33

Solaronix©(Abr 2014) Innovative solutions for solar professionals, Ver 050514DM © Solaronix SA. Acesso em: http://www.solaronix.com./

Agnaldo JS, Bastos JBV, Cressoni JC, Viswanathan GM (2006) Células Solares de TiO_2 Sensibilizado por corante. Rev. Brasileira de Ensino de Física 28(1):77–84

Alessandra LA, Andriely B, Anelise B, Cristiane V, Daniela V (2006) Espectroscopia Raman

Baiju K et al (2007) Effect of tantalum addition on anatase phase stability and photoactivity of aqueous sol–gel derived mesoporous titania. J Mol Catal A Chem 276(1–2:41–46

Baik JM et al (2008) High-yield TiO_2 nanowire synthesis and single nanowire field-effect transistor fabrication. Appl Phys Lett. 92: 242111.

Bavykin DV, Walsh FC (2009) Elongated titanate nanostructures and their applications. Eur J Inorg Chem 2009:977

Bavykin DV, Walsh FC (2010) Titanate and titania nanotubes synthesis, properties and applications. School of Engineering Sciences. University of Southampton, Southampton

Bavykin DV, Parmonb VN, Lapkina AA, Walshc FC (2004) J Mater Chem 14:3370

Bavykin DV, Walsh FC, Frank C (2006) Protonated titanates and TiO_2 nanostructured materials: synthesis, properties, and applications. Adv Mater. https://doi.org/10.1002/adma.200502696

Bavykin DV, Walsh FC, Frank C (2006) Protonated titanates and TiO_2 nanostructured materials: synthesis, properties, and applications. Adv Mater 18:2807–2824

Besra L, Liu M (2007) A review on fundamentals and applications of electrophoretic deposition (EPD). Prog Mater Sci 52:1–61

Campbell WM, Jolley KW, Wagner P, Wagner K, Walsh PJ, Gordon KC, Mende S, Nazeeruddin MK, Wang Q, Michael G, Officer DL (2007) Highly efficient porphyrin sensitizers for dye-sensitized solar cells. J Phys Chem C 111:11760

Cao G (2004) J Phys Chem B 108:921

Chagas FCM (1984) Células solares: Estrutura Semicondutor – Isolante – Semicondutor. 1984. 125 f. Dissertação (Mestrado em Engenharia Mecânica) Universidade Estadual de Campinas, Campinas

Chang YJ, Chow TJ (2011) Highly efficient triarylene conjugated dyes for sensitized solar cells. J Mater Chem 21:9523

Chen Q, Du GH, Zhang S, Peng LM (2002) Acta Cryst B 58:587

Chen Q, Du GH, Zhang S, Peng LM (2002b) Acta Cryst B58:587

Cid J, Yum J, Jang S, Nazeeruddin MK, Fererro M, Palomares E, Ko J, Michael G, Torres T (2007) Molecular cosensitization for efficient panchromatic dye-sensitized solar cells. Angewandte Chemie International Edition, vol 46, p 8358

Dittrich T, Weidmann J, Timoshenko VY, Petrov AA, Koch F, Lisachenko MG, Lebedev E (2000) Mater Sci Eng B 489:69–70

Dresselhaus MS, Lin YM, Rabin O, Jorio A, Souza FAG, Pimenta MA, Saito R, Samsonidze GG, Dresselhaus G (2003) Nanowires and nanotubes. Mater Sci Eng C Biomimetic Mater, Sensors Syst 23:129–140. Holanda.

Morgado E Jr, Bojan AM, Paula MJ, de Abreu Marco AS, Fernando CR (2009) Characterization and thermal stability of cobalt-modified 1-D nanostructured trititanates. J Solid State Chem 182:172–181

El-Mahdy GA, Atta M, Al-Lohedan HA (2013) Water soluble nonionic rosin surfactants as corrosion inhibitor of carbon steel in 1 M HCl. Int J Electrochem Sci 8:5052–5066

Ferreira OP (2006) Unveiling the structure and composition of titanium oxide nanotubes through ion exchange chemical reactions and thermal decomposition processes. J Braz Chem Soc 17:393–402

Ferreira OP (2006b) Nanotubos e Nanobastões de Óxidos e Sulfetos de Metais de Transição Obtidos via Sistemas Bidimensionais (Lamelares): Preparação. Caracterização e Propriedades, Campinas-SP

Ferreira Odair Pastor (2006c) Unveiling the structure and composition of titanium oxide nanotubes through ion exchange chemical reactions and thermal decomposition processes. J Braz Chem Soc 17:393–402

Ferreira OP (2006) Nanotubos e Nanobastões de Óxidos e Sulfetos de Metais de Transição Obtidos via Sistemas Bidimensionais (Lamelares): Preparação, Caracterização e Propriedades. Tese de Doutorado (2006). Campinas-SP

Giarola M, Sanson A, Monti F, Mariotto G, Bettinelli M, Speghini A, Salvino G (2010) Nature Mater Sci Phys Rev B 81:174305

Gil SK, Hyung KS, Godble VP, YoungSK, O-Bong Y, Hyung SS (2006) Electrophoretic deposition of titanate nanotubes from commercial titania nano-particles: application to dye-sensitized solar cells. Electrochem Commun 8:961–966

Hagfeldt A, Boschloo G, Sun L, Kloo L, Pettersson H (2010) Dye-sensitized solar cells. Chem Rev 110:6595–6663

Hara K, Sayama K, Ohga Y, Shinpo SA, Arakawa H (2001) A coumarin derivative dye sensitized nanocrystalline TiO_2 solar cell having a high solar-energy conversion efficiency up to 5.6%. Chem Commun 6:569

Hodes G et al (1992) J Electrochem Soc 139:3136

Hoyer P (1996) Formation of a titanium dioxide nanotube array. Langmuir. Am Chem Society 12(6):1411–1413

Iijima S (1991) Helical microtubules of graphitic carbon. Nature 354:56–58

Ito S et al (2009) Int J Photoenergy 517609

Kasuga T et al (1998) Formation of titanium oxide nanotube. Langmuir. https://doi.org/10.1021/la9713816b. Publication Date (Web): May 23, 1998 Copyright © 1998 American Chemical Society, vol 14, pp 3160–3163

Kasuga T et al (1998) Formation of titanium oxide nanotube. Langmuir 14:3160–3163

Kasuga T, Hiramatsu M, Hoson A, Sekino T, Niihara K (1999) Adv Mater 11:1307

Kim GS, Seo HK, Godble VP, Kim YS, Yang OB, Shin HS (2006) Electrophorect deposition of titanate nanotubes from commercial titania nanoparticles: application to dye-sensitized solar cells. Chonbuk National University, Korea. 8:961–966

Kolmakov A, Moskovits M (2004) Chemical sensing and catalysis by one-dimensional metal-oxide nanostructures. Ann Rev Mater Res. 34:151–180

Kukovecz Á, Hodos M, Horváth E, Radnóczi G, Kónya Z, Kiricsi I (2005) J Phys Chem B 109:17781

Law M, Goldberger J, Yang P (2004) Semiconductor nanowires and nanotubes. Ann Rev Mater Res. 34:83-122

Law M, Greene LE, Johnson JC, Saykally R, Yang PD (2005) Nature Mater 4:455–459

Lima FM (2013) Deposição de Dióxido de Estanho-Flúor (SnO_2:F) em Substrato Transparente para uso em Células Fotoeletroquímicas. Universidade Federal do Ceará Programa de Pós-Graduação em Engenharia Mecânica, Fortaleza

Lima FA (2015). Application of transition metal oxide based nanostructured thin films on third generation solar cells. Doctorate Thesis submitted to the Graduate Program in Materials Science and Engineering Fortaleza- Ce—Brasil, 2015

Liu Y et al (2011) Cobalt redox mediators for ruthenium-based dye-sensitized solar cells: a combined impedance spectroscopy and near-ir transmittance study. J Phys Chem C Madison 115, 38:18847–18855.

Ma R, Bando Y, Sasaki T (2004) J Phys Chem B 108:2115

Marszalek M et al (2011) Application of ionic liquids containing tricyanomethanide $[C(CN)_3]^-$ or tetracyanoborate $[B(CN)_4]^-$ anions in dye-sensitized solar cells. Inorg Chem Minnesota 50:11561–11567

Mende S, Bach U, Baker HR, Horiuchi T, Miura H, Ito S, Uchida S, Michael G (2005) Organic dye for highly efficient solid-state dye-sensitized solar cells. Adv Mater 17:813

Morgado E Jr (2007a) Estudo de titanatos nanoestruturados obtido por tratamento hidrotérmico de óxido de titânio em meio alcalino. Universidade Federal do Rio Grande do Norte-UFRN

Morgado E Jr (2007b) Tese de Doutorado. Universidade Federal do Rio Grande do Norte

Mozia S, Borowiak Palen E, Przepiorski J, Grzmil B, Tsumura T, Toyoda M, Grzechulska Damszel J, Morawski AW (2010) Physico-chemical properties and possible photocatalytic applications of titanate nanotubes synthesized via hydrothermal method. J Phys Chem Solids 71:263–272

Nàdia KZ, João FMF (2012) Core–shells of $SrAl_2O_4$:Ce(III), Ln phosphor coated by TiO_2 nanotube and nanowire-shaped structures. J Nanoeng Nanomanuf 2:1–9
O'regan B, Gratzel M (1991) A low-cost, high-efficiency solar cell based on dyesensitized colloidal TiO_2 films. Nature 353:737–740
Pang Qi, Limin Leng, Lijuan Zhaoc, Liya Zhou, Chunjie Liang, Yuei Lan (2011) Dye sensitized solar cells using freestanding TiO_2 nanotube arrays on FTO substrate as photoanode. Mater Chem Phys 125:612–616
Pang A, Lun CX, Luo H, Li Y, Wei M (2013) Highly efficient indoline dyes co-sensitized solar cells composed of titania nanorods. Electrochimica Acta 94:92–97
Santos NM, Rocha MJ, Matos JME, Ferreira OP, Filho JM, Viana BC, Oliveira AC (2013) Alkali metal intercalated titanate nanotubes: Metal cations intercalated titanate nanotubes as catalysts for α,β unsaturated esters production. Appl Catal A Gen 454:74–80
Sauvage F et al (2011) Butyronitrile-based electrolyte for dye-sensitized solar cells. J Am Chem Soc Salt Lake 133:13103–13109
Shalan AE et al (2012) Controlling the microstructure and properties of titania nanopowders for high efficiency dye sensitized solar cells. Electrochim Acta 89(2013):469–478
Silva FLR (2012) Síntese e Caracterização de Nanoestruturas à Base de Dióxido de Titânio. Universidade Federal de Minas Gerais Instituto de Ciências Exatas Departamento de Física
Sivalingam G et al (2003) Photocatalytic degradation of various dyes combustion synthesized nano anatase TiO2. Appl Catal B Environ. 45(1):23–38
Sivalingam G et al (2004) Photocatalytic degradation of various dyes combustion synthesized nano anatase TiO_2. Appl Catal B Environ 48(2):83–93. https://doi.org/10.1016/j.apcatb.2003.09.013
Souza APS et al (2018) Performance evaluation of titanate nanotubes and nanoribbons deposited by electrophoresis in photoelectrodes of dye-sensitized solar cells. Mat Res [online] 21(4) [cited 24 Oct 2019]. e20180110. Available from: http://www.scielo.br/scielo.php?script=sci_arttext&pid=S151614392018000400221&lng=en&nrm=iso. Epub May 28, 2018. ISSN15161439. http://doi.org/10.1590/1980-5373-mr-2018-0110
Souza APS et al (2019) Photoelectrodes with titanate nanotubes sensitized by mesoporphyrin derivative from cashew nut shell. Matéria (Rio J.) v.24,No(1),e-12274.
Sun X, Li Y (2003) Chem Eur J 9:2229
Tanaka K et al (1997) Photocatalytic degradation of mono, di and trinitro-phenol in aqueous TiO_2 suspension. J Mol Catal 122:67–74
Tereza B, Daniela P, Wei W, Peter O, Stanislav D, Xiao DZ, Radomir K (2013) Study of titanate nanotubes by X-ray and electron diffraction and electron microscopy. Mater Charac 87:166–171
Tereza B, Daniela P, Wei W, Peter O, Stanislav D, Xiaodong Z, Radomir K (2013) Study of titanate nanotubes by X-ray and electron diffraction and electron microscopy. Mater Charact 87(2014):166–171
Thiago JF (2010) Especialista de Produto PerkinElmer do Brasil. São Paulo
Thorne A, Kruth A, Tunstall D, Irvine JTS, Zhou W (2005) J Phys Chem B 109:5439
Unger E (2012) Excitonic dye solar cells. PhD thesis, Uppsala University
Vanessa RB (2011) Síntese e Fotosensibilização de Nanotubos de Titanatos. ISEL Instituto Superior de Engenharia de Lisboa
Viana BC, Ferreira OP, Souza Filho AG, Angel AH, Filho JM, Alves OL (2011) Highlighting the mechanism of the titanate nanotubes to titanate nanoribbons transformation. J Nanopart 13:3259–3265. http://dx.doi.org/10.1007/ s11051-011-0240-3 (Research paper)
Wang YQ, Hu GQ, Duan XF, Sun HL, Xue QK (2002) Chem Phys Lett 365:427
Wang Q, Moser J, Grätzel M (2005) Electrochemical impedance spectroscopic analysis of dye-sensitized solar cells. J Photochem Photobiol B 109:14945–14953
William A, Vallejo L, Cesar A, Quiñones S, Johann A, Hernandez S (2011, Nov 9) The chemistry and physics of dye-sensitized solar cells, solar cells - Dye-Sensitized Devices, Leonid A. Kosyachenko, IntechOpen. https://doi.org/10.5772/20609. Available from: https://www.intechopen.com/books/solar-cells-dye-sensitized-devices/the-chemistry-and-physics-of-dye-sensitized-solar-cells

Wu D, Chen Y, Liu J, Zhao X, Li A, Ming N (2005) Appl Phys Lett 87:112–501
Xiao MW, Wang LS, Wu YD, Huang JJ, Dang Z (2008) Nanotechnology 19
Xiao MW, Wang LS, Wu YD, Huang JJ, Dang Z (2008b) Preparation and characterization of CdS nanoparticles decorated into titanate nanotubes and their photocatalytic properties. Nanotechnology 19(1):015706
Yamada M et al (2006) One-dimensional proton conductor under high vapor pressure condition employing titanate nanotube. Electrochem Commun. 8:1549-1552.
Yang J, Jin Z, Wang X, Li W, Zhang J, Zhang S, Guo X, Zhang Z (2003) Dalton Trans 20:3898
Yao BD, Chan YF, Zhang XY, Zhang WF, Yang ZY, Wang N (2003) Appl Phys Lett 82:281
Yella A et al (2011) Porphyrin-sensitized solar cells with cobalt (ii/iii)–based redox electrolyte exceed 12 percent efficiency. Science Wash 334(4):629–633
Yella A, Lee HW, Tsao HN, Yi C, Chandiran AK, Nazeeruddin MK, Diau EW, Yeh CY, Zakeeruddin SM, Michael G (2014) Porphyrin-sensitized solar cells with cobalt (II/III) -based redox electrolyte exceed 12 percent efficiency. Science, 629
Ylhäinen EK, Nunes MR, Silvestre AJ, Monteiro OC. "Titania-free synthesis of titanates nanostructures". (2008) Nature Mater submitted to Sol Stat Sci
Ylhäinen EK, Nunes MR, Silvestre AJ, Monteiro OC (2008) Nat Mater Sol Stat Sci
Zeit FT (1931) The Kubelka-Munk theory of reflectance. Kubelka and Munk Physik 12:593
Zhang et al (2003) TEM study on the formation process of TiO_2 nanotubes. Chin Chem Lett 14(4):419–422
Zhang Q, Guozhong C (2011) Nanostructured photoelectrodes for dye-sensitized solar cells. Nano Today 6(91–109):2010
Zhang S, Peng LM, Chen Q, Du GH, Dawson G, Zhou WZ (2003) Phys Rev Lett 91:256103
Zhao L, Yu J Fan J, Peng CZ, Wang S (2009) Dye-sensitized solar cells based on ordered titanate nanotube films fabricated by electrophoretic deposition method. Electrochem Commun 11:2052–2055
Zhitomirsky I (1998) Cathodic electrosynthesis of titanium and ruthenium oxides. Mater Lett 33(5–6):305–310
Zhitomirsky I (1998) Cathodic electrosynthesis of titanium and ruthenium oxides. Mater Lett

Chapter 12
Interaction of Nanomaterials with Biological Systems

Thaiz Batista Azevedo Rangel Miguel, Sergimar Kennedy de Paiva Pinheiro, and Emilio de Castro Miguel

1 Plenty of Room at the Bottom

In 1959, Richard Feynman presented a lecture entitle "*There is plenty of room at the bottom – an invitation for a new field of Physics*" (Feynman 1960) during the American Physical Society meeting at Caltech. On this lecture, he introduced the concept of manipulating matter at the atomic level, trying to lead a new field of physics.

The lecture was structured in several sessions, that include the problem of manipulating and controlling things on a small scale, and that there may be certain advantages to making elements smaller. Despite the possible miniaturization it would still be necessary to read the information, so the author spends a lot of time on microscopy techniques.

Despite being at the forefront of scientific thinking of the time, Feynman's ideas were influenced by current ideas in the academic world of the time. These include the elucidation of DNA structure (Watson and Crick 1953) and Arthur von Hippel's article entitled "*Molecular Engineering*" (Von Hippel 1956), published three years before Feynman's lecture. In this paper, von Hippel does not make connection with miniaturization of information, however, in a book published three years later, articles

T. B. A. R. Miguel
Laboratório de Biotecnologia—Labiotec, Departamento de Engenharia de Alimentos, Universidade Federal do Ceará, Fortaleza, Brazil
e-mail: thaizrangel@gmail.com

S. K. de P. Pinheiro · E. de Castro Miguel (✉)
Laboratório de Biomateriais—BioMat, Departamento de Engenharia Metalúrgica e de Materiais, and Central Analítica da UFC, Universidade Federal do Ceará, Fortaleza, Brazil
e-mail: emiliomiguel@ufc.br

S. K. de P. Pinheiro
e-mail: sergimarkennedy@hotmail.com

R. F. do. Nascimento et al. (eds.), *Nanomaterials and Nanotechnology*, Materials Horizons: From Nature to Nanomaterials,
https://doi.org/10.1007/978-981-33-6056-3_12

on computer memory and molecular engineering in vehicles of the future were noted, recalling the motivation of miniaturization to the space race (Schulz 2018). The idea of "*manipulating and controlling things at atomic scale*" was therefore already in the air.

Although there is controversy about Feynman's real influence on the conception of nanotechnology, it is undoubtedly that his speech in 1959 changed the direction of this science, still incipient at the time. It is impossible not to recognize Feynman as a visionary, in fact anticipating various aspects of what we today call nanotechnology (Schulz 2018). It is also possible to read the lecture as a manifesto for physicists to take control of both the physical and biological sciences (Editorial 2009).

Even 60 years after Feynman's lecture, new optoelectronic and photonic technologies outperformed their casual conjectures (Daukantas 2019) shows that nanotechnology is still a frontier field and far from its limit.

2 Nanotechnology and Beyond: The Past, the Present and the Future

The science known as nanotechnology is defined as the understanding and control of matter at dimensions between 1 and 100 nm, where unique phenomena enable novel applications (Hulla et al. 2015). Such peculiarity has been explored for a long time.

The history or nanoparticles is very long (Heiligtag and Niederberger 2013; Jeevanandam et al. 2018). Nanostructured material was exploited to reinforcement of ceramic matrixes by adding natural asbestos nanofibers 4500 years ago (Heiligtag and Niederberger 2013). Egyptians also used nanomaterials more than 4000 years ago based on a synthetic chemical process to synthesize PbS nanoparticles for hair dye (Walter et al. 2006).

Another hair dyeing formula, "*Egyptian blue,*" was the first synthetic pigment which was prepared using a sintered mixture nanometer-sized glass and quartz around third century BC (Walter et al. 2006). These nanoparticles are quite similar to PbS quantum dots synthesized by recent materials science techniques (McDonald et al. 2005). Egyptian blue represents a mixture of $CaCuSi_4O_{10}$ and SiO_2. In ancient geographical regions of the Roman Empire, including countries such as Egypt, Mesopotamia, and Greece, the extensive use of Egyptian blue for decorative purposes has been observed during archaeological explorations (Heiligtag and Niederberger 2013).

Metallic nanoparticles synthesis by chemical methods dates back to the fourteenth century and thirteenth century Before Christ. The metallic nanoparticles era started with the Egyptians and Mesopotamians making glass using metals (Schaming and Remita 2015). These materials may be the earliest examples of synthetic nanoparticles in a practical application.

From the late Bronze Age (1200–1000 BC), red glass has been found in Frattesina di Rovigo (Italy) that is colored by surface plasmon excitation of Cu NPs (Colomban 2009).

Likewise, the Celtic red enamels originating from the 400–100 BC period have been reported to contain copper nanoparticles and cuprous oxide (Brun et al. 1991). Still, a Roman glass workpiece is the most famous example of ancient metallic nanoparticles application (Heiligtag and Niederberger 2013; Jeevanandam et al. 2018). The Lycurgus Cups are a fourth-century Roman glass cup presently in the British Museum, made of a dichroic glass that displays different colors: red when a light passes from behind and green when a light passes from the front (Leonhardt 2007). This dichroic coloring is attributed to nanocrystals of an Au/Ag alloy dispersed throughout the glass matrix (Colomban 2009; Schaming and Remita 2015). More recent studies found that the Lycurgus Cups contain Ag–Au alloy nanoparticles, with a ratio of 7:3 in addition to about 10% Cu (Freestone et al. 2007).

Nowadays, the application of nanostructured materials and nanotechnology is almost unlimited. Engineered nanomaterials are used in different scientific and industrial disciplines including medicine (Arca-Lafuente et al. 2020; Arruebo et al. 2007; McDonald et al. 2005; Qiu et al. 2019; Rupar et al. 2019); agriculture (Qiu et al. 2019; Torrent et al. 2019; Zhang et al. 2019); food (Du et al. 2019; Jiménez-López et al. 2020; Kritchenkov et al. 2019; Santos et al. 2019); cosmetic (Khezri et al. 2018; de Stavale et al. 2019; Swain et al., 2019); and electrical transformers (Contreras et al. 2017).

Despite the almost unlimited application of nanostructured materials (almost), all have the potential not only perform the desired functions but causing adverse impacts. Thus, the future of these materials points to "safe-by-design" nanomaterials (Hwang et al. 2018).

In this extremely diverse, extensive, and complex scenario, this chapter is not intended to exhaust the subject; on the other hand, it attempts to address it in a straightforward manner by punctually showing the main classes of nanostructured materials and state of the art of nanomaterials/biological systems interaction.

3 Main Classes of Nanostructured Materials

Nanoparticles (NPs) can be found in the environment naturally or through artificial synthesis processes (Kaur and Luthra 2016). Naturally occurring NPs are inorganic (silicates, phosphates, metal oxides, and sulfides) or organic (macromolecules and biocolloids). Synthetic or engineered NPs are introduced intentionally or unintentionally into the environment (Bhatt and Tripathi 2011). Figure 1 shows the main categories of nanoparticles that can be introduced into the environment.

There are many types of intentionally produced nanomaterials, and a variety of others is expected to appear in the future, nevertheless, most current nanomaterials could be organized into four types:

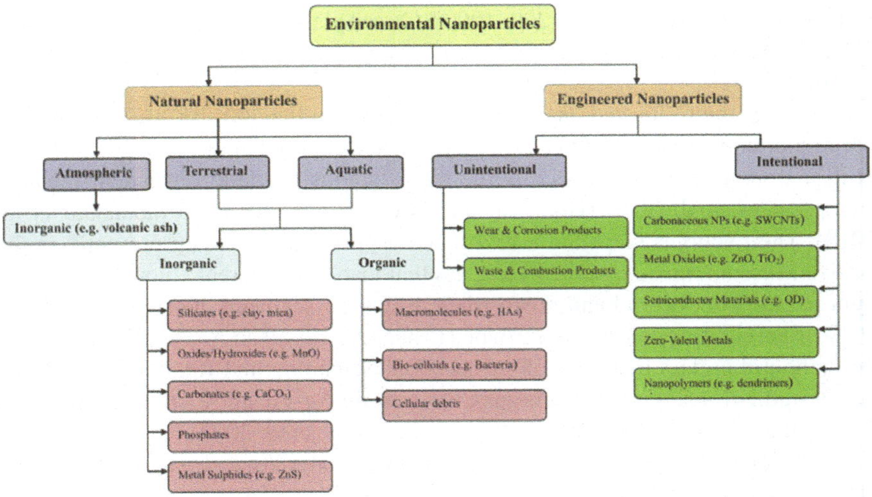

Fig. 1 Categories of nanoparticles present in the environment. Reproduced with permission from Bhatt and Tripathi (2011). Copyright (2011) Elsevier

- Carbon-based materials
- Metal-based materials
- Dendrimers
- Composites.

3.1 *Carbon-Based Materials*

Carbon-based nanomaterials (CBNs) have shown tremendous impact in the biomedical field with their ability to deliver therapeutic molecules and allow visualization of cells and tissues, which are necessary for the cure and treatment of diseased and damaged tissues. In addition, these nanomaterials are promising candidates in the field of cancer therapy (Hinzmann et al. 2014). CBN includes fullerenes, carbon nanotubes (CNTs), graphene (G), and its derivatives (graphene oxide (GO), nanodiamonds (NDs), and carbon-based quantum dots (CQDs) (Patel et al. 2019).

These nanomaterials are composed mostly of carbon, commonly taking the form of a hollow spheres, ellipsoids, or tubes. Spherical and ellipsoidal carbon nanomaterials are referred to as fullerenes, while cylindrical ones are called nanotubes. These CBNs have many advantageous properties, including excellent separation properties, large surface areas, high mechanical strength, unique electrical and thermal conductivity properties, and superior antibacterial properties (Wu et al. 2020b), being used in supplementary fuels and high-end materials (Kalman et al. 2019). Furthermore, these particles have many potential applications, including improved films and coatings, stronger and lighter materials, and applications in electronics (Kalman et al. 2019).

The carbon-based nanoparticles can impact the biological activity of species exposed to these NPs. For example, materials from the graphene family can lead to toxic effects in the aquatic environment (Yao et al. 2019). Even at low doses, toxicity was noted in Zebrafish embryos (Bangeppagari et al. 2019). Other carbon-based nanoparticles such as graphene oxide or multi-walled carbon nanotubes caused cellular damage and lower survival rate on *Streptomyces* sp. (Liu et al. 2019). Although most studies point to the toxicity of CBNs, the toxic effects for higher life forms are not conclusive (Liu et al. 2019).

3.2 Metal-Based Materials

These nanomaterials include quantum dots, nanogold, nanosilver, and metal oxides, such as titanium dioxide. These metal oxide nanoparticles have been used in a wide range of applications such as energy storage devices, catalysis, electrochemistry lubrication, sensors, coatings, and environmental remediation (Narayanan and El-sayed 2005; Zhang et al. 2016a, b, c). A quantum dot is a closely packed semiconductor crystal comprised of hundreds or thousands of atoms, and whose size is on the order of a few nanometers to a few hundred nanometers. The quantum dots (QDs) have shell structure and discrete energy levels, like real atoms, and have attracted considerable attention because of their potential applications in nanotechnology (Yakar et al. 2020). Gold nanoparticles have been used in several biomedical applications including imaging, diagnosis, and as drug delivery system within cells (Boisselier and Astruc 2009).

Nanosilver is a generic term that refers to nanoscale silver materials that are at least less than 100 nm in size. This material is generally in the form of silver nanoparticles (AgNPs) (Rezvani et al. 2019). Silver nanoparticles have received significant attention due to their antimicrobial activity and prevention of biofilm formation, as well as their physical, chemical, biological properties, and their applicability in electronics, optics, and medicine (Ansari et al. 2014), in addition to being widely applied in medical devices, cosmetics, clothing, household items, water treatments, and detergents (Marambio-Jones and Hoek 2010).

Nano- and micro-TiO_2 particles are used in several areas including sunscreen, cosmetics, paints, food additives, medicine as well as decontamination of air, water, and soil (Ozkan et al. 2016).

3.3 Dendrimers

These nanomaterials are nanosized polymers built from branched units. The surface of a dendrimer has numerous chain ends, which can be tailored to perform specific chemical functions. These structural characteristics make dendrimers different from linear and branched polymers (Lyu et al. 2019). Because three-dimensional dendrimers contain interior cavities into which other molecules could be placed, they may be useful for drug delivery (Kurtoglu et al. 2010; Mignani et al. 2014).

The structure of dendrimers allows its modification by adding amino acids or small units of peptides to its structure, being classified into three distinct categories according to the position of the amino acids (Crespo et al. 2005). This structure called the dendrimer peptide acquires desirable characteristics for a wide range of applications (Mirakabad et al. 2019). The dendrimer peptide can be used as diagnostic tools, drug delivery, and therapeutic applications (Mirakabad et al. 2019).

3.4 Composites

Composites are generally a mixture of constituents that differ in properties made to make a single material with integrated properties (Jaspal and Malviya 2020). Composites combine nanoparticles with other nanoparticles or with larger, bulk-type materials. Nanoparticles, such as nanosized clays, are already being added to products ranging from auto parts to packaging materials, to enhance mechanical, thermal, barrier, and flame-retardant properties.

The class of composites is divided into composites based on oxides, hybrids, biosorbents, nanocomposites, activated carbon, and polymeric composites (Jaspal and Malviya 2020). Oxide-based composites are mainly used in water treatment. In a study conducted by Thakur and Kandasubramanian (2019), composites based on graphene oxide were used to remove pollutants in the textile industry and agriculture. Hybrid composites are used efficiently to remove heavy metals from the environment. The composite prepared from papaya seed with clay was efficient in absorbing metal ions from the water (Unuabonah et al. 2013).

Activated carbon composite has been used mainly in the treatment, purification, and discoloration of water, including tap water, groundwater, and wastewater (Azari et al. 2014). Finally, polymers have been a viable option in water treatment and environmental remediation (Jaspal and Malviya 2020).

4 General Characteristics of Nanoparticles

The study of nanoparticles (NPs) is essential to assess their interaction with the environment and their effects on organisms, since NPs have different physical and

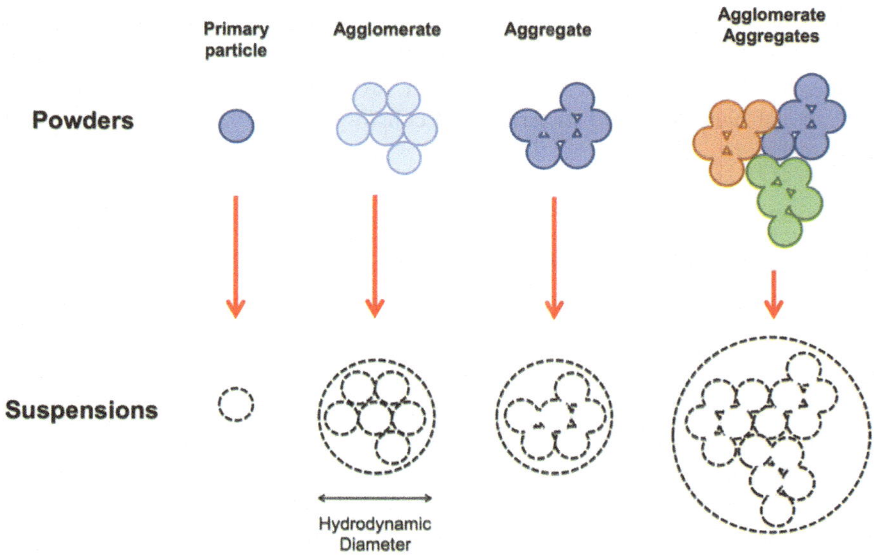

Fig. 2 Schematic representation of the different forms of particles is primary particle, aggregate, agglomerate, and agglomerate aggregates. Note the difference in real diameter and hydrodynamic diameter

chemical characteristics from their source material (Bulk) (Christian et al. 2008). Studies show that the surface energy increases as particle size decreases, which stimulates hydrophobic and/or electrostatic interaction between NPs (Loosli et al. 2015; Nur et al. 2015).

The intensity of the NPs interaction generates different aggregation states resulting in aggregates and clusters that are related to their hydrodynamic radius as shown in Fig. 2. These clusters depend on the surface chemistry of the NPs, on the concentration and composition of the medium to form loose structures (Clément et al. 2013). On primary particles and agglomerates, repulsive forces are dominant; however, weak repulsive forces in liquid result in aggregates and/or agglomerated aggregates (Soares et al. 2018).

The particle characteristics contribute actively to the interactions with the medium through by promoting the adsorption of ions, proteins, natural organic materials, and detergents; double-layer formation; dissolution; or minimizing free surface energy by surface restructuring (Min et al. 2009; Nel et al. 2009). In general, the most important physicochemical features of nanomedicine are composition, structure, size, surface properties, porosity, charge, and stability (Kettiger et al. 2013).

Characteristics such as size, shape, composition, and hydrodynamic size directly control their ability to interact with toxic molecules or substances (Boran et al. 2016). The potential harmful effect of these NPs is on biota and the risk of contamination of agricultural land, water resources, and air (Tosco and Sethi 2018).

Understanding the interaction of nanostructured materials with biological systems will be very helpful to evaluate the risk and impact of health (Gehr 2018) and environmental (Bundschuh et al. 2018). However, this is not a simple mission given the diversity of nanostructured materials and cells/organisms that can be part of this relationship.

5 Nanomedicine: Predictable Interactions of Nanoparticles with Biological Systems

Nanomedicine can be defined as the application of nanotechnology for medical purposes on diagnosis, monitoring, control, prevention, and treatment of diseases (Tinkle et al. 2014). This science involves highly interdisciplinary science from a range of fields, including engineering, chemistry, physics, biology, medical sciences, and pharmaceutical sciences. Although considered still in its infancy as a scientific discipline, having first been coined in the 1990s, nanomedicine research has resulted in significant impact through a range of applications (Nance 2019).

The application of nanomaterials to medical problems has already demonstrated a clinical impact in terms of delivery strategies for a range of bioactive molecules, including therapeutic agents, nucleic acids, and imaging contrast agents (Sakamoto et al. 2010).

For successful applications of nanomaterial in medicine, it is needed precise control over surface modifications, size, shape, and other particle characteristics (Zhang et al. 2012), since the cell varies in many features. For example, cardiac cells contain numerous mitochondria while secretory cells have extensive networks of Golgi apparatus and endoplasmic reticulum. Therefore, it is critical to highlight the features associated with nanomaterials (NMs) that could affect the recognition and uptake in a manner dependent on cell type. As NMs have emerged as effective drug carriers to treat complex diseases, it has also become crucial to understand the mechanisms of nanoparticle endocytosis (Akhtar et al. 2018).

To enter cells via endocytosis (clathrin-dependent or independent), NMs must contact and attach to the cell surface, which forms a vesicle around the NM to be endocytosed (Akhtar et al. 2018). In each incident of endocytosis, a certain fraction of the membrane is consumed (Roy and Wrana 2005). There would come a time when the surface tension of the membrane will not allow more endocytosis to occur without efficient recycling of the vesicles in the plasma membrane (Roy and Wrana 2005; Steinman et al. 1983). Moreover, a larger cell membrane surface area would be required for the endocytosis of bigger NMs and vice versa. It is therefore not surprising that professional phagocytes are significantly bigger in size than non-phagocytes (Steinman et al. 1983). For the foregoing, it is reasonable to conclude that microparticles would be internalized in a particular cell in fewer numbers than their nanosized counterparts, a reason partly explaining why micron-sized particles are lesser toxic than their nanosized counterparts, however, there are dramatic differences between normal metabolically driven endocytosis and the endocytosis induced by NMs at high concentrations during in vitro studies (Behzadi et al. 2017).

Once endocytosed, nanomaterials are internalized and remain entrapped in transport vesicles which traffic along the endolysosomal scaffold, thereby exerting key effects on subcellular organelles. Intracellular trafficking and the fate of nanomaterials are linked to their physicochemical properties and endocytic pathways (Miller et al. 2009). For example, nanoparticles taken up by clathrin-dependent receptor-mediated endocytosis are typically destined for lysosomal degradation; on the other hand, clathrin-independent receptor-mediated endocytosis internalization leads to endosomal accumulation and sorting to a nondegradative path (Bareford and Swaan 2007). While some drug delivery systems aim to avoid lysosomal degradation (Whitehead et al. 2009), understanding the key intracellular interactions of nanoparticles has allowed researchers to engineer nanoparticles for highly specialized delivery. Appropriate design and engineering of nanocarriers could therefore allow for controlled intracellular delivery of therapeutics to individual intracellular compartments, which provides benefits to therapies associated with these unique organelles, including cancer therapy, gene therapy, and lysosomal storage disease treatments.

The investigation of two large scientific databases over the years reveals that only after 2004 articles with the keyword "nanomedicine" were noticed (although Liposomal Doxorubicin has been used since 1995 (Porche 1996). The number of publications on the subject is growing every year (Fig. 3).

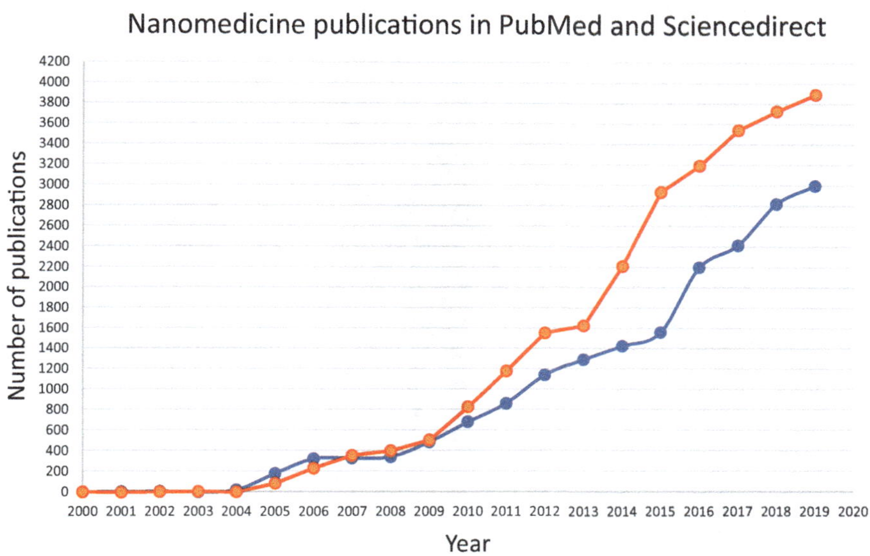

Fig. 3 Graph showing the rise of nanomedicine publications (2004/2019) in two databases. For this research, we used PubMed (red line) (https://www.ncbi.nlm.nih.gov/pubmed) and Science Direct (blue line) (https://www.sciencedirect.com) databases. The search was done filtering by keyword using the keyword "*nanomedicine.*"

Nanomedicine was been used in diagnosis and treatment of numerous diseases. Among these diseases, cancer deserves to be highlighted due to the difficulty (and sometimes inefficiency) of treating some of its variations. In this sense, nanostructured materials are promoting a new point of view.

6 Nanomedicine in Cancer Diagnostic

A wide variety of platforms is being investigated as nanocarriers for cancer diagnosis and treatment, including lipid-based, polymer-based, inorganic, viral, and drug-conjugated nanoparticles (Tran et al. 2017). Some examples of these applications can be observed in Fig. 4. Several of these platforms have also been approved for clinical use.

Nanoparticles developed for imaging have contributed to the development of new possibilities for diagnosis and design of drugs aimed at clinical applications (Weissleder and Pittet 2008). Several types of nanomaterials have been developed using light-responsive components, composed mainly of organic materials (fluorophores, photosensitizers, and carbon-based NPs) and inorganic NPs (gold NPs, quantum dots, and upconversion NPs) (Choi and Frangioni 2010). These NPs have

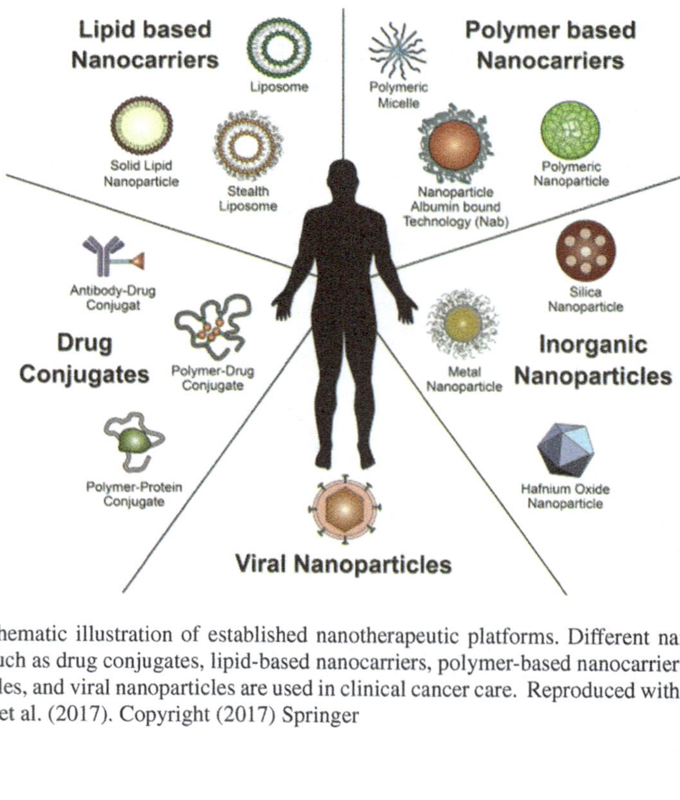

Fig. 4 Schematic illustration of established nanotherapeutic platforms. Different nanomedicine products such as drug conjugates, lipid-based nanocarriers, polymer-based nanocarriers, inorganic nanoparticles, and viral nanoparticles are used in clinical cancer care. Reproduced with permission from Tran et al. (2017). Copyright (2017) Springer

been used as fluorescent markers, as an alternative to conventional techniques such as the requirement of color-matched lasers, fluorescence bleaching and lack of discriminatory capacity of multiple dyes. Fluorescent NPs can greatly overcome these problems (Parveen et al. 2012). These NPs are generally multifunctional, with the administration of medication inside the cell being used in diagnostic imaging (Son et al. 2019).

Graphene NPs have characteristics such as high thermal conductivity, large surface area, and flexibility which allow its use in the acquisition of optical imaging (Chen et al. 2018a, b). In addition, graphene can be functionalized with quantum dots, making it more stable, becoming appropriate for cell imaging (Jaleel et al. 2017). According to Zhang et al. (2016a, b, c) graphene quantum dots (GQDs) exhibited good stability, low level of cytotoxicity, and easy entry into cells. These GQDs produced clear images of cells with remarkable image stability. Zhu et al. (2011) synthesized GQDs that can be used as a bioimaging agent due to their high stability and solubility in many solvents. Other NPs such as upward-converting photon nanoparticles (UCNPs) have also been used as high contrast imaging agents (Lv et al. 2017). UCNPs have emerged as a promising new class of nanomaterials due to their ability to convert near-infrared light into visible luminescence (Muhr et al. 2014). Other commonly used NPs are composed by oxide metal, including titanium oxide, zinc oxide, and manganese oxide (La Van et al. 2003; Rasmussen et al. 2010; Zhen and Xie 2012).

The synthesis of the fluorescence nanobiosensors has been performed for the detection of pancreatic adenocarcinomas. This nanobiosensor was synthetized with water-dispersible Fe/Fe_3O_4 nanoparticles with two tethered fluorescent dyes, the TCPP (Tetrakis(4-carboxyphenyl)porphyrin) and cyanine, as shown in Fig. 5. The fluorescent nanobiosensor is sensitive to proteases and arginase, which are precursors in the formation of malignant tumors related to pancreatic cancer. These proteases and arginase are present in the biopsy fluid and contain key signatures specific to this type of cancer (Mahmoodzadeh et al. 2018).

In a study carried out in patients with histologically proven pancreatic cancer, an immunoassay was developed that uses surface-enhanced Raman scattering (SERS) effect to detect this type of cancer. The SERS platform is based on NPs to detect cancer biomarkers. Modified AuNPs were adsorbed on a mica surface and the gold interface was coated with octadecanethiol (ODT) and specific anti-MUC4 8G7 monoclonal antibody. This assay was able to detect and quantify MUC4 levels in individual serum samples, distinguishing healthy patients, and patients with pancreatic cancer (Krasnoslobodtsev et al. 2015). In addition to NPs (inorganic, organic, and biological) that are used to locate and diagnose diseases related to cancer, these NPS can also act together in the diagnosis and therapy of the disease.

Nanomedicine growth drove a new area which combines diagnostics, and therapy termed theranostics is emerging and is a promising approach which holds in the same system both the diagnosis/imaging agent and the medicine (Soares et al. 2018). Theranostics develops methods to diagnose and treat diseases (especially various types of cancer) in their initial stage, minimizing side effects due to the targeting of drugs to the specific target (Muthu et al. 2014).

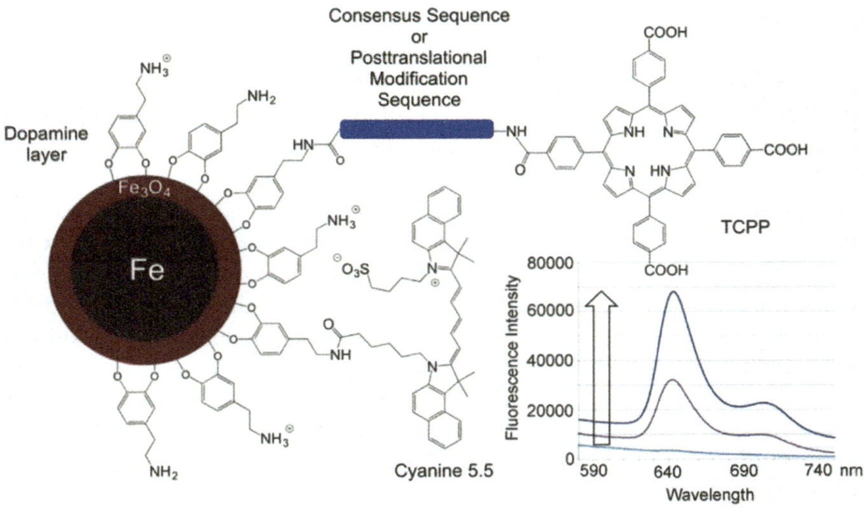

Fig. 5 Chemical structure of the nanobiosensors for protease and arginase detection. The core of the nanobiosensor consists of dopamine-coated a Fe/Fe_3O_4 core to which 50 ± 4 cyanine 5.5 and 35 ± 3 TCPP molecules are bound, following a random deposition-based modeling approach. The consensus sequences experience either proteolytic cleavage by their respective proteases or the chemical constitution of the post-translational modification sequence is changed. For instance, arginases I + II convert arginine to ornithine without proteolytic cleavage of the oligopeptide. Inset: The fluorescence occurring from the nanobiosensors increases with incubation time. This enables fluorometric detection of protease/arginase activities. Reproduced with permission from Kalubowilage et al. (2018). Copyright (2018) Elsevier

Mahmoodzadeh et al. (2018) developed a new method of theranostic based on photothermal chemotherapy (chemophotothermal therapy) for solid tumors (addenocarcinoma MCF7 cell line). In this article, the authors synthetized gold nanoparticles (GNPs) conjugated to the HS-PCL-b-PNIPAAm-b-PAA triblock copolymer and GNPs@polymer. This new compound was loaded with doxorubicin hydrochloride (DOX) as an anticancer drug through electrostatic interactions to afford GNPs@polymer-DOX theranostic nanomedicine. In this approach, the compound is endocitated by the cell and the heat generated by the irradiation of the lasers on the compound promotes the release of the drug (DOX) in the cell, directed to specific locations as shown in Fig. 6.

In another research related to cancer, the compound Co-57-labeled MTX-LDH was evaluated for anticancer activity in mouse colon-carcinoma CT-26 cells. This compound was synthesized using layered double hydroxide (LDH) with incorporation of an anticancer drug, methotrexate (MTX), and a radioisotope, Co-57. The results suggest that MTX-LDH radiolabeled in addition to marking tumor cells had a high cancer-cell suppression effect on CT-26 cells (Kim et al. 2020).

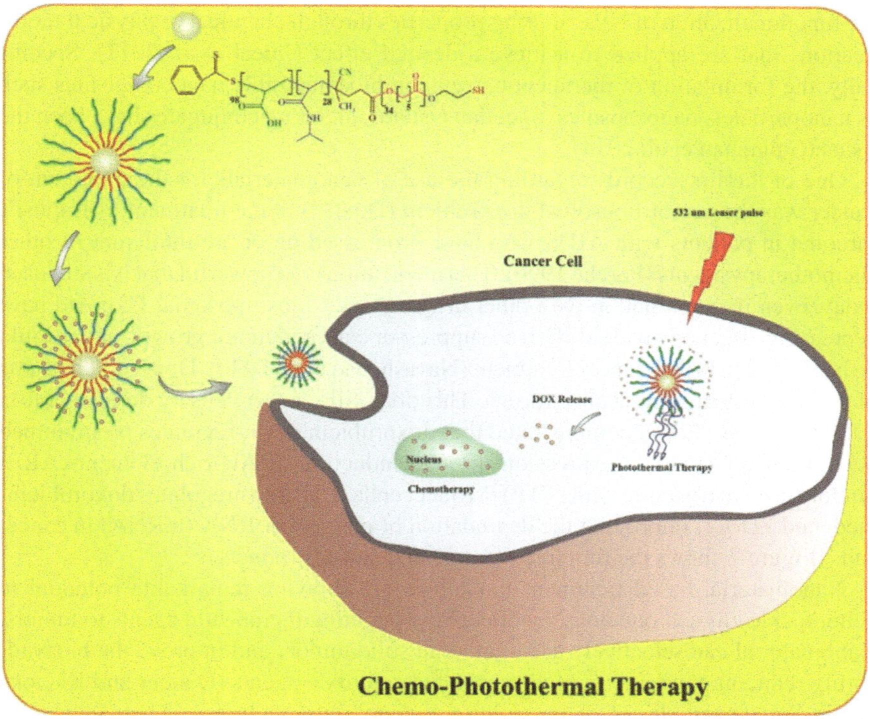

Fig. 6 Overall strategy for development of GNPs@polymer-DOX theranostic nanomedicine for chemo-photothermal therapy of solid tumors. Reproduced with permission from Mahmoodzadeh et al. (2018). Copyright (2018) Elsevier

7 Nanomedicine in Cancer Treatment

The range of diseases generically called cancer is characterized by of uncontrolled cell proliferation and spread. Therapeutic approaches used ordinarily in the cancer treatment include surgery, chemotherapy, radiation therapy, immunotherapy, and hormone therapy (Awasthi et al. 2018). These therapeutic approaches have improved patient survival and treatment outcomes; however, for most of these, they remain challenged by a number of limitations. Among these limitations, drug targeting and delivery are particularly challenging due to the non-selective tissue toxicity and presence of highly organized physiological, physical, and enzymatic barriers, which limit drug partitioning and distribution to the target site (Callaghan et al. 2014). Major advances in drug targeting and delivery have been the focus of research efforts in recent years. Nanotechnology-based drug delivery platforms offer a viable means of delivering small molecules and macromolecules in a localized or targeted manner (Akhtar et al. 2018). One of the most interesting capabilities in nanomedicine is the functionalization of NPs, altering properties through chemical or physical modifications that are applied to achieve a desired effect (Jaleel et al. 2017). Specifically, the formulation of therapeutic agents in biocompatible nanocomposites such as nanoparticles, nanocapsules, micellar systems, and drug conjugates have been the focus (Callaghan et al. 2014).

One of the first records regarding the use of nanomaterials for the treatment of cancer was the use of liposomal doxorubicin (Doxil®) in the treatment of Kaposi's sarcoma in patients with AIDS who have progressed on or are intolerant of other chemotherapy agents (Porche 1996). This medication is so powerful that it is still used today, even in combination with other drugs. Doxil® plus liposomal P5 could have decreasing effect on myeloid-derived suppressor cells and tumor growth, and it could be beneficial in breast cancer treatment (Navashenaq et al. 2020). Doxorubicin is one of the most effective anticancer drugs. This drug kills cancer cells by direct cytotoxicity. Kim et al. (2020) demonstrated that doxorubicin down-regulates programmed death-ligand 1 (PD-L1) expression through induction of AU-rich element (ARE) binding protein tristetraprolin (TTP) in cancer cells. TTP downregulates doxorubicin-mediated PD-L1, improving the degradation of messenger RNA (mRNA) in cancer cells. Figure 7 shows the timeline of some anticancer products.

Nanomaterial-based treatment modalities are showing remarkable potential to better tackle clinical oncology by effectively targeting therapeutic agents to tumors. Nanomaterial can selectively accumulate in solid tumors and improve the bioavailability reducing the toxicity of encapsulated cytotoxic agents (Quader and Kataoka 2017). Some examples of successful use of nanodrugs can be found below.

The pegylated liposomal doxorubicin (Doxil/caelyx) proved to be efficient in the treatment of patients with Kaposi's sarcoma (KS). The drug in this formulation localizes better to the tumor and has higher efficacy (Lichterfeld et al. 2005). Overall, 81.5% of study patients developed a response to the combined treatment, with a total of 55.5% and 26% of individuals reaching partial or complete responses, respectively. The treatment of KS in HIV-infected patients has been considerably improved by the use of Doxil/caelyx; however, response to therapy must be evaluated carefully and regularly to prevent excessive dosages of chemotherapy and long-term negative consequences (Cainelli and Vallone 2009).

The ferrocene emerges as an effective anticancer drug against ovarian cancer cells A2780 and SK-OV-3. The ferrocene has several unique properties such as thermal and hydrolytic stability, simple preparation, accessibility to a large variety of derivatives, and favorable electrochemical properties that predispose these compounds for biomedical applications. The family of ferrocene derivatives of the general formula [Fe{(η5-C5H4CH2 (p–C6H4) CH2 (N–het)} 2] as illustrated in Fig. 8 (Hodík et al. 2017).

The ferrocene cellular uptake mechanism involves receptor and transporter proteins, indicating that ferrocenes are transferred through interactions with trasferin–trasferin receptors. Within the cell, the ferrocenes induced an increase of reactive oxygen species levels in SK-OV-3 cells, indicating that the generation of reactive oxygen species is tightly associated with activation of the apoptotic program

Fig. 7 Timeline with some examples of nanoparticles for cancer treatment

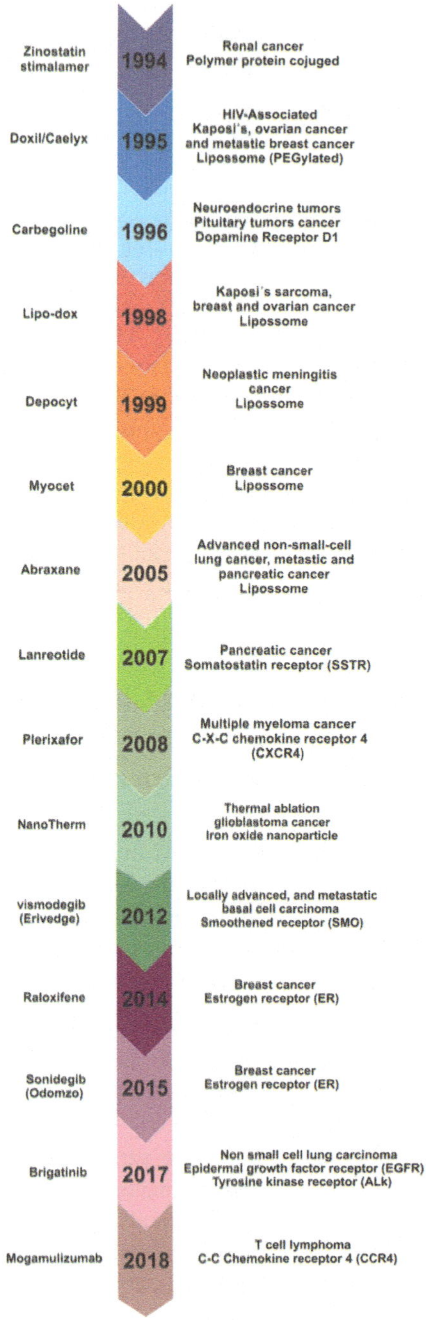

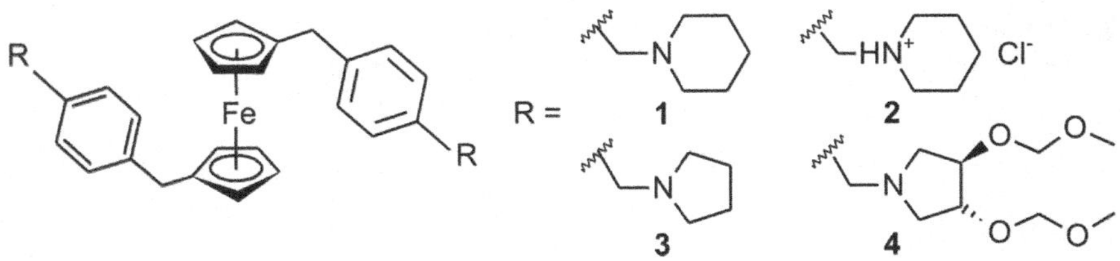

Fig. 8 Chemical formula of ferrocene derivates with anticancer potential. Reproduced with permission from Skoupilova et al. (2020). Copyright (2020) Elsevier

(Skoupilova et al. 2020). These organometallics show cytotoxic activity comparable with cisplatin (a classic drug for ovarian cancer treatment) presenting strong anticancer efficacy due to their intracellular accumulation caused by transferrin receptor-mediated endocytosis (Skoupilova et al. 2020).

Despite many studies conducted experimentally for new anticancer drugs, the disparity between preclinical and clinical studies is remarkable (Quader and Kataoka 2017). Even if nanomedicine shows us success stories in the areas cited above, this new area still has major challenges and limitations that must be overcome to achieve the desired outcome (Wu et al. 2020b).

8 Limitations and Challenges of Nanomedicine

8.1 Biological Barriers and Specific Targeting

Nanostructured materials that act as drugs must successfully reach the target site (Wu et al. 2020a). For this, they must cross several biological barriers such as reticuloendothelial system, endothelial/epithelial membranes, complex networks of blood vessels, abnormal flow of blood, and interstitial pressure gradients (Alonso 2004; Dilnawaz et al. 2018; Sanhai et al. 2008; Wolfram and Ferrari 2019).

- The reticuloendothelial system consists of both cellular and noncellular components. Phagocytic cells may bind nanoparticles and cause a release of cytokines, increasing nanoparticle clearance from the bloodstream and local inflammation of tissue (von Roemeling et al. 2017). Proteins, lipids, and other macromolecules may also bind to the surface of the nanoparticles and create a "biomolecular corona" around the nanoparticles (Tran et al. 2017). This effect may increase clearance from the bloodstream via recognition by the immune system or disrupt the ability of the nanoparticles to be internalized by the tissue of interest (García et al. 2014; Miele et al. 2009). Surface modifications of nanoparticles may permit escape from this mechanism (García et al. 2014).
- The blood–brain barrier is a challenge for treating brain cancers. Only 2% of molecules go through it, including ions, nutrients, specific peptides and proteins,

and leukocytes (Pardridge 2005). The barrier consists of endothelial cells joined by tight junctions and enclosed by astrocytic cells, basal lamina, pericytes, and microglia. Current methods for increasing penetration may involve direct introductions into the brain such as intraventricular or intracerebral injection, infusion, and implantation and may increase toxicity risks and non-uniform drug dispersal (von Roemeling et al. 2017).
- Tumor tissue is often characterized by leaky vasculature rich in fenestrations and poor in pericyte coverage. This characteristic has been utilized for targeting of nanoparticles to tumor tissue, however deeper penetration into the tumor is frequently restricted due to heterogeneity of the tumor microenvironment (Chauhan et al. 2012). Methods to increase nanoparticle penetration into the tumor bed are currently being investigated. Use of smaller nanoparticles may allow enhanced passage through the vasculature and deeper penetration into the tumor (Blanco et al. 2015).

In addition to the limitations mentioned above (and others not explored), there is still a great challenge for oral delivery, because of the wide range of pH variation and enzymatic degradation in gastrointestinal tract (Evans et al. 1988). Thus, orally applied nanoparticles need to have high stability in the gastrointestinal tract, the ability to penetrate intestinal epithelium to keep the high systemic bioavailability of drugs after crossing several barriers (Wu et al. 2020a) as noted in chitosan-coated zein nanoparticles for oral delivery of resveratrol (Pauluk et al. 2019). The same protein (zein) was also used in the form of nanofibers as drug carriers for controlled delivery of L-3, 4-dioxyphenylalanine levodopa for treatment of forms of Parkinson's syndrome (Ansari et al. 2019).

For the reasons cited above, the careful characterization of nanotechnology-related products requires a combination of different techniques to understand the physicochemical features and how these affect efficacy and product safety (Marques et al. 2019).

9 Regulatory Challenges

The Food and Drug Administration (FDA) and European Medicines Agency (EMA) approved a number of nanomedicine products for cancer therapy; however, there are not specifically implemented guidelines for drug products containing soft materials yet (Tinkle et al. 2014; Wu et al. 2020a). The lack of guidance in the examination of nanomedicine and nanomedicine therapeutics can only make regulatory decisions based on individual assessment of benefits and risks (Desai 2012).

The same scenario is found on Brazil. The Brazilian Health Surveillance Agency (ANVISA) regulates or support regulation of medicines, medical devices, sanitizers, pesticides, food, cosmetics, blood, tissues and organs and tobacco but has no clear regulation on nanomedicine. ANVISA also formed a committee which comprised of experts from the nanomaterials niche area with focus on drugs, medical devices, food, hygiene products as well as diagnostic equipment and supplies. The aim was to develop a questionnaire for manufacturers willing to register products that contain nanomaterials. However, the initiative ended in 2016 without further actions. Up till now, all medicinal products containing engineered nanomaterials are evaluated on a case-by-case basis (Marques et al. 2019).

10 Nanomaterials and Environment: Nanotoxicology

The technological advance allowed the manipulation of matter on nanoscale, bringing new possibilities related to the production of materials with unique characteristics. Thus, nanoparticles, which are the result of this manipulation, could be introduced in products used daily such as cosmetics, paints, packaging, pharmaceuticals, electronic devices, and most industrialized products (Marambio-Jones and Hoek 2010). The increasing use of this input also succumbs to the concern with the release of these particles into the environment and the potential adverse effects on ecosystems, as well as human health (Eckhardt et al. 2013).

Engineered nanoparticles can be highly toxic, because due to their small size they acquire new physicochemical characteristics and properties that potentially cause adverse effects on organisms (Auffan et al. 2010). The inherent risk of the disposition of these NPs in the environment and their effects are attended by nanotoxicology. Recently, the term nanotoxicology has gained space, including the increase in the number of journals that address this topic (Table 1).

Nanotoxicology is a field of toxicology dedicated to the study of the toxicological effects of nanomaterials in different biological systems, including cells, tissues, and living organisms (Selvaraj et al. 2018). The interaction of nanomaterials with the environment is quite complex and evaluating the impacts of these nanostructures on cells, animals, and humans is a challenge, considering that the nanoparticles differ from their original material (bulk). Despite the significant number of studies related to the toxicity of nanostructures in biological systems (Hastings et al. 2015) estimate the general mechanism of action of NPs is a complicated task. This difficult occurs because even small modifications in the characteristics of the particles can drastically change their toxicity (Pikula et al. 2020).

Furthermore, toxicity of a NP is directly related to the dose at which the organism is receiving, and the way that nanomaterial/organism interactions occur (Wang et al. 2018; Zhang et al. 2016a, b, c). Characteristics such as particle size, surface charge, morphology, chemical composition, synthesis process, size/surface ratio must be considered for toxicity assessment (Barrena et al. 2009; Kettler et al. 2014; Murdock et al. 2008). In the subsequent topics, the interaction of these nanomaterials in biological systems will be demonstrated, with an emphasis on the most studied nanoparticles in the last few years (Fig. 9).

Table 1 Main journals that presented articles with the keyword "*nanotoxicology*."

Journals publishing articles on nanotoxicology		
Publication vehicle	Publisher	Frequency
ACSNano	ACS	Monthly
Aquatic Toxicology	Elsevier	Monthly
Archives of Toxicology	Springer	Monthly
Biomaterials	Elsevier	Weekly
Chemistry and Ecology	Taylor and Francis	Mensal
Colloids and Surfaces B: Biointerfaces	Elsevier	Monthly
Current Opinion in Toxicology	Elsevier	Monthly
Environmental Pollution	Elsevier	Monthly
Food and Chemical Toxicology	Elsevier	Monthly
NanoImpact	Elsevier	Semiannual
Nanomaterials	MDPI	Monthly
Nature Nanotechnology	Nature Pub. Group	Monthly
Toxicology and Applied Pharmacology	Elsevier	Monthly
Toxicology Letters	Elsevier	Monthly
Toxicology Reports	Elsevier	Yearly

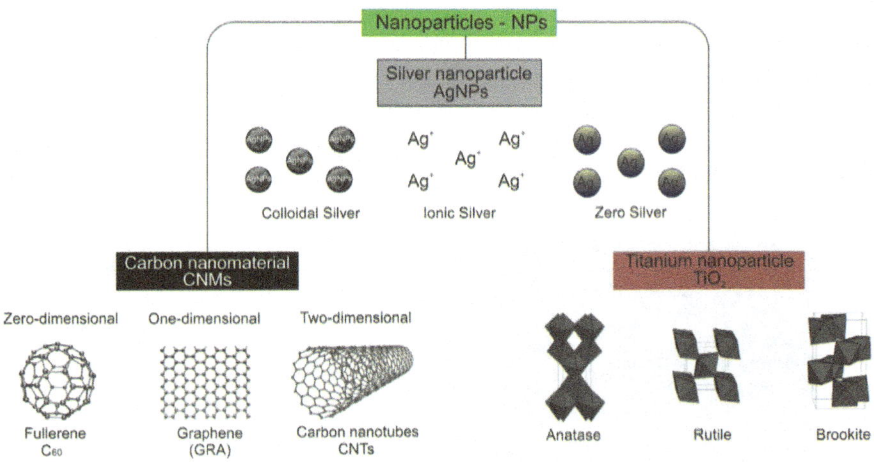

Fig. 9 Different shapes and nature of nanoparticles. These variations are related to their chemical composition and spatial geometry of atoms

11 Interactions of Silver Nanoparticles with Biological Systems

Silver nanoparticles (AgNPs) are recognized to have antimicrobial activity and are currently one of the most studied and industrial used nanomaterials. This fact directly reflected in the vertiginous increase of its production (Durán et al. 2019; Vance et al. 2015).

According to the synthesis method, AgNPs can vary in shape and size, giving different physical–chemical characteristics, directly affecting their interaction and toxicity to biological systems (Durán et al. 2019). An example of this variation can be seen in Fig. 10. This figure shows scanning and transmission electron micrographs of AgNPs synthesized by citrate method. These particles have different sizes and irregular morphology.

According to Tombuloglu (Tombuloglu et al. 2018), different results can be obtained through the interaction of NPs and plants. These results depend on the size of the NPs, coating, and type of species used as a parameter to evaluate the results. Figure 11 illustrates the general mechanism of absorption of nanoparticles in plants.

The absorption of nanoparticles in plants occurs basically through the roots (Cifuentes et al. 2010; Geisler-Lee et al. 2013; Pariona et al. 2017); however, if NPs are suspended in the air they can be absorbed by the leaves (Anjum et al. 2016). As soon as they are absorbed, the NPs reach the conducting vessels and are transported to the aerial part via apoplast or symplast and interact with the plant cells. The plant cellulosic cell wall acts as a barrier that allows the passage of only small particles and restricts the larger ones, so that smaller nanoparticles can pass through this layer relatively easily in relation to larger nanoparticles (Rastogi et al. 2017).

For most authors, nanoparticles larger than 20 nm are adsorbed and accumulate on the cell wall. This adsorption can occur by electrostatic attraction, hydrogen bonding, and specific coordination (Wang et al. 2018). The adsorption of nanoparticles leads

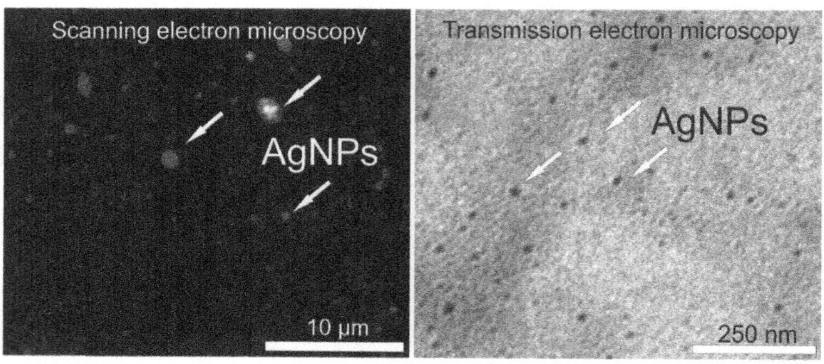

Fig. 10 Micrograph of silver nanoparticles (AgNPs) (arrows) showing morphology and irregular surface. AgNPS can have different sizes due to agglomeration resulting from the increased surface load

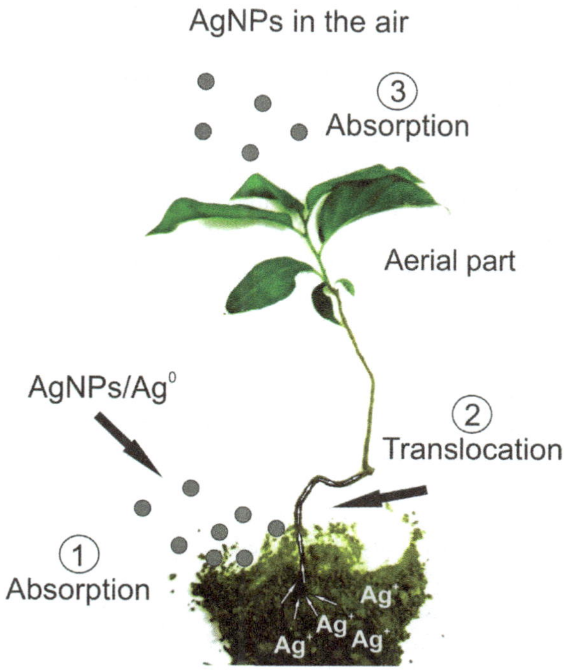

Fig. 11 Absorption of nanoparticles in plants. 1—AgNPs and Ag^0 accumulate in the soil, coming into contact with oxidizing agents (H_2O/O_2); 2—AgNPs and Ag^0, in addition to being absorbed by the roots, dissociate in Ag^+. Inside Ag^+ cells it induces the production of intracellular ROS (strong oxidizer) causing damage to the cell wall structure; 3—Otherwise, if the AgNPs are suspended in the air, they can be absorbed by the aerial part of the plant

to denaturation and oxidation of the cell wall in both plants and microorganisms, in addition to promoting the dissolution of AgNPs and Ag^0 in Ag^+ ionic silver (Le Ouay and Stellacci 2015; Piccapietra et al. 2012; Xiu et al. 2012); however, Ag^+ is more toxic to cells (Sendra et al. 2018). The oxidation of the cell wall allows the passage of large amounts of AgNPs, remaining Ag^0 and Ag^+.

Inside the cell, the dissolution of the remaining AgNPs and Ag^0 normally occurs in the presence of an oxidizing agent, which can be oxygen, water or reactive oxygen species (ROS). Consequently, the production of ROS in the cell is triggered by the interaction of Ag^+ with the cellular content. The main target of Ag^+ is proteins and thiol groups present in DNA, fundamental molecules for the proper functioning of cellular machinery. This entire cycle of nanoparticles/cell interaction tends to produce more oxidizing agents that denature the cell wall leading to the death of the organism. Plants, microorganisms, and animals are susceptible to the effects of AgNPs. Many studies have been conducted using different species to assess the potential risk of these nanomaterials in the environment (Yu et al. 2013).

Several plant species were used as a model to assess the toxicity of AgNPs and different responses were observed as a result of this interaction. In the pioneering work conducted by Barrena (Barrena et al., 2009) in plants of *Cucumis sativus* (cucumber) and *Lactuca sativa* (lettuce), it was observed that AgNPs inhibited germination, however the roots of exposed plants were longer. In *Arabidopsis thaliana,* the interaction of AgNPs and Ag^+ also inhibited germination and the length of exposed plant roots was shorter (Geisler-Lee et al. 2013). In *Triticum aestivum* (wheat) seedlings, no inhibition in germination rate was observed, only reduction in root length (Vannini et al. 2014). *Lycorpesicon esculentum* (tomato) delayed seed germination without changing the germination percentage. AgNPs reduced root development of seedlings with undeveloped stems and absence of root hair in the hypocotyl region (Gonçalves et al. 2016). It is likely that the presence of AgNPs induced the activity of catalases twice as high, indicating that there was an increase in the production of oxygen species, leading to oxidative stress. The oxidative stress would justify the delay in root development and other anomalies such as the absence of root hair, development unusual hypocotyl, and underdeveloped plant.

In bacteria, AgNPs showed a broad bactericidal effect in both gram positive and gram negative bacteria (Amato et al. 2011; Bapat et al. 2018). The antibacterial action of AgNPs is related to the release of Ag^+ ions, which is greater when smaller nanoparticles are used (Roco 2004). *Staphylococcus aureus* growth was inhibited in the presence of Ag^+ (Loza et al. 2014; Mirzajani et al. 2011), recurrent result in several studies using different species of bacteria as *Nitrosomonas europaea* (Radniecki et al. 2011; Yuan et al. 2013), *Escherichia coli* (Cho et al. 2005; Sondi and Salopek-Sondi 2004; Xu et al. 2012) and *Streptococcus mutans* (Dutra-Correa et al. 2018).

Studies were also conducted using organisms present in the aquatic environment as a model (Wang et al. 2018) using *Scenedesmus obliquus* (green algae) observed that the increase in the concentration of AgNPs directly affected the growth of these algae. In *Chlorella vulgaris* and *Dunaliella tertiolecta*, exposure to NPs led to the accumulation of AgNPs in the cells, negatively affecting the development of algae (Oukarroum et al. 2012). Silver nanomaterials (silver nanospheres, AgNPs; silver nanowires, AgNWs; silver nanoplates) with different sizes and surfaces inhibited the development of *Chlorococcum infusionum*.

Silver nanoplates are the more toxic among the nanomaterials cited (Nam and An 2019). In *Danio rerio* (zebrafish), embryos exposed to AgNPs and Ag^+ changes in neural circuits were observed, leading to dysfunctional responses to touch and locomotor changes (Zhao et al. 2019). As in aquatic environment, the soil receives a large amount of nanomaterials present in consumer products. In the soil, AgNPs tend to accumulate, being harmful to small animals. In a study using *Eisenia fetida* (earthworms), AgNPs were toxic with an increased mortality rate (Garcia-Velasco et al. 2019).

In rat neural cells, exposure of cells to AgNPs for 24 h caused neural function disorders, suggesting that this exposure may progress to the progression of degenerative diseases (Lin et al. 2016).

All of these results suggest a toxic effect of AgNPs on organisms in contact with these nanomaterials. The effects can be differentiated considering the characteristics of the nanoparticles and the interaction environment.

12 Interactions of Titanium Nanoparticles with Biological Systems

Titanium dioxide (TiO_2) nanoparticles are present in many commercial products, such as paints, cosmetics, and sunscreens (Nakata and Fujishima 2012; Servin et al. 2013). In addition to the use of TiO_2 in product consumer, these nanoparticles can be used in environmental remediation (Aitken et al. 2006) once many of them have photocatalytic activity resulting in the photochemical degradation of substances present in the environment (Hund-Rinke and Simon 2006). Photocatalytic activity is activated by ultraviolet radiation and releasing radical (OH^-) (Bundschuh et al. 2011). This radical causes damage to microorganisms showing antimicrobial activity, being desirable characteristics in medical, food, and water treatment industries (Ravishankar Rai and Jamuna Bai 2011).

TiO_2 has three distinct shapes as spatial geometry varies between atoms. The rutile phase and anatase phase have tetragonal geometry and brookite orthorhombic geometry (Costa et al. 2006). These variations give different characteristics for each conformational state of TiO_2. The rutile phase is used in the composition of sunscreens because it is a more stable structure in relation to the other two structures.

The daily use of sunscreens increases the possibility of these nanoparticles entering the environment and contaminating soil and water. The expectation over the next 10 years is that the accumulation of TiO_2 will reach 6,000,000 tons (Robichaud et al. 2009). These data are alarming, once these NPs can be toxic to microorganisms, plants and animals.

The interaction of TiO_2 nanoparticles in plants, microorganisms, and animals can have different effects since the toxicity of these NPs is associated with geometry and stability structure. In a study using *Linum usitatissimum* (linseed) the anatase phase contributed to a positive effect on germination and root growth (Clément et al. 2013). This fact can be attributed to the antimicrobial properties of the anatase structure, which increased the plant's resistance to stress. A similar result was described by (Kibbey and Strevett 2019) studying the interaction of TiO_2 nanoparticles in *L. sativa* plants. However, in *Spirodela polyrrhiza,* the interaction of TiO_2 nanoparticles significantly reduced the growth parameter (root length) (Movafeghi et al. 2018).

Exposition of aquatic organism (*Chlorella vulgaris*, *Dafnia magna* and *Phaeodactylum tricornutum*) proved that the anatase form is more toxic than the rutile phase, since the latter formed more aggregates when interacting with the tested organisms (Clément et al. 2013). Another study using three species of algae (*Synedra ulna, Scenedesmus quadricauda,* and *Stigeoclonium tenue*) indicated that exposure to TiO_2 nanoparticles did not affect the growth of these species, indicating that there was no toxicity (Kulacki et al. 2012). The toxic effect of TiO_2 nanoparticles is associated with their composition, but this effect may vary according to the studied species.

In different bacterial species, in general, the accumulation of TiO_2 nanoparticles showed toxicity observed by growth inhibition and metabolic changes. In a study using samples of the bacterial community that makes up the soil, changes in the activity, composition, and diversity of the microbial community were observed

(Nogueira et al. 2012). A similar result was observed by (Kibbey and Strevett 2019). TiO_2 nanoparticles were effective against *Staphylococcus aureous* and *Pseudomonas aeruginosa* showing an inhibition rate in 100% of microorganisms (Haider et al. 2017).

13 Interactions of Nanostructured Carbons with Biological Systems

Carbon-based nanomaterials (CNMs) are basically formed by the carbon element with one of its dimensions smaller than 100 nm. The main forms of these nanomaterials include carbon nanotubes (CNTs) that can be single-walled carbon nanotubes (SWCNTs) and double-walled nanotubes (MWCNTs), graphene (GRA) with planar structure and fullerene (C_{60}) with spherical shape (Chichiriccò and Poma 2015). These carbon-based nanomaterials have different characteristics from their macroscale material, acquiring new properties of industrial interest (Jiang et al. 2019).

Fullerenes have a unique structure and property, and for this reason they can be applied in areas of photovoltaic energy and photocatalysis (Mroz et al. 2007; Zhong et al. 2016). This material also has high conductivity, magnetism and catalytic activity (Shan et al. 2017), optical transparency, and mechanical flexibility (Mohajeri et al. 2018). Among the carbon-based materials, graphene (GRA) has a larger surface area, providing exceptional optical and thermal properties (Jiang et al. 2019). Graphene atoms are connected by carbobo-carbon bond and their mechanical properties are superior to those of other materials, so this material has been used in the manufacture of compounds with high mechanical resistance (Peng et al. 2020).

CNTs have been widely used for environmental remediation and wastewater treatment due to their characteristics such as large surface area, greater pore volume, and mechanical flexibility (Zhou et al. 2018); furthermore, it is also used as drug carriers, bioimaging, and biosensors (Shim et al. 2002).

The variety of applications of CNMs in the most varied areas increased the risk of these nanomaterials accumulating in the environment (Chen et al. 2018a, b). Species of plants, microorganisms, and animals would come into direct contact with CNMs, generating harmful effects for these organisms.

The interaction of CNMs in plants directly affects morphology, resulting in variation in germination rate and root length. These toxic effects are common for plant species. The phytotoxicity is related to the production of reactive oxygen species (ROS), causing oxidative stress, lipid peroxidation, and plant DNA damage (Tripathi et al. 2017). In cabbage plants (*Brassica oleracea*), CNTs accumulated mainly in the roots and leaves, inhibiting growth and decreasing biomass (Awad et al. 2017). The interaction of C_{60}, CNTs, MWCNTs, and graphene (GRA) with seedlings of *Oriza sativa* (rice) inhibited the absorption of nutrients by the roots, altering the morphology of the cells and increasing the antioxidant enzymes activity (Hao et al.

2018). Similar result was reported by (Chen et al. 2017) using graphene oxide (OG) in plants of *Triticum aestivum* (wheat).

Ghosh et al. investigated the interaction of MWCNTs with *Allium cepa* (onion). MWCNTs caused changes and damage to DNA, resulting in morphological changes of plants (Ghosh et al. 2015).

In microorganisms, the results were similar to that observed in plants. CNMS induced lipid peroxidation, DNA damage, protein denaturation, and cell death (Chen et al. 2018a, b). In microflora, MWCNTs caused damage at the cellular level, generating toxicity (Yadav et al. 2016). Chen et al. observed that the interaction of CNTs in a microbial community reduced the amount of carbon in the medium, decreasing the diversity of bacteria (Chen et al. 2015).

In animals, interaction with CNTs alters metabolic pathways and causes cell damage, genotoxicity, and apoptosis. The administration of SWCNts in mice (*Mus musculus*) caused damage to mitochondria and cellular apoptosis due to the accumulation of these CNTs in the lungs, spleen, and kidneys (Principi et al. 2016). Fujita et al. (2016) administered doses of SWCNTs and MWCNTs in *Rattus novergicus* (rat) and observed pleural and pulmonary inflammation.

14 Conclusions

Nanostructured materials have been used throughout human history; however, the concept of manipulating matter at the atomic level had great impetus in the 1960s.

Nanoparticles can be found in the environment naturally or through artificial synthesis processes. The study of nanoparticles is essential to assess their interaction with the environment and their effects on organisms, since they have different physical and chemical characteristics from their bulk material.

The technological advance allowed the manipulation of matter on nanoscale, bringing new possibilities related to the production of materials with unique characteristics. Thus, nanoparticles, which are the result of this manipulation, could be introduced in products used daily such as cosmetics, paints, packaging, pharmaceuticals, electronic devices, and most industrialized products. In the other hand, the increasing use of nanostructured materials also succumbs to the concern with the release of these particles into the environment and the potential adverse effects on ecosystems, as well as human health.

Undoubtedly, nanostructured materials will occupy an even more prominent place in different areas of science in the next years.

Acknowledgements The authors would like to thank the Central Analítica-UFC/CT-INFRA/MCTI-SISANO/Pró-Equipamentos CAPES for the support. TBARM acknowledge funding from CNPq (grant 350023/2020-4) and Central Analítica-UFC/CT-INFRA-FINEP/Pro-Equipamentos-CAPES/CNPq-SisNano-MCTI 2019 (Grant 442577/2019-2). SKPP is a doctoral student in the Postgraduate Program in Biotechnology of Natural Resources.

References

Aitken RJ, Chaudhry MQ, Boxall ABA, Hull M (2006) Manufacture and use of nanomaterials: current status in the UK and global trends. Occup Med (Chic Ill) 56:300–306

Akhtar MJ, Ahamed M, Alhadlaq HA (2018) Challenges facing nanotoxicology and nanomedicine due to cellular diversity. Clin Chim Acta 487:186–196

Alonso MJ (2004) Nanomedicines for overcoming biological barriers. Biomed Pharmacother 58:168–172

Amato E, Diaz-Fernandez YA, Taglietti A, Pallavicini P, Pasotti L, Cucca L, Milanese C, Grisoli P, Dacarro C, Fernandez-Hechavarria JM, Necchi V (2011) Synthesis, characterization and antibacterial activity against gram positive and gram negative bacteria of biomimetically coated silver nanoparticles. Langmuir 27:9165–9173

Anjum NA, Rodrigo MAM, Moulick A, Heger Z, Kopel P, Zítka O, Adam V, Lukatkin AS, Duarte AC, Pereira E, Kizek R (2016) Transport phenomena of nanoparticles in plants and animals/humans. Environ Res 151:233–243

Ansari AQ, Ansari SJ, Khan MQF, Khan MQF Qureshi UA, Khatri Z, Ahmed F, Kim IS (2019) Electrospun Zein nanofibers as drug carriers for controlled delivery of Levodopa in Parkinson syndrome. Mater Res Express 6:075405

Ansari MA, Khan HM, Khan AA, Cameotra SS, Pal R (2014) Antibiofilm efficacy of silver nanoparticles against biofilm of extended spectrum β-lactamase isolates of *Escherichia coli* and *Klebsiella pneumoniae*. Appl Nanosci 4:859–868

Arca-Lafuente S, Martínez-Román P, Mate-Cano I, Madrid R, Briz V (2020) Nanotechnology: areality for diagnosis of HCV infectious disease. J Infect 80(1):8–15

Arruebo M, Fernández-Pacheco R, Ibarra MR, Santamaría J (2007) Magnetic nanoparticles for drug delivery applications. J Nanosci Nanotechnol 2:22–32

Auffan M, Bottero J-YY, Chaneac C, Rose J, Ros J (2010) Inorganic manufactured nanoparticles: how their physicochemical properties influence their biological effects in aqueous environments. Nanomedicine 5:999–1007

Awad YM, Vithanage M, Niazi NK, Rizwan M, Rinklebe J, Yang JE, Ok YS, Lee SS (2017) Potential toxicity of trace elements and nanomaterials to Chinese cabbage in arsenic- and lead-contaminated soil amended with biochars. Environ Geochem Health 41:1777–1791

Awasthi R, Roseblade A, Hansbro PM, Rathbone MJ, Dua K, Bebawy M (2018) Nanoparticles in cancer treatment: opportunities and obstacles. Curr Drug Targets 19:1696–1709

Azari A, Babaei AA, Rezaei-Kalantary R, Esrafili A, Moazzen M, Kakavandi B (2014) Nitrate removal from aqueous solution using carbon nanotubes magnetized by nano zero-valent iron. J Maz Univ Med Sci 23:14–27

Bangeppagari M, Park SH, Kundapur RR, Lee SJ (2019) Graphene oxide induces cardiovascular defects in developing zebrafish (Danio rerio) embryo model: in-vivo toxicity assessment. Sci Total Environ 673:810–820

Bapat RA, Chaubal TV, Joshi CP, Bapat PR, Choudhury H, Pandey M, Gorain B, Kesharwani P (2018) An overview of application of silver nanoparticles for biomaterials in dentistry. Mater Sci Eng C 91:881–898

Bareford LM, Swaan PW (2007) Endocytic mechanisms for targeted drug delivery. Adv Drug Deliv Rev 59:748–758

Barrena R, Casals E, Colón J, Font X, Sánchez A, Puntes V (2009) Evaluation of the ecotoxicity of model nanoparticles. Chemosphere 75:850–857

Behzadi S, Serpooshan V, Tao W, Hamaly MA, Alkawareek MY, Dreaden EC, Brown D, Alkilany AM, Farokhzad OC, Mahmoudi M (2017) Cellular uptake of nanoparticles: journey inside the cell. Chem Soc Rev 46:4218–4244

Bhatt I, Tripathi BN (2011) Interaction of engineered nanoparticles with various components of the environment and possible strategies for their risk assessment. Chemosphere 82:308–317

Blanco E, Shen H, Ferrari M (2015) Principles of nanoparticle design for overcoming biological barriers to drug delivery. Nat Biotechnol 33:941–951

Boisselier E, Astruc D (2009) Gold nanoparticles in nanomedicine: preparations, imaging, diagnostics, therapies and toxicity. Chem Soc Rev 38:1759–1782

Boran H, Boyle D, Altinok I, Patsiou D, Henry TB (2016) Aqueous Hg^{2+} associates with TiO_2 nanoparticles according to particle size, changes particle agglomeration, and becomes less bioavailable to zebrafish. Aquat Toxicol 174:242–246

Brun N, Mazerolles L, Pernot M (1991) Microstructure of opaque red glass containing copper. J Mater Sci Lett 10:1418–1420

Bundschuh M, Filser J, Lüderwald S, McKee MS, Metreveli G, Schaumann GE, Wagner S (2018) Nanoparticles in the environment: where do we come from, where do we go to? Environ Sci Eur 30:1–17

Bundschuh M, Zubrod JP, Englert D, Seitz F, Rosenfeldt RR, Schulz R (2011) Effects of nano-TiO_2 in combination with ambient UV-irradiation on a leaf shredding amphipod. Chemosphere 85:1563–1567

Cainelli F, Vallone A (2009) Safety and efficacy of pegylated liposomal doxorubicin in HIV-associated Kaposi's sarcoma. Biol Targets Ther 3:385–390

Callaghan R, Luk F, Bebawy M (2014) Inhibition of the multidrug resistance P-glycoprotein: time for a change of strategy? Drug Metab Dispos 42:623–631

Chauhan VP, Stylianopoulos T, Martin JD, Popovic Z, Chen O, Kamoun WS, Bawendi MG, Fukumura D, Jain RK (2012) Normalization of tumour blood vessels improves the delivery of nanomedicines in a size-dependent manner. Nat Nanotechnol 7:383–388

Chen F, Hableel G, Zhao ER, Jokerst JV (2018a) Multifunctional nanomedicine with silica: role of silica in nanoparticles for theranostic, imaging, and drug monitoring. J Colloid Interface Sci 521:261–279

Chen L, Wang C, Li H, Qu X, Yang ST, Chang XL (2017) Bioaccumulation and toxicity of 13C-skeleton labeled graphene oxide in wheat. Environ Sci Technol 51:10146–10153

Chen M, Zhou S, Zhu Y, Sun Y, Zeng G, Yang C, Xu P, Yan M, Liu Z, Zhang W (2018b) Toxicity of carbon nanomaterials to plants, animals and microbes: recent progress from 2015-present. Chemosphere 206:255–264

Chen Q, Wang H, Yang B, He F, Han X, Song Z (2015) Responses of soil ammonia-oxidizing microorganisms to repeated exposure of single-walled and multi-walled carbon nanotubes. Sci Total Environ 505:649–657

Chichiriccò G, Poma A (2015) Penetration and toxicity of nanomaterials in higher plants. Nanomaterials 5:851–873

Cho KH, Park JE, Osaka T, Park SG (2005) The study of antimicrobial activity and preservative effects of nanosilver ingredient. Electrochim Acta 51:956–960

Choi HS, Frangioni JV (2010) Nanoparticles for biomedical imaging: fundamentals of clinical translation. Mol Imaging 9:291–310

Christian P, Von der Kammer F, Baalousha M, Hofmann T (2008) Nanoparticles: structure, properties, preparation and behaviour in environmental media. Ecotoxicology 17:326–343

Cifuentes Z, Custardoy L, de la Fuente JM, Marquina C, Ibarra MR, Rubiales D, Pérez-de-Luque A (2010) Absorption and translocation to the aerial part of magnetic carbon-coated nanoparticles through the root of different crop plants. J Nanobiotechnol 8:1–8

Clément L, Hurel C, Marmier N (2013) Toxicity of TiO_2 nanoparticles to cladocerans, algae, rotifers and plants—effects of size and crystalline structure. Chemosphere 90:1083–1090

Colomban P (2009) The use of metal nanoparticles to produce yellow, red and iridescent colour, from bronze age to present times in lustre pottery and glass: solid state chemistry, spectroscopy and nanostructure. J Nano Res 8:109–132

Contreras JE, Rodriguez EA, Taha-Tijerina J (2017) Nanotechnology applications for electrical transformers—a review. Electr Power Syst Res 143:573–584

Costa ACFM, Vilar MA, Lira HL, Kiminami RHGA, Gama L (2006) Síntese e caracterização de nanopartículas de TiO_2. Ceramica 52:255–259

Crespo L, Pons M, Giralt E, Royo M, Albericio F (2005) Peptide and amide bond-containing dendrimers. Chem Rev 105:1663–1681

Daukantas P (2019) Still plenty of room at the bottom. Opt Photonics News 30:24–31

Desai N (2012) Challenges in development of nanoparticle-based therapeutics. AAPS J 14:282–295

de Stavale AA, Fonseca GO, Duarte PS, Macedo LC, Percebom AM (2019) Nanoparticles of surfactant and block copolymers with high uptake of oily ingredients for cosmetic formulations. Colloids Surfaces A Physicochem Eng Asp 581:123779

Dilnawaz F, Acharya S, Sahoo SK (2018) Recent trends of nanomedicinal approaches in clinics. Int J Pharm 538:263–278

Du J, Hu Z, Yu Z, Li H, Pan J, Zhao D, Bai Y (2019) Antibacterial activity of a novel *Forsythia suspensa* fruit mediated green silver nanoparticles against food-borne pathogens and mechanisms investigation. Mater Sci Eng C 102:247–253

Durán N, Rolim WR, Durán M, Fávaro WJ, Seabra AB (2019) Nanotoxicologia de nanopartículcas de prata: Toxicidade em animais e humanos. Químiva Nov 42:206–213

Dutra-Correa M, Leite AABV, de Cara SPHM, Diniz IMA, Marques MM, Suffredini IB, Fernandes MS, Toma SH, Araki K, Medeiros IS (2018) Antibacterial effects and cytotoxicity of an adhesive containing low concentration of silver nanoparticles. J Dent 77:66–71

Eckhardt S, Brunetto PS, Gagnon J, Priebe M, Giese B, Fromm KM (2013) Nanobio silver: Its interactions with peptides and bacteria, and its uses in medicine. Chem, Rev

Editorial (2009) "Plent of room" revised. Nat Nanotechnol 4:781

Evans DF, Pye G, Bramley R, Clark AG, Dyson TJ, Hardcastle JD (1988) Measurement of gastrointestinal pH profiles in normal ambulant human subjects. Gut 29:1035–1041

Feynman R (1960) There's plenty of room at the bottom. An invitation to enter a new field of physics, Eng Sci

Freestone I, Meeks N, Sax M, Higgitt C (2007) The lycurgus cup—a roman nanotechnology. Gold Bull 40:270–277

Fujita K, Fukuda M, Endoh S, Maru J, Kato H, Nakamura A, Shinohara N, Uchino K, Honda K (2016) Pulmonary and pleural inflammation after intratracheal instillation of short single-walled and multi-walled carbon nanotubes. Toxicol Lett 257:23–37

Garcia-Velasco N, Irizar A, Urionabarrenetxea E, Scott-Fordsmand JJ, Soto M (2019) Selection of an optimal culture medium and the most responsive viability assay to assess AgNPs toxicity with primary cultures of Eisenia fetida coelomocytes. Ecotoxicol Environ Saf 183:109545

García KP, Zarschler K, Barbaro L, Barreto JA, O'Malley W, Spiccia L, Stephan H, Graham B (2014) Zwitterionici-coated "stealth" nanoparticles for biomedical application: Recent advances in coutering biomolecular corona formation and uptake by the mononuclear phagocyte system. Nano Small Micro 10:2516–2529

Gehr P (2018) Interaction of nanoparticles with biological systems. Colloids Surfaces B Biointerfaces 172:395–399

Geisler-Lee J, Wang Q, Yao Y, Zhang W, Geisler M, Li K, Huang Y, Chen Y, Kolmakov A, Ma X (2013) Phytotoxicity, accumulation and transport of silver nanoparticles by *Arabidopsis thaliana*. Nanotoxicology 7:323–337

Ghosh M, Bhadra S, Adegoke A, Bandyopadhyay M, Mukherjee A (2015) MWCNT uptake in Allium cepa root cells induces cytotoxic and genotoxic responses and results in DNA hypermethylation. Mutat Res Fundam Mol Mech Mutagen 774:49–58

Gonçalves SPC, Strauss M, Delite FS, Clemente Z, Castro VL, Martinez DST (2016) Activated carbon from pyrolysed sugarcane bagasse: silver nanoparticle modification and ecotoxicity assessment. Sci Total Environ 565:833–840

Haider AJ, Al-Anbari RH, Kadhim GR, Salame CT (2017) Exploring potential environmental applications of TiO_2 nanoparticles. Energy Procedia 119:332–345

Hao Y, Ma C, Zhang Z, Song Y, Cao W, Guo J, Zhou G, Rui Y, Liu L, Xing B (2018) Carbon nanomaterials alter plant physiology and soil bacterial community composition in a rice-soil-bacterial ecosystem. Environ Pollut 232:123–136

Hastings J, Jeliazkova N, Owen G, Tsiliki G, Munteanu CR, Steinbeck C, Willighagen E (2015) eNanoMapper: harnessing ontologies to enable data integration for nanomaterial risk assessment. J Biomed Semant 6:1–15

Heiligtag FJ, Niederberger M (2013) The fascinating world of nanoparticle research. Mater Today 16x:262–271

Hinzmann M, Jaworski S, Kutwin M, Jagiełło J, Koziński R, Wierzbicki M, Grodzik M, Lipińska L, Sawosz E, Chwalibog A (2014) Nanoparticles containing allotropes of carbon have genotoxic effects on glioblastoma multiforme cells. Int J Nanomed 9:2409–2417

Hodík T, Lamač M, Červenková Šťastná L, Cuřínová P, Karban J, Skoupilová H, Hrstka R, Císařová I, Gyepes R, Pinkas J (2017) Improving cytotoxic properties of ferrocenes by incorporation of saturated N-heterocycles. J Organomet Chem 846:141–151

Hulla JE, Sahu SC, Hayes AW (2015) Nanotechnology: history and future. Hum Exp Toxicol 34:1318–1321

Hund-Rinke K, Simon M (2006) Ecotoxic effect of photocatalytic active nanoparticles (TiO_2) on algae and daphnids. Environ Sci Pollut Res 13:225–232

Hwang R, Mirshafiee V, Zhu Y, Xia T (2018) Current approaches for safer design of engineered nanomaterials. Ecotoxicol Environ Saf 166:294–300

Jaleel JA, Sruthi S, Pramod K (2017) Reinforcing nanomedicine using graphene family nanomaterials. J Control Release 255:218–230

Jaspal D, Malviya A (2020) Composites for wastewater purification: a review. Chemosphere 246:125788

Jeevanandam J, Barhoum A, Chan YS, Dufresne A, Danquah MK (2018) Review on nanoparticles and nanostructured materials: history, sources, toxicity and regulations. Beilstein J Nanotechnol 9:1050–1074

Jiang BP, Zhou B, Lin Z, Liang H, Shen XC (2019) Recent advances in carbon nanomaterials for cancer phototherapy. Chem A Eur J 25:3993–4004

Jiménez-López J, Llorent-Martínez EJ, Ortega-Barrales P, Ruiz-Medina A (2020) Graphene quantum dots-silver nanoparticles as a novel sensitive and selective luminescence probe for the detection of glyphosate in food samples. Talanta 207:120344

Kalman J, Merino C, Fernández-Cruz ML, Navas JM (2019) Usefulness of fish cell lines for the initial characterization of toxicity and cellular fate of graphene-related materials (carbon nanofibers and graphene oxide). Chemosphere 218:347–358

Kalubowilage M, Covarrubias-Zambrano O, Malalasekera, AP, Wendel SO, Wang H, Yapa AS, Chlebanowski L, Toledo Y, Ortega R, Janik KE, Shrestha TB, Culbertson CT, Kasi A, Williamson S, Troyer DL, Bossmann SH (2018) Early detection of pancreatic cancers in liquid biopsies by ultrasensitive fluorescence nanobiosensors. Nanomed: Nanotechnol, Biol Med 14(6):1823–1832

Kaur P, Luthra R (2016) Silver nanoparticles in dentistry: an emerging trend. J Res Dent Sci 7:162

Kettiger H, Schipanski A, Wick P, Huwyler J (2013) Engineered nanomaterial uptake and tissue distribution: from cell to organism. Int J Nanomed 8:3255–3269

Kettler K, Veltman K, van de Meent D, van Wezel A, Hendriks AJ (2014) Cellular uptake of nanoparticles as determined by particle properties, experimental conditions, and cell type. Environ Toxicol Chem 33:481–492

Khezri K, Saeedi M, Maleki Dizaj S (2018) Application of nanoparticles in percutaneous delivery of active ingredients in cosmetic preparations. Biomed Pharmacother 106:1499–1505

Kibbey TCG, Strevett KA (2019) The effect of nanoparticles on soil and rhizosphere bacteria and plant growth in lettuce seedlings. Chemosphere 221:703–707

Kim HJ, Lee JY, Kim TH, Gwak GH, Park JH, Oh JM (2020) Radioisotope and anticancer agent incorporated layered double hydroxide for tumor targeting theranostic nanomedicine. Appl Clay Sci 186:105454

Krasnoslobodtsev AV, Torres MP, Kaur S, Vlassiouk IV, Lipert RJ, Jain M, Batra SK, Lyubchenko YL (2015) Nano-immunoassay with improved performance for detection of cancer biomarkers. Nanomed Nanotechnol Biol Med 11:167–173

Kritchenkov AS, Egorov AR, Dubashynskaya NV, Volkova OV, Zabodalova LA, Suchkova EP, Kurliuk AV, Shakola TV, Dysin AP (2019) Natural polysaccharide-based smart (temperature sensing) and active (antibacterial, antioxidant and photoprotective) nanoparticles with potential application in biocompatible food coatings. Int J Biol Macromol 134:480–486

Kulacki KJ, Cardinale BJ, Keller AA, Bier R, Dickson H (2012) How do stream organisms respond to, and influence, the concentration of titanium dioxide nanoparticles? a mesocosm study with algae and herbivores. Environ Toxicol Chem 31:2414–2422

Kurtoglu YE, Mishra MK, Kannan S, Kannan RM (2010) Drug release characteristics of PAMAM dendrimer-drug conjugates with different linkers. Int J Pharm 384:189–194

LaVan DA, McGuire T, Langer R (2003) Small-scale systems for in vivo drug delivery. Nat Biotechnol 21:1184–1191

Le Ouay B, Stellacci F (2015) Antibacterial activity of silver nanoparticles: a surface science insight. Nano Today 10:339–354

Leonhardt U (2007) Invisibility cup. Nat Photonics 1:207–208

Lichterfeld M, Qurishi N, Hoffmann C, Hochdorfer B, Brockmeyer NH, Arasteh K, Mauss S, Rockstroh JK (2005) Treatment of HIV-1-associated Kaposi's sarcoma with pegylated liposomal doxorubicin and HAART simultaneously induces effective tumor remission and CD4+ T cell recovery. Infection 33:140–147

Lin HC, Huang CL, Huang YJ, Hsiao IL, Yang CW, Chuang CY (2016) Transcriptomic gene-network analysis of exposure to silver nanoparticle reveals potentially neurodegenerative progression in mouse brain neural cells. Toxicol Vitr 34:289–299

Liu X, Tang J, Wang L, Liu Q, Liu R (2019) A comparative analysis of ball-milled biochar, graphene oxide, and multi- walled carbon nanotubes with respect to toxicity induction in Streptomyces. J Environ Manage 243:308–317

Loosli F, Vitorazi L, Berret JF, Stoll S (2015) Towards a better understanding on agglomeration mechanisms and thermodynamic properties of TiO_2 nanoparticles interacting with natural organic matter. Water Res 80:139–148

Loza K, Diendorf J, Sengstock C, Ruiz-Gonzalez L, Gonzalez-Calbet JM, Vallet-Regi M, Köller M, Epple M (2014) The dissolution and biological effects of silver nanoparticles in biological media. J Mater Chem B 2:1634–1643

Lv R, Yang P, Hu B, Xu J, Shang W, Tian J (2017) In situ growth strategy to integrate up-conversion nanoparticles with ultrasmall CuS for photothermal theranostics. ACS Nano 11:1064–1072

Lyu Z, Ding L, Huang AYT, Kao CL, Peng L (2019) Poly(amidoamine)dendrimers: covalent and supramolecular synthesis. Mater Today Chem 13:34–48

Mahmoodzadeh F, Abbasian M, Jaymand M, Salehi R, Bagherzadeh-Khajehmarjan E (2018) A novel gold-based stimuli-responsive theranostic nanomedicine for chemo-photothermal therapy of solid tumors. Mater Sci Eng C 93:880–889

Marambio-Jones C, Hoek EMV (2010) A review of the antibacterial effects of silver nanomaterials and potential implications for human health and the environment. J Nanoparticle Res 12:1531–1551

Marques MRC, Choo Q, Ashtikar M, Rocha TC, Bremer-Hoffmann S, Wacker MG (2019) Nanomedicines—tiny particles and big challenges. Adv Drug Deliv, Rev

McDonald SA, Konstantatos G, Zhang S, Cyr PW, Klem EJD, Levina L, Sargent EH (2005) Solution-processed PbS quantum dot infrared photodetectors and photovoltaics. Nat Mater 4:138–142

Miele Evelina, Spinelli GP, Miele Ermanno, Tomao F, Tomao S (2009) Albumin-bound formulation of paclitaxel (Abraxane® ABI-007) in the treatment of breast cancer. Int J Nanomed 4:99–105

Mignani S, El Kazzouli S, Bousmina MM, Majoral JP (2014) Dendrimer space exploration: An assessment of dendrimers/dendritic scaffolding as inhibitors of protein-protein interactions, a potential new area of pharmaceutical development. Chem Rev 114:1327–1342

Miller K, Erez R, Segal E, Shabat D, Satichi-Fainaro R (2009) Targeting bone metastases with a bispecific anticancer and antiangiogenic polymer-alendronate-taxane conjugate. Med Chem 48:2949–2954

Min Y, Akbulut M, Kristiansen K, Golan Y, Israelachvili J (2009) The role of interparticle and external forces in nanoparticle assembly. Nanosci Technol A Collect Rev Nat J 7:38–49

Mirakabad FST, Khoramgah MS, Keshavarz K, Tabarzad M, Ranjbari J (2019) Peptide dendrimers as valuable biomaterials in medical sciences. Life Sci 233:116754

Mirzajani F, Ghassempour A, Aliahmadi A, Esmaeili MA (2011) Antibacterial effect of silver nanoparticles on *Staphylococcus aureus*. Res Microbiol 162:542–549

Mohajeri M, Behnam B, Sahebkar A (2018) Biomedical applications of carbon nanomaterials: drug and gene delivery potentials. J Cell Physiol 234:298–319

Movafeghi A, Khataee A, Abedi M, Tarrahi R, Dadpour M, Vafaei F (2018) Effects of TiO_2 nanoparticles on the aquatic plant *Spirodela polyrrhiza*: evaluation of growth parameters, pigment contents and antioxidant enzyme activities. J Environ Sci (China) 64:130–138

Mroz P, Tegos GP, Gali H, Wharton T, Sarna T, Hamblin MR (2007) Photodynamic therapy with fullerenes. Photochem Photobiol Sci 6:1139–1149

Muhr V, Wilhelm S, Hirsch T, Wolfbeis OS (2014) Upconversion nanoparticles: from hydrophobic to hydrophilic surfaces. Acc Chem Res 47:3481–3493

Murdock RC, Braydich-Stolle L, Schrand AM, Schlager JJ, Hussain SM (2008) Characterization of nanomaterial dispersion in solution prior to in vitro exposure using dynamic light scattering technique. Toxicol Sci 101:239–253

Muthu MS, Leong DT, Mei L, Feng SS (2014) Nanotheranostics—application and further development of nanomedicine strategies for advanced theranostics. Theranostics 4:660–677

Nakata K, Fujishima A (2012) TiO_2 photocatalysis: design and applications. J Photochem Photobiol C Photochem Rev 13:169–189

Nam SH, An YJ (2019) Size- and shape-dependent toxicity of silver nanomaterials in green alga *Chlorococcum infusionum*. Ecotoxicol Environ Saf 168:388–393

Nance E (2019) Careers in nanomedicine and drug delivery. Adv Drug Deliv Rev 144:180–189

Narayanan R, El-sayed MA (2005) Catalysis with transition metal nanoparticles in colloidal solution: nanoparticle shape. J Phys Chem 109:12663–12676

Navashenaq JG, Zamani P, Nikpoor AR, Tavakkol-Afshari J, Reza M, Pharmd J (2020) Doxil chemotherapy plus liposomal P5 immunotherapy decreased myeloid-derived suppressor cells in murine model of breast cancer. Nanomed Nanotechnol Biol Med

Nel AE, Mädler L, Velegol D, Xia T, Hoek EMV, Somasundaran P, Klaessig F, Castranova V, Thompson M (2009) Understanding biophysicochemical interactions at the nano-bio interface. Nat Mater 8:543–557

Nogueira V, Lopes I, Rocha-Santos T, Santos AL, Rasteiro GM, Antunes F, Gonçalves F, Soares AMVM, Cunha A, Almeida A, Gomes NNCM, Pereira R (2012) Impact of organic and inorganic nanomaterials in the soil microbial community structure. Sci Total Environ 424:344–350

Nur Y, Lead JR, Baalousha M (2015) Evaluation of charge and agglomeration behavior of TiO_2 nanoparticles in ecotoxicological media. Sci Total Environ 535:45–53

Oukarroum A, Bras S, Perreault F, Popovic R (2012) Inhibitory effects of silver nanoparticles in two green algae, *Chlorella vulgaris* and *Dunaliella tertiolecta*. Ecotoxicol Environ Saf 78:80–85

Ozkan Y, Altinok I, Ilhan H, Sokmen M (2016) Determination of TiO_2 and $AgTiO_2$ nanoparticles in *Artemia salina*: toxicity, morphological changes, uptake and depuration. Bull Environ Contam Toxicol 96:36–42

Pardridge WM (2005) The blood-brain barrier: bottleneck in brain drug development. NeuroRx 2:3–14

Pariona N, Martinez AI, Hdz-garcı HM, Cruz LA (2017) Effects of hematite and ferrihydrite nanoparticles on germination and growth of maize seedlings. Saudi J Biol Sci 24:1547–1554

Parveen S, Misra R, Sahoo SK (2012) Nanoparticles: a boon to drug delivery, therapeutics, diagnostics and imaging. Nanomed Nanotechnol Biol Med 8:147–166

Patel KD, Singh RK, Kim H-WW (2019) Carbon-based nanomaterials as an emerging platform for theranostics. Mater Horizons 6:434–469

Pauluk D, Padilha AK, Khalil NM, Mainardes RM (2019) Chitosan-coated zein nanoparticles for oral delivery of resveratrol: formation, characterization, stability, mucoadhesive properties and antioxidant activity. Food Hydrocoll 94:411–417

Peng Z, Liu X, Zhang W, Zeng Z, Liu Z, Zhang C, Liu Y, Shao B, Liang Q, Tang W, Yuan X (2020) Advances in the application, toxicity and degradation of carbon nanomaterials in environment: a review. Environ Int 134:105298

Piccapietra F, Allue CG, Sigg L, Behra R (2012) Intracellular silver accumulation in *Chlamydomonas reinhardtii* upon exposure to carbonate coated silver nanoparticles and silver nitrate. Environ Sci Technol 46:7390–7397

Pikula K, Zakharenko A, Chaika V, Kirichenko K, Tsatsakis A, Golokhvast K (2020) Risk assessments in nanotoxicology: bioinformatics and computational approaches. Curr Opin Toxicol 19:1–6

Porche DJ (1996) Liposomal Doxorubicin (Doxil). Treat Rev 7:55–59

Principi E, Girardello R, Bruno A, Manni I, Gini E, Pagani A, Grimaldi A, Ivaldi F, Congiu T, De Stefano D, Piaggio G, de Eguileor M, Noonan DM, Albini A (2016) Systemic distribution of single-walled carbon nanotubes in a novel model: alteration of biochemical parameters, metabolic functions, liver accumulation, and inflammation in vivo. Int J Nanomed 11:4299–4316

Qiu H, Ye M, Zeng Q, Li W, Fortner J, Liu LL, Yang L (2019) Fabrication of agricultural waste supported UiO-66 nanoparticles with high utilization in phosphate removal from water. Chem Eng J 360:621–630

Quader S, Kataoka K (2017) Nanomaterial-enabled cancer therapy. Mol Ther 25:1501–1513

Radniecki TS, Stankus DP, Neigh A, Nason JA, Semprini L (2011) Influence of liberated silver from silver nanoparticles on nitrification inhibition of *Nitrosomonas europaea*. Chemosphere 85:43–49

Rasmussen JW, Martinez E, Louka P, Wingett DG (2010) Zinc oxide nanoparticles for selective destruction of tumor cells and potential for drug delivery applications. Expert Opin Drug Deliv 7:1063–1077

Rastogi A, Zivcak M, Sytar O, Kalaji HM, He X, Mbarki S, Brestic M (2017) Impact of metal and metal oxide nanoparticles on plant: a critical review. Front Chem 5:1–16

Ravishankar Rai V, Jamuna Bai A (2011) Nanoparticles and their potential application as antimicrobials. Formatex:197–209

Rezvani E, Rafferty A, Mcguinness C, Kennedy J (2019) Acta biomaterialia adverse effects of nanosilver on human health and the environment. Acta Biomater 94:145–159

Robichaud CO, Uyar AE, Darby MR, Zucker LG, Wiesner MR (2009) Estimates of upper bounds and trends in nano-TiO_2 production as a basis for exposure assessment. Environ Sci Technol 43:4227–4233

Roco MC (2004) Nanoscale science and engineering: unifying and transforming tools. Am Inst Chem Eng J 50:890–897

Roy LC, Wrana JL (2005) Clathrin- and non- Clatrhrin-mediated endocytic regulation of cell signalling. Nat Cell Biol 6:112–126

Rupar MJ, Golusinski P, Golusinski W, Masternak MM (2019) Human Papillomavirus and the use of nanoparticles for immunotherapy in HPV-related cancer: a review. Reports Pract Oncol Radiother 24:544–550

Sakamoto JH, van de Ven AL, Godin B, Blanco E, Serda RE, Grattoni A, Ziemys A, Bouamrani A, Hu T, Ranganathan SI, De Rosa E, Martinez JO, Smid CA, Buchanan RM, Lee SY, Srinivasan S, Landry M, Meyn A, Tasciotti E, Liu X, Decuzzi P, Ferrari M (2010) Enabling individualized therapy through nanotechnology. Pharmacol Res 62:57–89

Sanhai WR, Sakamoto JH, Canady R, Ferrari M (2008) Seven challenges for nanomedicine. Nat Nanotechnol 3:242–244

Santos V, Ribeiro APB, Santana MHA (2019) Solid lipid nanoparticles as carriers for lipophilic compounds for applications in foods. Food Res Int 122:610–626

Schaming D, Remita H (2015) Nanotechnology: from the ancient time to nowadays. Found Chem 17:187–205

Schulz PA (2018) Há mais história lá embaixo - um convite para rever uma palestra. Rev. Bras, Ensino Física, p 40

Selvaraj C, Sakkiah S, Tong W, Hong H (2018) Molecular dynamics simulations and applications in computational toxicology and nanotoxicology. Food Chem Toxicol 112:495–506

Sendra M, Blasco J, Araújo CVM (2018) Is the cell wall of marine phytoplankton a protective barrier or a nanoparticle interaction site? Toxicological responses of Chlorella autotrophica and Dunaliella salina to Ag and CeO_2 nanoparticles. Ecol Indic 95:1053–1067

Servin AD, Morales MI, Castillo-Michel H, Hernandez-Viezcas JA, Munoz B, Zhao L, Nunez JE, Peralta-Videa JR, Gardea-Torresdey JL (2013) Synchrotron verification of TiO2 accumulation in cucumber fruit: a possible pathway of TiO_2 nanoparticle transfer from soil into the food chain. Environ Sci Technol 47:11592–11598

Shan SJ, Zhao Y, Tang H, Cui FY (2017) A mini-review of carbonaceous nanomaterials for removal of contaminants from wastewater. IOP Conf Ser Earth Environ Sci 68:012003

Shim M, Kam NWS, Chen RJ, Li Y, Dai H (2002) Functionalization of carbon nanotubes for biocompatibility and biomolecular recognition. Nano Lett 2:285–288

Skoupilova H, Bartosik M, Sommerova L, Pinkas J, Vaculovic T, Kanicky V, Karban J, Hrstka R (2020) Ferrocenes as new anticancer drug candidates: determination of the mechanism of action. Eur J Pharmacol 867:172825

Soares S, Sousa J, Pais A, Vitorino C (2018) Nanomedicine: principles, properties, and regulatory issues. Front Chem 6:1–15

Son J, Yi G, Yoo J, Park C, Koo H, Choi HS (2019) Light-responsive nanomedicine for biophotonic imaging and targeted therapy. Adv Drug Deliv Rev 138:133–147

Sondi I, Salopek-Sondi B (2004) Silver nanoparticles as antimicrobial agent: a case study on *E. coli* as a model for gram-negative bacteria. J Colloid Interface Sci 275:177–182

Steinman RM, Mellman IS, Muller WA, Cohn ZA (1983) Endocytosis and the recycling of plasma membrane. J Cell Biol 96:1–27

Swain B, Park JR, Park KS, Lee CG (2019) Synthesis of cosmetic grade TiO_2-SiO_2 core-shell powder from mechanically milled TiO_2 nanopowder for commercial mass production. Mater Sci Eng C 95:95–103

Thakur K, Kandasubramanian B (2019) Graphene and graphene oxide-based composites for removal of organic pollutants: a review. J Chem Eng Data 64:833–867

Tinkle S, McNeil S, Muhlenbach S, Bawa R, Borchard G, Beranholz Y, Tamarkin L, Desai N (2014) Nanomedicines: addressing the scientific and regulatory gap. Ann New York Acad Sci 1313:1–41

Tombuloglu H, Tombuloglu G, Slimani Y, Ercan I, Sozeri H, Baykal A (2018) Impact of manganese ferrite ($MnFe_2O_4$) nanoparticles on growth and magnetic character of barley (*Hordeum vulgare* L.). Environ Pollut 243:872–881

Torrent L, Marguí E, Queralt I, Hidalgo M, Iglesias M (2019) Interaction of silver nanoparticles with mediterranean agricultural soils: lab-controlled adsorption and desorption studies. J Environ Sci (China) 83:205–216

Tosco T, Sethi R (2018) Human health risk assessment for nanoparticle-contaminated aquifer systems. Environ Pollut 239:242–252

Tran S, DeGiovanni P-J, Piel B, Rai P (2017) Cancer nanomedicine: a review of recent success in drug delivery. Clin Transl Med 6:1–21

Tripathi DK, Singh S, Singh S, Swati, Pandey R, Singh VP, Sharma NC, Prasad SM, Dubey NK, Chauhan DK (2017) An overview on manufactured nanoparticles in plants: uptake, translocation, accumulation and phytotoxicity. Plant Physiol Biochem 110:2–12

Unuabonah EI, Günter C, Weber J, Lubahn S, Taubert A (2013) Hybrid clay: a new highly efficient adsorbent for water treatment. ACS Sustain Chem Eng 1:966–973

Vance ME, Kuiken T, Vejerano EP, McGinnis SP, Hochella MF, Hull DR (2015) Nanotechnology in the real world: redeveloping the nanomaterial consumer products inventory. Beilstein J Nanotechnol 6:1769–1780

Vannini C, Domingo G, Onelli E, De Mattia F, Bruni I, Marsoni M, Bracale M (2014) Phytotoxic and genotoxic effects of silver nanoparticles exposure on germinating wheat seedlings. J Plant Physiol 171:1142–1148

Von Hippel AR (1956) Molecular engineering. Science 123:315–317

von Roemeling C, Jiang W, Chan CK, Weissman IL, Kim BYS (2017) Breaking down the barriers to precision cancer nanomedicine. Trends Biotechnol 35:159–171

Walter P, Welcomme E, Hallégot P, Zaluzec NJ, Deeb C, Castaing J, Veyssière P, Bréniaux R, Lévêque JL, Tsoucaris G (2006) Early use of PbS nanotechnology for an ancient hair dyeing formula. Nano Lett 6:2215–2219
Wang P, Zhang B, Zhang H, He Y, Ong CN, Yang J (2018) Metabolites change of Scenedesmus obliquus exerted by AgNPs. J Environ Sci (China)76:1–9
Watson JD, Crick FHC (1953) Molecular structure of nucleic acids. a structrue for deoxyribose nucleic acid. Nature 171:737–738
Weissleder R, Pittet MJ (2008) Imaging in the era of molecular oncology. Nature 452:580–589
Whitehead KA, Langer R, Anderson DG (2009) Knocking down barriers: advances in siRNA delivery. Nat Rev Drug Discov 8:129–138
Wolfram J, Ferrari M (2019) Clinical cancer nanomedicine. Nano Today 25:85–98
Wu L-P, Wang D, Li Z (2020a) Grand challenges in nanomedicine. Mater Sci Eng C 106:1–7
Wu Y, Xia Y, Jing X, Cai P, Igalavithana AD, Tang C, Tsang DCW, Ok YS (2020b) Recent advances in mitigating membrane biofouling using carbon-based materials. J Hazard Mater 382:120976
Xiu ZM, Zhang QB, Puppala HL, Colvin VL, Alvarez PJJ (2012) Negligible particle-specific antibacterial activity of silver nanoparticles. Nano Lett 12:4271–4275
Xu Hengyi, Qu F, Xu Hong, Lai W, Wang YA, Aguilar ZP, Wei H (2012) Role of reactive oxygen species in the antibacterial mechanism of silver nanoparticles on *Escherichia coli* O157:H7. Biometals 25:45–53
Yadav T, Mungray AA, Mungray AK (2016) Effect of multiwalled carbon nanotubes on UASB microbial consortium. Environ Sci Pollut Res 23:4063–4072
Yakar Y, Çakır B, Özmen A (2020) Polarizability and electric field gradient of two-electron quantum dots. J Phys Chem Solids 137
Yao J, Wang H, Chen M, Yang M (2019) Recent advances in graphene-based nanomaterials: properties, toxicity and applications in chemistry, biology and medicine. Microchim, Acta, p 186
Yu S, Yin Y, Liu J (2013) Silver nanoparticles in the environment. Silver Nanoparticles Environ 15:78–92
Yuan Z, Li J, Cui L, Xu B, Zhang H, Yu CP (2013) Interaction of silver nanoparticles with pure nitrifying bacteria. Chemosphere 90:1404–1411
Zhang H, Wang X, Chen C, An C, Xu Y, Dong Y, Zhang Q, Wang Y, Jiao L, Yuan H (2016a) Facile synthesis of diverse transition metal oxide nanoparticles and electrochemical properties. Inorg Chem Front 3:1048–1057
Zhang J, Ma Y, Li N, Zhu J, Zhang T, Zhang W, Liu B (2016b) Preparation of graphene quantum dots and their application in cell imaging. J Nanomater 31:337–344
Zhang L, Dong H, Zhu Y, Zhang J, Zeng G, Yuan Y, Cheng Y, Li L, Fang W (2019) Evolutions of different microbial populations and the relationships with matrix properties during agricultural waste composting with amendment of iron (hydr)oxide nanoparticles. Bioresour Technol 289:121697
Zhang L, He Y, Goswami N, Xie J, Zhang B, Tao X (2016c) Uptake and effect of highly fl uorescent silver nanoclusters on Scenedesmus obliquus. Chemosphere 153:322–331
Zhang X-Q, Xu X, Bertrand N, Pridgen E, Swami A, Farokhzad OC (2012) Interactions of nanomaterials and biological systems: implications to personalized nanomedicine. Adv Drug Deliv Rev 64:1363–1384
Zhao G, Wang ZY, Xu L, Xia CX, Liu JX (2019) Silver nanoparticles induce abnormal touch responses by damaging neural circuits in zebrafish embryos. Chemosphere 229:169–180
Zhen Z, Xie J (2012) Development of manganese-based nanoparticles as contrast probes for magnetic resonance imaging. Theranostics 2:45–54
Zhong Y, Munir R, Balawi AH, Sheikh AD, Yu L, Tang MC, Hu H, Laquai F, Amassian A (2016) Mesostructured fullerene electrodes for highly efficient n-i-p perovskite solar cells. ACS Energy Lett 1:1049–1056

Zhou B, Li Y, Zheng G, Dai K, Liu C, Ma Y, Zhang J, Wang N, Shen C, Guo Z (2018) Continuously fabricated transparent conductive polycarbonate/carbon nanotube nanocomposite films for switchable thermochromic applications. J Mater Chem C 6:8360–8371

Zhu S, Zhang J, Qiao C, Tang S, Li Y, Yuan W, Li B, Tian L, Liu F, Hu R, Gao H, Wei H, Zhang H, Sun H, Yang B (2011) Strongly green-photoluminescent graphene quantum dots for bioimaging applications. Chem Commun 47:6858–6860

Chapter 13
Creating Smart and Functional Textile Materials with Graphene

J. H. O. Nascimento, B. H. S. Felipe, J. M. T. C. Dias, A. G. F. Souza, A. P. S. Júnior, F. M. F. Galvão, R. L. B. Cabral, B. R. Carvalho, J. P. S. Morais, and Awais Ahmad

The research and development of graphene-based materials are happening at an intense pace due to their extraordinary physical and chemical properties. Their high electronic mobility, high thermal conductivity, mechanical properties, photoluminescence, among others, made them a wonder material in several research fields. The textile community is aware of this evolution and, therefore, is also taking advantage of the properties of these graphene-based materials for the development of textiles

J. H. O. Nascimento (✉) · B. H. S. Felipe
Programa de Pós Graduação em Engenharia Mecânica, Universidade Federal do Rio Grande do Norte, Centro de Tecnologia, Natal, Rio Grande do Norte, Brazil
e-mail: joseheriberto@ct.ufrn.br

J. H. O. Nascimento · J. M. T. C. Dias · F. M. F. Galvão · R. L. B. Cabral
Programa de Pós Graduação em Engenharia Têxtil, Centro de Tecnologia, Universidade Federal do Rio Grande do Norte, Av. Sen. Salgado Filho, 3000, Natal, Rio Grande do Norte 59072-970, Brazil

J. H. O. Nascimento · B. H. S. Felipe · J. M. T. C. Dias · A. G. F. Souza · A. P. S. Júnior · F. M. F. Galvão · R. L. B. Cabral
Grupo de Pesquisa de Inovação em Micro e Nanotecnologias, Centro de Tecnologia, Universidade Federal do Rio Grande do Norte, Natal, Rio Grande do Norte, Brazil

B. R. Carvalho
Departamento de Física, Universidade Federal do Rio Grande do Norte, Natal, Rio Grande do Norte, Brazil

J. P. S. Morais
Embrapa Algodão: Campina Grande, Campina Grande, Paraíba, Brazil

A. Ahmad
Department of Chemistry, University of Lahore, Lahore 54590, Pakistan

Department of Applied Chemistry, Government College University Faisalabad, Faisalabad 38000, Pakistan

R. F. do. Nascimento et al. (eds.), *Nanomaterials and Nanotechnology*,
Materials Horizons: From Nature to Nanomaterials,
https://doi.org/10.1007/978-981-33-6056-3_13

with functionalities that include UV protection, antistatic, antibacterial, photoluminescent finishes, improvement of mechanical properties, flexible supercapacitors, sensors, etc. In this context, this chapter aims to address the main concepts, applications, and perspectives on the application of graphene-based material in the textile area. We will mainly focus on the progressing research using graphene oxide (GO), reduced graphene oxide (rGO), and graphene quantum dots (GQDs).

1 The Main Materials Based on Graphene

Graphene's history starts in 1947 by Philip Wallace by developing a theory to describe its electronic band structure. Landau and Peierls argued that 2D crystals could not exist since it was thermodynamically unstable. Although some experimental works on 'graphite' oxide were already being in progress, it was only in 2004 by Geim and Novoselov that a one-atom thick of 2D graphite was isolated. This came to be known as graphene where the prefix 'graph' (derived from graphite) and the suffix 'ene' (derived from polycyclic aromatic hydrocarbons, such as naphthalene and anthracene). Therefore, graphene consists of a monolayer of sp^2 hybridized carbon atoms packed in a 2D hexagonal structure (Inagaki et al. 2014). Later on, Geim and Novoselov received the Nobel Prize in Physics in 2010 for graphene's isolation and outstanding applications (Bassalo and Farias 2017).

Graphene in its pure form has remarkable properties such as fracture strength with about 200 times greater than that of structural steel (130 GPa), high values of its Young's modulus ($\sim$1 TPa) (Lee et al. 2008), thermal conductivity five times greater than diamond (~5000 W $(m K)^{-1}$) (Khanafer and Vafai 2017), high mobility value of cargo conveyors (~200,000 $cm^2 (V s)^{-1}$) (Bolotin et al. 2008), high transparency for a single-layer film (97.7% transmittance) (Cole et al. 2015), complete impermeability to any gases (Bunch et al. 2008; Yu et al. 2015), and hydrophobicity (Munz et al. 2015). The main synthesis routes are based on silicon carbide thermal decomposition, graphite exfoliation, and chemical vapor deposition processes (Abbasi et al. 2016).

Graphene oxide (GO), one of the most important graphene derivatives, has many functional groups containing oxygen in its basal plane and at the edges in the form of epoxy, hydroxyl, and carboxyl groups, in addition to a high index of defects in the carbon structure. Thus, this material has its properties distinct from pure graphene, such as the electrical insulating and hydrophilic behavior, which allows the formation of suspensions in aqueous medium. The main synthesis techniques are based on the controlled exfoliation of graphite.

Reduced graphene (rGO) is produced by GO reduction reactions. The resultant product of this reaction is a material of carbonic structure with low oxygen index and defects in the basal plane of graphene. So far, it is not possible to state that rGO is completely oxygen free, as functional oxygen groups form highly stable carbonyl and ether groups that are difficult to remove, so that rGO contains a residual portion of oxygen (between 15 to 25%) (Bagri et al. 2010). Even so, rGO becomes an alternative material to pure graphene, at a lower cost, but its electronic and mechanical properties have substantially reduced values when compared (Voiry et al. 2016).

Graphene quantum dots (GQDs) are small fragments of graphene, where the effect of the confinement of excitons in the three spatial dimensions is observed. Usually, the effect occurs when the particles are less than 30 nm in size. Consequently, GQDs have a band gap value greater than zero and depend on the synthesis method, dopants, and average particle size. They present good conductivity; high chemical stability; environmentally friendly; low toxicity; and strong photoluminescence emission. In addition, they can be easily obtained and low in cost production, depending on the synthesis method (Bacon et al. 2014). Carboxylic functional groups can be present on the surface of GQDs, resulting in hydrophilic and biocompatible characteristics. GQDs are also suitable for surface passivation and chemical modification with various polymeric, inorganic, organic, or biological materials, which can result in new or improved properties for certain applications (Poortavasoly et al. 2014; Qu et al. 2015).

Doping of GQDs with heteroatoms leads to better efficiency of properties such as photoluminescence and catalytic activities for oxidation–reduction reactions, which are a prerequisite for energy storage/conversion. The introduction of heteroatoms (e.g., boron, nitrogen, oxygen, phosphorus, and sulfur) into the carbon network can properly disorient the electron network, which is homogeneously conjugated. This network regulates the properties of the surface by adjusting the load distribution of doped domains. For example, in an N-GQD (quantum dot of graphene doped with nitrogen), the electron states of the network can be intrinsically altered by *N*-doping and, subsequently, give the nanoparticle a greater photoluminescent activity (Kaur et al. 2018).

A comparison between the characteristics of the main classes of graphene-based materials (GO, rGO, and GQDs) is presented in Table 1.

A survey of the last ten years was carried out with the objective of obtaining quantitative data from patent registrations and publications in scientific journals worldwide covering graphene, graphene oxide, reduced graphene oxide, graphene quantum dots involving textile materials, be they fibers, yarns, knits, woven, and non-woven.

The survey of the number of patents was carried out through the WIPO website in the Field Combination bar in search option. In similar algorithmic writing, the keyword graphene and codes equivalent to textile materials were inserted in the Coverage Page tab: D01—natural or synthetic fibers or yarns; spinning; D02—yarns; mechanical finishing of yarns or ropes; D03—weaving; D04—braiding; lace-making; knitting; trimmings; non-woven fabrics; D05—sewing; embroidering; tufting; D06—textile treatment or finishing; laundering; flexible materials not otherwise provided for; D07—ropes; cables other than electric; D10—indexing scheme associated with subclasses of section d, relating to textiles.

To obtain data from publications in scientific journals, the metric system of the Web of Science platform was used, inserting the keywords in the search tab graphene and fabrics.

Table 1 A comparison of the characteristics of graphene and its derivatives (GO, rGO, and GQDs)

	Graphene	Graphene oxide (GO)	Reduced graphene oxide (rGO)	Graphene quantum dots (GQDs)
Typical molecular structure	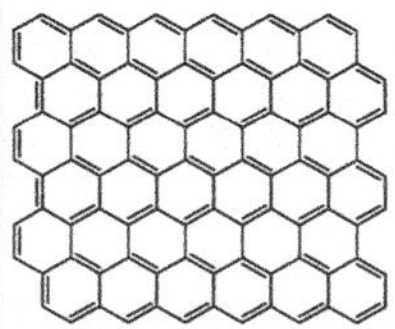	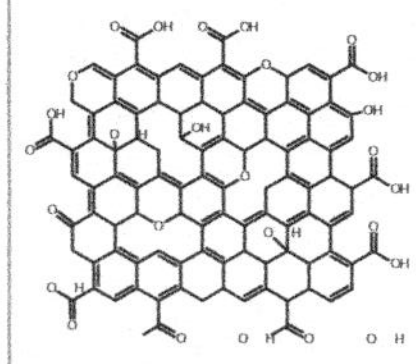	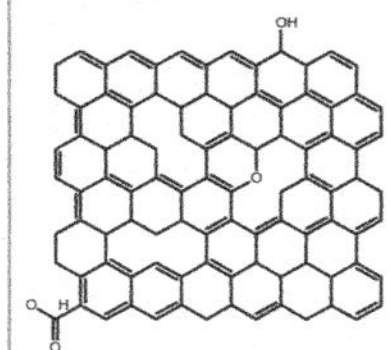	
Main synthesis routes	Thermal decomposition of SiC; Chemical vapor deposition; Graphite exfoliation	Controlled exfoliation of graphite	Reduction of graphene oxide	Controlled carbonization of organic compounds (e.g., citric acid, cellulose); breaking of larger carbon-based structures (e.g., graphite, activated carbon)
C:O ration	There is no oxygen.	High	Low	Variable
Size	High	>30 nm	>30 nm	≤30 nm
Doping	B and N	B, N, O, P, and S	B, N, P and S	B, N, O, P, and S
References	Rao et al. (2014), Perreault et al. (2015), Yi and Shen (2015), Wang et al. (2016), Chen et al. (2017), Yazdi et al. (2018)	Putri et al. (2015), Muzyka et al. (2017)	Abdolhosseinzadeh et al. (2015), Bag et al. (2015), Thakur and Karak (2015), Yuan et al. (2016), Guex et al. (2017)	Ogi et al. (2014), Lim et al. (2015), Kaur et al. (2018), Chen et al. (2019)

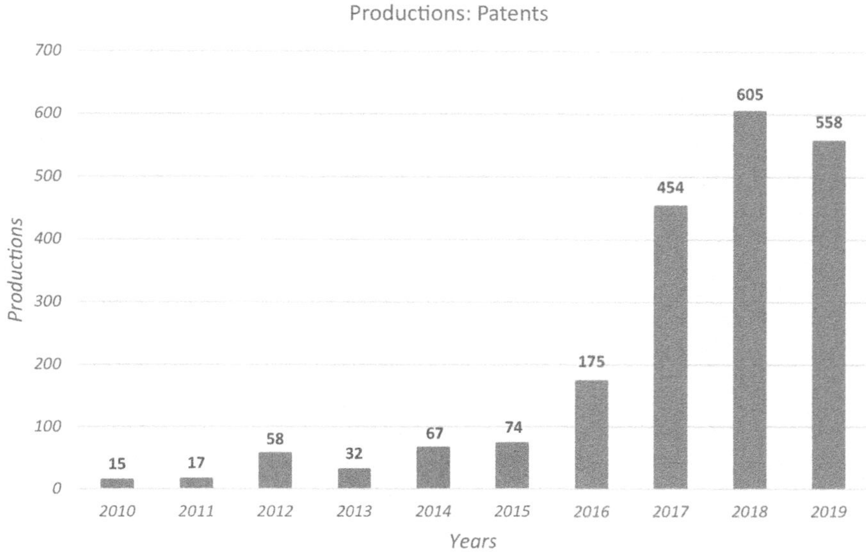

Fig. 1 Graphene Patents registered in the last ten years worldwide. Data from WIPO (World Intellectual Property Organization)

Figure 1 presents the graph containing data obtained through the WIPO portal (World Intellectual Property Organization), where it shows that in the last ten years there has been a marked increase in the registration of patents with graphene. Only the year 2016 registered approximately the same number of patents when compared to the first five years. In the year 2017, there was an increase close to 62% compared to the previous year. In turn, 2019 saw a slight reduction of approximately 8% compared to 2018, which in turn led to 605 registrations.

In Fig. 2, one can observe the number of scientific articles published in the last ten years worldwide with the term graphene applied to textile materials. It is observed that the data corroborate with the data from patent registrations; that is, there has been a significant increase in the last five years, especially in the year 2017, when there was a jump of approximately 46% in scientific publications when compared with the year 2016. It is also observed that the year 2019 was responsible for more publications than the sum of the years 2010 to 2016, making it a prominent year.

Figure 3 shows the registration of patents by countries containing graphene applied to textile materials in the last ten years. It is observed that the sum of patent registrations from all countries in this period does not equal 10% of the total registrations in China, which alone corresponds to 1.684 registrations followed by the USA with 130 registrations.

When checking by region of deposits shown in Fig. 4, the Patent Corporation Treaty (PCT) and the Eurasian Patent Organization (EPO) are included, which are independent centers for patent registrations, but together account for only 7% of the total registration number.

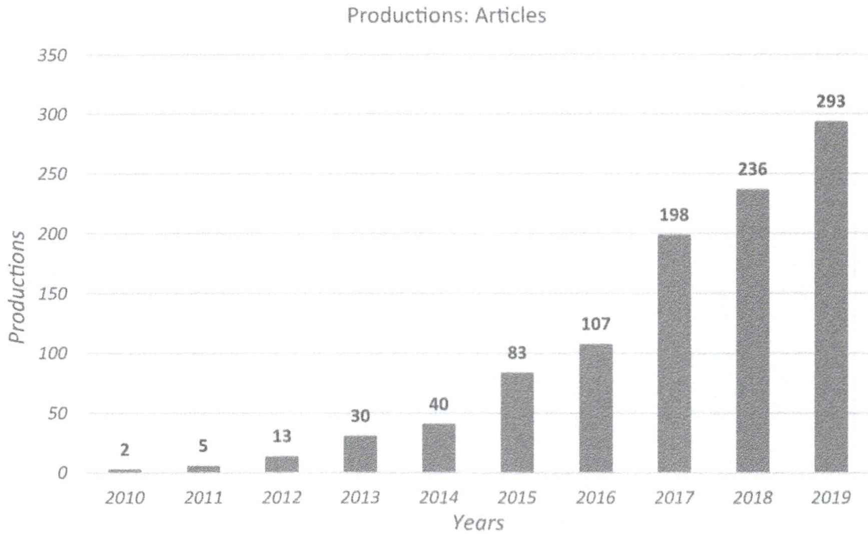

Fig. 2 Scientific articles published in the last ten years worldwide with the term graphene applied to textile materials. Data from the Web of Science

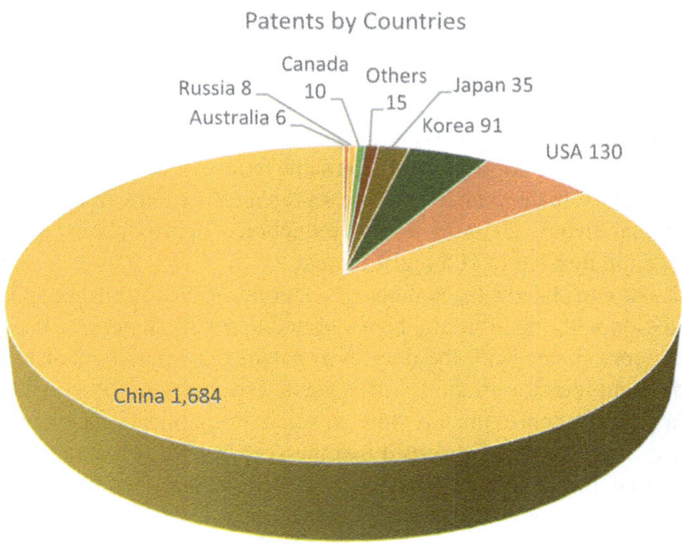

Fig. 3 Patents registered by countries with graphene applied to textile materials in the last ten years. Data from WIPO (World Intellectual Property Organization)

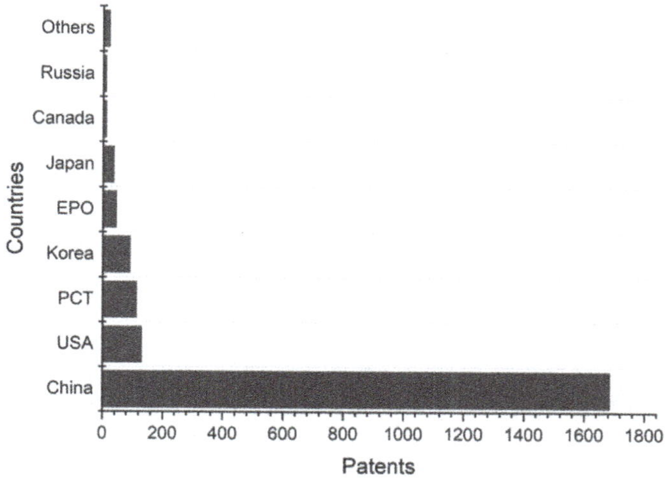

Fig. 4 Patents filed by country with graphene on textile materials in the last ten years. Data from WIPO (World Intellectual Property Organization)

Figure 5 shows the data of publications of articles in scientific journals divided by countries. As with the registration of patents, China stands out positively in comparison with other countries. This country alone accounts for 56% of all scientific publications in the past ten years, followed by the USA with only 9% of all publications.

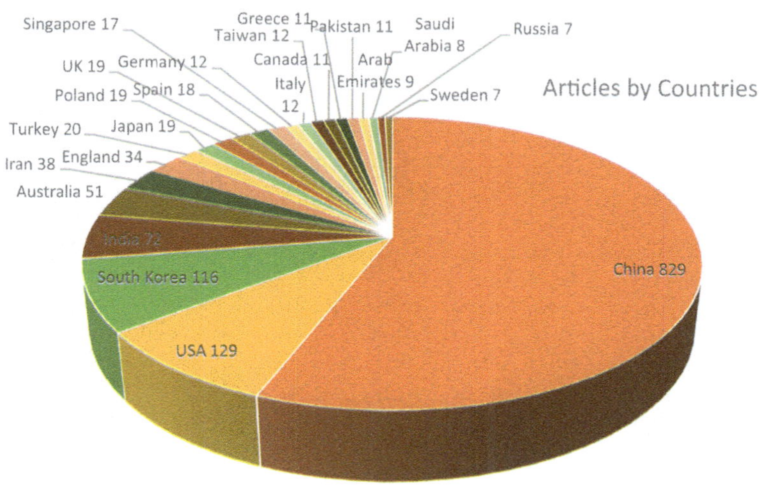

Fig. 5 Scientific articles by country with graphene applied to textile materials. Data from Web of Science

2 The Applications of Graphene in the Textile Area

Due to the excellent properties that are involved with graphene and its derivatives, there is a great development of research and patents applied to the most diverse areas of technology with the use of these materials. The textile industry aims to take advantage of these materials to improve functionality or incorporate new characteristics to fibers, yarns, and fabrics.

As shown in Table 2, each graphene structure can result in characteristics that can be used in specific applications in textile materials. For applications involving the development of sensors, structures with conductivity or capacitance, pure graphene or rGO are widely used. For self-cleaning treatments, rGO can be used both to enhance the photocatalytic activity of semiconductors (such as ZnO and TiO_2) and to make the fabric liquid-repellent. Pure graphene is also very hydrophobic; however, in most cases, the use of rGO is prioritized for this application due to the greater ease of obtaining. Antimicrobial treatments have already been applied to obtain this GO and rGO property associated with semiconductors. Graphene and its derivatives normally

Table 2 Applications of graphene and derivatives on textile materials

Application	Graphene	GO	rGO	GQDs	References
Photoluminescence	x	x	x	xx	Pang et al. (2018)
Self-cleaning/photocatalysis	x	x	xx	x	Kumbhakar et al. (2018)
Conductivity	xxx	x	xxx	x	Shateri-Khalilabad and Yazdanshenas (2013), Karimi et al. (2014), Cai et al. (2017), Zhao et al. (2018)
Capacitance	xxx	x	xxx	x	Shao et al. (2016), Sun et al. (2016), Yang et al. (2017)
Flame retardancy	x	xx	xx	xx	Huang et al. (2012), Some et al. (2015), Khose et al. (2018), Ji et al. (2019)
UV protection	xx	xxx	xxx	xx	Tian et al. (2015), Cai et al. (2017), Pan et al. (2018), Zhao et al. (2018), Ji et al. (2019)
Water repellence	xxx	xx	xxx	x	Shateri-Khalilabad and Yazdanshenas (2013), Tissera et al. (2015), Cai et al. (2017), Yang et al. (2017), Masae (2018), Zhao et al. (2018)
Antimicrobial	x	xx	xx	x	Zhao et al. (2013), Sun et al. (2014), Masae (2018), Pan et al. (2018), Stan et al. (2019)
Flexible LEDs	xx	x	x	x	Lee et al. (2017)

Source Author
Subtitle x = not very relevant, xx = reasonably relevant, xxx = very relevant

have a high absorption of radiation in the ultraviolet range of the electromagnetic spectrum, which makes them excellent materials with properties to increase the UV protection of fabrics. Treatments that involve flame repellency may be involved with GO, rGO, or GQDs, often in synergy with other components with repellent functions. There are studies that prove the antimicrobial activity of GQDs; however, few studies have been applied to fabrics (Ristic et al. 2014; Kuo et al. 2017; Wong 2018). Photoluminescent textiles can be obtained through treatment with GQDs; however, there are also few studies that apply this property to fabrics despite the good property of the material (Lin et al. 2014). There are several ways to apply graphene and its derivatives to textile materials. In this chapter, applications involving depletion, dip coating, layer-by-layer, spray coating, inkjet print, and extrusion printing will be discussed.

2.1 Exhaustion

In the beginning of exhaustion, a certain amount of textile material is loaded into the machine and brought into equilibrium with a solution containing chemicals, such as dyes and textile auxiliaries, for a period of minutes to hours. The exhaustion process involves the adsorption and absorption of dyes and textile auxiliaries from dye baths (or any liquor) due to the substantivity of the chemicals to the textile materials. Many textile materials, such as fibers, yarns, knitted, and fabrics, can be finished by exhaustion process. In general, the stages of an exhaustion process for fabrics include pretreatments, such as desizing, scouring, bleaching, and mercerization. The steps can be followed by rinsing and washing several times, depending on the concentration, solubility, and affinity of the chemical not fixed on the textile fabric (Cay et al. 2009). Usually, the use of water is much higher when compared to techniques such as dip coating or spray coating.

Exhaustion is one of the most widely or partially used methods to immobilize graphene and its derivatives on textile materials due to the ease and versatility of the technique. Miankafshe et al. (2019) used the technique to pre-modify PET fabric and then apply rGO to obtain a textile with suitable conductive properties. The pre-modification of the polyester initially consisted of an alkalization from a sodium hydroxide bath (1:50/60 °C/90 min). Then, the fabric was washed with hot water and acetone to remove remaining impurities and was oven-dried. The surface load of the fabric was modified by immersion in a solution of chitosan or hexadecylpyridinium chloride, which was stirred at room temperature for 24 h. The fabric was washed and dried. The fabric sample with the modified surface was treated by immersion in an aqueous dispersion of GO for 30 min/40 °C and then dried in an oven. Finally, the fabric functionalized with GO was chemically reduced using an aqueous solution of sodium hydrosulfite (1:100/30 min/95 °C), and the samples were rinsed and dried.

Afroj et al. (2019) applied the exhaustion process to dye textile threads with graphene-based inks. Thus, these threads will be suitable for multifunctional applications in next-generation electronic textiles, that is, wearable devices associated with health, sports, fashion, household items, and military items. To simulate an ultra-fast, industrial yarn dyeing process, a laboratory-scale Mathis LABOMAT dyeing machine was used to dye (coat) ~2 g of bleached cotton yarn using a batch dyeing technique. A 1:50 ratio of liquor materials (M: L) was used to dye a batch of cotton yarn with a rGO ink at 60 °C for 30 min. The batch of yarn dyed with rGO was subsequently dried at 100 °C for 15 min. This process can potentially be expanded in a high-speed industrial facility to produce tons (~1000 kg/h) of electroconductive textile yarns for next-generation wearable electronic applications.

2.2 Dip Coating

Dip coating is one of the most used processes for dyeing, printing, or applying various chemical finishes on textile materials such as fabrics, knits, yarns, and non-woven. In a typical process, the substrate is immersed in a chemical solution and then squeezed. During squeezing, part of the solution is forced in the opposite direction and returns to the bath tank, another part is forced into the fabric, and a small portion is dragged over the substrate. Then, the textile material is subjected to controlled heating processes, so that the curing of the finish and drying occurs. It is important to control the time and temperature at this stage to avoid the formation of stains on the fabric. The fabric is then washed to remove residual chemicals and dried. Dip coating is very advantageous industrially as there is the possibility of adding other finishing treatments to the production line continuously, such as pre-treatments, washing, and drying processes. In this way, the processes become more compact, faster, and cheaper. In addition, dip coating provides considerable water savings, as it is necessary to use solutions with low volumes and high concentrations. Figure 6 outlines the typical principle of a continuous dip-coating process using industrially to treat textile materials.

Various chemical and physical parameters must be controlled in a dip-coating process to result in an efficient treatment for textile materials, that is, with good uniformity, pick-up values, and optimized solids loading for a good cost-benefit ratio. During immersion and coating, the thickness and morphology of the deposition of

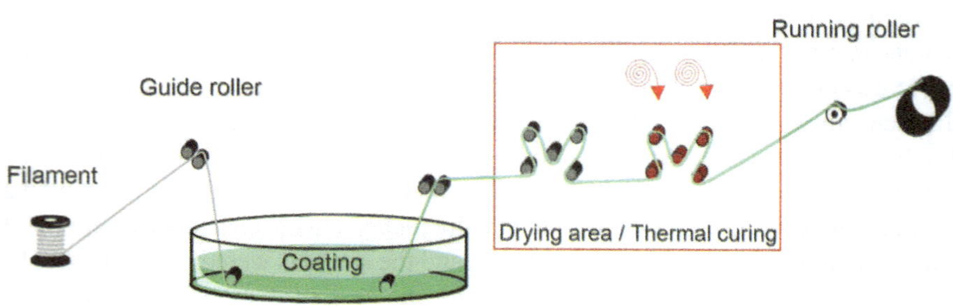

Fig. 6 Typical scheme of a continuous dip coating using industrial process. *Source* Author

thin films are determined by parameters, such as immersion time, withdrawal speed, immersion coating cycles, density and viscosity, surface tension, substrate surface and substrate conditions, evaporation of coating solutions, structural conditions of the textile substrate used (e.g., weight and pattern), chemical composition of the substrate, and pre-treatments carried out.

Several studies have been developed to understand the mechanisms that involve dip-coating processes and theoretical formulas have been established to model mechanisms that involve dip-coating processes (Tang and Yan 2017), such as the famous Landau–Levich theory, which predicts the thickness of the deposited films through the following equation:

$$t_1 = 0.944 C_a^{\frac{1}{6}} \left(\frac{\eta U}{\rho g} \right)^{1/2}$$

where C_a is the capillary number given by $C_a = \eta U/\delta$; η, δ, and ρ denote viscosity, surface tension, and density of the coating solutions, respectively; U is the withdrawal speed; g is the gravitational acceleration constant.

Groenveld built a model to estimate the thickness that emphatically considered the flow (J) using the following equation (Tang and Yan 2017):

$$t_1 = J \left(\frac{\eta U}{\rho g} \right)^{1/2}$$

Note that both equations are valid for $C_a < 10^{-3}$ and for fluid Newtonian.

The dip-coating technique is widely used to immobilize graphene and its derivatives on textile materials. Stan et al. (2019) used the technique to perform a treatment on cotton fabric based on TiO_2 and rGO to obtain self-cleaning and antimicrobial properties. First, powder dispersions based on rGO and TiO_2 were synthesized. Then, cotton fabrics were immersed in the dispersions prepared above, kept at 40 °C for 30 min, and then dried at 100 °C for 2 min. The fabrics were immersed in perfluorocarbon polymer, dried at 110 °C, and thermoset at 170 °C for 40 s. Gan et al. (2015) used the technique to immobilize graphene nanoribbon on cotton fabric to obtain conductive properties, in addition to an improvement in thermal stability and mechanical properties (tensile stress and Young's modulus). For the development of this material, an aqueous dispersion of graphene nanoribbon was prepared. A cotton cloth was dipped in the solution, removed, and dried. This cycle was repeated several times until the resistance of the cotton fabric became constant.

Polyamide monofilaments of 3000 tex were the object of studies by Tas et al. (2019) with the purpose of nanocoating with graphene oxide using the dip-coating method. Soon after, the graphene layer was reduced by green reduction method using ascorbic acid. It was observed that samples coated with 7 layers of rGO presented a sheet resistance of 3.09 kΩ/sq and that the number of graphene oxide coatings adhering to the surface of the filaments is the most important parameters that affect post-reduction conductivity. It was determined that the contact angles obtained on polyamide monofilaments nanocoated with graphene oxide are superior to the uncoated samples, thus decreasing the hydrophilic properties of the monofilaments. Thus, their optical properties were also altered with the increase in the number of nanocoatings, there was a reduction of about 40% of light transmission, and with this, it is believed that these structures can be used as electrodes in solar cells in the form of fibers, as well as sensors, actuators, and supercapacitors.

In addition to the number of nanocoatings on textile substrate by dip coating with oxidized or reduced graphene oxide being important in the results of sheet resistance, Wang et al. (2020) studied the influence of textile structure as fabric, knitting, or non-woven on the properties of capacitive materials nanocoated with rGO. Electrical and electrochemical properties were systematically studied and it was proved that the textile fabric, knitting, and non-woven structures presented higher amount of rGO sheet mass but lower sheet resistance. And the woven fabric structure presented better result of specific capacitance (40.5 F g^{-1}) with potential application in wearable electronic textiles.

2.3 *Spray Coating*

Various chemical finishes, including those based on graphene, can also be applied to fabrics using the spray coating technique. Through it, a low-viscosity chemical compound is contained in a pressurized vessel and transported to a substrate by a high-pressure air stream through an injection nozzle, followed by a drying process. A wide variety of nozzle system models allows varying spray patterns to be achieved, depending on product requirements. This technique can be done using a robotic spray gun (shown in Fig. 7) or manual. Coatings can be applied to one or both sides of the fabric. The amount of chemical applied is governed by the concentration of chemicals, the diameter of the nozzles, and pressure in the collector. Care must be taken to avoid overlapping spray patterns that could lead to uneven and unacceptable finish distribution. In addition, viscosity control is important to ensure that chemicals penetrate the fabric (Joshi and Butula 2013; Conway 2016). The biggest advantages of this technique compared to the previous cited techniques are (Choudhury 2017):

- The fact that there is no generation of effluents that could cause damage to the environment and treatment costs.
- It is a continuous finishing process, which makes it more compact, faster, and cheaper.
- Possibility to control process parameters with ease.

Zahid et al. (2017) applied PEDOT: PSS (poly(3,4-ethylenedioxythiophene): polystyrene sulfonate) and graphene to cotton fabric using the spray coating technique to develop conductive properties. Conductive cotton fabrics can be suitable for applications in wearable bionics and in elastic electronic sensors, such as measuring body movement and biomedical monitoring. Briefly, a solution of PEDOT: PSS, graphene, and dimethyl sulfoxide in isopropanol was prepared (a total solid content

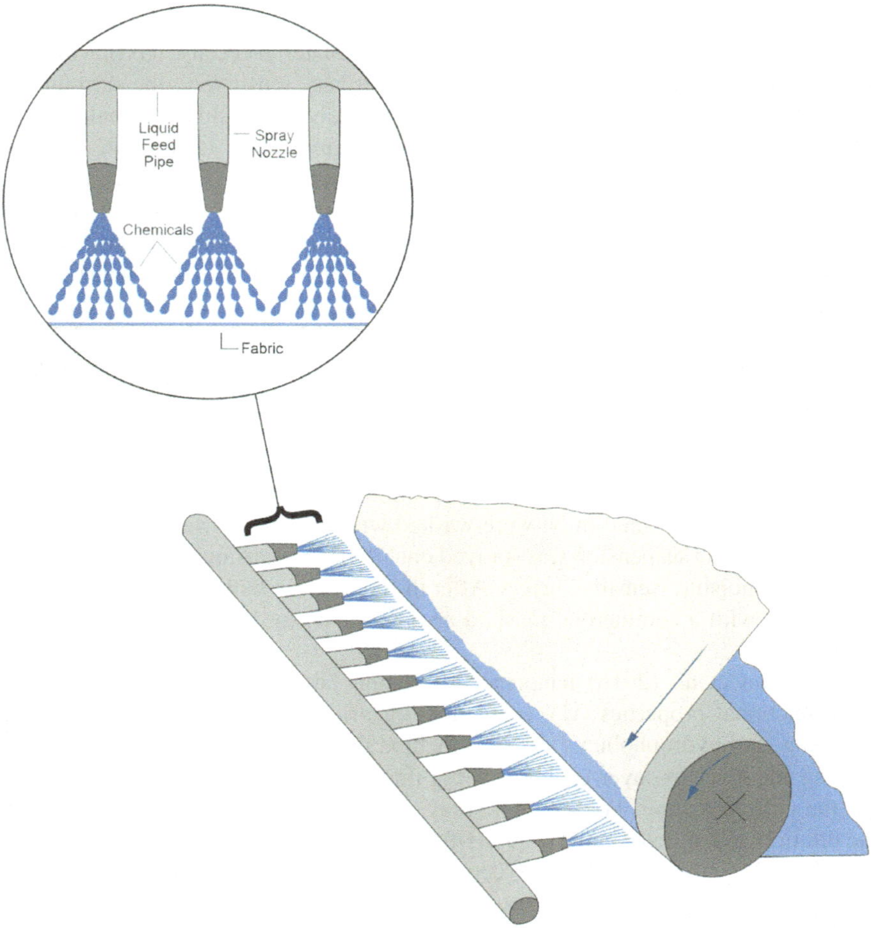

Fig. 7 Scheme of a chemical product application on fabric via spray coating. *Source* Author

of 1% by mass was maintained). The solution was applied to both sides of a mercerized cotton fabric using an airbrush spray system with a 0.73-mm nozzle, pressure of 2.5 at a distance of 15 cm. After the application, 2 spray cycles were performed with a hot air gun, and a dry uptake of 10–12% of the fabric weight was measured. The dry fabric was thermally cured in an oven for 30 min at 130 °C and then conditioned for 24 h at 23 ± 1 °C and 65 ± 2% relative humidity. Conductivity levels corresponding to sheet resistance of 25 Ω/sq have been achieved (~1.6 S/cm). The conductive fabrics were breathable, light, and resistant to laundry cycles.

Li et al. (2019) developed a flexible moisture sensor based on silk fabrics to monitor human breathing. With real-time breathing monitoring, it is possible to obtain more reliable data for the assessment of personal health conditions, considering that the frequency of breathing can indicate the progress of acute diseases, such as pneumonia, asthma, obstructive sleep apnea, cardiac arrest, and lung cancer. Graphene oxide was used in the construction of this material because it is rich in functional groups containing oxygen, which makes it very sensitive to the surrounding atmosphere, including humidity and temperature. Under a humid environment, the GO's functional surface groups react with water molecules to generate protons, further leading to a decrease in electrical impedance. Thus, the GO can be used as a sensor element to monitor human breathing, differentiating the changes in humidity during the inhalation–expiration cycles. For the manufacture of this sensor, interdigital nickel electrodes were prepared using the electrolyte-free coating technique on immaculate silk fabrics (Li et al. 2018). Briefly, Kapton tapes with a pre-cut interdigital pattern were used to seal both sides of the silk fabrics. Then, the sealed tissues were treated successively with a sensitization solution (tin chloride and hydrochloric acid) and an activation solution (palladium chloride and hydrochloric acid) at room temperature for 10 min. After rinsing with water, the tissues were immersed in an aqueous solution (pH = 10.0) containing nickel (II) sulfate, sodium citrate, boric acid, sodium hypophosphite, and ammonium chloride for electroplating (80 °C, 15 min). The silk fabrics deposited in Ni were washed with water, followed by drying in an oven. Then, a GO suspension was sprayed onto the fabrics, using a portable air spray to produce moisture-sensitive layers. After the connection with the copper wires and integration with a commercial mask, a respiration sensor based on silk fabric was obtained.

Nooralian et al. (2016) nanocoated a cotton fabric with graphene to obtain multifunctional properties (UV protection, electrical conduction, electromagnetic shielding, and hydrophobicity) using a method that involved the concepts of spray coating and layer-by-layer techniques. For this, two colloidal sprays were prepared for the cotton coating. For the first spray, a colloidal dispersion was made from the dilution in distilled water of COOH-functionalized graphene powder (FGN) and azo-bis-isobutyronitrile and *N*,*N*,*N*-cetyltrimethylammonium bromide (CTAB). The colloidal dispersions were then treated with ultrasonic waves at 30 °C for 4 h to produce stable colloids. For the second spray, vinylphosphonic acid was mixed with azo-bis-isobutyronitrile in deionized water under stirring conditions at 30 °C for 4 h. Then, a layer-by-layer coating was made on cotton fabric following the order of application:

1. Application of FGN colloidal dispersion (via spray);
2. Immersion of the fabric in deionized water;
3. Application of the mixture of vinylphosphonic acid and azo-bis-isobutyronitrile (via spray);
4. Immersion of the fabric in deionized water.

The cycle was repeated until the formation of 10 layers of graphene/polyvinylphosphonic acid in the cotton. Finally, the fabric was then dried at 60 °C and cured at 100 °C for 60 min.

2.4 Layer-by-Layer

The layer-by-layer (LBL) technique is a method of coating thin films on substrates through self-assembling layers by a cyclic process where a polyelectrolyte is adsorbed, and after washing, a material of the opposite charge is adsorbed by electrostatic interaction over the previously deposited layer, forming a thick bilayer in the order, in large part, nanometric (Fu et al. 2019). The LBL assembly process can be repeated until to produce a multilayer coating assembled to the desired thickness. In addition, interaction by hydrogen bonds (Xie et al. 2019) and covalent bonds (Li et al. 2019) is also established for this methodology. Most of the polyelectrolytes used for the functionalization of the fabrics and other textiles substrates are polymeric bases such as poly(diallyldimethylammonium chloride) (PDDA) (Zhang et al. 2014), poly(allylamine hydrochloride) (PAH) (Liu et al. 2018), poly(sodium styrenesulfonate) (PSS) (David et al. 2015), chitosan (Patil et al. 2016), poly(dopamine) (PDA) (Yang et al. 2019), and bovine serum albumin (Yun et al. 2013).

Ller was the first to use it in 1966 in his work entitled 'Multilayers of colloidal particle' where he studied the deposition of silica and alumina on a smooth surface like glass (Iler 1966). But, the scientific community did not give importance to the discovery of this method (Inagaki et al. 2013). Only in the mid-1990s, Decher and collaborators suggested the production of ultrathin films using a technique to coat substrates with polymeric materials, colloids, biomolecules, and biological material (Bolotin et al. 2008; Lee et al. 2008; Cole et al. 2015; Khanafer and Vafai 2017). Although many of the works deal with coating materials on substrates, its use does not make it necessary for certain applications that use self-assembly, as found in membranes (Sha'rani et al. 2019), microcapsule (Kolesnikova et al. 2010), and nanoparticles (Ramasamy et al. 2014).

The procedures used in the LBL assembly technique are simple and versatile in relation to other methods of depositing thin films, such as Langmuir–Blodgett (LB), which requires a gas–liquid interface for the production of nanostructured films in monolayers (Sharma et al. 2011). In this case, its application in the various fields of science expanded to the development of different assembly technologies, including creaming (Shim et al. 2007), electrodeposition (Cranston and Gray 2006), spray (Heo et al. 2020), roll-to-roll (Razza et al. 2016), lamina (Sun et al. 2020), spin (Thomas 1987), magnetic assembly (Hong et al. 2004), fluidics (Picart et al. 2001), saturation (Hoogeveen et al. 1996), molecular (Qi et al. 2011), high gravity (Ma et al. 2012), centrifugation (Nishiguchi et al. 2011), dipping (Gill et al. 2020), and conical fluidized beds (Noi et al. 2015). In essence, these processes are constituted on the principle that materials are input to promote the coating of layers. Although there are numerous techniques for deposition of layers via LBL, it is important to note that the thickness, homogeneity, and interaction between the layers of the produced films are directly related to the applied assembly method and this is a determining factor for the physical–chemical properties of the material obtained (Chen et al. 2017; Yazdi et al. 2018).

As the deposition method adjusts the composition of nano- and microstructural films, the LBL assembly technique provides a powerful tool for employing a variety of materials such as polymers (Estillore and Advincula 2011), metallic nanoparticles (Nath and Chilkoti 2004), nucleic acids (Nault et al. 2010), and semiconductors (Srivastava and Kotov 2008), in the formation of new hybrids. Among them, 2D graphene-based nanosheets have aroused great interest in the field of new flexible textile hybrids due to their extraordinary properties. Therefore, different technologies for depositing films by LBL can be considered under the aspect of five different categories: immersive, spray, spin, electromagnetic, and fluid assembly.

2.4.1 Immersion

The LBL immersion assembly technique is the most widely used form for coating graphene-based materials. Generally, the deposition of films is performed by manually immersing a polymeric matrix in the nanoparticle solution, followed by washing (Gill et al. 2020) (Fig. 8). Furthermore, there are also mechanisms that speed up the LBL assembly process changing the deposition mechanism in an automated way through steps using immersion machines to transfer the substrate (Dubas and Schlenoff 1999; Shiratori and Yamada 2000). In the case where the substrate is particulate, the material must be collected between the deposition and washing steps, where it needs centrifugation (Ye et al. 2005). Tian et al. (2015) used this approach to deposit layers of conductive poly (3,4-ethylenedioxythiophene) polymer bound to the polyanion PSS doped with graphene adsorbed on chitosan layers (polycation), revealing excellent electronic properties and a hybrid of the UPF 312.

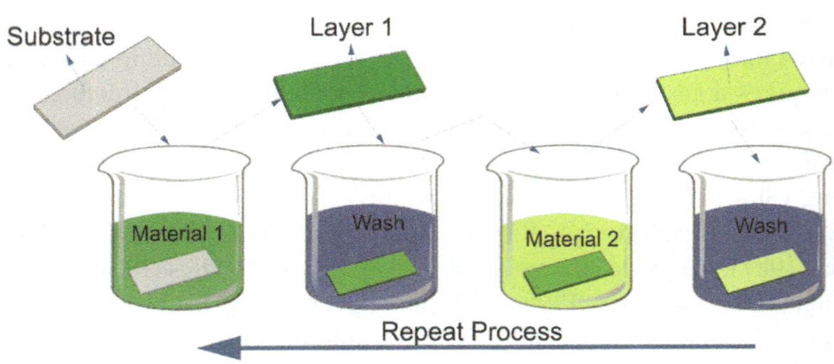

Fig. 8 Scheme of the immersion LBL assembly coating process

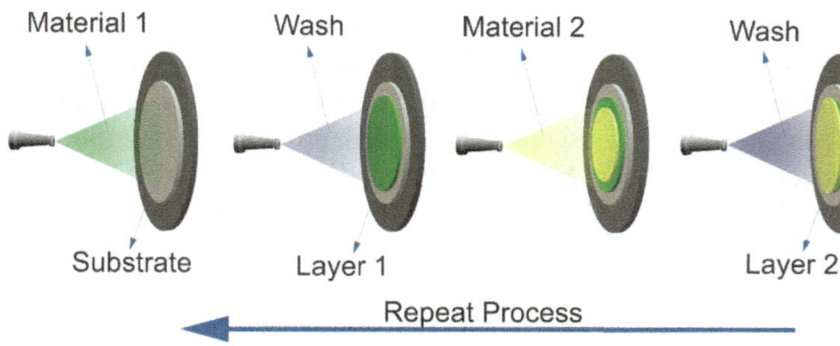

Fig. 9 Scheme of the spray LBL assembly coating process

2.4.2 Spray

In this type of LBL assembly (Fig. 9), the films are formed from polycationic and polyanionic solutions in aerosol and spray them on substrate, following a previously defined sequence (Heo et al. 2020). The great advantage of this method is the speed at which the layer-by-layer deposition process is carried out. In contrast to immersion, the spray coating cycle is faster, reaching 6 s (Izquierdo et al. 2005), which is an interesting feature that approaches industrial levels. Nooralian et al. (2016) produced multifunctional nanocomposite of cotton fabric coated with graphene/polyvinylphosphonic acid (anion) bilayer assembled by spraying, contributing to increased hydrophobicity, thermal stability, thermal conductivity, UV absorption, and electromagnetic shielding. Heo et al. (2019) used the same procedure to coat multilayer graphene oxide on polyethylene films, a potential route for the production of food packaging.

2.4.3 Spin

Through the LBL spin assembly method (Fig. 10), the substrate is rotated during the material deposition process (Chiarelli et al. 2001) or the suspension containing

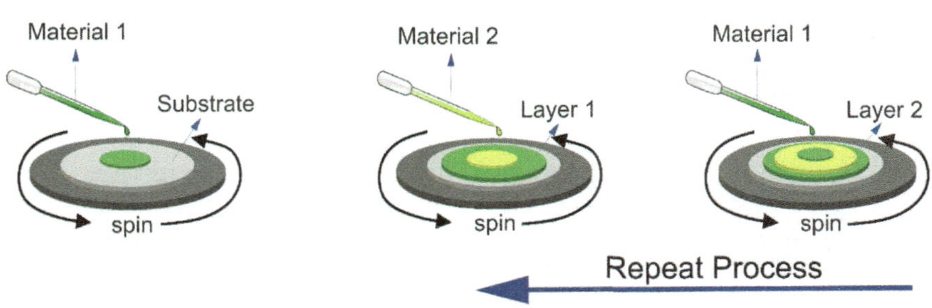

Fig. 10 Scheme of the spin LBL assembly coating process

the polyelectrolytes that will compose the films is molded to the substrate and then to be rotated (Cho et al. 2001). Thus, a set of forces, including electrostatic interactions, contributes to the formation of more homogeneous layers because they are centrifugal, air-shearing, and viscous forces, which promote the desorption of polymers (polyelectrolyte) that have been weakly bound to the substrate (Cho et al. 2001). Furthermore, shear forces also guarantee the production of thinner and more orderly films, when compared to immersion coatings that produce thicker and interpenetrated films (Patel et al. 2007). The automated assembly of films by immersion promotes thicker layers, unlike automated assembly of films by spin results in the deposition of thinner films (Seo et al. 2008).

Most of the works that involve LBL assembly by spin use flat substrates and are not capable of more complex forms, as in the immersion method. However, the LBL assembly techniques by spray and spin to immobilize multi-walled carbon nanotubes (MWCNTs) and graphene on fibrous carbon cloth substrate for flexible supercapacitor applications (Shakir 2014). Rajasekar et al. (2013) constructed a polymeric composite with polyethylene terephthalate (PET) coated with PDDACl bilayers (positive charge) containing graphene oxide and sulfonated polyvinylidene fluoride matrix (negative charge) for high hydrogen barrier application.

2.4.4 Electromagnetic Assembly

In this type of approach, materials are deposited using an electric or magnetic field to make layers on substrates that act as electrodes, by electrochemical deposition (Fig. 11). For the electric field, the generation of a voltage causes an electric current to be applied to electrodes immersed in polymer solutions; then, the electrodes are raised and immersed in polymer solution of opposite charges, to again reverse the polarity, forming multilayers by electroplating (Richardson et al. 2016); therefore, current density and voltage are important parameters in the construction of electrodeposited layers (Kulandaivalu and Sulaiman 2019). Zhang and Cui (2011) developed an ultrasensitive flexible substrate sensor where graphene films were self-assembled with

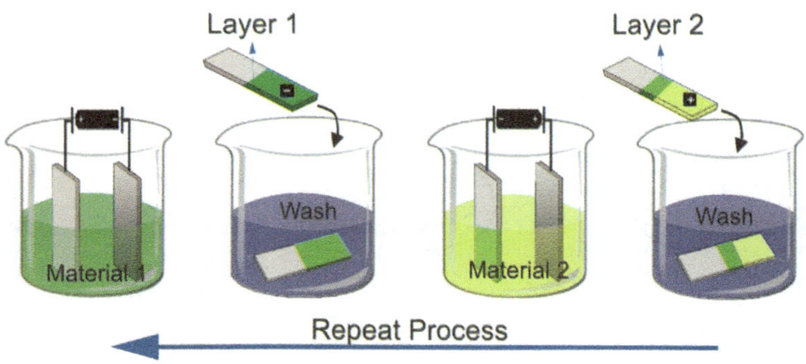

Fig. 11 Scheme of the electromagnetic LBL assembly coating process

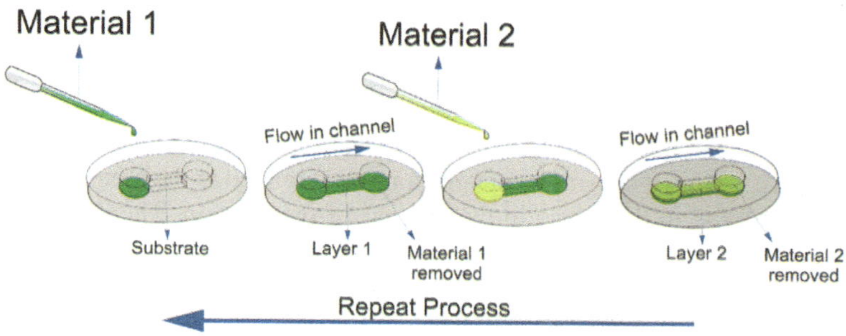

Fig. 12 Schematization of the fluidic LBL assembly coating process

layers of PDDA and PSS, capable of detecting concentrations of prostate antigens around three orders of magnitude lower than sensors based on carbon nanotubes under same manufacturing and measurement conditions. This approach is capable of controlling thickness, structure, surface area, and deposition rate by applying the current potential and density (Kulandaivalu and Sulaiman 2019).

2.4.5 Fluidic Assembly

The fluid assembly method involves alternating polymeric solutions and washing to deposit multilayers with fluid channels that enable the coating on the channels themselves or on substrates inserted or immobilized in the channels, such as tubes or capillaries, or microfluidic networks designed (Wang et al. 2011) using pressure or vacuum to move sequentially solutions (Fig. 12). Contact time is a crucial factor in determining the amount of polymer adsorbed. Furthermore, higher concentrations of polymeric solutions normally produce thicker films (Richardson et al. 2015).

In general, fluidic assembly is a technology that enables the coating of multifilms to be obtained on difficult surfaces by other assembly methods. Although there is no work regarding coating on textile substrates, one possibility of implementing this technique is based on the production of continuous fibers with geometric complexity, through the manufacture of microfluidic fibers (MFF). In this system, a continuous flow of a polymeric solution that will constitute the microfiber is imbibed by a second fluid in a microchannel, where the core undergoes a physical or chemical transformation until its solidification (Daniele et al. 2015). This method was also used to obtain alginate microfiber hydrogels for application in tissue engineering (Yamada et al. 2012).

2.5 Silk Screen Printing

Screen printing is a very old printing technique, but widely used by the textile industry today. In this technique, a mesh is used to transfer a pigmented paste to the textile fabric surface, giving it color. The pattern formed is related to the areas that have not been waterproofed in the fabric in the process of preparation. Therefore, at points where the paste must not pass through, a light-sensitive resin is applied. A squeegee is responsible for dragging and forcing the paste through the mesh. For the printing of colored patterns, it is necessary to use multiple screens, each corresponding to a pigmented paste that must be directed to specific points on the textile surface. The mesh can be supported on a flat frame, known as a frame (wood or metal), featuring the screen printing. The mesh can also be supported on a cylindrical frame, known as rollers (metal), featuring rotary printing. Screen printing can be applied manually or automatically. It is the most flexible to the use of various patterns; however, it is slower when compared to rotary screen printing. After printing the paste on the fabrics, thermal processes (thermofixation or vaporization) are used to fix the prints. Figure 13 outlines the main components of a manual screen printing, while Fig. 14 outlines the components of a rotary screen printing.

The mesh is a porous structure that can be formed by silk, nylon, polyester, phosphor bronze, or stainless steel. The density of threads can vary from 19 to 200 threads per centimeter. Higher densities result in more defined prints, but the characteristics of the paste, such as viscosity, must be controlled to avoid flaws. Silk knits have good stability and considerable resistance. Nylon fabrics are the most used because they are up to 30 times more resistant to breakage than silk fabrics, have a high resistance to abrasion, and have a high resistance to most chemicals that can be added to stamping paste. Polyester is slightly less resistant to nylon, chemical agents, and wear; however, it remains very suitable for use. There are also metallic

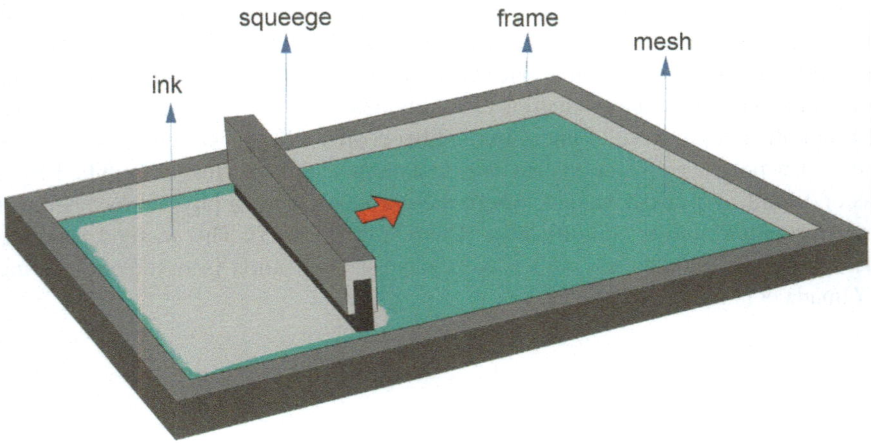

Fig. 13 Manual picture screen printing scheme

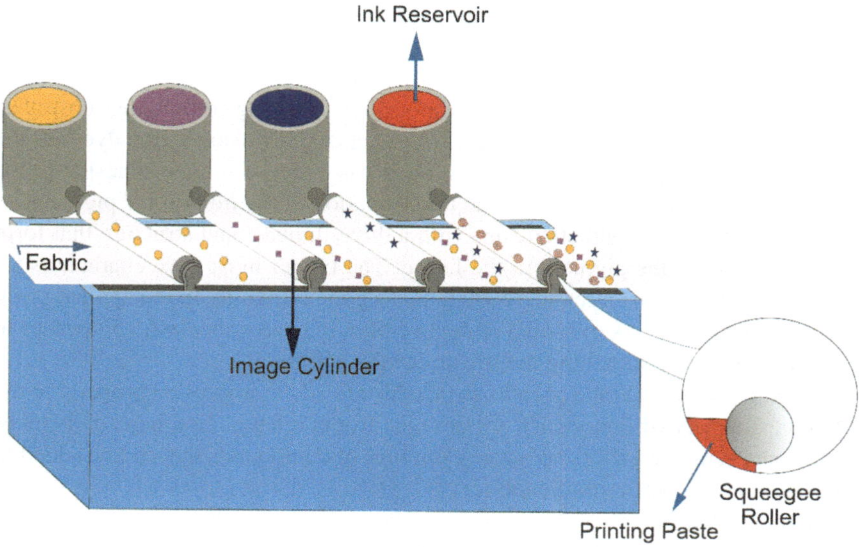

Fig. 14 Rotary screen printing scheme

meshes, phosphorous bronze, or stainless steel, which offer extraordinary strength, optimal dimensional stability, resistance to heat and alkalis, and exceptional fineness. Yet, they are expensive, difficult to manually tension, and vulnerable to shocks. Its classification ranges from 29 to 129 threads per centimeter. Metallic screens are widely used in industrial screen printing for large runs, and mainly in the printing of printed circuits, where absolute precision is required.

The stamping paste is composed of a mixture of products that will be necessary for the paste's own stability or for fixation on the textile substrate, among them are:

- Thickeners: They are used to guarantee the ideal viscosity of the paste, preventing failures due to the diffusion of the chemical components of the paste through the fabric. It is important that these materials have no reactivity with the other components present in the formulation. The thickeners commonly used are starches and gums, sodium alginate, carboxymethylcellulose, methylcellulose, and methylcellulose.
- Hygroscopic agents: Usually, urea or glycerin is used, as they absorb enough water and help to dissolve other components that are present in the formulation of the paste.
- Acid/alkali: They are used to promote certain reactions. Alkali, for example, is used in conjunction with reactive dyes to facilitate covalent attachment to the substrate.
- Dispersing agents: They are used to form a dispersion and prevent sedimentation of low solubility components.
- Defoaming agents: Emulsion-based defoaming agents are generally used.

- Catalysts: They are used to accelerate the rate of a specific reaction. Some of the common catalysts are potassium ferrocyanide, copper sulfide, and ammonium vanadate.
- Dyes/pigments: They are the agents responsible for giving color to the substrates. While dye-based printing pastes facilitate localized diffusion of the dye into the textile substrate, the pigment-based printing paste focuses on the adhesion of the pigment to the substrate surface and the formation of the film on the pigment.
- Binders: These are synthetic resins with a three-dimensional structure that form a film that fixes the pigment in the fabric. The binder must have characteristics such as high washing strength to guarantee a good durability to the applied print, the ability to crosslink to form a protective film, to have viscoelastic properties to allow folding and stretching the substrate after printing.
- Finishing products: Some agents can be added to add or improve a property to the fabric, such as hydrophobicity, softness, or aroma release. Graphene derivatives and some semiconductors are some examples of compounds that can be added to give special properties to the fabric.
- Photoinitiating agents: They are substances that absorb UV radiation and are directly involved in the production of free radicals necessary to promote polymerization reactions.
- Evaporation inhibitors: They are used to prevent the screen from clogging.
- Oxidizing agent: It prevents 'frosting' of prints or reduction of dye during steaming—compounds such as sodium m-nitrobenzene sulphonate.
- Solvents: They are used to dissolve dyes and pigments and prevent precipitation, e.g., water, ethylene glycol, thiodiethylene glycol, diethylene glycol.

Although the screen printing technique is widely used for the application of prints on textiles, it can also be used to apply chemical finishes such as softener, antimicrobial finishes (Ocepek et al. 2012), and aroma releasers (Golja et al. 2013). The technique has also been shown to be promising for the use of graphene in textiles. Abdelkader et al. (2017) applied a graphene-based supercapacitor on a textile substrate using the screen printing technique. A suspension of GO in water with a concentration of 5 mg/ml was used to print the supercapacitor device, whose viscosity was modified using an acrylate thickener to achieve the proper viscosity. This mixture was stirred for 15 min using a mechanical stirrer to produce a viscous printing paste. The fabric was printed using a manual 62/cm mesh screen and 10 mm squeegee width. A prepared screen was used to print a circuit with the GO paste prepared on a textile substrate. After printing, the GO was reduced to rGO using an electrochemical method. The sample was carefully washed with water and dried for a few hours at 70 °C under vacuum. The conductive network functioned as a current collector and an active supercapacitor material. The obtained capacitance was as high as 2.5 mF cm^{-2} and remained at 95.6% of this value when tested under bending conditions. The supercapacitor device maintained 97% of its original capacitance even after 10,000 cycles.

Song et al. (2019) performed a rGO-based cotton fabric treatment using a screen printing technique to obtain UV protection properties. RGO is a material that is difficult to apply to textile materials due to the low amount of hydrophilic groups in its structure, which makes it poorly soluble in water. For this reason, most of the routes covered in the literature, including the one used by the authors, part of the application of GO in fabric for a subsequent conversion to rGO from a reduction reaction (Karim et al. 2017b). Briefly, in the procedure performed, a paste based on GO and waterborne polyurethane (WPU) was prepared. In addition, thickener, binder, and a photoinitiator were also added. The paste was applied to a cotton fabric using the manual screen printing technique. The fabric was dried at 80 °C for 15 min, followed by curing the WPU under a UV lamp. The fabrics printed in GO/WPU composite paste were immersed in an aqueous solution containing 5 mg/ml of sodium hydrosulfite, at about 90 °C for 1 h to obtain fabrics printed in rGO/WPU composite paste. Finally, the fabric was carefully rinsed and dried at 80 °C for 30 min, followed by curing at 150 °C for 3 min. Screen printing techniques are very advantageous to guarantee the adhesion of particles that naturally have little chemical reactivity with textiles, as was the case with rGO.

2.6 *Inkjet Print*

Inkjet printing has been one of the most modern techniques for applying prints on textile substrates. The technique provides advantages in process efficiency, flexibility, ease of use, and reduced environmental impact. Digital fabric printing has evolved rapidly in recent decades, which has generated faster speeds, more economical processes, and greater durability in sensitive printing components. The challenge of current technological development is no longer to improve the printing of color images, but to print electronic circuits for the development of wearable electronic textiles (e-textiles). E-textiles have become a focus of significant research interest due to their possible applications in sportswear, military uniforms, environmental monitoring, and health care. Metallic paints based on Ag, Cu, or Au are the materials commonly used due to their high electrical conductivity. However, inks based on graphene or rGO have the advantages of being environmentally friendly, biocompatible, and not requiring a high sintering temperature (Tawiah et al. 2016; Karim et al. 2017a).

Unlike more conventional techniques such as screen printing, the inks used by inkjet printing must have a low viscosity and a high purity in terms of dye content. For inkjet printing, there is no need to manufacture printing screens, as the design is a digital image. Inkjet printing involves the deposition of small jets (like ink drops) of colored ink on the substrate. The inkjet is controlled by a computer, as required by the digital image. The quality of the printed image is defined by the print resolution in dots per inch (dpi). The technologies of this technique are classified as continuous inkjet technology (a), thermal inkjet printing (b), and piezo inkjet printing (c) (Fig. 15), as can be seen in the following diagrams (Wardman 2018).

In 'Continuous Ink Jet Technology,' a continuous flow of ink is fired out of the print head nozzle. An electrical charge is then used to direct individual droplets to the

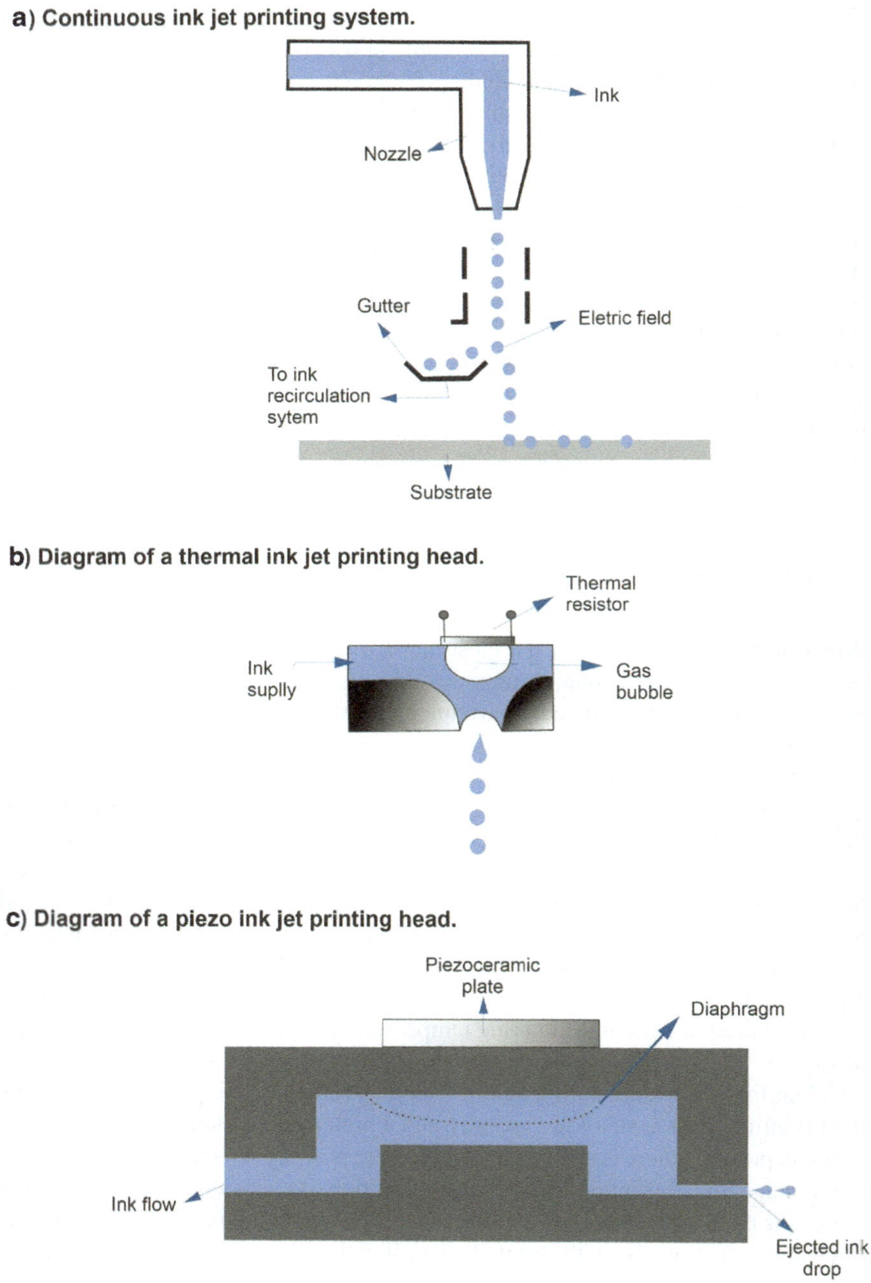

Fig. 15 Scheme of continuous inkjet technology (**a**), thermal inkjet printing (**b**) and piezo inkjet printing (**c**)

substrate or a collection chute where the ink is recirculated back to the ink reservoir. In the 'Thermal Ink Jet Printing' system, controlled heating of the ink occurs, causing pressure that forces the ink out of the nozzle. In the 'Piezo Ink Jet Printing' system as an electrical charge is applied, a piezoelectric component expands and contracts. This movement forces the diaphragm to open and close, generating pressure in the ink, forcing it out of the nozzle on the substrate. As there is no heating, there is the possibility of using a greater amount of paints, but it is a more expensive system.

Karim et al. (2017a, b) team developed a technique to optimize the printing of graphene on textile substrates. The technique is based on the pre-printing of a surface treatment based on organic nanoparticles; thus, the surface becomes more suitable for receiving the rGO ink impression, minimizing problems such as circuit breakage failures and significantly increasizng the electrical conductivity.

2.7 Extrusion

Textile fibers derived from chemical processes are quite versatile in terms of the incorporation of the most diverse particles in their structure, considering that it is possible to immobilize them during the fiber synthesis process in the spinneret or in treatments after the fiber formation. Basically, there are three techniques for the production of synthetic fibers (Bonaldi 2018; Sanes et al. 2020).

(a) Melt spinning—a process in which the polymer is heated above its melting point and, once melted, is forced to pass through a spinneret, whose holes can have different diameters and shapes. The jet of polymer emerging from each orifice is guided to a cooling zone, where the polymer solidifies, to complete the fiber formation process, e.g., polyamides and polyesters.
(b) Dry spinning—a process by which the polymer is dissolved in a suitable solvent and the resulting solution is extruded through a spinneret. The jet of the polymer solution is guided to a heating zone, where the solvent evaporates and the filament solidifies, e.g., cellulose acetates and polyacrylonitriles.
(c) Wet spinning—a process by which the polymer is also dissolved and the solution is forced to pass through a spinneret immersed in a coagulant bath. As the polymer solution emerges from the nozzle holes in the bath, the polymer is precipitated or chemically regenerated, e.g., viscose and polyacrylonitriles (Fig. 16).

Basically, the polymer mass is guided by means of distribution lines and metering pumps, which guarantee a constant flow rate for the spinning positions, composed of a series of filters that purify and distribute the polymer, which are coupled with plates perforated with variable thickness and size. The holes (capillaries) in the plate, vary depending on the type of fiber, may have circular, special, or hollow cross sections, as shown in Table 3. The geometry selected for the fiber can influence the characteristics of the fabric formed by them, such as shine, touch, and permeability (Ramsden 2016).

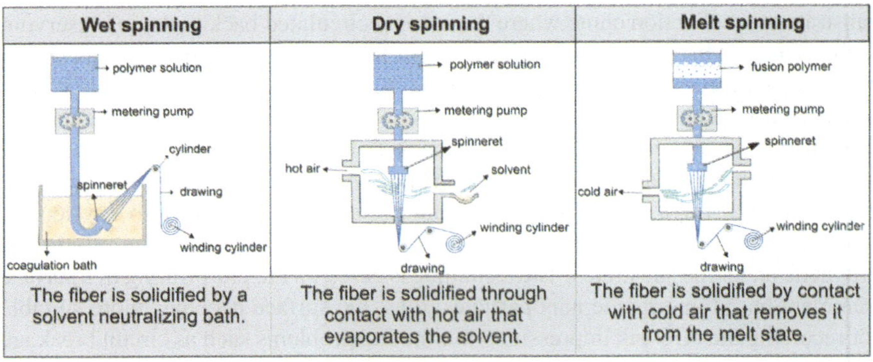

Fig. 16 Synthetic spinning processes. *Source* Author

When synthetic fibers containing graphene and its derivatives are prepared by one of these three spinning processes, it is important to disperse the graphene in the fiber evenly, forming a well-dispersed system that makes graphene have its optimized properties, among them the conductivity, thermal stability, mechanical reinforcement, etc. Generally, the dispersion of graphene in the synthetic fibers is evaluated through analyzes involving high-resolution transmission electron microscopy (HRTEM) and X-ray diffraction (XRD). Covalent or non-covalent modifications can be used to increase the dispersion of graphene in the polymeric matrix. The modifications can also alter the microstructure of graphene, resulting in graphene with variation in the atomic proportion of carbon/oxygen, functional groups, electrical conductivity, and solubility in polar or nonpolar solvents (Ji et al. 2016).

Khan et al. (2012) produced graphene/polyester fibers with improved mechanical properties over pure polyester filament. A melt spinning process was carried out. Briefly, a PET sheet weighing 1 g was cut into pieces and heated to ~200 ° C for 10 min. A composite foundry was produced by adding a certain volume of graphene dispersed in *N*-methyl-2-pyrrolidone (NMP) to the molten PET. The mixture was stirred constantly to ensure homogeneity. After 10 min of heating, almost all of the NMP evaporated. The fibers were formed from the ejection of the molten compound through a fine needle at a rate of ~40 cm s^{-1}. The fibers were cooled under ambient conditions for a minimum period of 1 h. Then, the fibers were stretched until breaking. It was observed that the pure PET fibers could only be stretched up to ~200%, while the composite fibers could be stretched up to 400%. The fibers were pulled while they were cold and released immediately afterward. In all cases, they kept the entire length. It was observed that with an addition of up to 2% by weight of graphene load, there was an increase in Young's modulus twice its initial value and its resistance has quadrupled in value.

Xu and Gao (2010) produced graphene/nylon-6 fibers with improved mechanical properties over pure nylon 6 filament. With a load of 0.1% by weight of graphene oxide in the polymer matrix, there was an increase of more than 2 times the tensile strength and Young's modulus. The procedure involved the addition of GO amid

Table 3 Round, trilobal, and hexagonal cross section, respectively, and reflection of light on its surface

Form	Longitudinal direction	Transversal section
Serrated	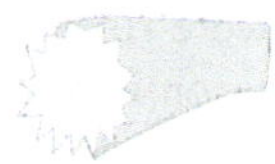	
Square whit void		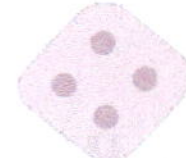
Dog bone	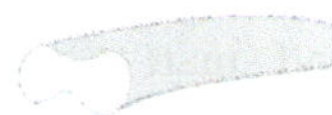	
Round		
Flat, oval whit convolutions		
Trilobal		
Hollow-core	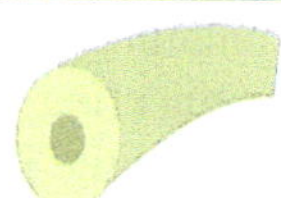	

Source Author

the precursors used in the polymerization reaction of nylon 6 (caprolactam and aminocaproic acid). The fibers were manufactured using a melt spinning process.

He et al. (2012) produced fibers reinforced with sodium alginate and graphene oxide for applications in the field of biomedical engineering. Fibers made of sodium alginate have important properties for application in dressings, such as non-toxicity, biocompatibility, and relatively low cost. In addition, alginate is highly soluble in water and can form hydrogels in the presence of divalent cations (such as Ca^{2+}), due to ionic cross-linking by calcium bridges between the residues of glucuronic acid in adjacent chains. While the addition of graphene oxide was responsible for a considerable increase of up to two times more in the maximum tensile strength, it is up to three times more than Young's modulus, allowing for better handling. The fibers were prepared using a wet-spinning method, in which a calcium chloride solution was used as a coagulation bath. Briefly, an aqueous suspension of GO was prepared by ultrasonic treatment and then added to a certain volume of sodium alginate solution, and the resulting solution was stirred constantly for about 3 h to form a sodium alginate/GO solution. The solution was squeezed through a needle and dragged to the coagulation bath at room temperature. The spun fibers were collected and soaked in water for 10 min, and then, they were rinsed several times and dried. The cured fibers were stretched for a certain reason before being put back in the coagulation bath to cure for several minutes, and then, they were rinsed several times and dried.

3 Concluding Remarks

The development of functional and smart textiles using graphene, one of the most important and attractive nanomaterials of the present time, is in progress. Multifunctional textile with properties supercapacitive, conductive, antimicrobial, UV protection, and photocatalytic and with high mechanical resistance has been in development in the industry. For the application of the graphene to textiles, several types of processes have been studied and optimized, and it is clear that the methods that use digital printing have been the most efficient due to the low consumption of water and energy in the textile industry. Although several so-called nanotextiles with other nanoparticles are already available on the market, there are still no textile products available with graphene, but there has been a growing demand for patents and scientific articles in the last 10 years, with China leading the largest quantity of studies and research involving the application of graphene in the textile industry. There are still several manufacturing processes in the research stage and the production of multifunctional textile products containing graphene. As a result, graphene has been extensively studied in textile engineering, thus replacing some conventional textiles, thus promoting a change in the textile industry, and with a better prospect of modernization for the 4th industrial revolution.

References

Abbasi E et al (2016) Graphene: synthesis, bio-applications, and properties. Artif Cells Nanomed Biotechnol 44(1):150–156. https://doi.org/10.3109/21691401.2014.927880

Abdelkader AM et al (2017) Ultraflexible and robust graphene supercapacitors printed on textiles for wearable electronics applications. 2D Mater 4(3):35016

Abdolhosseinzadeh S, Asgharzadeh H, Seop Kim H (2015) Fast and fully-scalable synthesis of reduced graphene oxide. Sci Rep 5(1):10160. https://doi.org/10.1038/srep10160

Afroj S et al (2019) Engineering graphene flakes for wearable textile sensors via highly scalable and ultrafast yarn dyeing technique. ACS Nano 13(4):3847–3857

Bacon M, Bradley SJ, Nann T (2014) Graphene quantum dots. Part Part Syst Charact 31(4):415–428. https://doi.org/10.1002/ppsc.201300252

Bag S et al (2015) Nitrogen and sulfur dual-doped reduced graphene oxide: synergistic effect of dopants towards oxygen reduction reaction. Electrochim Acta 163:16–23. https://doi.org/10.1016/J.ELECTACTA.2015.02.130
Bagri A et al (2010) Structural evolution during the reduction of chemically derived graphene oxide. Nat Chem 2(7):581–587. https://doi.org/10.1038/nchem.686
Bassalo JMF, Farias RF (2017) Prêmio Nobel de 2016: Química e Física. Caderno Brasileiro de Ensino de Física 34(2):479. https://doi.org/10.5007/2175-7941.2017v34n2p479
Bolotin KI et al (2008) Temperature-dependent transport in suspended graphene. Phys Rev Lett 101(9):096802. https://doi.org/10.1103/PhysRevLett.101.096802
Bonaldi RR (2018) Functional finishes for high-performance apparel. In: High-performance apparel, pp 129–156
Bunch JS et al (2008) Impermeable atomic membranes from graphene sheets. Nano Lett 8(8):2458–2462. https://doi.org/10.1021/nl801457b
Cai G et al (2017) Functionalization of cotton fabrics through thermal reduction of graphene oxide. Appl Surf Sci 393:441–448. https://doi.org/10.1016/J.APSUSC.2016.10.046
Cay A, Tarakçıoğlu I, Hepbasli A (2009) Assessment of finishing processes by exhaustion principle for textile fabrics: an exergetic approach. Appl Therm Eng 29(11–12):2554–2561
Chen M, Haddon Robert C et al (2017) Advances in transferring chemical vapour deposition graphene: a review. Mater Horiz 4(6):1054–1063. https://doi.org/10.1039/C7MH00485K
Chen W et al (2019) Green synthesis of graphene quantum dots from cotton cellulose. ChemistrySelect 4(10):2898–2902. https://doi.org/10.1002/slct.201803512
Chiarelli PA et al (2001) Controlled fabrication of polyelectrolyte multilayer thin films using spin-assembly. Adv Mater 13(15):1167–1171
Cho J et al (2001) Fabrication of highly ordered multilayer films using a spin self-assembly method. Adv Mater 13(14):1076–1078
Choudhury AKR (2017) Principles of textile finishing. Woodhead, Kidlington
Cole RM et al (2015) Evanescent-field optical readout of graphene mechanical motion at room temperature. Phys Rev Appl 3(2):024004. https://doi.org/10.1103/PhysRevApplied.3.024004
Conway R (2016) Coating of textiles. In: Horrocks AR, Anand SC (eds) Handbook of technical textiles. Elsevier
Cranston ED, Gray DG (2006) Formation of cellulose-based electrostatic layer-by-layer films in a magnetic field. Sci Technol Adv Mater 7(4):319
Daniele MA et al (2015) Microfluidic strategies for design and assembly of microfibers and nanofibers with tissue engineering and regenerative medicine applications. Adv Healthc Mater 4(1):11–28
David M et al (2015) Acidic and basic functionalized carbon nanomaterials as electrical bridges in enzyme loaded chitosan/poly (styrene sulfonate) self-assembled layer-by-layer glucose biosensors. Electroanalysis 27(9):2139–2149
Dubas ST, Schlenoff JB (1999) Factors controlling the growth of polyelectrolyte multilayers. Macromolecules 32(24):8153–8160
Estillore NC, Advincula RC (2011) Stimuli-responsive binary mixed polymer brushes and free-standing films by LbL-SIP. Langmuir 27(10):5997–6008
Fu J et al (2019) Layer-by-Layer electrostatic self-assembly silica/graphene oxide onto carbon fiber surface for enhance interfacial strength of epoxy composites. Mater Lett 236:69–72
Gan L et al (2015) Graphene nanoribbon coated flexible and conductive cotton fabric. Compos Sci Technol 117:208–214
Gill AK et al (2020) Mussel-inspired UV protective organic coatings via layer-by-layer assembly. Eur Polymer J 124:109455
Golja B, Šumiga B, Forte Tavčer P (2013) Fragrant finishing of cotton with microcapsules: comparison between printing and impregnation. Color Technol 129(5):338–346
Guex LG et al (2017) Experimental review: chemical reduction of graphene oxide (GO) to reduced graphene oxide (rGO) by aqueous chemistry. Nanoscale 9(27):9562–9571. https://doi.org/10.1039/C7NR02943H

He Y et al (2012) Alginate/graphene oxide fibers with enhanced mechanical strength prepared by wet spinning. Carbohyd Polym 88(3):1100–1108. https://doi.org/10.1016/j.carbpol.2012.01.071

Heo J et al (2020) Spray-assisted layer-by-layer self-assembly of tertiary-amine-stabilized gold nanoparticles and graphene oxide for efficient CO_2 capture. J Membr Sci 117905

Heo J, Choi M, Hong J (2019) Facile surface modification of polyethylene film via spray-assisted layer-by-layer self-assembly of graphene oxide for oxygen barrier properties. Sci Rep 9(1):1–7

Hong X et al (2004) Fabrication of magnetic luminescent nanocomposites by a layer-by-layer self-assembly approach. Chem Mater 16(21):4022–4027

Hoogeveen NG et al (1996) Formation and stability of multilayers of polyelectrolytes. Langmuir 12(15):3675–3681

Huang G et al (2012) Thin films of intumescent flame retardant-polyacrylamide and exfoliated graphene oxide fabricated via layer-by-layer assembly for improving flame retardant properties of cotton fabric. Ind Eng Chem Res 51(38):12355–12366

Iler RK (1966) Multilayers of colloidal particles. J Colloid Interface Sci 21(6):569–594. https://doi.org/10.1016/0095-8522(66)90018-3

Inagaki M et al (2013) Advanced materials science and engineering of carbon. Butterworth-Heinemann

Inagaki M et al (2014) Advanced materials science and engineering of carbon. Elsevier. https://doi.org/10.1016/C2012-0-03601-0

Izquierdo A et al (2005) Dipping versus spraying: exploring the deposition conditions for speeding up layer-by-layer assembly. Langmuir 21(16):7558–7567

Ji X et al (2016) Review of functionalization, structure and properties of graphene/polymer composite fibers. Compos A Appl Sci Manuf 87:29–45. https://doi.org/10.1016/j.compositesa.2016.04.011

Ji Y, Chen G, Xing T (2019) Rational design and preparation of flame retardant silk fabrics coated with reduced graphene oxide. Appl Surf Sci 474:203–210

Joshi M, Butula B (2013) Application technologies for coating, lamination and finishing of technical textiles. In: Gulrajani M (ed) Advances in the dyeing and finishing of technical textiles. Elsevier Science, p 448

Karim N, Afroj S, Malandraki A et al (2017a) All inkjet-printed graphene-based conductive patterns for wearable e-textile applications. J Mater Chem C 5(44):11640–11648

Karim N, Afroj S, Tan S et al (2017b) Scalable production of graphene-based wearable e-textiles. ACS Nano 11(12):12266–12275

Karimi L et al (2014) Using graphene/TiO_2 nanocomposite as a new route for preparation of electroconductive, self-cleaning, antibacterial and antifungal cotton fabric without toxicity. Cellulose 21(5):3813–3827

Kaur M, Kaur M, Sharma VK (2018) Nitrogen-doped graphene and graphene quantum dots: a review onsynthesis and applications in energy, sensors and environment. Adv Coll Interface Sci 259:44–64. https://doi.org/10.1016/j.cis.2018.07.001

Khan U et al (2012) High strength composite fibres from polyester filled with nanotubes and graphene. J Mater Chem 22(25):12907–12914. https://doi.org/10.1039/c2jm31946b

Khanafer K, Vafai K (2017) Analysis of the anomalies in graphene thermal properties. Int J Heat Mass Transfer 104:328–336. https://doi.org/10.1016/J.IJHEATMASSTRANSFER.2016.07.103

Khose RV et al (2018) Novel approach towards the synthesis of carbon-based transparent highly potent flame retardant. Carbon 139:205–209. https://doi.org/10.1016/j.carbon.2018.06.049

Kolesnikova TA et al (2010) Nanocomposite microcontainers with high ultrasound sensitivity. Adv Func Mater 20(7):1189–1195. https://doi.org/10.1002/adfm.200902233

Kulandaivalu S, Sulaiman Y (2019) Recent advances in layer-by-layer assembled conducting polymer based composites for supercapacitors. Energies 12(11):2107

Kumbhakar Partha et al (2018) In-situ synthesis of rGO-ZnO nanocomposite for demonstration of sunlight driven enhanced photocatalytic and self-cleaning of organic dyes and tea stains of cotton fabrics. J Hazard Mater 360:193–203

Kuo W-S et al (2017) Graphene quantum dots with nitrogen-doped content dependence for highly efficient dual-modality photodynamic antimicrobial therapy and bioimaging. Biomaterials 120:185–194. https://doi.org/10.1016/J.BIOMATERIALS.2016.12.022

Lee C et al (2008) Measurement of the elastic properties and intrinsic strength of monolayer graphene. Science 321(5887):385–388. https://doi.org/10.1126/science.1157996

Lee KS et al (2017) Transparent nanofiber textiles with intercalated ZnO@ graphene QD LEDs for wearable electronics. Compos B Eng 130:70–75

Li B et al (2018) A flexible humidity sensor based on silk fabrics for human respiration monitoring. J Mater Chem C 6(16):4549–4554. https://doi.org/10.1039/c8tc00238j

Li X et al (2019) Design and development of layer-by-layer based low-pressure antifouling nanofiltration membrane used for water reclamation. J Membr Sci 584:309–323

Lim SY, Shen W, Gao Z (2015) Carbon quantum dots and their applications. Chem Soc Rev 44(1):362–381. https://doi.org/10.1039/C4CS00269E

Lin L et al (2014) Luminescent graphene quantum dots as new fluorescent materials for environmental and biological applications. TrAC Trends Anal Chem 54:83–102

Liu H et al (2018) Differently-charged graphene-based multilayer films by a layer-by-layer approach for oxygen gas barrier application. Compos B Eng 155:391–396

Ma L et al (2012) Layer-by-layer self-assembly under high gravity field. Langmuir 28(25):9849–9856

Masae M (2018) Hydrophobic and antibacterial activity of silk textile surfaces using reduced graphene oxide (RGO) and TiO_2 coating. J Mater Sci Appl Energy 7(3):307–316

Miankafshe MA, Bashir T, Persson N-K (2019) The role and importance of surface modification of polyester fabrics by chitosan and hexadecylpyridinium chloride for the electrical and electro-thermal performance of graphene-modified smart textiles. New J Chem 43(17):6643–6658

Munz M et al (2015) Thickness-dependent hydrophobicity of epitaxial graphene. ACS Nano 9(8):8401–8411. https://doi.org/10.1021/acsnano.5b03220

Muzyka R et al (2017) Oxidation of graphite by different modified hummers methods. New Carbon Mater 32(1):15–20. https://doi.org/10.1016/S1872-5805(17)60102-1

Nath N, Chilkoti A (2004) Label-free biosensing by surface plasmon resonance of nanoparticles on glass: optimization of nanoparticle size. Anal Chem 76(18):5370–5378

Nault L et al (2010) Cell transfection using layer-by-layer (LbL) coated calixarene-based solid lipid nanoparticles (SLNs). Chem Commun 46(30):5581–5583

Nishiguchi A et al (2011) Rapid construction of three-dimensional multilayered tissues with endothelial tube networks by the cell-accumulation technique. Adv Mater 23(31):3506–3510

Noi KF et al (2015) Assembly-controlled permeability of layer-by-layer polymeric microcapsules using a tapered fluidized bed. ACS Appl Mater Interfaces 7(50):27940–27947

Nooralian Z, Parvinzadeh Gashti M, Ebrahimi I (2016) Fabrication of a multifunctional graphene/polyvinylphosphonic acid/cotton nanocomposite via facile spray layer-by-layer assembly. RSC Adv 6(28):23288–23299. https://doi.org/10.1039/c6ra00296j

Ocepek B et al (2012) Printing of antimicrobial microcapsules on textiles. Color Technol 128(2):95–102

Ogi T et al (2014) Transient nature of graphene quantum dot formation via a hydrothermal reaction. RSC Adv 4(99):55709–55715. https://doi.org/10.1039/C4RA09159K

Pan N et al (2018) Fabrication of cotton fabrics through in-situ reduction of polymeric N-halamine modified graphene oxide with enhanced ultraviolet-blocking, self-cleaning, and highly efficient, and monitorable antibacterial properties. Colloids Surf A 555:765–771

Pang Y et al (2018) Facile preparation of N-doped graphene quantum dots as quick-dry fluorescent ink for anti-counterfeiting. New J Chem 42(20):17091–17095

Patel PA, Dobrynin AV, Mather PT (2007) Combined effect of spin speed and ionic strength on polyelectrolyte spin assembly. Langmuir 23(25):12589–12597

Patil PO et al (2016) Facile green synthesis of reduced graphene oxide and fabrication of layer by layer self-assembled rGO@ chitosan@ rGO@ folic acid nanocomposite for possible biosensing application. J Bionanosci 10(2):150–157

Perreault F, Fonseca de Faria A, Elimelech M (2015) Environmental applications of graphene-based nanomaterials. Chem Soc Rev 44(16):5861–5896. https://doi.org/10.1039/C5CS00021A
Picart C et al (2001) Buildup mechanism for poly (L-lysine)/hyaluronic acid films onto a solid surface. Langmuir 17(23):7414–7424
Poortavasoly H, Montazer M, Harifi T (2014) Simultaneous synthesis of nano silver and activation of polyester producing higher tensile strength aminohydroxylated fiber with antibacterial and hydrophilic properties. RSC Adv 4(86):46250–46256. https://doi.org/10.1039/C4RA04835K
Putri LK et al (2015) Heteroatom doped graphene in photocatalysis: a review. Appl Surf Sci 358:2–14. https://doi.org/10.1016/J.APSUSC.2015.08.177
Qi A et al (2011) Template-free synthesis and encapsulation technique for layer-by-layer polymer nanocarrier fabrication. ACS Nano 5(12):9583–9591
Qu D et al (2015) Formation mechanism and optimization of highly luminescent N-doped graphene quantum dots. Sci Reports 4(1):5294. https://doi.org/10.1038/srep05294
Rajasekar R et al (2013) Electrostatically assembled layer-by-layer composites containing graphene oxide for enhanced hydrogen gas barrier application. Compos Sci Technol 89:167–174
Ramasamy T et al (2014) Layer-by-layer assembly of liposomal nanoparticles with PEGylated polyelectrolytes enhances systemic delivery of multiple anticancer drugs. Acta Biomater 10(12):5116–5127
Ramsden J (2016) Nanotechnology: an introduction. William Andrew
Rao CNR, Gopalakrishnan K, Govindaraj A (2014) Synthesis, properties and applications of graphene doped with boron, nitrogen and other elements. Nano Today 9(3):324–343. https://doi.org/10.1016/J.NANTOD.2014.04.010
Razza S et al (2016) Research update: large-area deposition, coating, printing, and processing techniques for the upscaling of perovskite solar cell technology. APL Mater 4(9):91508
Richardson JJ, Björnmalm M, Caruso F (2015) Technology-driven layer-by-layer assembly of nanofilms. Science 348(6233):aaa2491
Richardson JJ et al (2016) Innovation in layer-by-layer assembly. Chem Rev 116(23):14828–14867. https://doi.org/10.1021/acs.chemrev.6b00627
Ristic BZ et al (2014) Photodynamic antibacterial effect of graphene quantum dots. Biomaterials 35(15):4428–4435. https://doi.org/10.1016/J.BIOMATERIALS.2014.02.014
Sanes J et al (2020) Extrusion of polymer nanocomposites with graphene and graphene derivative nanofillers: an overview of recent developments. Materials 13(3):549
Seo J et al (2008) Effect of the layer-by-layer (LbL) deposition method on the surface morphology and wetting behavior of hydrophobically modified PEO and PAA LbL films. Langmuir 24(15):7995–8000
Sha'rani SS et al (2019) Improving the redox flow battery performance of low-cost thin polyelectrolyte membranes by layer-by-Layer Surface assembly. J Power Sources 413:182–190
Shakir I (2014) High energy density based flexible electrochemical supercapacitors from layer-by-layer assembled multiwall carbon nanotubes and graphene. Electrochim Acta 129:396–400
Shao F et al (2016) Fabrication of polyaniline/graphene/polyester textile electrode materials for flexible supercapacitors with high capacitance and cycling stability. Chem Asian J 11(13):1906–1912
Sharma D et al (2011) Synthesis of ZnO nanoparticles using surfactant free in-air and microwave method. Appl Surf Sci 257(22):9661–9672
Shateri-Khalilabad M, Yazdanshenas ME (2013) Preparation of superhydrophobic electroconductive graphene-coated cotton cellulose. Cellulose 20(2):963–972
Shim BS et al (2007) Nanostructured thin films made by dewetting method of layer-by-layer assembly. Nano Lett 7(11):3266–3273
Shiratori SS, Yamada M (2000) Nano-scale control of composite polymer films by mass-controlled layer-by-layer sequential adsorption of polyelectrolytes. Polym Adv Technol 11(8–12):810–814
Some S et al (2015) Phosphorus-doped graphene oxide layer as a highly efficient flame retardant. Chem Eur J 21(44):15480–15485

Song W et al (2019) Graphene oxide/waterborne polyurethane composites for fine pattern fabrication and ultrastrong ultraviolet protection cotton fabric via screen printing. Appl Surf Sci 463:403–411
Srivastava S, Kotov NA (2008) Layer-by-layer (LBL) assembly with semiconductor nanoparticles and nanowires. In: Semiconductor nanocrystal quantum dots, pp 197–216
Stan MS et al (2019) Reduced graphene oxide/TiO_2 nanocomposites coating of cotton fabrics with antibacterial and self-cleaning properties. J Ind Text 49(3):277–293
Sun H et al (2014) Graphene quantum dots-band-aids used for wound disinfection. ACS Nano 8(6):6202–6210. https://doi.org/10.1021/nn501640q
Sun H et al (2016) Large-area supercapacitor textiles with novel hierarchical conducting structures. Adv Mater 28(38):8431–8438
Sun R et al (2020) A layer-by-layer architecture for printable organic solar cells overcoming the scaling lag of module efficiency. Elsevier, Joule
Tang X, Yan X (2017) Dip-coating for fibrous materials: mechanism, methods and applications. J Sol-Gel Sci Technol 81(2):378–404
Tas M et al (2019) Graphene and graphene oxide-coated polyamide monofilament yarns for fiber-shaped flexible electrodes. J Text Inst 110(1):67–73
Tawiah B, Howard EK, Asinyo B (2016) The chemistry of inkjet inks for digital textile printing—review. BEST 4:61–78
Thakur S, Karak N (2015) Alternative methods and nature-based reagents for the reduction of graphene oxide: a review. Carbon 94:224–242. https://doi.org/10.1016/J.CARBON.2015.06.030
Thomas IM (1987) Single-layer TiO_2 and multilayer TiO_2–SiO_2 optical coatings prepared from colloidal suspensions. Appl Opt 26(21):4688–4691
Tian M et al (2015) Robust ultraviolet blocking cotton fabric modified with chitosan/graphene nanocomposites. Mater Lett 145:340–343
Tissera ND et al (2015) Hydrophobic cotton textile surfaces using an amphiphilic graphene oxide (GO) coating. Appl Surf Sci 324:455–463
Voiry D et al (2016) High-quality graphene via microwave reduction of solution-exfoliated graphene oxide. Science (New York, N.Y.) 353(6306):1413–1416. https://doi.org/10.1126/science.aah3398
Wang Y et al (2011) Coupling electrodeposition with layer-by-layer assembly to address proteins within microfluidic channels. Adv Mater 23(48):5817–5821
Wang L et al (2016) Mechanism of boron and nitrogen in situ doping during graphene chemical vapor deposition growth. Carbon 98:633–637. https://doi.org/10.1016/J.CARBON.2015.11.058
Wang L et al (2020) Effects of three fabric weave textures on the electrochemical and electrical properties of reduced graphene/textile flexible electrodes. RSC Adv 10:6249–6258
Wardman RH (2018) An introduction to textile coloration: principles and practice. Wiley, Pondicherry
Wong KK (2018) Synthesis of nitrogen-doped graphene quantum dots and its antibacterial property. The Hong Kong Polytechnic University
Xie M et al (2019) Layer-by-layer modification of magnetic graphene oxide by chitosan and sodium alginate with enhanced dispersibility for targeted drug delivery and photothermal therapy. Colloids Surf, B 176:462–470
Xu Z, Gao C (2010) In situ polymerization approach to graphene-reinforced nylon-6 composites. Macromolecules 43(16):6716–6723. https://doi.org/10.1021/ma1009337
Yamada M et al (2012) Controlled formation of heterotypic hepatic micro-organoids in anisotropic hydrogel microfibers for long-term preservation of liver-specific functions. Biomaterials 33(33):8304–8315
Yang Y et al (2017) Waterproof, ultrahigh areal-capacitance, wearable supercapacitor fabrics. Adv Mater 29(19):1606679
Yang X et al (2019) Surface modification of poly (p-phenylene terephthalamide) fibers by polydopamine-polyethyleneimine/graphene oxide multilayer films to enhance interfacial adhesion with rubber matrix. Polym Testing 78:105985

Yazdi GR, Iakimova T, Yakimova R (2018) Fabrication of graphene by thermal decomposition of SiC. In: Epitaxial graphene on silicon carbide. Jenny Stanford Publishing, pp 63–109, https://doi.org/10.1201/9781315186146-3

Ye S et al (2005) Multilayer nanocapsules of polysaccharide chitosan and alginate through layer-by-layer assembly directly on PS nanoparticles for release. J Biomater Sci Polym Ed 16(7):909–923

Yi M, Shen Z (2015) A review on mechanical exfoliation for the scalable production of graphene. J Mater Chem A 3(22):11700–11715. https://doi.org/10.1039/C5TA00252D

Yu S et al (2015) Sorption of radionuclides from aqueous systems onto graphene oxide-based materials: a review. Inorg Chem Frontiers 2(7):593–612. https://doi.org/10.1039/C4QI00221K

Yuan B et al (2016) Boron/phosphorus doping for retarding the oxidation of reduced graphene oxide. Carbon 101:152–158. https://doi.org/10.1016/J.CARBON.2016.01.080

Yun YJ et al (2013) A novel method for applying reduced graphene oxide directly to electronic textiles from yarns to fabrics. Adv Mater 25(40):5701–5705

Zahid M et al (2017) Strain-responsive mercerized conductive cotton fabrics based on PEDOT:PSS/graphene. Mater Des 135:213–222. https://doi.org/10.1016/j.matdes.2017.09.026

Zhang B, Cui T (2011) An ultrasensitive and low-cost graphene sensor based on layer-by-layer nano self-assembly. Appl Phys Lett 98(7):73116

Zhang D, Tong J, Xia B (2014) Humidity-sensing properties of chemically reduced graphene oxide/polymer nanocomposite film sensor based on layer-by-layer nano self-assembly. Sens Actuators B Chem 197:66–72

Zhao J et al (2013) Graphene oxide-based antibacterial cotton fabrics. Adv Healthc Mater 2(9):1259–1266

Zhao H et al (2018) Fast and facile graphene oxide grafting on hydrophobic polyamide fabric via electrophoretic deposition route. J Mater Sci 53(13):9504–9520

Chapter 14
Nanotechnology Systems for Biofuels Production

Francisco Thálysson Tavares Cavalcante, Katerine da Silva Moreira, Paula Jéssyca Morais Lima, Rodolpho Ramilton de Castro Monteiro, Bruna Bandeira Pinheiro, Carlos Alberto Chaves Girão Neto, Kimberle Paiva dos Santos, Maria Cristiane Martins de Souza, Rita Karolinny Chaves de Lima, and José Cleiton Sousa dos Santos

1 Introduction

Petroleum-based fossil fuels account for more than 90% of global primary energy consumption and are a major energy resource (Li et al. 2019). World energy consumption tends to increase over the years, energy demand is expected to be approximately 30 TW by 2050 and 46 TW by 2100 (Sadaf et al. 2018). One of the main reasons for environmental pollution is the generation of energy from fossil fuels, and the combustion of fossil fuels causes the emission of greenhouse gases, such as CO_2. In the last 150 years, the concentration of CO_2 in the environment has increased to 370 ppm (Li et al. 2019).

The development of renewable energy emerges as a solution for reducing greenhouse gas emissions (Pradhan et al. 2018). In this context, the main sources of renewable energy today are those from natural phenomena such as solar, wind, and biomass-derived bioenergy (Li et al. 2019). For the more, developments in technologies to convert biomass into sustainable fuel are being widely studied and implemented to reduce the damage caused by fossil fuels (Li et al. 2019).

In this regard, biofuels are classified into two groups, namely primary and secondary biofuels (Sekoai et al. 2019). Thus, first-generation biofuels are produced from sucrose, animal fats, plant oils, and crops such as corn, wheat, among others

F. T. T. Cavalcante · K. da Silva Moreira · P. J. M. Lima · R. R. de Castro Monteiro ·
B. B. Pinheiro · C. A. C. G. Neto · K. P. dos Santos
Departamento de Engenharia Química, Universidade Federal do Ceará, Campus do Pici, CEP, Fortaleza, CE 60455-760, Brazil

M. C. M. de Souza · R. K. C. de Lima · J. C. S. dos Santos (✉)
Instituto de Engenharias E Desenvolvimento Sustentável, Universidade da Integração Internacional da Lusofonia Afro-Brasileira, Rua José Franco de Oliveira, S/N, CEP, Redenção, CE 62790-970, Brazil
e-mail: jcs@unilab.edu.br

R. F. do. Nascimento et al. (eds.), *Nanomaterials and Nanotechnology*,
Materials Horizons: From Nature to Nanomaterials,
https://doi.org/10.1007/978-981-33-6056-3_14

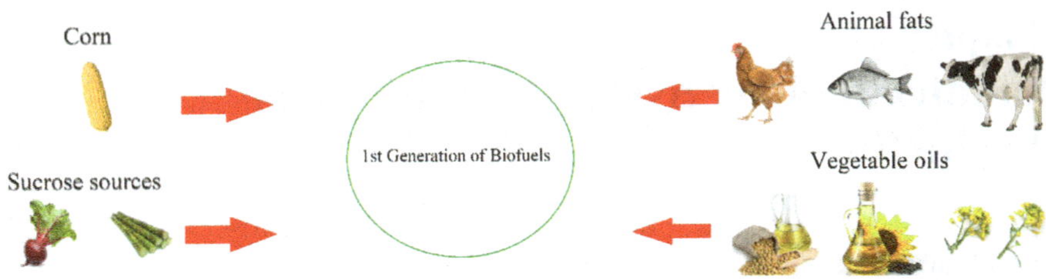

Fig. 1 Sources of the first generation of biofuels

(Fig. 1). Meanwhile, second-generation biofuels use non-food raw material as wood waste, agricultural waste, lignocellulosic biomass, among others (Leong et al. 2018; Zhang et al. 2018). In addition to these two types of biofuel generation, there is a third-generation employing processing microalgae to produce biofuel (Leong et al. 2018), and a fourth generation focused on modifying the metabolism of these organisms, lowering production costs and increasing production (Moravvej et al. 2019). Some examples of biofuels generated from biomass are biodiesel, bioethanol, biohydrogen, and biogas (Sekoai et al. 2019).

Biodiesel is a clean, biodegradable, non-toxic, and low pollutant biofuel. In addition, it can be obtained from various renewable sources (Nisar et al. 2017; Tian et al. 2017). Biodiesel has environmental technical and strategic advantages: It can reduce most exhaust emissions, it is biodegradable and has inherent lubricity (Rai et al. 2016). In recent years, the biodiesel industry has grown considerably, and globally the biodiesel industry will expand its production with an annual expansion of 7.3% to $54.8 billion in 2025 (Sekoai et al. 2019).

Other types of biofuels, such as bioethanol and biogas, are alternative candidates that can be used in the renewable energy market (Cesaro and Belgiorno 2015). Bioethanol is also considered a clean, renewable, and non-toxic fuel and can be obtained from various biomass raw materials such as sugar cane, cornstarch, and algae (Sirajunnisa and Surendhiran 2016). In 2017, global ethanol production was estimated at approximately 100 billion liters and is expected to double in the next decade (Aditiya et al. 2016). The technique for production of second-generation bioethanol is not yet stable compared to first-generation bioethanol. However, bioethanol obtained from non-food raw materials does not present socioeconomic questions because they are considered waste (Sekoai and Daramola 2017). On the other hand, biogas is a biofuel produced from anaerobic digestions of microorganisms (Bundhoo and Mohee 2016). It is mostly composed of 50–75% methane (CH_4), 25–45% carbon dioxide (CO_2), and other components in small quantities (Andritz Group 2013).

Biohydrogen is another biofuel, which has a high energy content that is considered three times higher than fossil fuels, uses various raw materials such as organic effluents and can be generated under operating conditions of pressure and ambient temperature, enabling the large-scale production (Sekoai et al. 2019). Biohydrogen production is increased by nanomaterials since these materials can potentiate the activity of microorganisms that possess physicochemical properties (Pugazhendhi et al. 2019).

Fortunately, the application of nanotechnology in recent years has intensified in various segments, such as the food, agricultural, cosmetic, pharmaceutical, and electronic industries, due to the ability to use various nanoscale materials, such as nanoparticles, in a range from 1 to 100 nm (Sekoai et al. 2018; Tyagi et al. 2018). This use of nanomaterials in various sectors is attributed to the properties of nanoparticles, which include high reactivity and structure/morphology (Sekoai et al. 2018; Tyagi et al. 2018). Nanoparticles have other characteristics such as high degree of crystallinity, catalytic activity, stability, durability, efficient storage, high recovery potential, reuse, and recycling make these materials exceptional candidates for biofuel systems (Nizami and Rehan 2018).

Due to these characteristics, many nanomaterials can be applied in biofuel production. For example, nanoparticles can be used mainly as catalysts because they perform an important job in electron transfer, improve anaerobic agent activities, and reduce inhibitory substances (de Vasconcellos et al. 2018). However, some nanosystems suffer from some technical issues and economic viability, but continuous researches are being developed and some of the future trends will be pointed at the end of this chapter, with recommendations for further studies.

2 Biofuels Production with Nanotechnology Systems

Application of nanotechnology in biofuel production, especially in biocatalysts, is being reported in several studies to increase process efficiency (Biswas 2019). The small size of nanoparticles and different forms in nature (metallic, semiconductor, or polymeric) imply in a big versatility and are an advantage in biocatalysis, due to higher surface area, which increases catalytic activity, better stability, and reusability (Ahmadi et al. 2019). They also present a high degree of crystallinity and could be synthesized using a top-down or bottom-up approach (Sekoai et al. 2019). This chapter will broach these and many other interesting characteristics and applications of nanotechnology in biofuels production, showing which barriers are being studied to be overcome in the future.

2.1 Biodiesel

Biodiesel is a very promising fuel as it is produced especially from renewable energy sources or biomass (Fig. 2), is biodegradable and contributes to the reduction of combustion emissions (Rosha et al. 2019; Sekoai et al. 2019). Biodiesel accounts for 82% of total biofuel production, making it the main alternative to diesel (Bozbas

Fig. 2 Biomass for biodiesel production

2008; Sekoai et al. 2019), and vegetable oils are the main source of biodiesel production. However, due to the great competition with their use in cooking and food production, their use is becoming increasingly expensive (Muniru et al. 2018; Zhang et al. 2013). Animal fats, soaps, greases, inedible oils, used frying oils are examples of low-cost biodiesel feedstock, however, their quantities available are not sufficient to meet the current requirements of this biofuel (Fingerman et al. 2018). Thus, microalgae, an inedible raw material that does not interfere with the global food economy, is gaining great importance today (Banerjee et al. 2019). Given the need to reduce environmental impacts and ensure the same level of performance of existing fuels, microalgae can contribute to a reduction in the need for land due to their energy yield, high growth rate in a short time, high production of biomass, and the lack of agricultural land (Lee et al. 2010). Microalgae also stand out for having higher photosynthetic efficiency when compared to plants (Banerjee et al. 2019).

Many studies have also been performed on lipase immobilization by nanomaterials (Bezerra et al. 2017; Costa et al. 2016; Galvão et al. 2018; Monteiro et al. 2019; Rios et al. 2019; Souza et al. 2017). Lipase from *Candida rugosa* was covalently linked to a magnetic microsphere on Fe_3O_4/Poly (styrene-methacrylic acid) for biodiesel production from soybean oil (Xie and Wang 2014). In this nanosystem, the developed immobilized enzyme converted methanol and soybean oil to 86% FAME and the biocatalyst remained active for 4 cycles without significant loss of activity (Xie and Wang 2014). In another investigation with a nanosystem (Kalantari et al. 2013), lipase from *Pseudomonas cepacian* was bound to the nanocomposite particles of non-porous silica-coated magnetite agglomerates covalently amino functionalized by glutaraldehyde as a coupling agent. The authors tested the FAME conversion with the free enzyme, where the conversion was only 34% and with the immobilized enzyme, obtaining 54% conversion, conserved after 5 times reused (Kalantari et al. 2013). In nanosystem presented by Tran et al. (2012) (Tran et al. 2012), the lipase of *Burkholderia* sp. was immobilized on a ferric silica nanocomposite for biodiesel production. They coated the Fe_3O_4 core with silica shell to synthesize core-shell nanoparticles, where they achieved 90% FAME yield in 30 h in a batch operation, further validating the new support and good selection of the immobilization method and immobilized lipase was used for 10 cycles without significant loss of activity.

In another study on nanosystem, Wang et al. (2011) designed a four packed bed reactors (PBR) system for repeated and highly efficient lipase use. Both the conversion rate and stability obtained using the four accumulated bed reactor system were much higher than that obtained using the single PBR, and the biodiesel conversion was maintained at a high rate of over 88% for 8 days (Wang et al., 2011). According to the authors due to the longer residence time of the reaction mixture in the reactor and the reduction in inhibition of lipase-nanoparticle biocomposite by-products, it was possible to obtain a high conversion rate and considerable stability in the four-bed reactor, which further contributes to reducing the cost of biodiesel production. Thus, the advantages of using an immobilized nanobiocatalytic system could outweigh the costs of the immobilization step (Wang et al. 2011). These reactor scale studies highlight the possibility of designing and operating even industrial scale enzyme systems for biodiesel production (Verma et al. 2013).

Therefore, considering the above, it is important to emphasize that the interdisciplinary combination of biotechnology and nanotechnology represents a very promising opportunity, increasing the amount and efficiency of biodiesel production. The possibility of reusing stable and efficient nanobiocatalytic systems could greatly improve the economic viability of biodiesel.

2.2 *Bio-Jet Fuels*

Air transport is necessary for world tourism and trade. The International Air Transport Association (IATA) reported that about 4.0 billion passengers and 64 million tons of cargo were transported by air in 2018, generating 65 billion jobs and underpinning $2.7 trillion (3.6%) of the world's gross domestic product (IATA 2019). Indeed, the global aviation industry is projected to expand, creating 90 million jobs and nearly $6 trillion in annual economic activities by 2034; besides, the number of air passengers will double in the next 20 years (Yang et al. 2019).

The growth in air traffic requires a big consumption of jet fuels; in fact, according to US Energy Information Administration (EIA), excluding electricity, jet fuel consumption is projected to grow more than any other transportation fuel, rising 35% from 2018 to 2050 (EIA 2019), as well as the average price of jet fuel is expected to grow at a 2.7% annual rate from 2016 to 2050 (EIA 2019). Jet fuel costs account for about 24% of all the airline's operating costs in 2018, and it is mainly related to oil prices (IATA 2019). Furthermore, the large consumption of jet fuel increases greenhouse gas emissions. As it was reported by Air Transport Action Group, worldwide, flights produced 895 million tons of CO_2 in 2018; thus, the global aviation industry produces more than 2% of all human-induced CO_2 emissions and 12% of the transport sector (ATAG 2019). As a result, the aviation industry aims to achieve a 50% reduction in CO_2 emission by 2050 as compared to 2005s level (Wei et al. 2019). Therefore, to ensure sustainable development of the aviation industry, to rely only on improving fuel efficiency is not enough, it is crucial to develop an alternative renewable and environmental innocuous fuel to meet the growing demand and reduce the dependency of fossil fuel (Wei et al. 2019).

Bio-based jet fuels or, for short, bio-jet fuel (also known as synthetic paraffinic kerosene) consists of renewable hydrocarbons which properties are almost identical or, in some cases, superior to those of fossil jet fuel (Gutiérrez-Antonio et al. 2017). The properties of bio-jet fuels are related to their chemical composition, which mainly defines the performance characteristics (low-temperature fluidity, thermal oxidation stability, combustion property, fuel compatibility with current aviation system, fuel volatility, and fuel metering and aircraft range) of bio-jet fuels in turbine aero-engines (Braun-Unkhoff and Riedel 2015).

For bio-jet fuel production, as for other biofuels and bioenergy field in general, nanotechnology has different applications, such as the modifications of feedstocks, development of more efficient catalysts, among others (Andrade et al. 2011).

In this sense, depending on the characteristic of the triglyceride's feedstocks, it is necessary to remove oxygenated compounds and obtain n-alkanes, which ensures the energy density of the fuel, by hydrodeoxygenation (Itthibenchapong et al. 2017). For upgrading bio-oils to produce bio-jet fuels by removal of oxygen using hydrogen, noble metal catalysts, which are environmentally friendly nanoparticles doped on porous or acid/base solid supports, such as Pt, Pd, Rh, and Ru have been used (Yang et al. 2019).

As an example, Gutierrez et al. (2009) studied the upgrading of a bio-oil by hydrodeoxygenation using zirconia-supported mono- and bimetallic noble metal (Rh, Pd, Pt) catalysts, which were active and selective in the hydrodeoxygenation guaiacol at 300 °C (Gutierrez et al. 2009). In addition, silica nanoparticles, an environmentally friendly catalyst support, have been widely used in the hydrodeoxygenation reactions. For instance, Duan et al. (2017) reported an efficient hydrodeoxygenation of water-insoluble bio-oil to alkanes by using a catalyst that combines highly dispersed palladium and ultrafine molybdenum phosphate nanoparticles on silica (Duan et al. 2017).

Besides bio-oils, bio-jet fuels can be produced from different kind of biomass feedstocks, like hemicellulose, cellulose, and lignin. Lignin is the only large-scale biomass source with an aromatic functionality; nevertheless, its unique aromatic structure is quite stable. Indeed, the transformation of lignin into C8–C15 cycloparaffins and aromatics provides a pathway for the development of aromatic components in bio-jet fuels using lignin (Yang et al. 2018). As an example, Bi et al. (2015) reported the transformation of lignin into jet and diesel fuel, involving directional production of C8–C15 aromatics with HZSM-5 nanocomposite and C8–C15 cycloparaffins by the hydrogenation of aromatics over palladium on carbon nanocatalyst (Bi et al. 2015).

By the above, bio-jet fuel has been emerging as an alternative and complementary to jet fuels, once it mitigates CO_2 emissions and helps to meet the increasing demand of fuels of aviation industry. In this regard, nanosystems have been mainly used to modify bio-jet fuels feedstock in order to upgrade the properties of the biofuel.

2.3 *Biohydrogen*

Considered as an efficient and promising "energy-carrier," hydrogen has gained global attention focus because it is a fuel with high energy content and carbon-free, therefore the only final combustion by-product is water, which makes it favorable for reduction in GHG emissions (Chezeau et al. 2019). It can be applied either in combustion engines or to generate electricity by using fuel cell technologies, since its similarity to electricity is greater than fossil fuels in terms of energy systems (Chandrasekhar et al. 2015; Ghimire et al. 2015). The major hindrance of its use is related to the production for commercial applications, which is performed mainly by using fossil fuels in a thermo-chemical process, and the storage (Al-Mohammedawi et al. 2019; Ghimire et al. 2015; Ren et al. 2019). Hydrogen biological generation is an alternative pathway that allows an eco-friendly and attractive renewable production (Fakhimi et al. 2019; García-Depraect et al. 2019).

Biohydrogen is a second-generation biofuel that can be obtained by biological pathways which requires a lower energy supply and is less harmful in terms of CO_2 emission. Its production can be performed by some process using the biomass of several categories (Dinesh Kumar et al. 2019; Ghimire et al. 2015; Ren et al. 2019). The use of lignocellulosic substrates presents several advantages for they are abundant and renewable materials cultivated in the no arable soils and thus it does not cause any competition for food production (Kumar et al. 2015; Sambusiti et al. 2015). The bioprocess could be even more attractive when the cost is reduced with the utilization of low-value material, as agro-industrial residues and organic municipal waste and wastewater (Ghimire et al. 2015).

Biohydrogen can be generally obtained by some process, as dark fermentation, photo-fermentation, water biophotolysis, and indirect biophotolysis which have their own positive and negative points (Seelert et al. 2015). Due to the possibility to use cheap organic substances as substrate, as food and industrial waste, photo and dark fermentation are the most explored and reported techniques in literature (Cai et al. 2019; Wimonsong et al. 2014). Figure 3 shows mechanisms for biohydrogen production.

Photo-fermentation is a widely studied method for biohydrogen production for it allows a relatively high substrate conversion and can make use of several substrates as carbon source. Besides that, it can be performed at ambient conditions, which allows a lower energy demand (Al-Mohammedawi et al. 2019; Mirza et al. 2019). In

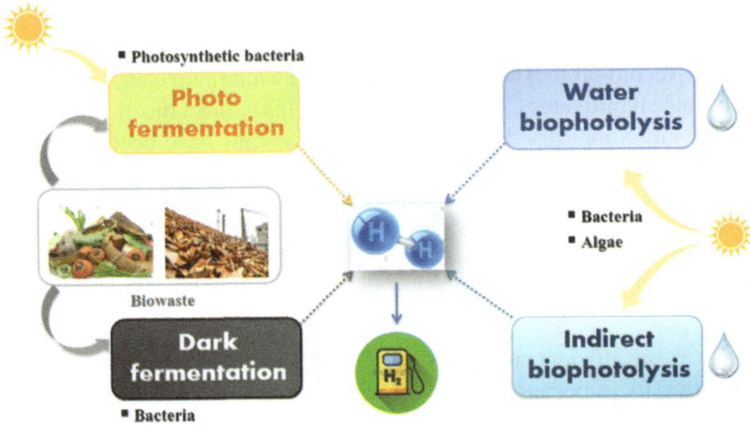

Fig. 3 Mechanisms and sources for biohydrogen production

this process, the production is regulated by two key enzymes, the membrane-bound hydrogenases and the nitrogenases (Dolly et al. 2015; Show et al. 2019)

In an anaerobic dark fermentation, the biohydrogen production is achieved by a redox reaction catalyzed by hydrogenase (Chezeau et al. 2019). The dark fermentation presents some good environmental and economic benefits for the biohydrogen production that makes it a viable way to that fuel obtaining (Malik et al. 2014; Palomo-Briones et al. 2019). The process can be operated under mild conditions of temperature and pressure, it presents a better resistance against contamination, it accepts different substrate species and generally presents a greater efficiency when compared to the other biological process for that production (García-Depraect et al. 2019; Malik et al. 2014).

However, several barriers remain a challenge to be overcome for successful biohydrogen production and commercialization, as the low substrate conversion and production rates, which decrease the processes efficiency (Chandrasekhar et al. 2015; Fakhimi et al. 2019; Show et al. 2019). In this context, nanotechnology systems have attracted attention because it may enhance several process of hydrogen production, both by dark and by photo fermentation (Malik et al. 2014; Wimonsong et al. 2014). Because of the high specific surface of nanoparticles, its simple addition influences directly on the reaction rate. Besides that, these particles might induce the interactions inside the microorganism improving the electrons transference which benefits the production kinetics (Jafari and Zilouei 2016).

In this point, organic and nonorganic nanoparticles have been studied as a carrier for cell immobilization aiming the improvement in biohydrogen production (Kumar et al. 2016). Several studies involving metallic nanoparticles have been realized because of their ability to react quickly with the electron donors, which enhance the reaction kinetic (Dolly et al. 2015). The addition of metal nanoparticles, like hematite, gold, nickel, and silver, to dark fermentation process has showed to improve the process efficiency, by increasing the hydrogen yield and microbes bioactivity, for example, and also by reducing the lag phase for the biohydrogen production (Hsieh et al. 2016; Pugazhendhi et al. 2018; Zaidi et al. 2019). Some of those studies attribute these results to stimulation of hydrogenase enzyme activity increase caused by the metal addition (Hsieh et al. 2016).

2.3.1 Iron and Nickel Nanoparticles as Nanosystems for Biohydrogen Production

Bulk and nanoparticles form of iron sulfate were studied as a supplement on photo-biohydrogen production and both forms achieved the highest cumulative hydrogen production at the same concentration in mg/L (Dolly et al. 2015). However, this maximum result obtained for the nano form was 1.2 fold higher than the bulk form, which testifies the enhancement caused by the utilization of nanosized particles (Dolly et al. 2015).

The addition of Fe and Ni to the fermentation has shown a direct influence on the activity of *hydrogenase*, the key enzyme in the hydrogen production by dark fermentation bioprocess, because these metals are present in its active site and many studies have demonstrated an improvement of results by adding iron and nickel salts (Elreedy et al. 2017; Gadhe et al. 2015). Besides that, the Fe and Ni nanoparticles utilization may enhance also the activity of the *ferredoxin oxidoreductase*—which can indirectly generate hydrogen in dark fermentation—for they can ameliorate the electron transfer rate due to surface increase, which makes this utilization attractive (Gadhe et al. 2015). In process of anaerobic digestion of an industrial wastewater containing mono-ethylene glycol (used in petrochemical productions), nanoparticles composed of Ni and graphene were added as a supplement to the process and there was an increase of more than 100% in the biohydrogen production (Elreedy et al. 2017).

Iron (II) oxide nanoparticles were added to an anaerobic process, conducted at a thermophilic condition, aiming the biohydrogen improvement using glucose as the substrate. The results showed a rate augmentation with no relevant modifications in the metabolic pathway and the authors considered this addition as a vital factor to the required enhancement achievement. Another nanosystem using molasses-based distillery wastewater to the biological production of hydrogen, iron oxide nanoparticles were added in the process and showed to make the hydrogen production increase (Malik et al. 2014). Ferrihydrite nanoparticles present an enhancement effect in a dark fermentation for hydrogen production with an increase of more than 60% in the production (Zhang et al. 2019a, b). The Fe supplementation has attracted attention also in photo-fermentation because it is part of the structure of the key enzymes—hydrogenase and nitrogenase—and it can work in the electron transfer chain (Dolly et al. 2015).

2.3.2 Hematite Nanoparticles for Biohydrogen Production

Hematite (Fe_2O_3) and nickel oxide (NiO) nanoparticles added to a process of biohydrogen generation from dairy wastewater presented an enhancement in the production by shortening the lag phase and increasing the microbial activity (Gadhe et al. 2015). Both individual and conjoint addition presented a positive effect, however best results were obtained when the conjoint addition was performed. A significant positive effect was obtained by the addition of TiO_2 and magnetic hematite (Fe) nanoparticle in a dark fermentation by *Clostridium pasteurianum* for the biohydrogen production (Hsieh et al. 2016). By performing gene expression measurements, the authors noted that the nanoparticles addition effect was not at the hydrogenase gene level, which was speculated in the previous literature, however, it was attributed to the enhanced electron transference.

2.3.3 Others Nanosystems for Biohydrogen Production

Literature reports that gold can be applied as the catalyst in a great variety of reactions. Usually, the activity of the gold particles is related to the size and other factors, as the support nature and the preparation method (Sekoai et al. 2019). Different metallic nanocatalysts (Au/Fe-Zn-Mg–Al-O) supported onto hydrotalcite were applied in dark fermentation with sucrose as the substrate for biohydrogen production (Wimonsong et al. 2014). The greatest achieved increase in the hydrogen yield (2 times higher than the control) was attributed to the Au, that can act as electron sinks, supported on a surface having Zn, which can work as the active site for the key enzyme hydrogenase.

Nano-spray dried particles of ferric citrate, ferrous sulfate, nickel chloride, and nickel acetate were added to biohydrogen process production by dark fermentation of crude glycerol generated by biodiesel industry (Sarma et al. 2014). It was observed that only ferric citrate particles showed an enhancement of hydrogen production, with a yield value more than 2 times higher than the standard process. Nano-titanium dioxide was applied in the pre-treatment of sugarcane bagasse for using in a dark fermentation process and the results showed an improvement of fermentation efficiency (Jafari and Zilouei 2016).

In this way, it can be noticed that nanosystems from different nature have shown an enhancement effect in different process for biohydrogen generation, by increasing different production parameters.

2.4 Algae-Derived Systems for Biofuels Production

Algae-derived biofuels present great potential. However, they suffer from high production costs and other technical barriers. For example, wastewater-algal biomass could have microbial contaminants compromising the production yield (Limayem et al. 2016). Nanotechnology and nanomaterials could be used to improve production efficiency and reduce processing cost.

2.4.1 Nanotechnology Applied to Algae Biofuel Production

Algae Cultivation

To cheap production costs and increase biomass growth, nanomaterials could be applied in light emitting diodes (LEDs) for artificial illumination of algae, manipulating illumination properties (Pompa et al. 2006). Metal nanoparticles can also be coupled to localized surface plasmon resonance and this technique could avoid photo-inhibition and optimize light frequencies preferred by algal species (Torkamani et al. 2010). For insurance of an adequate provision of CO_2 and nutrients for the growing cells, the use of nanosystems is also possible: using a mix of micro- and nano-bubbles in airlift loop bioreactors can increase mass transfer efficiency, the residence time of the bubbles and reduce energy consumption of the provision process (Zimmerman et al. 2010, 2011).

As it was presented, there is a problem with microbial contaminants of wastewater-algal biomass, but nanotechnology can overcome it. A combination of natural chitosan and zinc oxide nanoparticles was applied, with positive results. The nanoparticles were proven effective inhibitory agents against microbial lytic groups, like *Pseudomonas* spp., *Micrococcus luteus*, and *Bacilus pumilus*. Another advantage of the process was that did not compromise algae cells, showing great potential for in situ interventions (Limayem et al. 2016).

Algae Harvesting

Nanosystems are also being applied to extract oil from algae without breaking their cell. The spongiosum characteristic of some mesoporous nanoparticles enables an entrapping of lipids molecules and catalysts such as oxides of strontium and calcium can also be introduced into the pore structure and transesterify their molecules in situ (Akubude et al. 2019; Zhang et al. 2013).

In this way, the application of magnetophoretic separation technology using magnetite nanoparticles can provide rapid growth and high efficiency for algal harvesting. The process needs further studies to become economically practical, but a good strategy is coating the nanoparticles with a cationic polymer, reducing the costs and enabling reuse. The use of UV radiation increases algal harvesting efficiency and can also bring potential for large-scale production of algae-derived biofuels (Ge et al. 2014). A successful similar nanosystem process was also applied, obtaining 99% of harvesting efficiency of *Chlorella* sp. KR-1 using chitosan-Fe_3O_4 nanoparticles, without any adverse effects on microalgal growth, enabling medium recycling and therefore, lowering the cost of the employment of the biofuel from this source (Abo Markeb et al. 2019; Lee et al. 2013).

Biomass Transformation and Biofuel Employment

The use of enzymes to catalyze biomass transformations suffers from high cost and low stability of the catalysts. Using nanoscale structures can eliminate these problems, improving product quality, with high specific surface area, high catalytic activity, high resistance to saponification, and good rigidity (Akubude et al. 2019). Nanostructured supports are used to immobilize enzymes, like lipases, and this was already proved to increase transesterification yields for biodiesel production from different sources of oil (algae included), when comparing with traditional supports (Tahvildari et al. 2015; Teo et al. 2016). Besides that, nanomaterials can also be used as additives for the algae biodiesel, improving brake thermal efficiency and reducing exhaust emissions (Karthikeyan and Prabhakaran 2018; Karthikeyan and Prathima 2017).

2.5 *Nanosystems for Biogas Production*

The development of researchers to investigate the effects of nanomaterials (1–100 nm) on the production of biogas is currently intensified by the increasing demand for green energy (Baniamerian et al. 2019). As described by Zhang et al. (2016), biogas is a gaseous mixture containing mainly methane (CH_4) and carbon dioxide (CO_2) (Zhang et al. 2016). Sekoai et al. (2019) report CH_4 percentages of 50–75%, followed by 25–45% of CO_2 (Sekoai et al. 2019). Some impurities, like hydrogen sulfide (H_2S) and ammonia (NH_3), are also presented (Kadam and Panwar 2017). Moreover, typical biogas has some nitrogen (N_2), oxygen (O_2), carbon monoxide (CO), water vapor, siloxanes, dust particles, halogenated and aromatic compounds (Abdeen et al. 2016).

In this way, the anaerobic digestion of organic matter by microorganisms and enzymes, which results in biogas production, takes place in four complex steps: hydrolysis, acidogenesis, acetogenesis, and methanogenesis (Sahota et al. 2018). As stated by Kadam and Panwar (2017), the total biochemical process depends on different experimental factors, e.g., type of organic substrates, temperature changes, pH and inoculum concentration (Kadam and Panwar 2017). Several liquids and solids substrates can be treated by anaerobic digestion, including microalgae, agricultural residues, animal manure, wastewater, waste activated sludge, food waste, yard waste and solid organic waste (Mushtaq et al. 2016; Vasco-Correa et al. 2018).

Remarkably, the decomposition of municipal solid waste (MSW) in landfills is also interesting (Lima et al. 2018; Seman et al. 2019). According to Kormi et al. (2017), landfills are among the three largest anthropogenic sources of methane emissions in the world (Kormi et al. 2017). Thus, in addition to energy benefits, proper recovery of landfill biogas reduces local greenhouse gas emissions (Villanueva-Estrada et al. 2019). Xu et al. (2019) report that the amount of landfill produced biogas can be improved by controlling moisture conditions for MSW microorganisms (Xu et al. 2019). Besides, biogas production rates are affected by the type and age of landfill waste, properties of landfill leachate and environmental conditions (Plocoste et al. 2016). Recent works indicate that nanomaterials often found in landfills, due to the disposal of diverse commercial products, can also influence the biogas generation(Xu et al. 2019; Plocoste et al. 2016).

Temizel et al. (2017) investigated the effect of nano-ZnO on biogas generation from sanitary landfills (Temizel et al. 2017). Two conventional and two bioreactor landfills, operated at mesophilic temperature (35 °C) for about 1 year, were used in the experiments. The results suggest that the presence of nano-ZnO in MSW may decrease the potential of landfill biogas production. In the reported work, conventional and bioreactor landfills produced 15% less biogas in comparison with the control reactors. The authors suggested that the release of Zn^{2+} from nano-ZnO affects adversely methanogenesis phase by reducing the activity of key enzymes (Temizel et al. 2017). The same hypothesis was raised by Mu and Chen (2011) when they studied the impact of nano-ZnO on anaerobic digestion of waste activated sludge (Mu and Chen 2011). In another work, Mu et al. (2011) concluded that nano-ZnO has an inhibition effect on anaerobic digestion of waste activated sludge much lower than nano-TiO_2, nano-Al_2O_3, and nano-SiO_2 (Mu et al. 2011). Zhang et al. (2017) revealed that the Zn^{2+} released from nano-ZnO reduces the biogas production and it can be one of the main reasons for the positive impact on volatile fatty acids (VFAs) accumulation in sludge anaerobic digestion (Zhang et al. 2017).

In another approach, Luna-del Risco et al. (2011) studied the influence of copper and zinc oxide particles size on biogas production from mesophilic anaerobic digestion of cattle manure (Luna-del Risco et al. 2011). The tests were performed in gas-tight closed serum bottles during a period of 14 days. The results showed that both nano-Cu and nano-ZnO have inhibitory effects on biogas production. Relative to total biogas obtained in the control sample, a nano-CuO concentration of 15 mg/L reduced in 30% the biogas production. In the presence of 120 and 240 mg/L of nano-ZnO, the biogas production decreased by 43 and 74% (Luna-del Risco et al. 2011). Due to their toxicity on the methanogens, other nanomaterial additives, such as Mn_2O_3 (Gonzalez-Estrella et al. 2013) and Al_2O_3 (Alvarez and Cervantes 2012), are also associated with negative influence on biogas production.

On the other hand, application of nanosystems of supplementation of solid organic wastes (SOWs) with trace of metal nanoparticles is a promising way to enhance biogas production in view of the fact of many of the enzymes associated with methanogenesis step require a co-factor for its activation (Choong et al. 2016). As an example of nanoparticles commonly added into SOWs, Zhang et al. (2019a, b) quote zero-valent metals, metal oxides, nano-ash, and carbon-based materials (Zhang et al. (2019a, b). The work of Liu et al. (2015) demonstrated that zero-valent iron nanoparticles have great potential to promote significant enhancement in biogas yields obtained from waste activated sludge (Liu et al. 2015). In relation to such substrate, at mesophilic conditions, Su et al. (2013) found that the addition of 0.1 wt% zero-valent iron (ZVI) nanoparticles improves in 30.4% the biogas production rates (Su et al. 2013). About biogas production from anaerobic digestion of cattle dung slurry, Abdelsalam et al. (2017) reported that trace of metal nanoparticles such as 1 mg/L Co, 2 mg/L Ni, 20 mg/L Fe, and 20 mg/L Fe_3O_4 has demonstrated positive effects (Abdelsalam et al. 2017).

Abdallah et al. (2019) studied the impact of Ni-Ferrite and Ni-Co-Ferrite nanoparticles on biogas production from anaerobic digestion of cow manure (Abdallah et al. 2019). The experiments were carried out in 35 days, at mesophilic conditions. The results indicated that the use of Ni-Ferrite nanoparticles, at concentrations of 20, 70, and 130 mg/L, resulted in biogas enhancements of 30.8%, 28.5%, and 17.9%, respectively. Besides, Ni-Co-Ferrite nanoparticles, also at concentrations of 20, 70, and 130 mg/L, were able to increase the biogas production by 6.6%, 5.9%, and 32.9%, respectively (Abdallah et al. 2019). According to Sekoai et al. (2019), nanoparticles provide a large surface area for microorganisms to bind in active sites of molecules, which results in the increase of the substrate conversion by hydrolysis. In addition, the positive effects of nanomaterials on anaerobic digestion are attributed to their high reactivity and specificity (Sekoai et al. 2019).

Nanomaterials can also affect the biogas treatment for contaminant removal purposes. Ma and Zou (2018) investigated the effect of Cu and CuO nanoparticles on the removal of hydrogen sulfide in biogas using methyl diethanolamine (MDEA) as solvent (Ma and Zou 2018). The results showed that both nanoparticles tested, Cu and CuO can promote the gas-liquid mass transfer in the desulfurization process of biogas by MDEA. However, MDEA-based CuO nanofluids presented better absorption efficiency than MDEA-based Cu nanofluids (Ma and Zou 2018). All these works can conclude that nanoparticles could affect positively or negatively in biogas production, depending on the biogas source or in which step nanoparticles are employed.

2.6 Bioethanol

Modern use of nanosystems: enzymes-(magnetic nanoparticles) can increase ethanol production by optimizing the production of second-generation ethanol (E2G) from agricultural waste. The nanosystem for biofuels from lignocellulose biomass is still in its initial stage. The E2G ethanol has the advantages of not competing with food; advance technology still under development to reduce the cost of conversion and environmentally friendly (Rodríguez-Couto 2019). The biofuels industry has been looking for new biocatalysts to meet the growing demand for ethanol.

In this context, the prospect of using the cellulase enzyme complex to hydrolyze lignocellulosic biomass trapped in magnetic nanoparticles for E2G production is presented from the following characteristics: higher pH and temperature tolerance, higher storage stability, reusability in reaction cycles, and higher substrate affinity (Zhang et al. 2015).

Lupoi and Smith (2011) studied the hydrolysis of microcrystalline cellulose for cellulase-silica nanoparticles and obtained ethanol yields of 2.1 (P 1/4 0.06) to 2.3 (P 1/4 0.01) times higher in simultaneous saccharification and fermentation (SSF) compared to the enzyme in solution. These results indicate more technical and economic reliability for the use of nanosystems in the ethanol (E2G) production process (Lupoi and Smith 2011).

Different biomass was also studied to produce E2G ethanol using nanosystems. The jackfruit biomass and sugarcane leaves using the nanosystem cellulase supported on MnO_2 nanoparticles for E2G production (yield of 21.96 g/L) (Cherian et al. 2015). The biomass from sesbania aculeate using the nanosystem cellulase bound magnetic nanoparticles with E2G yields (5.31 g/L) (Baskar et al. 2016).

Carli et al. (2019) studied the use of nanosystems: enzymes β-glucosidase and endoglucanase supported on to functionalize magnetic nanoparticles for sugarcane bagasse hydrolysis. The hydrolysis of sugarcane bagasse alcohol insoluble residue treatment (AIR) and bagasse pretreated with dimethyl sulfoxide (DMSO) for incubation of nanosystems presented the total reducing sugar released with DMSO treated sugarcane were about 1.9-, 3.0-, and 2.1-fold higher than that released from AIR. The study also noted the improvement of the catalytic activity, high optimum temperature, with easy recovery using the magnetic field.

The hydrolysis feedstocks are available in bulk and in industrial facilities, resulting in an industrial facility, reducing costs compared to others biomass presented. Thus, the development of new technologies, nanosystems, should be designed in such a way that also encompasses solutions for the regional problems. Agribusiness brings different possibilities, especially in the use of energy residues, which even used, still does not have its full potential.

2.7 *Other Biofuels Production Using Nanotechnology Systems*

Several other biofuels have the potential to be obtained using nanosystems, among which we can mention biomethanol, biobutanol, green diesel, among others, and these are briefly discussed in the present topic (Fig. 4)

In addition to the biofuels already discussed, we can also mention another very important one, biomethanol. In general, the term biomethanol is used to describe methanol produced from two different methods, being Fischer–Tropsch reaction of syngas or biomethane (Minteer 2011). Methanol is the simplest alcohol with the chemical formula (CH_3OH) and compared to ethanol, is more volatile and more toxic. The raw materials that can be used for methanol production can be concentrated carbonaceous materials of any kind, such as biomass, coal, or even carbon dioxide and solid waste (Melikoglu et al. 2016). Among the main benefits of methanol, we can mention that it is a distributed energy source for power generation (Suntana et al. 2009) and after its combustion can easily be decomposed into carbon dioxide and water vapor (Shamsul et al. 2014).

In water resource recovery facilities (WRRFs), methanol is a largely used exogenous carbon source for denitrification, in order to achieve low total nitrogen levels

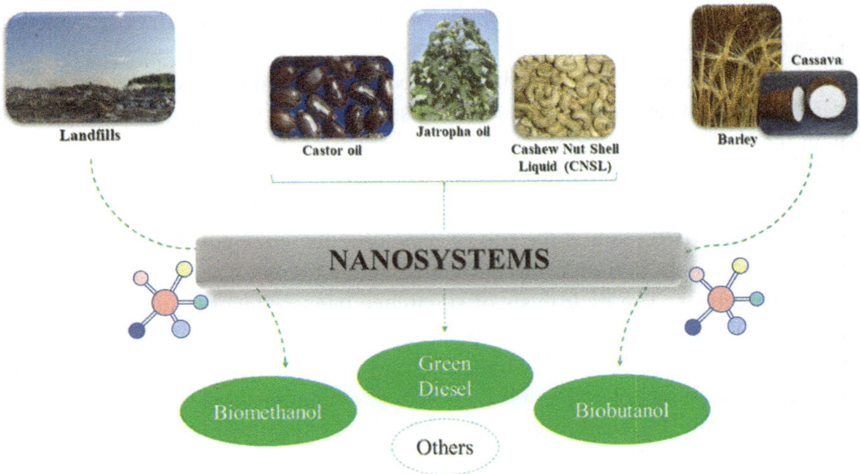

Fig. 4 Other examples of biofuels

(Cherchi et al. 2009). However, methanol used for denitrification is cost dependent on competing for demand from the chemical industries. Besides, there are certain safety concerns related to methanol storage and handling, making it difficult to purchase and transport commercial methanol to the facilities (Su et al. 2019). Biogenic methanol production can be an attractive solution to reduce dependence on external carbon sources and benefit the sustainability of these facilities. Biogenic conversion of methane to methanol has been previously demonstrated on a laboratory scale (Taher and Chandran 2013). However, the need for the use of high temperatures and the low conversion efficiency obtained made it more difficult to apply this technology to full scale (Shamsul et al. 2014).

Another biofuel that stands out is biobutanol. It is a straight-chain primary fuel of molecular formula C_4H_9OH with an alcohol (–OH) functional group included (Bharathiraja et al. 2017; Tigunova et al. 2013). Butanol which is produced from the alcoholic fermentation of natural, biodegradable, or renewable biomass is considered biobutanol (Pugazhendhi et al. 2019). It has great potential as a biofuel, as it has similar properties to gasoline and can be used directly in the car engine without modification (Malik et al. 2014; Xue et al. 2016). Besides, it can be mixed with gas, diesel, or gasoline in any percentage and normally used as a transportation fuel (Pugazhendhi et al. 2019). Biobutanol is considered superior to biomethanol, as it has some advantages such as a greater mixture with gasoline, higher energy efficiency, less need for engine modification and high performance, among others (Dürre 2008).

The use of biomass feedstock for biobutanol production via fermentation began in the early 1900s (Pugazhendhi et al. 2019). Since then, several cellulosic raw materials have been used for its production, such as cassava bagasse, changed grass and miscanthus, barley straw, maize, and wheat, as well as glucose and cornstarch (Huang et al. 2015). However, lignocellulosic biomass is the most promising raw material for biobutanol production (Jiang et al. 2019). Microorganisms of the genus clostridia are the main fermentative organisms used for biobutanol production, is known for their potential to ferment different types of renewable biomass in butanol through the fermentation pathway known as acetone-butanol-ethanol (ABE).

Although that certain microbial strains have intolerance to butanol accumulation in the fermentative medium, alternatives to solve this problem have been studied, such as increasing the resistance of these strains to butanol accumulation through the use of genetically modified or mutant strains with potential for production of biobutanol, searching for low-cost substrates for anaerobic fermentation, choosing the most suitable fermentation system, and also adapting the production system already used to nanosystems, thus seeking to improve this bioprocess (Pugazhendhi et al. 2019).

Biodiesel is a biofuel that has been previously discussed and despite having several advantages, it also has problems related to chemical stability, low calorific value, cold flow and filterability due to oxygen present in its structure (Knothe 2010; Santillan-Jimenez et al. 2013). As an alternative, green diesel consists of a drop-in biofuel, which has no oxygen in its composition and is different from those from petrochemical sources (Scaldaferri and Pasa 2019). Their present some features, like full compatibility with petro-diesel, has a high calorific value (44 MJ/kg), and low specific gravity (0.78), is very storage stable and promotes lower combustion emissions (Kalnes et al. 2009). Drop-in biofuels also have higher combustion heat compared to oxygenated biofuels (bioethanol and biodiesel) and are not hygroscopic (Scaldaferri and Pasa 2019).

For green diesel production, many raw materials can be applied, and the most used are cashew nut shell liquid (CNSL), jatropha oil, and castor oil (Orozco et al. 2017; Scaldaferri and Pasa 2019; Yenumala et al. 2019). Starting from these materials, the mentioned biofuel can be obtained from pyrolysis, catalytic cracking, and hydrodeoxygenation (HDO), commonly prepared under high pressure and in the presence of hydrogen, being widely accepted due to the high yield of product obtained (Yenumala et al. 2019). However, it is a less economic and ecological route, because it consumes a large amount of conventional hydrogen for the reaction to occur (Huber et al. 2007). Some alternative routes have been studied, such as the production under a free hydrogen atmosphere, better known as the deoxygenation process (DO). It is a route whose reaction conditions are lighter compared to the HDO process (Lestari et al. 2010).

3 Future Trends

Nanosystems have a wide range of applications for improving biofuel's production. However, some issues need to be solved to increase commercial availability for some of these systems. For this purpose, many studies are being developed, and this topic shows some of the prospective in this technology.

The use of immobilized enzymes is proven to be a powerful mechanism in biofuels processing, but it is impossible to predict and control enzyme stabilization with different supports and immobilization media (Kočar et al. 2015; Pavlidis et al. 2014). Although nanosystems have a potential stabilizing effect in the application of nanostructured materials, this mentioned effect is enzyme-specific, being difficult to understand the conformational changes as they interact with nanomaterials (Zeng et al. 2019). To expand studies of this type, it is necessary to lower the cost of the whole nanosystem, from the nanoparticle's synthesis to the immobilization protocol.

Another possible strategy is to combine enzyme immobilization with genetic modification. The last one is primarily used to produce soluble enzymes and these are not reusable and have low stability. The combination could bring novel biocatalysts with both qualities from the individuals' research fields and high efficiency for industrial applications (Bernal et al. 2018)

Biofuel cells were not mentioned in this chapter since they are on an early stage of research and more issues to be solved. However, nanomaterials are proven to increase electron transport between electrode surfaces and biocatalysts. Their exploration will increase due to their excellent electrocatalytic activity and that will enhance the current and power densities of biofuel cells (Zhao et al. 2017).

For nanosystems applying nanoengineered materials for bioelectrode development (metal nanoparticles carbon nanotubes and graphene), it is necessary to understand the interactions with the biomolecule/enzyme aggregated, making the appropriated modifications to their structure and immobilization and functionalization protocols to obtain practical nanobiocatalytic systems (Adeel et al. 2018). Those characteristics are linked and therefore need a lot of studies for optimization of the mentioned systems, providing better performance in the field of a biological fuel cell. The use of redox mediators to improve the communication between the enzyme's redox center and the electrode surface could be a way to increase the performance of the cells (Adeel et al. 2018).

The use of another type of nanosystems using simple synthetic methods to nanostructure biomimetic materials is also being studied, without the use of enzymes. These materials mimic the native environments of the biocatalysts, extending their longevity and therefore extending the cell's lifespan (Zhao et al. 2017). For biohydrogen cells, it is also necessary to improve the storage and distribution infrastructures and overcome other scale-up problems (Aricò et al. 2013).

For biodiesel production, the integration of nanosystems with biofuel production could be the key to improve their development to its full potential, and one strategy to increase the yield of the process is using co-immobilization technique, facilitating the contact of various enzymes in each step of the reaction or just improving the hydrolysis of complex substrates for biofuel synthesis (Lee et al. 2013).

On another step from biofuels production chain, the separation is primally made with inorganic materials, but nanosystems using nanoporous carbons can also be used as effective, selective, and economic absorbents for the separation of biomolecules (Peluso et al. 2019). More recently, biocompatible mesoporous nanoparticles with property to absorb hydrophobic molecules have been developed and investigated as a mechanism to harvest fatty acids from algal cultures (without killing them) and are subsequently used to produce biodiesel by esterification. The fact that the lipids are stored between the cell membrane and the cell wall makes this harvesting strategy possible without interfering with the cell membrane, putting mesoporous nanoparticles that withdraw oils from the cell and into the pores. To prevent transesterification in situ, strong oxides can be introduced into the pore structure. Nanofarming could provide another sustainable component in biodiesel production, produced directly without biocatalyst disruption (Biswas 2019)

Nanosystems could also bring revolutionizing changes to biogas production, improving the stability of cellulose enzymes, enhancing the catalytic capacity in biohydrogen production, and improving biological and chemical digestion with biocompatible nanoparticles. These topics need deeper comprehension and a cheaper production to increase the use of this technology in the future (Faisal et al. 2018). Anaerobic photo-fermentation reactors using visible-light photoactive metal oxides could be a way to increase methane production and turn this type of system more economically attractive (Ganzoury and Allam 2015).

4 Conclusion

Many researches had been made to increase biofuels production efficiency with the employment of nanosystems. For biocatalytic nanosystems, it is necessary to improve stability, biocompatibility, and economical availability. Strategies like co-immobilization of enzymes, nanofarming, synthesis of nanostructured biomimetic materials among others previously discussed could bring the full potential to biofuels production field, but many further researches are necessary to reach this point.

Acknowledgements We gratefully acknowledge the financial support of Brazilian Agencies for Scientific and Technological Development, Fundação Cearense de Apoio ao Desenvolvimento Científico e Tecnológico (FUNCAP), project number BP3-0139-00005.01.00/18, Conselho Nacional de Desenvolvimento Científico e Tecnológico (CNPq), project number 422942/2016-2, Coordenação de Aperfeiçoamento de Ensino Superior (CAPES).

References

Abdallah MS, Hassaneen FY, Faisal Y, Mansour MS, Ibrahim AM, Abo-Elfadl S, Salem HG, Allam NK (2019) Effect of Ni-Ferrite and Ni-Co-Ferrite nanostructures on biogas production from anaerobic digestion. Fuel 254:115673

Abdeen FRH, Mel M, Jami MS, Ihsan SI, Ismail AF (2016) A review of chemical absorption of carbon dioxide for biogas upgrading. Chin J Chem Eng 24:693–702

Abdelsalam E, Samer M, Attia YA, Abdel-Hadi MA, Hassan HE, Badr Y (2017) Effects of Co and Ni nanoparticles on biogas and methane production from anaerobic digestion of slurry. Energy Convers Manag 141:108–119

Abo Markeb A, Llimós-Turet J, Ferrer I, Blánquez P, Alonso A, Sánchez A, Moral-Vico J, Font X (2019) The use of magnetic iron oxide based nanoparticles to improve microalgae harvesting in real wastewater. Water Res 159:490–500

Adeel M, Bilal M, Rasheed T, Sharma A, Iqbal MN (2018) Graphene and graphene oxide: functionalization and nano-bio-catalytic system for enzyme immobilization and biotechnological perspective. Int J Biol Macromol 120:1430–1440

Aditiya HB, Mahlia TMI, Chong WT, Nur H, Sebayang AH (2016) Second generation bioethanol production: a critical review. Renew Sustain Energy Rev 66:631–653

Ahmadi MH, Ghazvini M, Alhuyi Nazari M, Ahmadi MA, Pourfayaz F, Lorenzini G, Ming T (2019) Renewable energy harvesting with the application of nanotechnology: a review. Int J Energy Res 43:1387–1410

Air Transport Action Group (ATAG), Facts & Figures (2019) [online]. Available at: https://www.atag.org/facts-figures.html. Accessed 25 Aug 2019

Akubude VC, Nwaigwe KN, Dintwa E (2019) Production of biodiesel from microalgae via nanocatalyzed transesterification process: a review. Mater Sci Energy Technol 2:216–225

Al-Mohammedawi HH, Znad H, Eroglu E (2019) Improvement of photofermentative biohydrogen production using pre-treated brewery wastewater with banana peels waste. Int J Hydrogen Energy 44:2560–2568

Alvarez LH, Cervantes FJ (2012) Assessing the impact of alumina nanoparticles in an anaerobic consortium: methanogenic and humus reducing activity. Appl Microbiol Biotechnol 95:1323–1331

Andrade F, Antunes F, Nascimento AV, Baptista S, Neves J, Ferreira D, Sarmento B (2011) Chitosan formulations as carriers for therapeutic proteins, 157–172

Andritz Group (2013) Biogas—an important renewable energy source. WBA Fact sheet

Aricò AS, Baglio V, Antonucci V (2013) Nanomaterials for fuel cell technologies. In: Nanotechnology for the energy challenge. Wiley-VCH Verlag GmbH & Co. KGaA, Weinheim, Germany, pp. 171–211

Banerjee Srijoni, Rout S, Banerjee Sanjukta, Atta A, Das D (2019) Fe2O3 nanocatalyst aided transesterification for biodiesel production from lipid-intact wet microalgal biomass: a biorefinery approach. Energy Convers Manag 195:844–853

Baniamerian H, Isfahani PG, Tsapekos P, Alvarado-Morales M, Shahrokhi M, Vossoughi M, Angelidaki I (2019) Application of nano-structured materials in anaerobic digestion: current status and perspectives. Chemosphere 229:188–199

Baskar G, Kumar RN, Melvin XH, Aiswarya R, Soumya S (2016) Sesbania aculeate biomass hydrolysis using magnetic nanobiocomposite of cellulase for bioethanol production. Renew Energ 98:23–28

Bernal C, Rodríguez K, Martínez R (2018) Integrating enzyme immobilization and protein engineering: an alternative path for the development of novel and improved industrial biocatalysts. Biotechnol Adv 36:1470–1480

Bezerra RM, Neto DMA, Galvão WS et al (2017) Design of a lipase-nano particle biocatalysts and its use in the kinetic resolution of medicament precursors. Biochem Eng J 125:104–115. https://doi.org/10.1016/j.bej.2017.05.024

Bharathiraja B, Jayamuthunagai J, Sudharsanaa T, Bharghavi A, Praveenkumar R, Chakravarthy M, Devarajan Y (2017) Biobutanol—an impending biofuel for future: a review on upstream and downstream processing techniques. Renew Sustain Energy Rev 68:788–807

Bi P, Wang J, Zhang Y, Jiang P, Wu X, Liu J, Xue H, Wang T, Li Q (2015) From lignin to cycloparaffins and aromatics: directional synthesis of jet and diesel fuel range biofuels using biomass. Bioresour Technol 183:10–17

Biswas A (2019) Nanotechnology in biofuels production: a novel approach for processing and production of bioenergy. In: Srivastava N, Srivastava M, Mishra PK, Upadhyay SN, Ramteke PW, Gupta VK (eds) Biofuel and biorefinery technologies. Springer International Publishing, Cham, pp 183–193

Bozbas K (2008) Biodiesel as an alternative motor fuel: production and policies in the European Union. Renew Sustain Energy Rev 12:542–552

Braun-Unkhoff M, Riedel U (2015) Alternative fuels in aviation. CEAS Aeronaut J 6:83–93

Bundhoo MAZ, Mohee R (2016) Inhibition of dark fermentative bio-hydrogen production: a review. Int J Hydrogen Energy 41:6713–6733

Cai J, Zhao Y, Fan J, Li F, Feng C, Guan Y, Wang R, Tang N (2019) Photosynthetic bacteria improved hydrogen yield of combined dark- and photo-fermentation. J Biotechnol 302:18–25

Carli S, Carneiro LABC, Ward RJW, Meleiro LP (2019) Immobilization of a β1032 glucosidase and an endoglucanase in ferromagnetic nanoparticles: a study of synergistic effects. Protein Expr Purif 160:28–35

Cesaro A, Belgiorno V (2015) Combined biogas and bioethanol production: opportunities and challenges for industrial application. Energies 8:8121–8144

Chandrasekhar K, Lee Y-J, Lee D-W (2015) Biohydrogen production: strategies to improve process efficiency through microbial routes. Int J Mol Sci 16:8266–8293

Cherchi C, Onnis-Hayden A, El-Shawabkeh I, Gu AZ (2009) Implication of using different carbon sources for denitrification in wastewater treatments. Water Environ Res 81:788–799

Cherian E, Dharmendirakumar M, Baskar G (2015) Immobilization of cellulase onto MnO_2 nanoparticles for bioethanol production by enhanced hydrolysis of agricultural waste. Chin J Catal 36:1223–1229

Chezeau B, Fontaine JP, Vial C (2019) Analysis of liquid-to-gas mass transfer, mixing and hydrogen production in dark fermentation process. Chem Eng J 372:715–727

Choong YY, Norli I, Abdullah AZ, Yhaya MF (2016) Impacts of trace element supplementation on the performance of anaerobic digestion process: a critical review. Bioresour Technol 209:369–379

Costa VM, De Souza MCM, Fechine PBA et al (2016) Nanobiocatalytic systems based on lipase-Fe3O4 and conventional systems for isoniazid synthesis: acomparative study. Brazilian J Chem Eng. https://doi.org/10.1590/0104-6632.20160333s20150137

de Souza MCM, dos Santos KP, Freire RM et al (2017) Production of flavor esters catalyzed by lipase B from *Candida antarctica* immobilized on magneticnanoparticles. Brazilian J Chem Eng 34:681–690. https://doi.org/10.1590/0104-6632.20170343s20150575

de Vasconcellos A, Miller AH, Aranda DAG, Nery JG (2018) Biocatalysts based on nanozeolite-enzyme complexes: effects of alkoxysilane surface functionalization and biofuel production using microalgae lipids feedstock. Colloids Surfaces B Biointerfaces 165:150–157

Dinesh Kumar M, Kaliappan S, Gopikumar S, Zhen G, Rajesh Banu J (2019) Synergetic pretreatment of algal biomass through H_2O_2 induced microwave in acidic condition for biohydrogen production. Fuel 253:833–839

Dolly S, Pandey A, Kumar B, Gopal R (2015) Process parameter optimization and enhancement of photo-biohydrogen production by mixed culture of *Rhodobacter sphaeroides* NMBL-02 and Escherichia coli NMBL-04 using Fe-nanoparticle. Int J Hydrogen Energy, 1–11

Duan H, Dong J, Gu X, Peng Y-K, Chen W, Issariyakul T, Myers WK, Li M-J, Yi N, Kilpatrick AFR, Wang Y, Zheng X, Ji S, Wang Q, Feng J, Chen D, Li Y, Buffet J-C, Liu H, Tsang SCE, O'Hare D (2017) Hydrodeoxygenation of water-insoluble bio-oil to alkanes using a highly dispersed Pd–Mo catalyst. Nat Commun 8:591

Dürre P (2008) Fermentative butanol production. Ann NY Acad Sci 1125:353–362

Elreedy A, Ibrahim E, Hassan N, El-dissouky A, Fujii M, Yoshimura C, Tawfik A (2017) Nickel-graphene nanocomposite as a novel supplement for enhancement of biohydrogen production from industrial wastewater containing mono-ethylene glycol. Energy Convers Manag 140:133–144

Energy Information Administration. Annual Energy Outlook (2019) [online]. Available at: https://www.eia.gov/outlooks/aeo/pdf/aeo2019.pdf. Accessed 25 Aug 2019

Faisal S, Hafeez FY, Zafar Y, Majeed S, Leng X, Zhao S, Saif I, Malik K, Li X (2018) A review on nanoparticles as boon for biogas producers-nano fuels and biosensing monitoring. Appl Sci 9:1–19

Fakhimi N, Dubini A, Tavakoli O, González-Ballester D (2019) Acetic acid is key for synergetic hydrogen production in Chlamydomonas-bacteria co-cultures. Bioresour Technol 289:121648

Fingerman KR, Sheppard C, Harris A (2018) California's low carbon fuel standard: modeling financial least-cost pathways to compliance in Northwest California. Transp Res Part D Transp Environ 63:320–332

Gadhe A, Sonawane SS, Varma MN (2015) Enhancement effect of hematite and nickel nanoparticles on biohydrogen production from dairy wastewater. Int J Hydrogen Energy 40:4502–4511

Galvão WS, Pinheiro BB, Golçalves LRB et al (2018) Novel nanohybrid biocatalyst: application in the kinetic resolution of secondary alcohols. J Mater Sci 53:14121–14137. https://doi.org/10.1007/s10853-018-2641-5

Ganzoury MA, Allam NK (2015) Impact of nanotechnology on biogas production: a mini-review. Renew Sustain Energy Rev 50:1392–1404

García-Depraect O, Rene ER, Gómez-Romero J, López-López A, León-Becerril E (2019) Enhanced biohydrogen production from the dark co-fermentation of tequila vinasse and nixtamalization wastewater: novel insights into ecological regulation by pH. Fuel 253:159–166

Ge S, Agbakpe M, Wu Z, Kuang L, Zhang W, Wang X (2014) Influences of surface coating, UV irradiation and magnetic field on the algae removal using magnetite nanoparticles

Ghimire A, Frunzo L, Pirozzi F, Trably E, Escudie R, Lens PNL, Esposito G (2015) A review on dark fermentative biohydrogen production from organicbiomass: process parameters and use of by-products. Appl Energy 144:73–95

Gonzalez-Estrella J, Sierra-Alvarez R, Field JA (2013) Toxicity assessment of inorganic nanoparticles to acetoclastic and hydrogenotrophic methanogenic activity in anaerobic granular sludge. J Hazard Mater 260:278–285

Gutierrez A, Kaila RK, Honkela ML, Slioor R, Krause AOI (2009) Hydrodeoxygenation of guaiacol on noble metal catalysts. Catal Today 147:239–246

Gutiérrez-Antonio C, Gómez-Castro FI, de Lira-Flores JA, Hernández S (2017) A review on the production processes of renewable jet fuel. Renew Sustain Energy Rev 79:709–729

Guo Z, Chen Y, Lu NL, Yang X, Tang K, Nasr A, Lin H (2018) The applications of nanocomposite catalysts in biofuel production. In: Multifunctional nanocomposites for energy and environmental applications. Wiley-VCH Verlag GmbH & Co. KGaA, Weinheim, Germany, pp 309–350

Hsieh P, Lai Y, Chen K, Hung C (2016) Explore the possible effect of TiO_2 and magnetic hematite nanoparticle addition on biohydrogen production by *Clostridium pasteurianum* based on gene expression measurements. Int J Hydrogen Energy, 1–7

Huang H, Singh V, Qureshi N (2015) Butanol production from food waste: a novel process for producing sustainable energy and reducing environmental pollution. Biotechnol Biofuels 8:147

Huber GW, O'Connor P, Corma A (2007) Processing biomass in conventional oil refineries: production of high quality diesel by hydrotreating vegetable oils in heavy vacuum oil mixtures. Appl Catal A Gen 329:120–129

International Air Transport Association, Annual Review (2017) [online]. Available at: https://www.iata.org/publications/Documents/iata-annual-review-2019.pdf. Accessed 25 Aug 2019

Itthibenchapong V, Srifa A, Faungnawakij K (2017) Heterogeneous catalysts for advanced biofuel production. In: Rai M, da Silva SS (eds) Nanotechnology for bioenergy and biofuel production, green chemistry and sustainable technology. Springer International Publishing, Cham, pp 231–254

Jafari O, Zilouei H (2016) Enhanced biohydrogen and subsequent biomethane production from sugarcane bagasse using nano-titanium dioxide pretreatment. Bioresour Technol 214:670–678

Jiang Y, Lv Y, Wu R, Sui Y, Chen C, Xin F, Zhou J, Dong W, Jiang M (2019) Current status and perspectives on biobutanol production using lignocellulosic feedstocks. Bioresour Technol Reports 7:100245

Kadam R, Panwar NL (2017) Recent advancement in biogas enrichment and its applications. Renew Sustain Energy Rev 73:892–903

Kalantari M, Kazemeini M, Arpanaei A (2013) Evaluation of biodiesel production using lipase immobilized on magnetic silica nanocomposite particles of various structures. Biochem Eng J 79:267–273

Kalnes TN, Koers KP, Marker T, Shonnard DR (2009) A technoeconomic and environmental life cycle comparison of green diesel to biodiesel and syndiesel. Environ Prog Sustain Energy 28:111–120

Karthikeyan S, Prabhakaran TD (2018) Environmental effects emission analysis of *Botryococcus braunii* algal biofuel using Ni-doped ZnO nano additives for IC engines. Energy Sources Part A Recover Util Environ Eff 00:1–8

Karthikeyan S, Prathima A (2017) Environmental effect of CI engine using microalgae methyl ester with doped nano additives. Transp Res Part D 50:385–396

Knothe G (2010) Biodiesel and renewable diesel: a comparison. Prog Energy Combust Sci 36:364–373

Kočar V, Božič Abram S, Doles T, Bašić N, Gradišar H, Pisanski T, Jerala R (2015) TOPOFOLD, the designed modular biomolecular folds: polypeptide-based molecular origami nanostructures following the footsteps of DNA. Wiley Interdiscip Rev Nanomed Nanobiotechnol 7:218–237

Kormi T, Bel Hadj Ali N, Abichou T, Green R (2017) Estimation of landfill methane emissions using stochastic search methods. Atmos Pollut Res 8:597–605

Kumar G, Bakonyi P, Periyasamy S, Kim SH, Nemestóthy N, Béla K (2015) Lignocellulose biohydrogen: practical challenges and recent progress 44:728–737

Kumar G, Mudhoo A, Sivagurunathan P, Nagarajan D (2016) Recent insights into the cell immobilization technology applied for dark fermentative hydrogen production. Bioresour Technol 219:725–737

Lee JY, Yoo C, Jun SY, Ahn CY, Oh HM (2010) Comparison of several methods for effective lipid extraction from microalgae. Bioresour Technol 101:S75–S77

Lee K, Yeun S, Na J, Goo S, Praveenkumar R (2013) Bioresource technology magnetophoretic harvesting of oleaginous *Chlorella* sp. by using biocompatible chitosan/magnetic nanoparticle composites. Bioresour Technol 149:575–578

Leong W-H, Lim J-W, Lam M-K, Uemura Y, Ho Y-C (2018) Third generation biofuels: a nutritional perspective in enhancing microbial lipid production. Renew Sustain Energy Rev 91:950–961

Lestari S, Mäki-Arvela P, Eränen K, Beltramini J, Max Lu GQ, Murzin DY (2010) Diesel-like hydrocarbons from catalytic deoxygenation of stearic acid over supported Pd nanoparticles on SBA-15 catalysts. Catal Lett 134:250–257

Li P, Sakuragi K, Makino H (2019) Extraction techniques in sustainable biofuel production: a concise review. Fuel Process. Technol

Lima RM, Santos AHM, Pereira CRS, Flauzino BK, Pereira ACOS, Nogueira FJH, Valverde JAR (2018) Spatially distributed potential of landfill biogas production and electric power generation in Brazil. Waste Manag 74:323–334

Limayem A, Gonzalez F, Micciche A, Haller E, Mohapatra S, Limayem A, Gonzalez F, Micciche A, Haller E (2016) Molecular identification and nanoremediation of microbial contaminants in algal systems using untreated wastewater, 1234

Liu Y, Wang Q, Zhang Y, Ni B-J (2015) Zero valent iron significantly enhances methane production from waste activated sludge by improving biochemical methane potential rather than hydrolysis rate. Sci Rep 5:8263

Luna-del Risco M, Orupõld K, Dubourguier H-C (2011) Particle-size effect of CuO and ZnO on biogas and methane production during anaerobic digestion. J Hazard Mater 189:603–608

Lupoi JS, Smith EA (2011) Evaluation of nanoparticle-immobilized cellulase for improved ethanol yield in simultaneous saccharification and fermentation reactions. Biotechnol Bioeng 108:2835–2843

Ma M, Zou C (2018) Effect of nanoparticles on the mass transfer process of removal of hydrogen sulfide in biogas by MDEA. Int J Heat Mass Transf 127:385–392

Malik SN, Pugalenthi V, Vaidya AN, Ghosh PC (2014) Kinetics of nano-catalysed dark fermentative hydrogen production from distillery wastewater. Energy Procedia 54:417–430

Melikoglu M, Singh V, Leu SY, Webb C, Lin CSK (2016) Biochemical production of bioalcohols. In: Handbook of biofuels production. Elsevier, pp 237–258

Minteer SD (2011) Biochemical production of other bioalcohols: biomethanol, biopropanol, bioglycerol, and bioethylene glycol. In: Handbook of biofuels production. Elsevier, pp 258–265
Mirza SS, Qazi JI, Liang Y, Chen S (2019) Growth characteristics and photofermentative biohydrogen production potential of purple non sulfur bacteria from sugar cane bagasse. Fuel 255:115805
Monteiro RRC, Lima PJM, Pinheiro BB et al (2019) Immobilization of lipase A from *Candida antarctica* onto chitosan-coated magnetic nanoparticles. Int J Mol Sci 20:4018. https://doi.org/10.3390/ijms20164018
Moravvej Z, Makarem MA, Rahimpour MR (2019) The fourth generation of biofuel, second and third generation of feedstocks. Elsevier Inc
Mu H, Chen Y (2011) Long-term effect of ZnO nanoparticles on waste activated sludge anaerobic digestion. Water Res 45:5612–5620
Mu H, Chen Y, Xiao N (2011) Effects of metal oxide nanoparticles (TiO_2, Al_2O_3, SiO_2 and ZnO) on waste activated sludge anaerobic digestion. Bioresour Technol 102:10305–10311
Muniru OS, Ezeanyanaso CS, Akubueze EU, Igwe CC, Elemo GN (2018) Review of different purification techniques for crude glycerol from biodiesel production. J Energy Res Rev 2:1–6
Mushtaq K, Zaidi AA, Askari SJ (2016) Design and performance analysis of floating dome type portable biogas plant for domestic use in Pakistan. Sustain Energy Technol Assess 14:21–25
Nisar J, Razaq R, Farooq M, Iqbal M, Khan RA, Sayed M, Shah A, Rahman I (2017) Enhanced biodiesel production from Jatropha oil using calcined waste animal bones as catalyst. Renew Energy 101:111–119
Nizami A-S, Rehan M (2018) Towards nanotechnology-based biofuel industry. Biofuel Res J 5:798–799
Orozco LM, Echeverri DA, Sánchez L, Rios LA (2017) Second-generation green diesel from castor oil: development of a new and efficient continuous-production process. Chem Eng J 322:149–156
Palomo-Briones R, Celis LB, Méndez-Acosta HO, Bernet N, Trably E, Razo-Flores E (2019) Enhancement of mass transfer conditions to increase the productivity and efficiency of dark fermentation in continuous reactors. Fuel 254:115648
Pavlidis IV, Patila M, Bornscheuer UT, Gournis D, Stamatis H (2014) Graphene-based nanobiocatalytic systems: recent advances and future prospects. Trends Biotechnol 32:312–320
Peluso A, Gargiulo N, Aprea P, Pepe F, Caputo D (2019) Nanoporous materials as H_2S adsorbents for biogas purification: a review. Sep Purif Rev 48:78–89
Plocoste T, Jacoby-Koaly S, Molinié J, Roussas A (2016) Effect of leachate recirculation on landfill methane production in a tropical insular area. Innov Energy Res 5:1–5
Pompa PP, Martiradonna L, Della Torre A, Della Sala F, Manna L, de Vittorio M, Calabi F, Cinagolani R, Rinaldi R (2006) Metal-enhanced fluorescence of colloidal nanocrystals with nanoscale control. Nat Nanotechnol 1:126–130
Pradhan P, Mahajani SM, Arora A (2018) Production and utilization of fuel pellets from biomass: a review. Fuel Process Technol 181:215–232
Pugazhendhi A, Shobana S, Duc D, Kumar V, Kumar G (2018) Application of nanotechnology (nanoparticles) in dark fermentative hydrogen production. Int J Hydrogen Energy 44:1431–1440
Pugazhendhi A, Shobana S, Nguyen DD, Banu JR, Sivagurunathan P, Chang SW, Ponnusamy VK, Kumar G (2019) Application of nanotechnology (nanoparticles) in dark fermentative hydrogen production. Int J Hydrogen Energy 44:1431–1440
Rai M, dos Santos JC, Soler MF, Franco Marcelino PR, Brumano LP, Ingle AP, Gaikwad S, Gade A, da Silva SS (2016) Strategic role of nanotechnology for production of bioethanol and biodiesel. Nanotechnol Rev 5:231–250
Ren HY, Kong F, Zhao L, Ren NQ, Ma J, Nan J, Liu BF (2019) Enhanced co-production of biohydrogen and algal lipids from agricultural biomass residues in long-term operation. Bioresour Technol 289:121774
Rios NS, Morais EG, dos Santos Galvão W et al (2019) Further stabilization of lipase from *Pseudomonas fluorescens* immobilized on octyl coated nanoparticlesvia chemical modification with

bifunctional agents. Int J Biol Macromol 141:313–324. https://doi.org/10.1016/j.ijbiomac.2019.09.003

Rodríguez-Couto S (2019) Green nanotechnology for biofuel production. In: Srivastava et al. (eds.) Sustainable approaches for biofuels production technologies, biofuel and biorefinery technologies, vol 7, p 1367. https://doi.org/10.1007/978-3-319-94797-6_4

Rosha P, Mohapatra SK, Mahla SK, Cho H, Chauhan BS, Dhir A (2019) Effect of compression ratio on combustion, performance, and emission characteristics of compression ignition engine fueled with palm (B20) biodiesel blend. Energy 178:676–684

Sadaf S, Iqbal J, Ullah I, Bhatti HN, Nouren S, Habib-ur-Rehman, Nisar J, Iqbal M (2018) Biodiesel production from waste cooking oil: an efficient technique to convert waste into biodiesel. Sustain Cities Soc 41:220–226

Sahota S, Shah G, Ghosh P, Kapoor R, Sengupta S, Singh P, Vijay V, Sahay A, Vijay VK, Thakur IS (2018) Review of trends in biogas upgradation technologies and future perspectives. Bioresour Technol Reports 1:79–88

Sambusiti C, Bellucci M, Zabaniotou A, Beneduce L, Monlau F (2015) Algae as promising feedstocks for fermentative biohydrogen production according to a biorefinery approach: a comprehensive review. Renew Sustain Energy Rev 44:20–36

Santillan-Jimenez E, Morgan T, Lacny J, Mohapatra S, Crocker M (2013) Catalytic deoxygenation of triglycerides and fatty acids to hydrocarbons over carbon-supported nickel. Fuel 103:1010–1017

Sarma SJ, Brar SK, Reigner J, Le Bihanc Y, Buelnac G (2014) Enriched hydrogen production by bioconversion of biodiesel waste supplemented with ferric citrate and its nano-spray dried particles. RSC Adv 4:49588–49594

Scaldaferri CA, Pasa VMD (2019) Green diesel production from upgrading of cashew nut shell liquid. Renew Sustain Energy Rev 111:303–313

Seelert T, Ghosh D, Yargeau V (2015) Improving biohydrogen production using *Clostridium beijerinckii* immobilized with magnetite nanoparticles. Bioenergy Biofuels, 4107–4116

Sekoai PT, Daramola MO (2017) The potential of dark fermentative bio-hydrogen production from biowaste effluents in South Africa. Int J Renew Energy Res 7:259–378

Sekoai PT, Awosusi AA, Yoro KO, Singo M, Oloye O, Ayeni AO, Bodunrin M, Daramola MO (2018) Microbial cell immobilization in biohydrogen production: a short overview. Crit Rev Biotechnol 38:157–171

Sekoai PT, Naphtaly C, Ouma M, Petrus S, Modisha P, Engelbrecht N, Bessarabov DG, Ghimire A (2019) Application of nanoparticles in biofuels : an overview. Fuel 237:380–397

Seman SZA, Idris I, Abdullah A, Shamsudin IK, Othman MR (2019) Optimizing purity and recovery of biogas methane enrichment process in a closed landfill. Renew Energy 131:1117–1127

Shamsul NS, Kamarudin SK, Rahman NA, Kofli NT (2014) An overview on the production of bio-methanol as potential renewable energy. Renew Sustain Energy Rev 33:578–588

Show KY, Yan Y, Zong C, Guo N, Chang JS, Lee DJ (2019) State of the art and challenges of biohydrogen from microalgae. Bioresour Technol 289:121747

Sirajunnisa AR, Surendhiran D (2016) Algae—a quintessential and positive resource of bioethanol production: a comprehensive review. Renew Sustain Energy Rev 66:248–267

Su L, Shi X, Guo G, Zhao A, Zhao Y (2013) Stabilization of sewage sludge in the presence of nanoscale zero-valent iron (nZVI): abatement of odor and improvement of biogas production. J Mater Cycles Waste Manag 15:461–468

Su Y-C, Sathyamoorthy S, Chandran K (2019) Bioaugmented methanol production using ammonia oxidizing bacteria in a continuous flow process. Bioresour Technol 279:101–107

Suntana AS, Vogt KA, Turnblom EC, Upadhye R (2009) Bio-methanol potential in Indonesia: forest biomass as a source of bio-energy that reduces carbon emissions. Appl Energy 86:S215–S221

Taher E, Chandran K (2013) High-rate, high-yield production of methanol by ammonia-oxidizing bacteria. Environ Sci Technol 47:3167–3173

Tahvildari K, Anaraki YN, Fazaeli R, Mirpanji S, Delrish E (2015) The study of CaO and MgO heterogenic nano-catalyst coupling on transesterification reaction efficacy in the production of biodiesel from recycled cooking oil. J Environ Heal Sci Eng 13:1–9

Temizel İ, Emadian SM, Di Addario M, Onay TT, Demirel B, Copty NK, Karanfil T (2017) Effect of nano-ZnO on biogas generation from simulated landfills. Waste Manag 63:18–26
Teo SH, Islam A, Taufiq-Yap YH (2016) Algae derived biodiesel using nanocatalytic transesterification process. Chem Eng Res Des 111:362–370
Tian X, Chen X, Dai L, Du W, Liu D (2017) A novel process of lipase-mediated biodiesel production by the introduction of dimethyl carbonate. Catal Commun 101:89–92
Tigunova OA, Shulga SM, Blume YB (2013) Biobutanol as an alternative type of fuel. Cytol Genet 47:366–382
Torkamani S, Wani SN, Tang YJ, Sureshkumar R (2010) Plasmon-enhanced microalgal growth in miniphotobioreactors. Appl Phys Lett 97:1–4
Tran D-T, Chen C-L, Chang J-S (2012) Immobilization of *Burkholderia* sp. lipase on a ferric silica nanocomposite for biodiesel production. J Biotechnol 158:112–119
Tyagi S, Rawtani D, Khatri N, Tharmavaram M (2018) Strategies for nitrate removal from aqueous environment using nanotechnology: a review. J Water Process Eng 21:84–95
Vasco-Correa J, Khanal S, Manandhar A, Shah A (2018) Anaerobic digestion for bioenergy production: global status, environmental and techno-economic implications, and government policies. Bioresour Technol 247:1015–1026
Verma ML, Barrow CJ, Puri M (2013) Nanobiotechnology as a novel paradigm for enzyme immobilisation and stabilisation with potential applications in biodiesel production. Appl Microbiol Biotechnol 97:23–39
Villanueva-Estrada RE, Rocha-Miller R, Arvizu-Fernández JL, Castro González A (2019) Energy production from biogas in a closed landfill: a case study of Prados de la Montaña. Mexico City Sustain Energy Technol Assessments 31:236–244
Wang X, Liu X, Yan X, Zhao P, Ding Y, Xu P (2011) Enzyme-nanoporous gold biocomposite: excellent biocatalyst with improved biocatalytic performance and stability. PLoS ONE 6:e24207
Wei H, Liu W, Chen X, Yang Q, Li J, Chen H (2019) Renewable bio-jet fuel production for aviation: a review. Fuel 254:115599
Wimonsong P, Nitisoravut R, Llorca J (2014) Application of Fe–Zn–Mg–Al–O hydrotalcites supported Au as active nano-catalyst for fermentative hydrogen production. Chem Eng J 253:148–154
Xie W, Wang J (2014) Enzymatic production of biodiesel from soybean oil by using immobilized lipase on Fe_3O_4/Poly(styrene-methacrylic acid) magnetic microsphere as a biocatalyst. Energy Fuels 28:2624–2631
Xu Q, Qin J, Ko JH (2019) Municipal solid waste landfill performance with different biogas collection practices: biogas and leachate generations. J Clean Prod 222:446–454
Xue C, Liu F, Xu M, Tang I-C, Zhao J, Bai F, Yang S-T (2016) Butanol production in acetone-butanol-ethanol fermentation with in situ product recovery by adsorption. Bioresour Technol 219:158–168
Yang J, Xin Z, He Q (Sophia), Corscadden K, Niu H (2019) An overview on performance characteristics of bio-jet fuels. Fuel 237:916–936
Yang X, Tang K, Nasr A, Lin H (2018) The applications of nanocomposite catalysts in biofuel production. Multifunctional nanocomposites for energy and environmental applications. Wiley, Weinheim, pp 309–350
Yenumala SR, Kumar P, Maity SK, Shee D (2019) Production of green diesel from karanja oil (*Pongamia pinnata*) using mesoporous NiMo-alumina composite catalysts. Bioresour Technol Reports 7:100288
Zaidi AA, Ruizhe F, Malik A, Khan SZ, Bhutta AJ, Shi Y, Mushtaq K (2019) Conjoint effect of microwave irradiation and metal nanoparticles on biogas augmentation from anaerobic digestion of green algae. Int J Hydrogen Energy 44:14661–14670
Zeng Z, Chen M, Liang J, Xiao R, Zeng G, Zhang J, Liu Z, Chen A, Zhou Y, Mo D (2019) Interaction of tetramer protein with carbon nanotubes. Appl Surf Sci 464:30–35

Zhang XL, Yan S, Tyagi RD, Surampalli RY (2013) Biodiesel production from heterotrophic microalgae through transesteri fi cation and nanotechnology application in the production. Renew Sustain Energy Rev 26:216–223

Zhang H, Zhao Y, Cao H, Mou G, Yin H (2015) Expression and characterization of a lytic polysaccharide monooxygenase from *Bacillus thuringiensis*. Int J Biol Macromol 79:72–75

Zhang Q, Hu J, Lee D-J (2016) Biogas from anaerobic digestion processes: research updates. Renew Energy 98:108–119

Zhang L, He X, Zhang Z, Cang D, Nwe KA, Zheng L, Li Z, Cheng S (2017) Evaluating the influences of ZnO engineering nanomaterials on VFA accumulation in sludge anaerobic digestion. Biochem Eng J 125:206–211

Zhang S, Su Y, Xu D, Zhu S, Zhang H, Liu X (2018) Assessment of hydrothermal carbonization and coupling washing with torrefaction of bamboo sawdust for biofuels production. Bioresour Technol 258:111–118

Zhang L, Loh K-C, Zhang J (2019a) Enhanced biogas production from anaerobic digestion of solid organic wastes: current status and prospects. Bioresour Technol Reports 5:280–296

Zhang Y, Xiao L, Wang S, Liu F (2019b) Stimulation of ferrihydrite nanorods on fermentative hydrogen production by *Clostridium pasteurianum*. Bioresour Technol 283:308–315

Zhao C, Gai P, Song R, Chen Y, Zhang J, Zhu J-J (2017) Nanostructured material-based biofuel cells: recent advances and future prospects. Chem Soc Rev 46:1545–1564

Zimmerman WB, Hewakandamby BN, Tesař V, Bandulasena HCH, Omotowa OA (2010) On the design and simulation of an airlift loop bioreactor with microbubble generation by fluidic oscillation. Int Sugar J 112:90–103

Zimmerman WB, Tesař V, Bandulasena HCH (2011) Towards energy efficient nanobubble generation with fluidic oscillation. Curr Opin Colloid Interface Sci 16:350–356

The manufacturer's authorised representative in the EU is Springer Nature Customer Service Centre GmbH, Europaplatz 3, 69115 Heidelberg, Germany. If you have any concerns regarding our products, please contact ProductSafety@springernature.com

Printed and bound by CPI Group (UK) Ltd, Croydon, CR0 4YY
30/09/2026
02225288-0014